IET HEALTHCARE TECHNOLOGIES SERIES 6

Active and Assisted Living

Active and Assisted Living: Technologies and Applications

Edited by Francisco Florez-Revuelta
and Alexandros Andre Chaaraoui

The Institution of Engineering and Technology

Published by The Institution of Engineering and Technology, London, United Kingdom

The Institution of Engineering and Technology is registered as a Charity in England & Wales (no. 211014) and Scotland (no. SC038698).

First published 2016

The Institution of Engineering and Technology
Michael Faraday House
Six Hills Way, Stevenage
Herts, SG1 2AY, United Kingdom

www.theiet.org

British Library Cataloguing in Publication Data
A catalogue record for this product is available from the British Library

ISBN 978-1-84919-987-2 (hardback)
ISBN 978-1-84919-988-9 (PDF)

Typeset in India by MPS Limited

Contents

Chapter 1

Technologies and applications for active and assisted living – current situation

Alexandros Andre Chaaraoui[1] *and Francisco Florez-Revuelta*[2]

1.1 Introduction

The world is addressing significant challenges due to the current and future demographic contexts. The number of people aged 65 years or over in Europe and the United States will almost double between 2015 and 2060 [1, 2]. This will be linked with an increase in people requiring long term care, i.e. a continuum of medical and social services designed to support the needs of people living with chronic health problems that affect their ability to perform everyday activities [3]. Currently, approximately 30% of people between 65 and 80 years of age require long-term care. This percentage reaches 50% for those over 80 [4]. Longevity of people combined with the decline in birth rate will also put pressure on the economic support of this care. The Statistical Office of the European Communities (EUROSTAT) projects that, in the next 30 years, the ratio between working and retired people, i.e. the old age support rate, will move from four-to-one to two-to-one in the EU [5]. Nowadays, EU Member States spend approximately a quarter of their GDP on social protection [6]. These demographic and economic situations raise significant challenges towards health and social care of the older population in terms of increased costs and lack of resources.

Parallel to this social need by citizens and governments, there is an economic opportunity arising from the expenditure related to the specific needs of the older population, what is called the Silver Economy [7]. According to Euromonitor [8], the spending power of this population group will reach $15 trillion by 2020 globally. Moreover, older people wish to keep their independence as long as possible, trying to fend for themselves by having a fulfilling life in terms of everyday activities and leisure.

Simultaneously to these changes in demography and attitude, faculties, and economics of older people, there have been great advances in information and

[1]Google, Inc., Zurich, Switzerland

[2]Faculty of Science, Engineering and Computing. Penrhyn Road, Kingston upon Thames, Surrey KT1 2EE, UK

communication technologies (ICT), with a rapid deployment and adoption by the general population. This has been mainly due to the introduction of the Internet and mobile telephony, and in the near future, generalisation of home automation systems and intelligent environments.

Technologies and services for Active and Assisted Living (AAL) stand out as a possible solution to address these challenges. AAL systems aim at improving the quality of life and supporting independent and healthy living of older or impaired people by using ICTs at home, at the workplace and in public spaces. AAL refers to the use of ICTs in the environments in which users are, so that these spaces are able to interact with people in a natural way, wherever and whenever they are needed, being aware of the context (situational, temporal, emotional, etc.) of the user or the environment, and to act proactively. An AAL environment requires the use of a distributed network of sensors and actuators to create a ubiquitous technological layer, able to interact transparently with the users, observing and interpreting their actions and intentions, learning their preferences and adjusting the parameters of the system to improve their quality of life and work.

AAL establishes a new paradigm of how people use technology. This is due to its holistic and person-centred conception, so that AAL could be usable, acceptable, useful and providing social value. Of course, these AAL technologies and services must be feasible and provide business value for the companies that develop them. In order to fulfil this holistic view of technology, the solutions developed must incorporate in their conception, design and development, the involvement of experts with different backgrounds: technology, health and care, social sciences, etc. What is paramount is users' engagement, in particular of those groups that may have a greater reliance on technology or who can receive greater support in their daily lives.

The objectives of AAL are diverse, as stated by the AAL Joint Programme [9]:

- to extend the time people can live in their preferred environment by increasing their autonomy, self-confidence and mobility;
- to support the preservation of health and functional capabilities of the elderly, promoting a better and healthier lifestyle for individuals at risk;
- to enhance security, preventing social isolation and supporting the preservation of the multifunctional network around the individual;
- to support carers, families and care organisations; and
- to increase the efficiency and productivity of used resources in the ageing societies.

The stakeholders addressed by these objectives also cover a wide spectrum [10]:

Primary Older persons, their families and informal caregivers;
Secondary Formal caregivers, service providers, charities and voluntary organisations;
Tertiary Organisations supplying AAL technologies and solutions; and
Quaternary Service procurers, policy makers, insurance companies.

The wide spectrum of objectives and stakeholders offers many possibilities for researchers and innovators in order to provide valuable services. This book aims at offering a comprehensive review of the technologies and applications for AAL. That is how this book is structured, incorporating some associated issues that intervene in the design and use of those solutions.

1.2 Technologies for AAL

In order to provide AAL services in their place of need, such as care centres or personal homes of older people, researchers, engineers and technicians are challenged by the wide scope of involved technological infrastructure. Depending on the specific application, several technological areas and research fields can be involved. However, a common underlying infrastructure usually exists regarding areas as ambient intelligence and ubiquitous computing, that are closely related to sensor and smart home technologies. The present book's first part digs deeply into these areas, chapter by chapter, highlighting and analysing specific works that have addressed these challenges recently in the state of the art, as well as presenting their own contributions. In the following, these chapters and the main goals which they address are briefly summarised.

In *Current State of the Art of Smart Environments and Labs from an AAL Point of View: Critical Analysis*, current contributions to the AAL technology development, and more specifically to smart environments and labs, are synthesised to reflect the current state of the field. *Crandall and Cook* focus on the engineering, including infrastructure such as sensors and middlewares, and user needs of AAL systems, along two of the major applications of smart homes, namely healthcare and home automation. The current challenges and ongoing projects of the state of the art of smart homes are then laid out based on summarising the work of the globally most active research groups and commercial smart home technologies available.

Next, in *Ambient and Wearable Sensors for Human Health Monitoring*, the authors *Rodgers et al.* study how monitoring of personal activity, vital signs and physiological measures can be enabled in a manner that minimises disruption to an individual's daily routine, while protecting their privacy at the same time. Ambient and wearable sensors that make this possible are reviewed, and special emphasis is made on favouring engagement of individuals to reduce the reliance on healthcare systems and improve self-care management of chronic conditions. Among other future challenges, the authors identify that both generating just-in-time actionable knowledge and creating a learning healthcare system are dependent of having scalable and interoperable systems, which can track behaviours concurrently and deal efficiently with the big data issues that result from a growing array of networking sensors and their data streams.

Visual sensors can be considered a very special type of sensors, since in most scenarios they are able to provide richer data about the environment than multiple other environmental sensors combined. The chapter *Computer Vision for Active and Assisted Living* analyses the state of the art of RGB cameras and depth sensors, detailing how pattern recognition and machine learning methods are applied to human motion and activity recognition and tracking. *Planinc et al.* use the traditional image processing pipeline to illustrate how current AAL projects take advantage of computer vision to approach a variety of applications from human behaviour analysis to physiological monitoring. The use of depth sensors, such as the popular Microsoft Kinect, for rehabilitation and robotics among others is reviewed, detailing skeletal pose estimation and tracking techniques, as well as methods based on depth maps, point clouds and plan-view maps. Finally, existing studies are synthesised related to the accuracy of infrared and time-of-flight depth sensors in comparison to stereo cameras and other marker-based tracking systems.

The different types of sensors and their readings, from binary to three-dimensional images, provide AAL systems with very heterogeneous information about the environment on which to act upon. The chapter entitled *A Data Fusion Approach for Identifying Lifestyle Patterns in Elderly Care* synthesises data fusion approaches which are aimed at combining information from multiple sensors of the same or different nature at varying levels of the system. A specific case of data fusion is considered, i.e. the combination of motion detectors and cameras, namely passive-infrared motion sensors, RGB cameras and depth sensors. *Eldib et al.* also introduce an in-home monitoring system which fuses a network of motion sensors and low-resolution cameras by combining the detected activities to identify the occupant's daily lifestyle pattern. Especially the deviations in these lifestyle patterns, such as *restful isolated days* or *busy social days*, can be very valuable health indicators.

Another challenge that arises from the heterogeneous sensors and actuators that can be employed in smart homes is their interoperability. Different underlying systems and protocols have to work together, since rarely one vendor can cover all in-home, wearable and mobile technologies, that users want to use transparently. This subject is addressed in *Towards Interoperable Enhanced Living Environments*, focusing on interoperability in Internet of Things infrastructures. *Spinsante et al.* summarise the basic challenges of interoperability as (1) device cooperation and discovery, (2) reduction of the number of gateways, and (3) user-centred personalisation. Specific design methodologies and existing projects and frameworks aimed at providing the necessary software architecture and the required network and data interoperability are reviewed.

AAL services may be as simple as reminding a person to take their prescription each eight hours, or as complex as recognising that the average daily amount of movement of an individual has been decreasing steadily and that this could imply a health risk due to their medical condition. This means that sophisticated long-term monitoring and detection may be necessary. In *Reasoning systems for AAL, Azkune et al.* distinguish three functional layers in reasoning systems for activity monitoring, activity modelling and activity inference. Monitoring is achieved through visual, environmental or wearable sensors, whereas modelling and inference requires learning techniques, either data-driven approaches that learn models from user data and infer the corresponding class, or knowledge-driven approaches that apply domain-specific knowledge and rules. Last but not least, the use of ontologies in AAL as support for reasoning systems is considered, as these can solve the modelling of complex data domains (e.g. user context) in an expressive and formal way.

Intelligent environments that enable AAL systems are not passive systems that continuously read data and eventually show results on a screen, but interactive systems that have to be able to proactively engage with the user when necessary. In the chapter named *Person–Environment Interaction*, the author *Caleb-Solly* presents an overview of human–computer interaction methods and design issues for person–environment interaction that are related to AAL. Special consideration is given to the accessibility needs of people with age-related physical and cognitive impairments, exploring different interaction modalities and technologies depending on the specific impairment and other additional variables, such as stress and social context.

The last chapter of this part of the book dedicated to AAL technologies, deals with telehealth. Among the valuable services AAL systems can provide, telehealth stands out as a viable option for enabling individuals to live at home safely and independently despite requiring periodic or continuous assistance, which is nowadays provided only in clinical environments. Health-oriented data analytics methods are reviewed in *Data Analytics for enabling Connected Health* by *Naguleswaran et al.* Furthermore, a novel data analytics method is presented. This decision support system relies on data from smart phones to identify human activity patterns and detect deviations based on individual health requirements. The system is also able to select the appropriate features for this purpose and ignore irrelevant inputs.

1.3 Applications for AAL

These technologies can be applied to offer a variety of services in different environments, i.e. at home, in public spaces, and at work. The AALIANCE2 Coordination Action presented several scenarios where AAL technologies can provide support to older people [10]: prevention of early degeneration of cognitive abilities, healthy living, management of chronic diseases, age-friendly and safe environments, fall prevention, management of daily activities and keeping control over own life, keeping social contact and having fun, outdoors mobility, avoiding caregivers isolation, and senior citizens at work. The second part of this book includes chapters discussing in detail some of these applications, the technologies employed in each case, and presenting their own work to provide valuable services.

Activities of Daily Living (ADLs) are those common tasks that people do every day. Basic activities such as bathing, dressing, toileting and eating; and instrumental activities, including, for instance, shopping, housekeeping, food preparation and ability to handle finances. In *Supporting Activities of Daily Living*, *Shewell et al.* review how AAL technologies, in particular smart environments, enable older people to perform some of these activities. They present several approaches to make environments intelligent: data driven, knowledge driven and context aware. Finally, the authors describe some challenges to design robust and reliable systems to support ADLs. These include data heterogeneity, interleaved activities, multiple occupancy, and ethical and security issues.

Gait is noticeably closely related to a person's well-being and personal autonomy, but it is also a very valuable metric for their physical and mental health state. In *Human Gait Analysis for Frailty Detection. Quantitative Techniques and Procedures*, *Fontecha et al.* study specifically the frailty syndrome considering gait activity as main predictor of the functional disorder of frailty. Multiple sensors, as accelerometer-enabled devices, pressure sensors and cameras, and analysis methods, based on different gait parameters and classification algorithms, are presented. Next, the authors present their own methods and results in three different case studies using accelerometer-enabled smartphones, wireless sensorised insoles, and a structured light sensor, respectively.

Increase in frailty may lead to a higher risk of having a fall incident. First, chapter *Fall Prevention and Detection* focuses on automatic techniques to assess fall risk, as a mechanism for prevention. *Baldewijns et al.* present different spatial and temporal parameters, which can be obtained from gait, in order to estimate an increase in fall risk. They also review the sensors that may be employed to calculate those parameters, next, different alternatives to visualisation and automatic interpretation of the gathered data. Finally, as the authors indicate, although these techniques may prevent falls, they cannot eliminate them completely. Therefore, they present different methods for fall detection using body-worn sensors (i.e. inertial measurement units) or sensors located in the environment (i.e. Doppler radars, cameras, and RGB-D devices).

Mobility is key to carry out many ADLs, for instance ensuring that older people can continue their interaction with others to avoid social isolation, that they are able to do their shopping by themselves, or that they can move independently at home. Most mobile devices incorporate pre-installed applications for location and navigation, however, commonly, they are not appropriate for being used by pedestrians or they miss valuable information required by people with physical or sensory impairments. *Hempel et al.* in their chapter *Outdoor Mobility Assistance – Technologies Helping on the Way* present different systems for localisation and positioning, as Global Navigation Satellite Systems (GNSS), infrastructure-based systems, dead reckoning methods, and hybrid methods. They also introduce other aspects that are relevant when designing navigation systems, as how to model the environment, and how to facilitate the interaction with the user. Finally, the authors review several products available in the market and relevant research projects in the area.

The chapter *Location and Orientation Technologies Based on WiFi Systems for People with Disabilities in Indoor Environment* complements the previous chapter, by studying the specific case of indoor systems, in which GNSS are not applicable, as the line of sight between satellites and the users is blocked. *Coret Gorgonio et al.* follow the Human Activity Assistive Technology (HAAT) model [11] to present the different aspects involved in mobility support systems for people with impairments.

The involvement of occupational therapists is very relevant at the moment of prescribing the appropriate AAL technologies and services to support independent living of older people. *Young* introduces the Person–Environment–Occupation Model (PEOM) [12] in chapter *Enabling Health, Well-being and Engagement in Life through Ambient Assisted Living Technologies: An Occupational Therapist's Lens*, as a framework for that prescription. Next, the principles for universal design are enumerated. Finally, the author reviews different examples of solutions to support older people in their well-being, health, social interaction, and engagement in life.

Other systems carry out a continuous monitoring of vital signs using wearable sensors, interpret the acquired data, support the medical staff or the carers to take a decision, and present the feedback in a meaningful and valuable way. *Baig et al.* review different decision support systems for inpatient care which provide data and feedback using tablets. The chapter *Tablet-based Clinical Decision Support System for Hospitalised Older Adults* analyses the current issues and challenges involved with the use of tablets in hospital care. Finally, the authors propose a decision system to provide individualised monitoring and interpretation of the vital signs in order to diagnose different health problems.

1.4 Associated issues and case studies

In the third and final part of the book, cross-cutting issues of AAL systems are tackled, which go beyond technology and applications and affect an AAL system's whole lifetime from design to result evaluation. These issues, such as the design of AAL services, accessibility, and privacy, have come up several times in this book, and it can be seen clearly how challenging and wide-scoped these subjects are when targeting in-home technology and healthcare services. In *Smart, Age-friendly Cities and Communities: The Emergence of Socio-technological Solutions in the Central and Eastern Europe*, smart cities and age-friendly communities are studied, focusing mainly on universal design, usability and accessibility. *Klimczuk and Tomczyk* present examples of such cities and communities and the concepts and premises that grounded these developments. Then, a closer look is taken to Central and Eastern Europe analysing the local needs of older adults, and how the challenge of a rapidly ageing population is handled along an often deficient existing infrastructure.

With all the analysed technologies and applications AAL compromises, one could ask the question of how a really accessible environment would look like that enables seamless multimodal interaction for older adults and impaired inhabitants, with both existing and emerging assistive technology solutions. *Stephanidis* attempts to answer this ambitious question in the chapter *Towards Accessible Ambient Assisted Living Environments*. By presenting such an accessible AAL environment, the emerging requirements are analysed, along the existing work of the state of the art and open challenges.

Perhaps one of the most clear issues associated with the adoption of AAL systems in private and semi-private environments is how the right to privacy can be protected without decreasing the quality of the service and, in particular, the recall of monitoring systems that detect situations of risk, such as falls. In *Privacy and Ethical Issues*, *Florez-Revuelta et al.* introduce privacy, ethical and legal issues related to healthcare technologies. With this context, design principles for privacy-aware technologies are presented. One particular case of technology is studied in further depth. This is visual monitoring with cameras, since this is one of the most intrusive technologies seen throughout this book. Different methods for visual privacy protection are reviewed, and an approach to offer different visualisations based on the user's context is presented.

End-users of AAL services have to be involved from very early design stages to avoid unexpected user experience conflicts. However, even when this is done, older or impaired adults are extremely heterogeneous, which makes it very difficult to find the right target group and approach the design stage appropriately. In *Human-Centred Design with Older Adults: Examples and Recommendations for Research, Ideation and Testing*, *Correia de Barros and Vasconcelos* go through the principles and practice of human-centred design with older adults reviewing existent literature and defining recommendations for the phases of ideation, testing and refinement. As a result, this chapter is very useful for both researchers and practitioners to support human-centred design processes.

Finally, the last chapter of this third part of the book presents a case study that goes from the initial design of a specific home platform to the final evaluation of experimental results obtained on an implemented proof of concept. In *Design and Implementation of a Smart Home Technological Platform for the Delivery of AAL Services: From Requirements to Field Experience*, *Spinsante et al.* first provide an overview of existing technology platforms used in research projects, and then introduce their platform for AAL service provisioning. Due to the close collaboration with several small medium enterprises which contributed their own technology, special attention is given to interoperability and product integration. During the evaluation of the experimental results, several aspects are addressed, not only limited to technology and the successful use of AAL services, but including also data management and analysis, and user interaction.

References

[1] European Union, *The 2015 Ageing Report. Underlying Assumptions and Projection Methodologies*, 2014.

[2] S. L. Colby and J. M. Ortman, *Projections of the Size and Composition of the US Population: 2014 to 2060*, 2015.

[3] N. McCall, *Who Will Pay for Long Term Care? Insights from the Partnership Programs*. Health Administration Press, 2001.

[4] F. Colombo, A. Llena-Nozal, J. Mercier, and F. Tjadens, *Help Wanted? Providing and Paying for Long-Term Care: Providing and Paying for Long-Term Care*, ser. OECD Health Policy Studies. OECD Publishing, 2011.

[5] OECD, *Pensions at a Glance 2015. OECD and G20 indicators*. OECD Publishing, 2015.

[6] European Commission, *Active Ageing Special Eurobarometer 378*, 2012.

[7] European Commission, *Growing the European Silver Economy*, 2015.

[8] Euromonitor, *Boomers as Consumers*, 2012.

[9] AAL Association, "Objectives – Active and Assisted Living Joint Programme," http://www.aal-europe.eu/about/objectives, (Accessed in March 2016).

[10] AALIANCE2 Consortium, "AALIANCE2 AAL Roadmap 2014," http://www.aaliance2.eu/system/files/drafts/AALIANCE2_Roadmap2014.pdf, (Accessed in March 2016).

[11] A. M. Cook and J. M. Polgar, *Assistive technologies: Principles and practice*. Elsevier Health Sciences, 2014.

[12] M. Law, B. Cooper, S. Strong, D. Stewart, P. Rigby, and L. Letts, "The person-environment-occupation model: A transactive approach to occupational performance," *Canadian Journal of Occupational Therapy*, vol. 63, no. 1, pp. 9–23, 1996.

Part 1

Technologies for Active and Assisted Living

Chapter 2

Current state of the art of smart environments and labs from an ambient assisted living point of view

Aaron S. Crandall[1] *and Diane J. Cook*[1]

The automated home was re-envisioned as the ubiquitous computing world where living spaces would adapt and support the daily lives of people inhabiting them. This vision leads to the rapid rise of integrating more computing, actuators, monitoring, and eventually artificial intelligence technologies into living spaces under the various banners of smart environments, ambient intelligence, pervasive computing, instrumented homes, etc. These new capabilities were quickly dragooned into use as a support tool for the medical community. Today, they have culminated in the concept of Ambient Assisted Living (AAL). AAL is primarily focused on helping various groups of people with day-to-day health difficulties live independently and to make better choices about their health needs.

To make this revolution of healthcare possible, many research groups, laboratories, companies, clinics, and others have risen up to the challenge. These researchers, engineers, social scientists, and healthcare professionals have sought to meld the capabilities of their disciplines together to deliver complex, but necessary, technologies to our communities. This process has accelerated over the last decade as the smart homes field began to incorporate more of the medical community into its work. This cross discipline effort was designed to address issues faced by our ageing populations, developmentally disabled adults, chronic health problems and more, especially in the face of rising healthcare costs and the projected lack of healthcare professionals. Keeping up to date on the developments in the state-of-the-art smart environments and laboratories involved in this field has become a significant effort.

Nearly every university, research hospital, and significant technology company has developed some kind of smart environments, health monitoring device, or AAL research initiative. This has only been accelerated by the rise of the Internet of Things (IoT) and combining it with cloud computing infrastructure as a serious commercial endeavour. These two concepts have led to a significant boom in available devices for

[1]Department of Electrical Engineering and Computer Science, Washington State University, Pullman, WA, USA

health monitoring an instrumented life. Evaluating the current technologies available, as well as developing new ones is a booming field today.

The work being done has pushed the state of the art in many fields. Notable advances have occurred in a wide range of fields such as sensors, biomarkers, cognition monitoring, mobility monitoring, fall detection, medicine compliance, behaviometrics, machine learning (ML), big data, privacy issues, wearables, social analysis, user experience (UX) issues, clinical metrics delivery, and more. This paper seeks to note current contributors to AAL technology development and critically analyse how the field is progressing. These facilities now span the globe, especially in communities with well-established issues with an ageing population and a clinician crush. Having a good grasp of the current participants in the AAL field is important for anyone seeking to contribute to AAL projects.

2.1 Introduction

The conversion of home automation to ubiquitous computing has generated many new applications for technology in everyday life. A significant application of these to healthcare has generated the Ambient Assisted Living (AAL) research and commercial fields. With the introduction of new low-cost sensors, low-power wireless protocols, cloud computing, and Internet of Things (IoT) technologies, the ability to deploy tools in our daily lives has never been more available.

Summing up and reviewing the AAL research and commercial landscape today is an ambitious goal. All aspects of the AAL world have advanced dramatically in the last 5–10 years. Ten years ago, there was a fractured group of researchers performing work in the ubiquitous computing world, medical biosensing was still transitioning from mechanical to digital systems, and the high cost of ambient sensors prohibited large-scale deployments in the real world. Today, a confluence of advances in these fields have created one of the fastest growing research and commercial areas available. New opportunities to help individuals live in our communities through technology advances seem to crop up on a nearly weekly basis. Tracking this deluge of groups involved in creating the AAL landscape is a notable task.

This chapter attempts to sample the current players in the various major topics of the AAL community, specifically in the field of smart homes with a strong emphasis in the areas of healthcare and energy efficiency in the home. It attempts to give the reader a sense of where different aspects of AAL smart homes are headed and where they can look to for new ideas to emerge.

To lay the foundation for this work, various themes of smart homes and their components are laid out. These areas of technology advancement are the necessary components for delivering smart home-based services with an AAL nature. Once these themes are defined, a group of active smart home research groups will be introduced along with their focus. To complete a review of the current AAL smart home space, companies with products addressing the various aspects of the smart home components will be covered.

Table 2.1 List of technology areas in this chapter

Infrastructure	User needs	Healthcare	Home automation
Sensors	User interfaces	Mobility/falls	Heating/cooling
Wearables	Social tracking	Behavioural	Preferences
Machine learning	Privacy concerns	Clinical metrics	Device control
Big data		Biostatus	User experience
Middlewares		Cognition info	

2.1.1 Areas of technology advancement

The abstract and introduction of this chapter note the diverse suites of technology necessary to make AAL smart homes possible and also note who is involved in creating them. The technology themes and topics for AAL development are outlined in Table 2.1 to demonstrate the various facets one should be looking at when evaluating smart home groups and commercial offerings.

The various themes contribute to our ability to deliver a comprehensive AAL solution. Detailed aspects of some of these themes will be addressed in other chapters of this book. The categories listed here are divided to address the engineering and user aspects of AAL systems, then features targeted by two of the major applications of smart homes: healthcare and home automation.

1. Infrastructure—The data collection and processing in any AAL
2. User needs—User facing considerations
3. Healthcare—Applications in the health space
4. Home automation—Applications in controlling AAL spaces, including energy efficiency

There are other areas this list does not address. Notably, security and regulatory aspects of smart homes. The discussion of security is a large area, and it has been a growing discussion with more and more technologies being attached to our homes with IoT aspects [1]. Having a person attack your appliances and home controls remotely is a very real danger with these systems [2].

AAL smart homes have notable regulatory hurdles to address privacy and liability issues. This is especially true with healthcare focused technologies. Being aware of local laws and getting involved in their updating to have a framework for AAL technologies is a must.

The first major theme of AAL smart homes is its infrastructure. Most research groups have contributions to this area to address the variability of their needs in developing their particular application. These hardware, software, algorithms, and policies have given new technologies the ability to have ever more complex models of the residents and their needs. Better sensors, lower power devices, and more capabilities in wearables have enabled the IoT revolution. Connecting such a dynamic and heterogeneous network of devices has continued to require advances in middleware and networking. Advances in ML have brought the artificial intelligence community into a new age of breakthroughs. ML is quickly advancing some of the most difficult problem areas, notably in vision, natural language processing, planning, health metrics, and behavioural modelling. The deployment of so many sensors, and the instrumented lifestyle, has generated a quantity of data unenvisioned by earlier technology developers. The need to move, store, and address this huge volume of information has founded a whole research area unto itself with big data and data science. Big data is a combination of information technology, algorithmics, ML, and traditional database fields traditionally applied to genomics, but is expanding quickly into smart homes. iControl has reported that their commercial platform has collected over 100 billion events last year [1]. Across the board, the capabilities these infrastructure oriented technologies enable have pushed the other themes to new heights.

Once the infrastructure is being established to underpin the smart home, the needs of the user should be considered and explored. The User Needs theme notes the kinds of new advances that contribute to our understanding of the needs of the individuals living within the AAL smart home space. This theme is heavily influenced by non-technical groups, such as sociologists, home design, and experimental psychology. These groups continue to address questions about how the user interfaces of smart home systems need to be engineered as to be effective, informative, subtle, and useful to the various stakeholders relying on the tools. Building user interfaces for whole home systems, mobile devices, voice controls, and ambient alerting needs notably more effort than many prior UX scenarios to contribute to the UX. At the same time as the UX systems are being refined, the look and feel of the systems influence social interactions between AAL users and others. These social stigmas or boons are relatively poorly understood, since the perception of monitoring oneself through technology is rapidly changing in our society. Additionally, the nature of privacy in such an instrumented community demands continued effort into controlling and reviewing data. Data privacy laws, digital rights management, and data access are a burgeoning area that combines legal, technological, and social research. People often do not wish to be monitored by third parties, but they do want or need services provided by AAL technologies. Finding a middle ground of what is both acceptable and legal when balanced against the need for privacy is an ongoing process. Research and commercial entities should not overlook the user needs aspects of AAL systems, though they are often given less priority than the other themes.

One of the major goals of smart home research is healthcare. Having an instrumented environment opens up possibilities to get a whole life, detailed understanding of how people are performing from a healthcare perspective.

The ability of an AAL system to calculate clinically relevant and actionable informatics is a powerful new tool for caregivers. Research and commercial efforts have focused heavily on these areas since they are very important to families and organisations. The ageing world population is leading quickly to community wide needs for more technology-based tools to help older adults live in their homes longer as they age [3, 4]. There are a number of major healthcare targets AAL smart homes try to hit, but there are only a few major topics that are pursued by most groups.

Tracking mobility and falls has given insights into rehabilitation, cognitive health, physical needs, and daily activities. The ability of people to move and how well they move is a key tool for clinicians. Of all the critical events in a person's life, falling down is often the most common instigator of major life changes. Falls cost older adults in the United States more than 19 billion USD per year [5] and the cost in quality of life is incalculable. Fall detection and fall prediction is a quickly advancing field with many approaches to this difficult problem. Behaviour and behaviour modelling are indicators of daily living and whole life help, especially in assistive care scenarios. Beyond behaviour, finding new approaches to generating established clinical metrics has been a successful research area. Mapping from AAL sensor-based data to clinical metrics gives researchers a tight framework with well-established boundaries and a strong success metric. This kind of bounded problem fits well with ML methodologies, so there is continued interest in the area. Biostatus sensing to feed AAL systems, such as blood pressure, glucose, heart rate, sleep, hydration, and many more has revolutionised how we can monitor the health of individuals. These biostatus sensors are now invaluable tools in AAL, home, and clinical spaces. While biostatus technologies can measure physiological metrics, the mapping of behaviour to cognitive status is a relatively unexplored field. Being able to track the cognitive health of individuals through their daily activities gives clinicians a much better view of the needs of AAL residents. These clinical themes are currently the most visible and active areas of the AAL research community. Given the size of healthcare budgets and the projected needs of our communities, significant commercial effort is spent here since it easily updates and advances the proven healthcare approaches in use today.

Another significant target for AAL smart homes is home automation [1]. Controlling our homes in an easier, even autonomous, fashion is the goal of many AAL projects, and really in the vision most people have of a "smart home." This takes a combination of UX design, preference learning, and controlling the devices around us. A big advantage of controlling the home is reducing energy usage. Our planet is facing a growing need for energy, but the environmental impact is beginning to show signs of long-term effects on the biosphere. If unchecked, global climate change could make Earth uninhabitable for future generations of mankind. Using smart home technologies to reduce our energy use has been an ongoing project for many years. Most homes have a handful of major energy-consuming devices. Controlling heating and cooling (HVAC) systems has the most impact on the home's energy footprint. Developing HVAC control systems and building preference models for the residents are the most obvious commercial projects.

All of these themes will have contributors in the research and commercial areas. Each group or organisation will be noted for contributions to these various fields.

2.2 AAL smart home research groups

This section intends to introduce some of the active players in the AAL smart home community. Their primary contribution areas and some of their commercial impacts are of interest. Given the explosion of AAL research contributors in the last five years, there is no way to encapsulate all of the players in the field today. This work intends to note the longstanding active members and some of the notable new ones to give readers a jumping off point when evaluating existing groups.

2.2.1 European network of living labs

The European Network of Living Labs (ENoLL) [6] is a consortium of AAL research groups across Europe, with almost 400 historically recognized Living Labs and more than 170 active members today. These research groups publish on all topics of AAL smart homes, as well as all other aspects of human environments with technology integration. The existence of ENoLL and the scale of its success demonstrate the pervasive scale and scope of AAL research.

2.2.2 TigerPlace—University of Missouri

TigerPlace [7] is a healthcare facility with a long history of advancements in AAL technologies. Their melding of nursing and technology established their work as a pioneer in AAL healthcare. The facility is an instrumented eldercare space with a focus on helping people age in place.

The TigerPlace contributions span many aspects of AALs. They have published work on many sensor issues [8] for deploying systems in healthcare scenarios. This has included the use of cameras [9] for resident identification and other 3D style systems for measuring position and patterns of residents [10]. These systems have led to significant studies in the areas of detecting illness [11] through behaviour, but also with clinical metrics [10]. TigerPlace's research team has contributed to behaviour analytics for detecting change [12]. A significant part of their research program contributes to detecting and predicting falls [13–16]. They have also investigated privacy and perceptions of smart monitoring technologies among older adults [17], which has led to methods of reducing video fidelity while still preserving the ability to run an analysis of the residents [18].

2.2.3 Aware home research initiative—Georgia Tech

The Aware Home Research Initiative (AHRI) [19] addresses a range of health and environmental issues. It includes a dedicated home, a suite of managed eldercare homes, and a laboratory focused on children's development. The systems include sensors: motion, light, temperature, humidity, door and window, and cameras and controls: lights, lamp and appliance modules, and door locks. These systems support work in the areas of health and wellbeing [20], entertainment [21], and sustainability. A notable contribution of the Aware Home group is to understanding and building AAL systems for individuals with autism [22, 23].

2.2.4 Future care lab—RWTH Aachen University

Future Care Lab hosts Project House HumTec, which is interested in interdisciplinary research between engineering and social sciences. The HumTec fields include eHealth for mobility and ageing, energy efficiency, and law and technology. Their contributions include areas such as clinical rehabilitation [24], UX issues [25], and older adult's needs from AAL smart homes [26].

2.2.5 Distributed systems group—ETH, Zurich

The Distributed Systems Group (DSG) of ETH Zurich works on many areas of ubiquitous computing. These areas are mostly centred and infrastructure needs in AAL smart homes such as: sensor networks [27], energy efficiency [28], IoT [29], UX design in smart environments, and security and privacy implications of AAL systems. DSG is also involved in commercial collaborations, such as M-Lab [30].

2.2.6 Smart environments research center—Washington State University

The Smart Environments Research Center (SERC) [31] is a large group on all aspects of AAL technologies. Their specialties include ML [32, 33], cognition analysis [34], engineering smart homes [35], wearable biosensors [36], and clinical metrics [37]. SERC includes the Center for Advanced Studies in Adaptive Systems (CASAS) [38], which is a longstanding AAL smart homes research group. One notable aspect of CASAS is their hosting of data sets from various AAL environments for public research use [39].

2.2.7 POSEiDON Project—Middlesex University

The POSEiDON Project is an AAL programme targeting individuals with Down Syndrome to provide a greater independence in life. Research topics include augmented reality tools for daily life [40], ethical considerations in engineering AAL systems [41], and personalizing the world for people with Down Syndrome [42].

2.2.8 ORCATECH—Oregon Health and Science University

ORCATECH Living Labs at OHSU is a longstanding and strong contributor to health-focused AAL smart homes. With over 480 homes and eight years of longitudinal data, their results have begun to contribute to many ecological validity studies on health metrics and therapies based on AAL technologies. Their publications have included infrastructure topics of sensors and systems [43], UX issues [44], clinical metrics [45], mobility in AAL spaces [46], and cognitive function [47].

2.2.9 Intelligent assistive technology and systems lab—University of Toronto

IATSL seeks to develop zero-effort technologies with adaptive AAL systems, notably for occupational therapy and gerontology. The group has been very successful with large-scale, inter-institutional research projects that have enabled significant contributions in many healthcare areas using AAL technologies. The group focuses on Alzheimer's disease, ageing in place [48], ambient healthcare [49], rehabilitation [50], and many other technology-based AAL topics.

2.2.10 Smart medical home—University of Rochester

The University of Rochester Smart Medical Home focuses on the use of AAL technologies to address a range of medical needs. They focus on patient education, gait monitoring, cognitive aids [51], and biosensing in the home [52]. Their current research programmes have been focused on mobile applications based on the Apple ResearchKit [53] for cognitive aids in conjunction with commercial groups.

2.2.11 mHealth research group—Northeastern University

The mHealth drive at NEU is a relative newcomer as a large AAL smart home group, but it has drawn notable talent to the programme. The group is interested in all aspects of AAL systems, from fixed in-home technologies [54], to wearable devices [55], to personal health informatics [56]. mHealth is the hub for two significant open source projects in the AAL world: Wockets [57] and BoxLab [58]. Wockets is a project to build open source sensors for smart phones to give a more capable and end-to-end model of the sensing platform for mobile AAL systems. BoxLab is a longstanding smart home in a package solution with significant contributions from Intel Corporation, Microsoft Research, and Mitsubishi Electric Research Laboratory (MERL). mHealth is active in releasing their datasets from their research for further analysis by other groups [59].

2.2.12 Research groups summary

This is only a sampling of significant research groups operating in the AAL smart home space. Today there are hundreds of public and private research organisations publishing in this field. Research centres with long-term contributions are somewhat rarer given the complexity and cost of extended projects, but the availability of public and private funding has never been better for AAL smart home projects.

2.3 Commercial AAL smart home technologies

The AAL smart home has become a significant commercial endeavour in the last five years or so. The availability of very low power processors and the latest in mesh networking has reduced the unit cost of sensors, the penetration of smart phones has brought mobile Internet computing to the market, and ML in the cloud has made large data set handling possible. This confluence of social and technological changes has allowed for a boom in devices, systems, and applications for AAL technologies. This section will cover notable offerings in the healthcare, energy efficiency, and general home automation device spaces.

2.3.1 In-home health care companies

Given the promise of health application from AAL technologies and the demographic shift towards an ageing population, there are many companies attempting to commercialize in-home AAL based technologies today. This is a very active space for

Table 2.2 List of health care AAL smart home companies

Company	Product	Tech summary
Life Alert	Life Alert	Wearable call button
Philips	LifeLine	Wearable call button
Alarm.com	BeClose	In home sensors for ADLs
HealthSense	eNeighbor	In home sensors for alerting
GrandCARE	GrandCARE	Ambient sensing for ADLs and alerting
Care innovations	QuietCare	Ambient sensing for ADLs
BehavioMetrics	Bevy360	Advanced ML-based in-home care support
GreatCall	MyLive!y	Ambient and wearables
SafeinHome	SafeinHome	Ambient sensing for alerting

startups, but there are many established players as well. Table 2.2 summarises some of the notable members of the In-home Heath Care AAL market. Most companies are seeking approaches to detect and leverage the activities of daily living (ADLs) to provide caregivers with a more informed view of how the residents in the home are doing on a daily basis and then enable more accurate alerts about possible issues. This list focuses on classic and whole home care technologies, and does not address offerings with a single device, such as only a medicine dispenser or a stand-alone biosensor.

The early companies into this space mostly focused on a combination of security and emergency call buttons. Life Alert [60], Philips LifeLine [61], and many others focus mostly on a wearable emergency call button. These sensors are designed to be a conduit for an individual to contact or alert an emergency responder in the face of an acute event, such as a fall.

Second generation offerings in this area were early ambient sensing with minimal ML backends for deep behavioural modelling. These companies such as GrandCARE [62], BeClose [63], and QuietCare [64] bring whole home sensing and early modelling to the marketplace. They also began to incorporate more of the security features to the home, such as digitally locking doors, to their offerings.

The latest generation of in-home companies differentiate themselves from earlier offerings through more advanced ML capabilities. These AAL systems use the latest in a variety of ML, cloud computing, and mobile platforms to determine the current behaviours and needs of the residents. Many of these newcomers are startups today, including MyLive!y [65], Bevy360 [66], SafeinHome [67], and eNeighbor [68]. The value add of more advanced ML to in-home health care technologies are still being established with new markets and business models under exploration.

2.3.2 Home automation devices

The home automation market is booming. Many people are interested in automating daily activities and bringing in security features to their homes [1]. New wireless devices and low power protocols have enabled more advanced versions of the kinds of automation available for many years. Companies will either focus on a few aspects

Table 2.3 List of AAL topics and related companies

Lighting	Security	Item tracking
Crestron	ADT	PROTAG Elite
Jasco	Vivint	BluTracker
Lutron	Frontpoint	StickNFind
Insteon	SkyCover	PebbleBee
Leviton	Xfinity	XY Find-It
Cyberhomes	Honeywell	BringTags
Synapse Wireless	Home Depot	Tile
Daintree Networks	Yale	Find'Em Tracking
Enlightened	Korner	Linquet
Philips Hue	Visonic	Lupo

of home control, try to integrate multiple other products, or seek out a whole home automation solution.

AAL technology device companies New gadgets and systems are coming onto the market continually. Table 2.3 lists some of the major topics and companies involved in specialised AAL solutions from a sampling of the fields out there.

With the focus on reducing lighting costs, the automatic lighting control market is growing rapidly and stands at around $322 million today [69]. Another note is companies building smart LED bulbs with embedded IoT devices such as WeMo [70] and GE Link [71]. These devices enable mesh networking throughout the home with always-on devices.

Security companies have quickly moved in to aspects of the AAL space. Their connection with customer's homes, large install bases, and professional installers position them well to deliver robust AAL smart home implementations. Most of the established companies added real time notifications, video feeds, and remote locking capabilities. Some have begun to push on the in-home healthcare fields, but not with the same drive as healthcare focused companies have. Special note goes to companies such as Korner [72], which seek to provide a do-it-yourself (DYI) security solution. These systems are much less expensive than a classic home security company, but still give a level of notification if a home is entered at the wrong time.

Tracking lost items is a big area right now. Most systems use a short range Bluetooth solution to communicate with a smart phone about the relative range to a battery powered tag. This kind of non-healthcare, general use device is a nice use of the AAL technologies available. Several of these companies have made a notable splash through crowdfunding such as Kickstarter and Indigogo.

Whole Home Automation Companies These companies focus on delivering a whole home solution for automation, security, and other features. A sampling of the major players is listed in Table 2.4. This is by no means an exhaustive list today, but it includes a range of offerings.

Most of these companies sell a complete solution for deployment in homes, offices, classrooms, and other spaces. Longstanding players such as Control4 [73]

Table 2.4 List of whole home automation companies

Company	Focus areas
AMX	Lighting, video, climate, security
Control4	Lighting, multimedia, mechanical systems
Home Tech Europe	Automation, lighting, A/V, video
Crestron	Lighting, classrooms, HVAC, audio
Elan Home Systems	A/V systems, lighting
Home Automation, Inc.	Lighting, EV cars, security
Exceptional Innovation	Hospitality, HVAC, security
iControl Networks	Security, HVAC, eldercare
THRONE	Lighting, climate, surveillance, fire, alarm

and AMX [74] focus on professionally installed enterprise grade implementations. The "consumer grade" sensors and products available today are often robust and easy enough to work with that more recent companies, such as iControl [75], seek to integrate existing offerings into a full service system.

Sensors and Controllers The cost of production in the sensors and controllers market has dropped dramatically. A simple wireless PIR motion sensor as a complete solution costs around 50 USD retail today. Many companies are selling products in the classic area of general home monitoring of occupancy and items. Sensors such as motion, contact switches, RFID, shake/gyro, and electricity usage are widely available from companies such as Samsung's SmartThings [76], Ademco, 2GIG, GE Interlox, and Optex.

Controlling the home has become easier, especially in the lighting space, but there are other options available such as smart appliances. An oft-watched example are connected refrigerators from Whirlpool, LG, and Samsung to help with our shopping and nutrition patterns. Haier has been particularly active in the smart appliance space with their full kitchen suite and U+ Smart Life Platform [77]. Hurdles still remain for natural lighting control due to the complexity and maintenance issues with automated blinds, but Polymer Dispersed Liquid Crystal (PDLC) showing promise for "smart glass" to dim windows at a reasonable price point [78].

User Interfaces To interact with the smart home, most systems use either mobile phones or tablets to provide a user interface with the AAL environment. This kind of interface works, but does not always provide an on-hand means of communication. A current push is for more natural language processing based solutions. The latest breakthroughs in voice recognition, language modelling for intent, big data analysis, and user preference building has opened up the field of NLP-based communication with computer systems. Historically, NLP voice control systems suffered from significant error rates leading to frustration and lack of adoption, but the latest algorithms are coming in at under eight per cent error rates, which is actually getting close to what humans themselves do [79]. Major player offerings such as Apple's Siri, Google's "OK Google," Amazon's Echo, and Microsoft's Cortana are joined by smaller offerings such as VoicePod, BMW voice control, Ubi, Athom Homey, CastleOS CastleHub, Insteon's HomeKit Hub, Ivee Sleek, Mycroft, VRing, and plenty of others.

Another route to the smart home is through the smart watch. This wearable computer is often connected with a Wi-Fi or bluetooth network and provides a combination of a screen, audio in/out, accelerometres, biosensors, cameras, and other sensors. The advantage of the smart watch is its relatively accessible positioning on the body and often available status, as opposed to a smart phone which will be in a pocket, bag, belt clip, or off the body most of the time. There is a deluge of new smart watches on the market. Earlier offerings were brought out with great acclaim such as the Apple Smart Watch, Samsung Watch, and Pebble. Currently some of the best offerings are from Sony SmartWatch, Moto 360, Asus ZenWatch, Huawei Watch, Omate Racer, Garmin Vivoactive, and LG Urbane. Smart watches are taken up quickly by people seeking to have their text messages early and often, as well as a more advanced fitness tracker solution, but as a user interface to an AAL smart home they still struggle to integrate and provide the level of control needed to quickly and easily manipulate your surroundings. With the rise in NLP voice recognition, this lack of control may begin to be addressed as the integrated home automated providers from Table 2.4 begin to bridge in these NLP-based technologies.

2.3.3 In-home energy management offerings

Aside from healthcare and general automation, one of the most exciting areas for commercial AAL technologies is energy management, notably in home HVAC controls. Keeping the home in the right temperature range to be comfortable for the residents while reducing energy has two needs competing with each other. Ever since the Neural Network House [80], there has been work to conform the home around the resident's need, but not spending energy heating or cooling whenever possible.

Today, there is a wave of available smart thermostats. These devices give the residents more flexible control over their HVAC systems. The most advanced ones today also learn about the home and the preferences of the residents to try and adapt in a more AAL way over time. To get information for their model building about resident preferences, the latest smart thermostats also integrate with sensors to be deployed around the home. For example, the EcoBee3 uses motion detectors with temperature sensors to learn when people are in rooms and to model the per-room heating throughout the house. NEST made a huge splash when it hit the market with its adaptive nature, and the new version has begun pulling data in from other sources. Honeywell has long been a vendor in the HVAC space, and their latest device also integrates with external sensors. Other options on the market today include the Emerson Sensi, Venstar Color Touch, Lennox iComfort, and Trane XL.

What matters most to researchers and technology integrators is whether the HVAC control is available through an API of some sort to both act as a sensor and controller in the AAL smart home space. Only some of these commercial offerings have such an interface, and some only offer it through their web APIs. The EcoBee solutions offer ZigBee for local integration with smart meters and devices. The Trane XL uses

Z-Wave and Nexia, while NEST requires using their web API for data access and custom controls.

Unfortunately, a smart thermostat can only do so much to balance the temperature in the home. A single point of control with a binary state means that rooms not in use will still have energy spent heating and cooling them. This inefficiency is significant in many homes, so finding solutions for per-room heating and cooling will be a big step forward for the AAL energy efficiency marketplace. The general term for this is "zoning," and is common in larger commercial spaces, but historically too expensive for most home installations. There are currently a few players in this space, such as EcoVent [81] and ZoneFirst [82]. Having this level of granual control over the heating and cooling throughout the home provides a powerful opportunity to use adaptive models within AAL environments [83], especially if there are people with radically different preferences for environmental controls present.

2.3.4 Commercial AAL smart home summary

This section gave a taste of the innovations going on in the AAL smart homes commercial space today. From inexpensive sensors, to cloud-based algorithms for modelling the infrastructure needed to build and control these tools is in place to support the IoT model of modern devices. The Healthcare, Security, and Home Energy fields are quickly adopting these kinds of AAL technologies to support their customers. As the user interfaces and more environmental controls are developed, the uptake of these kinds of smart homes will continue to be more prevalent in our daily lives.

2.4 Conclusion

Ambient Assisted Living (AAL) has begun to pervade our communities in all aspects. Notably, AAL has made great strides in healthcare, the instrumented lifestyle, and energy efficiency. There are many new players emerging as the ability to acquire and deploy Internet of Things (IoT) sensors, use the latest in machine learning (ML) libraries, leverage lessons in user experience (UX), and to refine the use cases our communities need. Tracking the groups and companies involved in this field is a difficult one in today's marketplace. Visiting a conference venue dedicated to these technologies such as the Consumer Electronics Show or IFA demonstrates the scale of effort going into "smart" devices and environments. Knowing the various AAL targets helps new practitioners locate the right resources as they begin to implement their own AAL projects.

References

[1] iControl Networks. "2015 state of the smart home", http://www.icontrol.com/blog/2015-state-of-the-smart-home-report/, (June 2015), accessed 2015.10.14.

[2] S. Yoon, H. Park, H. S. Yoo. "Security issues on smart home in IOT environment", *Computer Science and Its Applications*, pp. 691–696, (Springer Berlin Heidelberg, 2015).

[3] M. Chand, R. L. Tung. "The aging of the world's population and its effects on global business", *The Academy of Management Perspectives*, **28(4)**, pp. 409–429, (2014).
[4] U.S. Census Bureau. "Projections of the population by selected age groups and sex for the United States: 2010 to 2050 constant net international migration series (NP2009-T2-C)", Online, (December 2009), table 2-C.
[5] N. C. for Injury Prevention, C. (NCIPC). "FACTS preventing falls among older adults", http://www.cdc.gov/injury/pdfs/Falls2009CDCFactSheet-FINAL-a.pdf, (May 2009), [Online]. Available: http://www.cdc.gov/injury/pdfs/Falls2009CDCFactSheet-FINAL-a.pdf, accessed 2015.09.27.
[6] E. N. of Living Labs. http://www.openlivinglabs.eu/, accessed 2015.10.14.
[7] M. Fergenson. "Tigerplace: An innovative 'aging in place' community", *The American Journal of Nursing*, **113(1)**, pp. 68–69, (2013).
[8] M. J. Rantz, M. Skubic, S. J. Miller, *et al.* "Sensor technology to support aging in place", *Journal of the American Medical Directors Association*, **14(6)**, pp. 386–391, (2013).
[9] T. Banerjee, J. M. Keller, M. Skubic. "Resident identification using Kinect depth image data and fuzzy clustering techniques", *Engineering in Medicine and Biology Society*, pp. 5102–5105, (IEEE, San Diego, CA, 2012).
[10] T. Banerjee, M. Skubic, J. M. Keller, C. Abbott. "Sit-to-stand measurement for in-home monitoring using voxel analysis", *Biomedical and Health Informatics*, **18(4)**, pp. 1502–1509, (2014).
[11] M. J. Rantz, M. Skubic, R. J. Koopman, *et al.* "Automated technology to speed recognition of signs of illness in older adults", *Journal of Gerontological Nursing*, **38(4)**, p. 18, (2012).
[12] M. J. Rantz, M. Skubic, R. J. Koopman, *et al.* "Automated technology to speed recognition of signs of illness in older adults", *Journal of Gerontological Nursing*, **38(4)**, p. 18, (2012).
[13] Y. Li, K. Ho, M. Popescu. "Efficient source separation algorithms for acoustic fall detection using a Microsoft Kinect", *IEEE Transactions on Biomedical Engineering*, **61(3)**, pp. 745–755, (2014).
[14] L. Liu, M. Popescu, M. Skubic, M. Rantz, T. Yardibi, P. Cuddihy. "Automatic fall detection based on doppler radar motion signature", *Pervasive Computing Technologies for Healthcare*, PervasiveHealth, pp. 222–225, (IEEE, Dublin, Ireland, 2011).
[15] M. Rantz, T. Banerjee, E. Cattoor, S. Scott, M. Skubic, M. Popescu. "Automated fall detection with quality improvement", *"Rewind to Reduce Falls in Hospital Rooms", Journal of Gerontological Nursing Technology Innovations*, (2014).
[16] M. Rantz, M. Skubic, C. Abbott, *et al.* "Automated in-home fall risk assessment and detection sensor system for elders", *The Gerontologist*, **55(Suppl 1)**, pp. S78–S87, (2015).
[17] G. Demiris, M. J. Rantz, M. A. Aud, *et al.* "Older adults' attitudes towards and perceptions of 'smart home' technologies: a pilot study", *Informatics for Health and Social Care*, **29(2)**, pp. 87–94, (2004).

[18] E. E. Stone, M. Skubic. "Silhouette classification using pixel and voxel features for improved elder monitoring in dynamic environments", *Pervasive Computing and Communications Workshops*, PERCOM Workshops, pp. 655–661, (IEEE, Seattle, WA, USA, 2011).

[19] J. A. Kientz, S. N. Patel, B. Jones, E. Price, E. D. Mynatt, G. D. Abowd. "The georgia tech aware home", *extended abstracts on Human factors in computing systems*, CHI '08, pp. 3675–3680, (ACM, New York, NY, USA, 2008).

[20] C. B. Fausset, T. L. Mitzner, C. E. Price, B. D. Jones, B. W. Fain, W. A. Rogers. "Older adults' use of and attitudes toward activity monitoring technologies", *Proceedings of the Human Factors and Ergonomics Society Annual Meeting*, volume 57, pp. 1683–1687, (SAGE Publications, Thousand Oaks, CA, USA, 2013).

[21] C. B. Fausset, L. Harley, S. Farmer, B. Fain. "Older adults' perceptions and use of technology: A novel approach", *Universal Access in Human–Computer Interaction. User and Context Diversity*, pp. 51–58, (Springer, Berlin Heidelberg, 2013).

[22] J. Bidwell, I. A. Essa, A. Rozga, G. D. Abowd. "Measuring child visual attention using markerless head tracking from color and depth sensing cameras", *International Conference on Multimodal Interaction*, pp. 447–454, (ACM, New York, NY, USA, 2014).

[23] N. Nazneen, A. Rozga, C. J. Smith, R. Oberleitner, G. D. Abowd, R. I. Arriaga. "A novel system for supporting autism diagnosis using home videos: Iterative development and evaluation of system design", *JMIR mHealth and uHealth*, **3(2)**, (2015).

[24] L. Bremen, J. Kluge, M. Ziefle, A. Modabber, E. Goloborodko, F. Hölzle. " 'two faces and a hand scan' – pre-and postoperative insights of patients undergoing an orthognathic surgery", *HCI International 2014-Posters Extended Abstracts*, pp. 389–394, (Springer, Berlin Heidelberg, 2014).

[25] M. Ziefle, W. Wilkowska. "Why traditional usability criteria fall short in ambient assisted living environments", *Proceedings of the 8th International Conference on Pervasive Computing Technologies for Healthcare*, pp. 218–222, (ICST (Institute for Computer Sciences, Social-Informatics and Telecommunications Engineering, Gent, Belgium), 2014).

[26] S. Himmel, M. Ziefle, C. Lidynia, A. Holzinger. "Older users' wish list for technology attributes", *Availability, Reliability, and Security in Information Systems and HCI*, pp. 16–27, (Springer, 2013).

[27] S. Mayer, G. Sörös. "User interface beaming – seamless interaction with smart things using personal wearable computers", *International Conference on Wearable and Implantable Body Sensor Networks*, BSN '14, pp. 46–49, (Zurich, Switzerland, 2014).

[28] W. Kleiminger. "Occupancy sensing and prediction for automated energy savings", Ph.D. thesis, ETH Zurich, Zurich, Switzerland, (2015).

[29] M. Kovatsch. "Scalable web technology for the Internet of things", Ph.D. thesis, ETH Zurich, Zurich, Switzerland, (2015).

[30] E. Fleisch. "Business perspectives on ubiquitous computing", *M-Lab working paper*, **(4)**, pp. 83–87, (2001).

[31] S. E. R. C. (SERC). http://serc.wsu.edu/, [Online]. Available: http://serc.wsu.edu/, accessed 2015.10.14.

[32] K. D. Feuz, D. J. Cook. "Transfer learning across feature-rich heterogeneous feature spaces via feature-space remapping (FSR)", *ACM Transactions on Intelligent Systems and Technology (TIST)*, **6(1)**, p. 3, (2015).

[33] L. B. Holder, D. J. Cook. "Automated activity-aware prompting for activity initiation", *Gerontechnology: International Journal on the Fundamental Aspects of Technology to Serve the Ageing Society*, **11(4)**, p. 534, (2013).

[34] P. N. Dawadi, D. J. Cook, M. Schmitter-Edgecombe. "Automated cognitive health assessment using smart home monitoring of complex tasks", *Systems, Man, and Cybernetics: Systems, IEEE Transactions on*, **43(6)**, pp. 1302–1313, (2013).

[35] D. J. Cook, A. S. Crandall, B. L. Thomas, N. C. Krishnan. "Casas: A smart home in a box", *Computer*, **46(7)**, (2013).

[36] L. S. Evangelista, J.-A. Lee, A. A. Moore, *et al.* "Examining the effects of remote monitoring systems on activation, self-care, and quality of life in older patients with chronic heart failure", *The Journal of Cardiovascular Nursing*, **30(1)**, p. 51, (2015).

[37] G. L. Sprint, D. Cook, D. L. Weeks. "Towards automating clinical assessments: A survey of the timed up and go (TUG)", *Biomedical Engineering*, **8**, pp. 64–77, (2015).

[38] WSU CASAS. "Center for Advanced Studies in Adapative Systems (CASAS) at Washington State University", http://casas.wsu.edu/, (2014).

[39] W. CASAS. "CASAS public smart home datasets repository", http://casas.wsu.edu/datasets.html.

[40] D. Kramer, J. C. Augusto, T. Clark. "Context-awareness to increase inclusion of people with ds in society", *Workshops at the Twenty-Eighth AAAI Conference on Artificial Intelligence*, (AAAI, Palo Alto, CA, USA, 2014).

[41] S. Jones, S. Hara, J. Augusto. "eFRIEND: an ethical framework for intelligent environment development", *International Conference on Pervasive Technologies Related to Assistive Environments*, p. 58, (ACM, New York, NY, USA, 2014).

[42] E. Schulze, A. Zirk. "Personalized smart environments to increase inclusion of people with down's syndrome", *Computers Helping People with Special Needs*, pp. 144–147, (Springer, Berlin Heidelberg, 2014).

[43] Z. T. Beattie, C. C. Hagen, T. L. Hayes. "Classification of lying position using load cells under the bed", *Engineering in Medicine and Biology Society*, EMBC, pp. 474–477, (IEEE, New York, NY, USA, 2011).

[44] M. M. Davis, M. Freeman, J. Kaye, N. Vuckovic, D. I. Buckley. "A systematic review of clinician and staff views on the acceptability of incorporating remote monitoring technology into primary care", *Telemedicine and e-HEALTH*, **20(5)**, pp. 428–438, (2014).

[45] Z. T. Beattie, T. L. Hayes, C. Guilleminault, C. C. Hagen. "Accurate scoring of the apnea–hypopnea index using a simple non-contact breathing sensor", *Journal of sleep research*, **22(3)**, pp. 356–362, (2013).

[46] D. Austin, R. M. Cross, T. Hayes, J. Kaye. "Regularity and predictability of human mobility in personal space", *PloS one*, **9(2)**, p. e90256, (2014).

[47] J. Kaye, N. Mattek, H. H. Dodge, *et al.* "Unobtrusive measurement of daily computer use to detect mild cognitive impairment", *Alzheimer's & Dementia*, **10(1)**, pp. 10–17, (2014).

[48] B. Taati, J. Snoek, A. Mihailidis. "Towards aging-in-place: Automatic assessment of product usability for older adults with dementia", *Healthcare Informatics, Imaging and Systems Biology*, HISB, pp. 205–212, (IEEE, 2011).

[49] S. Blackman, C. Matlo, C. Bobrovitskiy, *et al.* "Ambient assisted living technologies for aging well: A scoping review", *Journal of Intelligent Systems*, (2015).

[50] A. Mihailidis, J. Boger, J. Hoey, T. Jiancaro. "Zero effort technologies: Considerations, challenges, and use in health, wellness, and rehabilitation", *Synthesis Lectures on Assistive, Rehabilitative, and Health-Preserving Technologies*, **1(2)**, pp. 1–94, (2011).

[51] J. Allen, G. Ferguson, N. Blaylock, *et al.* "Chester: Towards a personal medication advisor", *Journal of Biomedical Informatics*, **39(5)**, pp. 500–513, (2006), [Online]. Available: http://www.sciencedirect.com/science/article/pii/S1532046406000256, dialog Systems for Health Communications.

[52] K. Samson. "Wearing the detectives: Wristbands, smartwatches, and other wearable devices allow for more real-time monitoring of seizures and other neurologic symptoms–and, possibly, more precise treatment.", *Neurology Now*, **11(4)**, pp. 34–36, (2015).

[53] J. Jardine, J. Fisher, B. Carrick. "Apple's researchkit: smart data collection for the smartphone era?", *Journal of the Royal Society of Medicine*, **108(8)**, pp. 294–296, (2015).

[54] S. Intille, P. Kaushik, R. Rockinson. "Deploying context-aware health technology at home: Human-centric challenges", *Human-Centric Interfaces for Ambient Intelligence*, **22**, (2009).

[55] A. Mannini, S. S. Intille, M. Rosenberger, A. M. Sabatini, W. Haskell. "Activity recognition using a single accelerometer placed at the wrist or ankle.", *Medicine and science in sports and exercise*, **45(11)**, pp. 2193–2203, (2013).

[56] N. Saranummi, D. Spruijt-Metz, S. S. Intille, I. Korhone, W. J. Nilsen, M. Pavel. "Moving the science of behavior change into the 21st century: novel solutions to prevent disease and promote health", *IEEE Pulse*, **4(5)**, pp. 22–24, (2013).

[57] S. U. Northeastern University, MIT. "Wockets, open source accelerometers for phones", http://web.mit.edu/wockets/, accessed 2015.10.15.

[58] B. W. Group. "Boxlab", http://boxlab.wikispaces.com/, accessed 2015.10.15.

[59] mHealth Group. "mhealth publicly available datasets", http://mhealth.ccs.neu.edu/data/, accessed 2015.10.15.

[60] LifeAlert. "Lifealert web site", http://www.lifealerthelp.com/, accessed 2015.10.15.

[61] Philips. "Philips life line", http://www.lifeline.philips.com, accessed 2015.10.15.

[62] GrandCARE. "Grandcare web site", http://www.us.grandcare.com, accessed 2015.10.14.

[63] BeClose. "Beclose web site", http://beclose.com/, accessed 2015.10.15.

[64] C. Innovations. "Quietcare", http://resources.careinnovations.com/quietcare, accessed 2015.10.15.

[65] Live!y. "Mylive!y", http://www.mylively.com/, accessed 2015.10.15.

[66] BehavioMetrics. "Bevy360", http://bevy360.com, accessed 2015.10.15.

[67] SafeinHome. "Safeinhome web site", http://www.safeinhome.com/, accessed 2015.10.15.

[68] HealthSense. "Healthsense solutions", http://healthsense.com/, accessed 2015.10.15.

[69] Memoori. "Wireless lighting controls 2014 to 2018 market research", http://www.memoori.com/portfolio/smart-buildings-wireless-lighting-controls-2014-to-2018/, accessed 2015.10.15.

[70] Belkin. "WeMo web site", http://www.wemo.com, accessed 2015.10.15.

[71] GE. "Ge link bulbs web site", http://gelinkbulbs.com/, accessed 2015.10.15.

[72] Korner. "Korner safe web site", https://www.kornersafe.com/, accessed 2015.10.15.

[73] Control4. "Control4 home automation and smart home control web site", http://www.control4.com.

[74] AMX. "AMX residential solutions", http://www.amx.com, accessed 2015.10.15.

[75] iControl Networks. "icontrol one", http://www.icontrol.com/, accessed 2015.10.15.

[76] Samsung. "Smartthings", http://www.smartthings.com/, [Online]. Available: http://www.smartthings.com/, accessed on 2015.10.10.

[77] M. S. Home. "Haier launches U+ smart home vision", http://www.mysmahome.com/COMPANY/4364/haier-launches-u-smart-home-vision.aspx, accessed 2015.10.15.

[78] G. Apps. "Smart glass windows", http://www.glass-apps.com/products/smart-glass-windows/, accessed 2015.10.15.

[79] X. Huang, J. Baker, R. Reddy. "A historical perspective of speech recognition", *Communications of the ACM*, **57(1)**, pp. 94–103, (2014).

[80] M. C. Mozer. "The neural network house: An environment that adapts to its inhabitants", *Proceedings of the AAAI Spring Symposium on Intelligent Environments*, pp. 110–114, (AAAI Press, Menlo Park, CA, 1998), [Online]. Available: http://www.mendeley.com/research/the-neural-network-house-an-environment-that-adapts-to-its-inhabitants/.

[81] EcoVent. "Ecovent web site", https://www.ecoventsystems.com/, accessed 2015.10.15.

[82] ZoneFirst. "Zonefirst web site", http://zonefirst.com/existing-zone-systems/, accessed 2015.10.15.

[83] A. Rabbani, S. Keshav. "The Spot* system for flexible personal heating and cooling", *ACM Sixth International Conference on Future Energy Systems*, pp. 209–210, (ACM, New York, NY, USA, 2015).

Chapter 3
Ambient and wearable sensors for human health monitoring

Mary Rodgers[1], *Vinay Pai*[1], *and Richard Conroy*[1]

3.1 Introduction

The world's population is aging, and with age comes the potential for an increase in chronic health conditions, a decrease in mobility and impaired senses. This population shift and the desire to provide quality care has resulted in an increasing interest in developing technology-based solutions for enabling and enhancing healthy independent living among older adults, while improving the effectiveness of disease-prevention strategies and access to health care. This interest is paired with the increased availability of new wearable and ambient technologies and an ever-improving health information technology infrastructure. Advances in bioengineering and computing, coupled with demographic shifts, suggest a ripe opportunity for the design and development of appropriate ambient and mobile technologies that enable functional independence and improve quality of life for older adults. Home-health and mobile-health technologies are expected to function not only as monitoring devices, but as essential components in the delivery of health services.

The present-day health-care system relies on detection and characterization of a person's physical and mental states and of significant changes to those states being identified and treated by trained health-care providers. Currently, health-care systems assess health indicators by relying on methods that are constrained by the care provider's schedule, payer models, and geosocial factors and are based on the assumption that observations during an office visit represent typical function. Detection and status are also often identified by burdensome tests that either lack ecological validity or depend heavily on patient recall. Alone, each of these factors may result in missing significant acute events or important signals of declining function, or may poorly characterize detected events. Together, these factors combine to create synergies that result in multiple missed opportunities for preventive or pre-emptive intervention.

Over the last decade, there has been a steady revolution in the development of wearable and ambient sensors for human health monitoring. These technologies strive

[1]National Institute of Biomedical Imaging and Bioengineering, 6707 Democracy Boulevard, Bethesda, MD 20892, USA

to enable monitoring of personal motion, vital signs, and physiological measures in a manner that minimizes disruption to an individual's daily routine and at all times protects their privacy, dignity, and comfort. These systems ideally should integrate, process, analyze, communicate, and present data so that the individuals are engaged and empowered in their own health and reduce their reliance on the health-care system. Effective application of home-based and mobile technologies has the potential to increase adherence to rehabilitation and medical regimens, reduce incidence of avoidable post-acute complications, and improve self-care management of chronic conditions. Additionally, they may be able to capture rare, irregular, or transient events; symptoms that are difficult for a patient to report; and/or changes in condition that evolve slowly over time. These improvements, in turn, could yield more accurate and earlier detection of changes that may interfere with healthy and independent living.

In this chapter, we review the state of the art in wearable mobile and ambient technologies that can be used to track movement behavior. This chapter is not meant to be a comprehensive survey of the field, but highlights some recent research advances in sensors for tracking and analyzing gait and other movements, development of smart home environments, data analysis that can identify people at high risk, and human factors important to the long-term effectiveness of these technologies.

3.2 Wearable sensor technologies

An increase in global population, lifespan, and aging demographics is forcing a rapid rise in health-care costs. The health-care system is also going through a technology and organizational transformation in which continuous monitoring and interventions can be conducted outside the clinical setting. The advancement of sensing technologies, embedded systems, wireless communication technologies, nanotechnologies, and miniaturization make it possible to develop smart systems to monitor activities continuously, pervasively, and nonobtrusively.

Wearable sensors detect abnormal or unforeseen situations by monitoring physiological parameters along with other parameters important for establishing context. There have been several successful cases where technologies have moved out of the clinic to monitor patients going about their day-to-day life over extended periods. Perhaps the most notable of these is the ambulatory electrocardiography device (ECG Holter monitor) for detecting arrhythmias [1]. Wearable sensor systems are progressively becoming more unobtrusive and more powerful, enabling the monitoring of patients for longer periods of time in their normal environment.

Current commercially available systems are compact, enclosed in durable packaging, and utilize either portable local storage or low-power radios to transmit data to remote servers [1, 2]. The development and refinement of novel fabrication techniques, sustainable power sources, inexpensive storage capacity, and more efficient communication strategies are critical to continue this trend toward "wear and forget."

Current wearable sensors typically fall in the categories of accelerometers, pulse oximeters, temperature sensors, bend sensors, bio impedance sensors, electrocardiogram (ECG) sensors, and position sensors. These sensors are primarily used to monitor three types of signals: activity, physiological, and environmental context. Data from these sensors can be collected, analyzed, and made available to the wearers, caregivers, or health-care professionals with the goal of improving the management and delivery of care, engaging patients, and encouraging independent living.

3.2.1 Activity monitors

The analysis of movement can provide many insights into well-being, rehabilitation, and fitness. Noncontact devices such as pedometers have been widely available for many decades. The concept of taking 10,000 steps daily representing the activity energy expenditure to balance the average calorific intake has been developed and refined over the past three decades and embraced by several public health campaigns. It has been the development of low-cost inertial sensors utilizing micro-electromechanical systems (MEMS), and sophisticated software for accurately detecting steps that has resulted in a dramatic rise in the availability and use of the personal activity monitors that is now helping to inform whether this goal is being met on a widespread basis and is giving us a better quantitative measure of how physical exercise correlates with health outcomes. For instance, many personal electronic devices, including some smartphones, music players, and electronic pedometers, can track movement with sufficient sensitivity and specificity. The most accurate sensors under ideal circumstances, and calibrated for healthy adults, are typically accurate to ±3% [3].

The incorporation of multiple sensors including accelerometers, gyroscopes, goniometers, force sensors, and pressure sensors can provide more detailed insight into movement characteristics such as gait [4], falls [5], tremor, and dyskinesia [6]. A number of recent reviews have been published describing the technology involved in these devices, placement considerations, the measurements which they are capable of, and the validation of these measurements [7–9].

The intersection of flexible electronics and activity monitoring provides a rich area of research opportunity. The conversion of mechanical to electrical energy using flexible polymers may prove attractive for energy harvesting [10], though energy density, comfort, and durability are challenges for creating fully self-sustained systems. Nanotechnology, in the form of fiber strain-gauges [11] and carbon nanotubes for detecting strain [12], can be readily included into clothing and other conformal designs, and can be prepared in transparent forms [13]. However, capturing the full range of human motions in a stretchable, nonrestrictive design, and seamless integration with other system components are challenges that need to be addressed.

One area of great promise is the integration of therapeutic and diagnostic systems into theranostic devices. Recently, a multifunctional wearable device has been developed that records muscle activity and is integrated with a controlled transdermal delivery system for releasing nanoparticles [14]. There are also many intriguing technologies being developed that may impact future device design. Spray-on sensors [15] and self-healing polymers [16] are two examples of advances that may address some of the challenges with existing approaches. However, in the shorter term, the focus will potentially remain on integration and durability of technologies, improved algorithms for motion detection, and validation of results against other technologies.

3.2.2 Physiological monitors

For many health-care use cases, it is highly desirable to have sensors capable of directly monitoring the physiology of the wearer in real time. When in contact with the skin, these sensors can also measure biological, chemical, or physical phenomena to assess physiology. The primary technical challenge however is maintaining consistent contact for extended periods and under different conditions, while the primary health-care challenge is how to achieve a high sensitivity and specificity for detecting abnormal events in real time so that these tools can be usefully incorporated into care management.

Maintaining consistent contact with the body is a significant challenge when exposed to the varying conditions of daily life, and there are different strategies being utilized to achieve this. Traditionally adhesive-based and elastic-band approaches have been effective in many circumstances, though epidermal electronics and temporary tattoo approaches are increasingly being investigated [17]. When a sensor is in long-term contact with the body, a number of physical and electrical measurements can be made, including heart rate, breathing rate, blood oxygen saturation, ballistocardiography, blood pressure, electroencephalography (EEG), electrocardiography (ECG), electromyography (EMG), and skin temperature [18]. For EEG, ECG, and EMG, capacitive sensors are typically used to measure bio potentials. In contrast, for vital sign measurements such as heart rate, respiration rate and blood pressure, optical detection techniques such as photoplethysmography or piezoelectric strain sensors are generally used.

The commercial availability of patch-based wearable sensors represents a significant advancement in personal monitoring device design, functionality, and wear time. Examples include HealthPatch™ and Metria™, which are patch-based wearable biometric sensors that adhere to the user's skin and continuously gather physiological, lifestyle information, and other indicators for up to 7 days. These devices contain multiple sensors that enable monitoring of key health indicators such as heart rate, breathing rate, skin temperature, posture, steps taken, activity, and sleep patterns. Data from the patch-based wearable sensors can be uploaded to a computer or mobile device, providing a snapshot of the user's health. The information can be used to guide performance or behavioral modifications in support of sports and fitness or health and wellness goals.

Flexible, temporary transfer tattoo-based sensors, or "electronic skin" or "epidermal electronics," as illustrated in Figure 3.1, show great promise for analyzing metabolites [19] and many other potential applications [20]. The direct printing of multifunctional devices onto the skin may increase durability and mitigate some surface interface issues which is very attractive [21, 22], though the natural skin exfoliation process still limits the lifetime of these devices to a couple of weeks.

For chemical analysis, one of the long-term goals is to realize lab-on-a-chip approaches that can analyze multiple analytes in bodily secretions, such as sweat

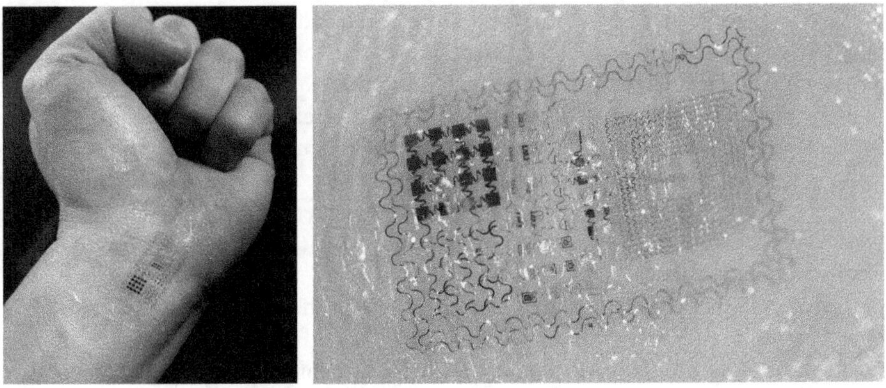

Figure 3.1 Skin-adhering wearable sensor. Photo shows components of a $1 \times 2\ cm^2$ array including transistors, an antenna, power coils, and temperature sensors (Photo courtesy of John Rogers, University of Illinois at Urbana-Champaign)

and saliva. Real-time sweat analysis can provide information on pH, electrolytes and hydration, and there have been efforts to build flexible, textile-based systems [23, 24]. Electrochemical approaches are more amenable to flexible designs than more sensitive or multiplexed optically based technologies, though the tradeoff between complexity of measurement and complexity of system implementation has yet to be fully explored. Biosensors that identify biological molecules or pathogenic organisms are of particular interest for studying complex processes. The ability to detect hormones, enzymes, or lipids would greatly assist in monitoring organ function, viral or bacterial infections, and metabolic disorders. Technologies which may open up new types of measurements in this area include ionogels and organic electrochemical transistors [25]. These approaches have yet to be integrated in fully functional devices deployed in the field, though there is an unmet need to be able to track biochemistry and molecular biomarkers on a routine or continuous basis.

Electrodermal activity (EDA), the change in conductance at the skin surface, is considered to represent activity within the sympathetic nervous system (SNS) [26, 27]. Typically, electrodes are used in either wired form or in combination with conductive gels to measure skin conductance (SC) at various locations on the skin surface, and the conductance values provide a measure of the SNS activity [26]. An interactive, continuous, autonomic logging and monitoring (iCalm) wearable system was proposed as a low power, low cost, and comfortable approach to tracking the SC [28, 29]. The commercial successor to the iCalm EDA sensor is the Affective Q™ Sensor, a wrist-worn sensor for measuring SC [30]. It has been used to monitor SC on the inner wrist (ventral forearm) with 1 cm diameter Ag–AgCl dry electrodes, and quantitatively evaluating the EDA sleep patterns [27].

Chetelat *et al.* [31] have explored the development of "cableless" approaches to measure bio potentials and body impedances. They propose using a high impedance interface layer to electrically isolate the interior of the body, with a conductive garment being used as a single galvanic line. In this case, physiological monitoring including bio potential and impedance measurements can be performed by sensors spread over the body, and embedded in a conductive garment. The electrical connection between each sensor and the conductive garment is achieved simply by contact with the sensor housing. For sensing ECG or other bio potentials, at least one galvanic line is required to serve as reference for biopotential and current return path for impedance measurements. However, since this line does not need to be isolated from the body, the conductive garment can address this need.

Rapin *et al.* [32] have used cooperative dry-electrode sensor concepts to simultaneously measure bio potential and bio impedance signals. In a cooperative approach, the electronics is split into separate electronic sensors, comprising each of two electrodes, one for voltage measurement, and one for current injection. While this approach requires one power supply per sensor, and synchronization among the sensors, it eliminates the need for shielded or multiwire cables. Additionally, since cooperative sensors directly sense and store the measured signals on-site, this eliminates the need for a central recording device.

Reliably detecting and alerting wearers and caregivers to abnormal physiological conditions with sufficiently high sensitivity and specificity will be critical to achieving wider spread adoption, and will be needed before acceptance of semi-automated or closed-loop systems. The challenge in developing a reliable, noninvasive blood glucose sensor illustrates the gap that can exist between demonstrating what is possible and translating that to a universally useful system for providing accurate and reproducible information [33].

3.2.3 Environmental monitors

Monitoring environmental variables around the subject is essential to determine context for physiological and chemical measurements. For example, many people experience short-term exposure to diesel exhaust, however the health effects of such exposures are only recently being measured. Recent studies have shown that short-term exposure to urban roadside diesel traffic may cause statistically significant reduction in lung function (as measured by lung function indices such as forced expiratory volume in the first second), and also lead to acidification of the airways and inflammation of the neutrophils [34].

As per the Centers for Disease Control and Prevention (CDC), during the 8 year period from 2001 to 2009, the proportion of persons of all ages with asthma in the United States increased significantly, going from 7.3% (20.3 million persons) to 8.2% (24.6 million persons), with prevalence increasing significantly among all populations (by age groups and demographic groups). The CDC suggested that one way to ensure that people with asthma can stay symptom-free is by “modifying their environment to reduce or eliminate exposure to allergens and irritants” [35]. Having personal sensor systems to track pollutant exposure at a personal level can not only help subjects who are sensitive to exposure to allergens and irritants but could also help precisely quantify the level of pollutants that subjects are exposed to.

Sensors embedded in textiles have been proposed, and wearable sensors built on flexible substrates have shown promise of detecting very low levels of various chemicals [36], such as volatile organic compounds (VOCs) in the environment. A tuning fork sensor platform approach has been presented as a wearable and wireless sensor system for real-time monitoring of VOCs [37]. This system uses tuning fork sensors coated with polymers or functional molecules to detect different target analytes. As the coating material interacts with analytes in the atmosphere, the resonant frequency of the tuning fork changes; a properly calibrated sensor can then be used to determine the concentration of the analyte that the sensor has been exposed to. While wearable, these devices still tend to be fairly bulky and about the size of two smartphones stacked vertically together. However, the researchers were able to use the system to detect toxic VOCs (such as benzene) in real time; moreover, the system could distinguish these VOCs from highly concentrated common interferents (such as Eau de toilette and mouthwash). For the latter study, the researchers used dual-sensing elements, one responsive to VOC and interferent, and another responsive only to the interferent. Field-effect transistors (FETs) fabricated by using multilayer films made from MoS_2, a transition metal dichalcogenide, and deposited on Si/SiO_2 have been shown to have stable and sensitive response to nitrous oxide (NO) down to a concentration of 0.8 ppm [38]. However, these FET devices do not appear to have been used for wearable sensors to date, though they are likely to be tested for these purposes in the near future.

The use of room temperature ionic-liquid (RTIL) interfaces to build electrochemical gas analysis systems as wearable sensor systems have also been demonstrated. Electrochemical (EC) gas sensors typically have ultra-low-power consumption, low cost, good selectivity, and can sense a wide range of gases [39]. A multimode EC sensory array approach has been proposed within the intelligent electrochemical gas analysis system (iEGAS) to address some of the concerns related to EC sensors (namely, slow response time and high maintenance cost of replacing liquid electrolytes). This system utilizes RTILs-based gas sensors. RTILs are thermally stable to up to 200°C, and gas sensors using RTILs have been shown to have wide detection limits, high sensitivity, and excellent reproducibility [40, 41]. The iEGAS system combines gas sensors, instrumentation electronics, sensor array processing algorithms and system controls in a miniature, light weight, and battery-powered unit (dimensions: 2.6 inch × 2 inch × 1 inch; power: 6.4 mW) [39]. This system is being used to simultaneously track SO_2, NO_2, methane, and oxygen and uses amperometry and impedance spectroscopy to detect these gases.

3.3 Ambient sensors

Ambient sensors embedded in the environment comprise a second major class of sensors that do not, in general, have the same size, power, and other constraints of wearable sensors. While miniaturized wearable sensors can provide detailed mobility information about a specific person wearing the device in most environments, they usually require direct interaction of the wearer since placement of the sensor

Table 3.1 Ambient and wearable sensors: summary of main characteristics relevant to human health monitoring

Type	Sensor	Measurement	Advantages	Limitations
Ambient	Video cameras	Human movement detection	Precise data	Privacy, computation expense, acceptability
Ambient	Microphones	Sound detection	Sound information	Implementation, computation cost, acceptability
Ambient	Binary	User-object interaction, location and movement detection	Inexpensive, low-maintenance, easy installation and replacement, less privacy sensitivity, minimal computation requirements	Simple and limited information for composite and multi-user activity monitoring
Ambient	RFID	Object and user identification	Small size Low cost	Reader and tag collision, range
Ambient	Environmental	Temperature, light, humidity	Intuitive monitoring of environment and object	Not suited for activity monitoring
Ambient	Movement	Ultrasound, radar, infrared	High accuracy, rich data stream	Privacy concern, multiple algorithms require significant infrastructure support
Wearable	Inertial	Acceleration, orientation	Compact size, low cost, non-intrusive, high accuracy, unique identification of users, easy location tracking of user	Cumbersome, uncomfortable, insufficient context information
Wearable	Physiologic	EEG, ECG, temperature, pulse, blood pressure	High sensitivity, accuracy in emergency situation detecting	Reliability, security, comfort with long-term skin attachment
Wearable	Chemical	Sweat, bodily fluids, implantable	Measurement of multiple biomarkers simultaneously	Complexity, limited sampling, reliability, comfort, requires intimate contact

influences accuracy of the measurement. While ambient sensors have several advantages over wearable sensors, they do introduce new layers of complexity, particularly if continuous monitoring is desired or if the environment cannot be easily monitored or controlled. See Table 3.1 for an overview of sensor advantages and limitations.

There has been an explosion of ambient sensors in the past decade, broadly categorized into positioning systems and environmental monitoring sensors. Positioning systems are used for monitoring high-value physical objects as well as for tracking people, animals, and vehicles. Environmental sensors can track indoor and outdoor conditions and monitor a range of phenomena such as climate to the presence of chemicals and biologics. Automation of networks of these sensors has become increasingly practical as their costs have come down and appropriate networking protocols and analysis tools have become available enabling scalability and greater complexity. In addition, these systems are becoming more robust, accurate, and reliable. Complementing these technical advances, the potential health-care challenges associated with an aging population are prompting a search for technologies that can improve or at least maintain wellness and quality of life for older adults.

The embedding and networking of these ambient sensors into personal living spaces has complemented the rise of the "smart home" concept [42]. Smart homes have a very broad definition, and may include home automation, assistive robots, calendar and event management, and networked fixtures and fittings with several goals including reduced energy consumption, multimedia access, improved security, or convenience. For the purpose of this chapter, however, we will focus on assisted living for the elderly using pervasive, intelligent, context-aware systems of sensors. The increasing focus of smart home test-beds on supporting independent elderly living is driven by demographics, costs, and the limited number of family members and health-care professionals available for caregiving. Assessing the mobility of those with impairments directly within their home environment could potentially prevent acute incidents and enable the provision of care or rehabilitation according to an individual needs. It could also reduce the burden on people being monitored if they are not capable or cognizant of using wearable sensors and where privacy is a significant concern.

A number of smart home test-beds for technology-assisted elderly living have been developed in the United States and Europe. The Gator Tech and Matilda Smart Houses at the University of Florida [43], the PlaceLab at MIT [44], TigerPlace at University of Missouri [45], and the Center for Advanced Studies in Adaptive Systems home project at Washington State University [46] are examples of projects that have been used to develop machine learning and activity recognition algorithms for assisted living. Several of these projects report positive effects on the system of health care, both for residents and caregivers.

A significant challenge is that older adults have many specific needs and health issues that need to be accounted for in ambient-assisted living ecosystems. High rates of chronic conditions, neurodegenerative diseases and movement disorders, and physical limitations present special challenges in designing and testing networks of sensors. On the sensor side, technologies must be robust, reliable, continuous and unobtrusive, while algorithms must be sensitive and specific to detecting health emergencies, disease progression, and variable context. Furthermore, these systems must be secure, provide privacy, and help the delivery of better health care at a lower cost. Perhaps one of the most overlooked areas of study is consideration of human factors and the application of human-centered design principles. Acceptability is perhaps the most important factor in getting a deployed technology integrated into daily life and usability is important in getting complete and accurate data. Integration into electronic and personal health records and the workflow of caregivers, allied health-care professionals, and clinicians will be critical to acceptance by the health-care system. Ultimately, a business and social case needs to be made for these networks of sensors and that will require reimbursement, a management plan, a sustainability model, and stakeholder benefit.

There have been several recent reviews on progress toward monitoring technologies that promote health and independent living in the elderly [47–51]. Other recent reviews have examined information challenges associated with smart homes [52] and some emerging business models for ambient-assisted living [53].

Ideally, in the future, these systems would also be prognostic and capable of automated intervention, in addition to detecting and responding to abnormal events. Although there have been significant advances on the development of sensors, the generalization, personalization, and scaling of the informatics tools remain a significant challenge. Building sensor measurements into robust activity classifiers and determining unusual behavior that may have health implications is a significant challenge. Furthermore, training the algorithms for individual preferences, capabilities and behavior as part of a large population is not practical at this stage. Therefore, much of the work to date has focused on acute event detection in small populations, such as falls, trying to develop system-level models, or by making the assumption that inter-individual and intra-individual variability is small compared to the mean signal. The field is evolving rapidly, however, and there has been significant progress toward context-aware activity recognition that is part of a semantic reasoning system with some degree of domain knowledge.

3.3.1 Motion sensors

Ambient motion sensors are sensors that can directly detect position and movement of people in the home environment. Examples of these sensors include passive infrared (IR) arrays, radar, laser range finding, ultrasound, video cameras, and microphones. Liu *et al.* [54] published a detailed survey of the engineering principles and technologies underlying indoor positioning systems and their application in automation, tracking, and control. One of the challenges with these sensors is that the more detailed the information is that they can capture, the higher are the concerns about privacy and security. Unless there is built-in event detection, these sensors also generally produce continuous, high-bandwidth data which needs to be fed through multiple algorithms, resulting in the need for significant infrastructure support. On the other hand, this rich data stream can be mined for many purposes, including gait analysis, interaction times, body position and individual identification, gestures, and facial expressions in a complex environment.

Video cameras are the most commonly used technology for tracking motion due to their nonintrusive nature, rich information, and broad applicability. Several reviews detail the development and progress of video-based techniques in smart homes [55–57]. Ideally vision systems can identify who is in the field of view, where they are specifically within the field of view, what they are interacting with, what they are doing and their movement characteristics. These parameters can be fed into event recognition algorithms to develop state-based models of behavior against which anomalies can be detected [56–58]. One of the challenges of these state-based systems is how to deal with unexpected events, such as significant scene changes, unexpected objects or movement, or unexpected behavior [56]. However these systems can be trained to recognized specific events such as falls [59, 60]. In a long-term retrospective study of video footage over three years in two long-term care facilities, Robinovitch *et al.* [61] identified that more than 40% of resident falls were caused by incorrect weight shifting, more than 20% from trips or falls, and 11% from bumps or lack of support. Such insights can help tune video-based algorithms to maximize their prognostic effectiveness for fall detection.

Other approaches to implementing unobtrusive geriatric assessment in the domestic environment include ultrasonic sensors, laser range sensors, and radar and RGB-depth sensors such as Microsoft Kinect. Yorozu *et al.* [62] recently used a laser range sensor for a multitarget stepping task to evaluate risk of falling. Their leg detection method tracked both legs and measured general walking parameters such as stride length and walking speed and accounted for cases in which one leg was hidden from the sensor or the legs were close, leading to tracking error. Nishiguchi *et al.* [63] developed a device using an IR laser range finder to assess stepping performance through detection of leg motion. They developed a stepping-response score using their device that was able to discriminate fallers and nonfallers in 152 community-dwelling elderly individuals. Kinect has emerged as a popular tool for exergaming (a portmanteau of "exercise" and "gaming") to increase balance and coordination in the elderly [64], but its capabilities of facial recognition, motion analysis, and gesture recognition have also led to it being used for studying gait characteristics in the elderly [65], and as an interactive assessment tool for assessing fall risk [66].

Passive IR sensors are another technology for monitoring movement indoors. They can detect thermal radiation wavelengths longer than 5 microns emitted by humans against background signals from inanimate objects. Their nominal sensing range is 1–12 m with a response time of less than a second, and with the duration of the output signal dependent on speed of movement and temperature difference. Passive IR sensors come in a number of formats from spot detectors to reflector geometries and horizontal scans, and can operate with low power consumption. They have a number of limitations that have hampered their deployment including difficulty with distinguishing animals from humans, detecting motionless heat sources, and working well in a changing or hot environment. Multielement thermopile array sensors can overcome some of these limitations and can detect temperature gradients, multiple objects and even gestures, gait and body shape [67]. Yun and Lee [68] showed that by using three PIR sensors aligned orthogonally in combination with instance-based learning, they were able to demonstrate 92% accuracy in classifying direction and speed from one sensor and 94% accuracy from two sensors.

Radar technologies have also been used for monitoring the elderly [69]. Some researchers prefer to focus on narrow-band solutions, especially those using the Doppler principle [70, 71]. Others opt for broadband solutions, primarily those using pulse type signals [72]. In unscripted environments, these systems have a sensitivity for detecting walking in an accuracy of approximately 92% [73], can distinguish walking with and without a walking aid [74], and have been used for fall detection

[75]. Radar has also been used as a noncontact approach for measuring vital signs [76] and for detecting disordered breathing [77].

3.3.2 Proximity and location-based sensors

Movement can also be inferred from simple location-based and proximity sensors. For example, time-stamped data from door and window switches, pressure pads, light barriers, and binary proximity sensors can be used to infer a person's walking direction or velocity [78] and provide insights into behavioral patterns and changes that may be indicative of stress, decline of faculties, or other important considerations. A complementary approach is the integration of power sensors into the fuse box that monitors appliance usage; this enables evaluation of the self-care ability of adults, and tracking of behaviors such as eating, sleeping, and television watching [79].

Proximity and location-based sensors have very different data properties compared to the other sensors considered in this chapter. These are primarily event-driven sensors that only provide 1-bit information, with generally higher accuracy and reliability. Typically, locations can be inferred with these binary systems because most proximity sensor systems used fixed locations and are triggered within a limited range or on contact, though who or what was at the location and the context of what they are doing can be much more difficult to identify. This limitation is, of course, an advantage if anonymization is important and where there are concerns about intrusion of technologies such as microphones or cameras.

There are a wide variety of technologies that can act as binary location sensors, including acoustic, ultrasound, pressure, contact, radar, millimeter-wave, and IR devices. Often these technologies can operate in multiple modes, though the power consumption is many-fold lower in a binary mode. To achieve binary operation, either a simple threshold can be set or a more elaborate onboard signal processing algorithm can be used, though these predetermined approaches can be limiting in environments where there is a wide dynamic range in scenes. As an example, one project to demonstrate their utility for monitoring movement was a study of stair use at the CDC. Proximity sensors were used to study the use of stairs by employees in response to four interventions, with the study finding that motivational signs and music significantly increased the usage of stairs [80]. Additionally, the practicality of proximity sensors [81, 82], pressure sensors [83], and contact switches [83] has all been demonstrated in smart home test-beds.

Despite the minimal information available from 1-bit sensors, within certain limits, networks can provide target tracking using combinatorial, greedy algorithms [84]. Alternatively, if targets move relatively smoothly, then probabilistic particle filter algorithms, such as ClusterTrack, can be used to identify and track different targets [85]. These sensor area networks (SANs) are becoming increasingly complex with validation of signal strengths, energy availability and buffering capacity, and there are general algorithms for analyzing traffic independent of sensor topology [86]. Arrays and networks of simple sensors can also be used as an assessment tool to give regular feedback on mobility [87], on monitoring sleep [88], or for guiding-assisted mobility devices [89], and can be scaled up to provide complete coverage in a building [90].

Although proximity sensors are normally embedded in the infrastructure, this is not a requirement. Tracking of a network of moving proximity sensors has also been demonstrated. Interactions between 119 participants in a pediatric ward involving health-care workers, patients, and caregivers were studied over a week-long period to understand how nosocomial infections may spread [91]. Radio-frequency identification tags (RFIDs) with a range of 1–1.5 m sensed more than 16,000 contacts, 95% of which lasted less than 4 min, with ward assistants having the most daily contacts at nearly 100 every day. From modeling this interaction data, the authors concluded that caregivers are probably not favoring the spread of potential infections to a large number of individuals, but that long and frequent interactions with nurses were more likely the cause. The authors also noted some of the limitations of using the RFID proximity sensors, including not being able to differentiate proximity from contact, and the potential for losing data because of dropouts in the communication infrastructure. A study looking at the activity of 135 hens in a noncage housing system validated proximity sensor data against video data, showing overall agreement of at least 84% for a hen's proximity to specific resources such as water, feeder, and nest-box [92]. Although the investigators were not able to identify a social hierarchy or whether the sensors modified behavior, they were able to demonstrate the feasibility of automated activity classification for some tasks, such as egg laying, using the sensor data. These systems can also be employed in geo-fencing applications and trajectory tracking algorithms, where there may be a concern that wandering is a sign of cognitive decline [93].

3.3.3 Environment sensors

Environmental information can help provide important contextual information for improving the sensitivity and specificity of signals derived from the sensors that directly measure motion. For example, high levels of pollen, dander, mold, or dust may lead to exacerbation of asthma that can change mobility patterns [94]. More broadly, environmental sensors can detect anomalous situations that may give rise to unusual behavior. Sleep patterns in older adults with neurodegenerative disorders can be impacted by high-intensity environments [95] where there are changes in lighting, from beeping of a battery in a fire alarm that needs replaced to rapid changes in the temperature and humidity, and can in turn change daily mobility patterns.

Environmental monitoring in the home is a key component of the smart home concept and has led to intelligent thermostats, networked smoke, and fire alarms and automated humidity control. The emergence of standards for low-cost, low-power communication, such as Zigbee, Z-wave, and Bluetooth Low Energy, has helped drive access to the internet of things for the consumer market [96]. Smart and connected environmental sensors are increasingly being embedded in consumer products to adapt operating characteristics to a changing environment to increase convenience and reduce power consumption. A number of studies have been published showing the implementation of integrated, scalable systems for monitoring the environment in the home [97, 98].

Many chemical species can be detected using a wide range of sensors, from inexpensive home monitors to emergency response and military grade hardware. There has been great interest in detecting and differentiating VOCs in the home environment, and progress has been made to develop compact sensors for formaldehyde [99], benzene [100], and reactive oxygen species [101]. There is particular concern for understanding how childhood exposure affects human development [102]. These sensors may also have applications in sensing food spoilage or contamination [103] and for breath analysis of VOCs that may be indicative of physiological changes [104] or unusual levels of drugs or alcohol [105].

For monitoring aerosols and particulate matter, a wide range of environmental field sensors have been developed, including low-cost, battery-powered systems [106]. Much attention has been directed toward monitoring personal exposure to air pollution with high spatial and temporal resolution [107], and for tracking exposure to smoke [108].

Outside the home, a number of groups have started trying to build integrated sensor platforms for environmental monitoring. Historically, this has focused on collecting weather data, and several internet community groups have sprung up to support sharing of weather data and improving weather models, such as the Citizen Weather Observer Program (CWOP) [109]. The emergence of inexpensive networkable, multisensor platforms such as Arduino, Raspberry Pi, and TI's SensorTag has led to reduction of barriers to participation and integration of other sensor types [110]. For example, the Smart Citizen project, a crowd-funded initiative started in Spain in 2012, has developed an Arduino shield that measures CO, NO_2, light, sound, temperature, and humidity, and provides a network of more than 6000 nodes across Europe and the Americas [111]. This bottom-up approach is starting to integrate with the top-down concept of smart cities and is likely to change how individuals, communities, and institutions communicate, cooperate, and collaborate [112]. A number of small businesses have sprung up to commercialize the knowledge that can be gained from networks of environmental sensors such as Aclima [113].

3.3.4 Sensor informatics

The informatics behind converting sensor data to activity recognition is perhaps the biggest area of opportunity over the next decade. Converting data streams into patterns and then actionable information requires reasoning, context awareness, ability to choose between event detection algorithms, and to carry out sensitive and specific, just-in-time pattern recognition. Historically much of the focus has been on image-based sensing where a single camera can capture subject as well as context information. However, the emergence of inexpensive sensor networks, better algorithms, and privacy concerns has led to more significant interest in fusing and analyzing one-dimensional data streams.

Activity recognition algorithms generally fall into two approaches: (1) machine learning and (2) logic modeling. Machine learning approaches have focused on supervised techniques, using expert labeled data as training, and unsupervised techniques that use probabilistic and statistical reasoning. Many algorithms and models have been developed for supervised learning, including Bayes networks, support vector machines, and Hidden Markov Models [52]. Variants of these algorithms can be used for unsupervised machine learning, where the most significant difference is that the model dynamically learns the parameters of the model through an iterative process keeping track of previous observations. The advantages of these machine learning approaches are that they somewhat robust against noisy and missing data, can give an estimate of uncertainty, and can be integrated with heuristic models of activity. The disadvantages of these approaches are that they are not easily scalable and reusable to many different individuals and behaviors and can have a high computational overhead, as in the case of supervised systems which require significant expert input.

Logic-based activity recognition algorithms represent patterns in sensor data as logical terms that are then reasoned as a logic formula in a library of activity models. The advantages of these approaches are that they are semantically clear, and can readily incorporate domain knowledge and heuristics. Disadvantages include their inability to estimate uncertainty, difficulty in differentiating between different models of behavior, and difficulty with adapting and learning over time. Ontologies for these logic models that are independent of algorithms are increasingly attractive for algorithm comparison, data sharing, and interoperability. A number of smart home studies have been published that incorporate these ontology-driven approaches [114–118]. Dynamically annotating continuous data streams from sensors, modeling complex activities, and automatically capturing domain knowledge for an individual's specific activities are outstanding research challenges in this area and are the subject of on-going research [52].

3.4 Conclusions

The growth of interest in the "Internet of Things" is highlighted by the rapid increase in the number of intelligent sensors embedded in our environment. Many futurists predict that we will surpass more than a trillion sensors within the next 5–10 years, emphasizing how pervasive they are predicted to become [119]. These sensors will transform our homes, workplace, and society and change how we monitor health and wellness. In order for these sensors to be viewed as beneficial, a multidisciplinary approach involving behavioral scientists, computer scientists, and physical scientists working together with human factors engineers, clinicians, public health, and allied health professionals to demonstrate the efficacy, applicability, reliability, and security of sensor systems will be required. The wider hope is that these sensors can reduce costs, improve quality and experience of care for individual, and engage all citizens in improving the health of the nation through reducing disparities and increasing the use of health information technology.

With 20% of the world's population expected to be age 60 or older by 2050 [120], there is a natural convergence between elderly health care and the use of these sensors to monitor health. Increasing health-care costs, a shortage of caregivers and many elderly living with multiple chronic conditions is putting the traditional health-care system under stress. Given the costs and loss of independence associated with nursing home care, 89% of older adults prefer to maintain an independent life in the comfort of their own home [121] though this will only be possible if technology tools are available that can monitor, alert, and predict events and engage the elderly in management of their health. The appropriate design of this technology is crucial so that it empowers older adults rather than overwhelms them [122].

To date, most sensor deployments for research have focused on a small (<100) number of sensor systems deployed with limited number of sensors on broad and periodic sampling. These systems are also maintained by experts and a costly infrastructure. However, a growing number of citizen science projects and commercial entities are trying to realize large-scale, pervasive sensor networks that have the potential to transform the way in which we understand our environment.

Technical and cost-benefit challenges remain for wider use of these sensors beyond these interested and motivated groups. Energy, reliability, communication, and computational limitations are all significant, inter-related technical challenges that need systems-level solutions and can be case-specific. For example, the question of whether to locally store data or transmit it is one of many complex issues challenging the scalability of these technologies.

The big data issue associated with a growing array of networked sensors is perhaps the greatest challenge and opportunity moving forward. The variety, volume, velocity, and veracity of these data streams, the open world in which these sensors operate, and the desire to generate just-in-time actionable knowledge as well as realizing a learning health-care system highlight the complexity of the challenge and the high expectations. Many of the test-beds work in a more controlled, predictable environment, with significant computational and expert resources on hand and limited numbers of serial activities being tracked. However, for real-world applications, these systems need to be scalable, track concurrent behaviors and multiple users with low resource overheads and ensuring privacy and interoperability. Access to information from these systems and the decision-making process that this knowledge triggers, will also lead to many political, economic, legal, and social discussions and shape how this technology is adopted and used for health care.

3.4.1 Future directions

Ambient and wearable sensor technologies are predicted to become increasingly pervasive in our daily lives. With the lengthening lifespan of our population, we are being challenged to address chronic disease management and to improve quality of life in the final years. There is a natural convergence between sensor technology, mobile communications, wellness management, and elderly care that is on one hand being pulled by the desire of older adults to remain mobile and live independent lives, and on the other hand being pushed by the personnel and fiscal limitations of the health-care system. The evidence for the efficacy and practicality of these ambient and wearable sensors to address these challenges is expanding, however with pervasive sensing come many political, economic, social, and technical challenges and opportunities.

The most prominent concerns about sensors networks are around security, privacy, and misuse; however, there are many other political and legal challenges to more widespread use. On the legal side, the questions of liability, malfeasance, and malpractice related to information and actions derived from assisted living systems are yet to be fully answered. With health-care management moving out of the clinic and into the home, the rights of individuals to be informed when data is being collected about them, who will have access to it, and how it will be used are being actively debated in the bioethics community. In the United States, in particular, there is concern around who owns and has access to an individual's health-related data and the decisions this may lead to. When this emerged as an issue for genetic information, the US government passed the Genetic Information Nondiscrimination Act in 2008 to prevent discrimination based on genetic information when it comes to health insurance and employment, however such protections do not yet exist for health-related sensor data. The rapid emergence of inexpensive sensors for monitoring physical exercise is blurring the lines between medical grade equipment for diagnostic and monitoring purposes and consumer grade equipment for lifestyle enhancement and entertainment. The question of how to regulate the use of sensors as medical devices and the associated underlying informatics networks is not fully resolved.

With health-care costs dramatically rising, there are increasing efforts to promote the use of cost-effective approaches. In the United States, 90% of adults own a cell phone, 71% of which are smart phones and the increasing use of them for lifestyle enhancement and entertainment provides an irresistible platform on which to build economically efficient health-care solutions [123]. The proliferation of sensors that are either embedded in cell phones or that make use of the same communication networks have dramatically expanded over the past five years; the sensor market is predicted to expand dramatically from a $10 billion industry in 2012 to $1 trillion in the next 10 years. This market will be extremely diverse, from municipal infrastructure management to gaming, but mobile and home-based health care is expected to play a significant role in driving demand. To realize this demand will require feasible investment and maintenance strategies either through out-of-pocket expenditure and insurance coverage or modifying of reimbursement packages and bundling of sensor services with other health-care management approaches.

The success of technologies for assisted living depends on social acceptability, access, and engagement. Chronic conditions (including impaired mobility, failing eyesight and hearing loss, higher rates of social isolation, and polypharmacy) increase the complexity and challenges for real-life implementation of sensor systems. Cultural differences, individual preferences, and community values can significantly affect adoption, adherence, and health impact. These social factors extend beyond the elderly person, because there are often multiple care-givers supporting their needs and these systems need to engage them and integrate with their daily lives. With older adults at greater risk of being economically disadvantaged, and lower socioeconomic status correlating with lower levels of care, worse psychological health and higher rates of morbidity and mortality, it is important that the barriers to access and use of these technologies are minimized so that these systems can be used to address some of our current health disparities.

Some futurists forecast that we are entering a third industrial revolution with the fusion of computing, communication, and sensing and greater accessibility to rapid prototyping, bioengineering, and energy sources. Many businesses have developed visions for a connected, smart world that is capable of supporting trillions of nodes in a heterogeneous global-scale network for applications ranging from climate monitoring to supply chain management and smart homes. Many analogies from swarms to central nervous systems have been used to describe this pervasive internet of things though it is clear that the lifecycle of technologies is becoming dramatically shorter, with more rapid development and adoption on a global scale. Networks of sensors are now capable of measuring and recording everything from temperature, light, and motion, to biohazards and physical indicators from the body. In the future, we are likely to see the emergence of more complex sensors such as ingestible and implantable sensors that monitor the body from within, mood measurement and enhancement, sensors for tracking and intervening with social interactions and advances in miniaturization, networking and self-powered that will make smart dust swarms practical. One ultimate outcome for this direction of technology development may be better integration of biological and semiconductor systems to enhance our lives and improve health. Declining costs and thoughtful user-centered design of sensor technology will make these developments accessible, widely used and an integral part of the digital ecosystem in the upcoming decade.

References

[1] P. Bonato, 'Wearable sensors/systems and their impact on biomedical engineering,' *Engineering in Medicine and Biology Magazine, IEEE*, vol. 22, pp. 18–20, May–June 2003.

[2] P. Bonato, 'Wearable sensors and systems,' *Engineering in Medicine and Biology Magazine, IEEE*, vol. 29, pp. 25–36, 2010.

[3] M. Karabulut, S.E. Crouter, and D.R. Bassett, Jr., 'Comparison of two waist-mounted and two ankle-mounted electronic pedometers,' *European Journal of Applied Physiology*, vol. 95, pp. 335–343, Oct 2005.

[4] W.J. Tao, T. Liu, R.C. Zheng, and H.T. Feng, 'Gait analysis using wearable sensors,' *Sensors*, vol. 12, pp. 2255–2283, Feb 2012.

[5] S. Chaudhuri, H. Thompson, and G. Demiris, 'Fall detection devices and their use with older adults: A systematic review,' *Journal of Geriatric Physical Therapy*, vol. 37, pp. 178–196, Oct–Dec 2014.

[6] B. Cole, S. Roy, C. De Luca, and S. Nawab, 'Dynamical learning and tracking of tremor and dyskinesia from wearable sensors,' *IEEE Transactions on Neural Systems and Rehabilitation Engineering*, vol. 22, pp. 982–991, Mar 19, 2014.

[7] C.C. Yang and Y.L. Hsu, 'A review of accelerometry-based wearable motion detectors for physical activity monitoring,' *Sensors*, vol. 10, pp. 7772–7788, 2010.

[8] D.R. Bassett, Jr., A. Rowlands, and S.G. Trost, 'Calibration and validation of wearable monitors,' *Medicine and Science in Sports and Exercise*, vol. 44, pp. S32–38, Jan 2012.

[9] N.F. Butte, U. Ekelund, and K.R. Westerterp, 'Assessing physical activity using wearable monitors: Measures of physical activity,' *Medicine and Science in Sports and Exercise*, vol. 44, pp. S5–12, Jan 2012.

[10] F.R. Fan, Z.Q. Tian, and Z. L. Wang, 'Flexible triboelectric generator!,' *Nano Energy*, vol. 1, pp. 328–334, Mar 2012.

[11] C. Pang, G.Y. Lee, T.I. Kim, *et al.*, 'A flexible and highly sensitive strain-gauge sensor using reversible interlocking of nanofibres,' *Nature Materials*, vol. 11, pp. 795–801, Sep 2012.

[12] T. Yamada, Y. Hayamizu, Y. Yamamoto, *et al.*, 'A stretchable carbon nanotube strain sensor for human-motion detection,' *Nature Nanotechnology*, vol. 6, pp. 296–301, May 2011.

[13] D.J. Lipomi, M. Vosgueritchian, B.C. Tee, *et al.*, 'Skin-like pressure and strain sensors based on transparent elastic films of carbon nanotubes,' *Nature Nanotechnology*, vol. 6, pp. 788–792, Dec 2011.

[14] D. Son, J. Lee, S. Qiao, *et al.*, 'Multifunctional wearable devices for diagnosis and therapy of movement disorders,' *Nature Nanotechnology*, vol. 9, pp. 397–404, May 2014.

[15] S. Schreml, R.J. Meier, K.T. Weiss, *et al.*, 'A sprayable luminescent pH sensor and its use for wound imaging in vivo,' *Experimental dermatology*, vol. 21, pp. 951–953, Dec 2012.

[16] B.C.K. Tee, C. Wang, R. Allen, and Z.N. Bao, 'An electrically and mechanically self-healing composite with pressure- and flexion-sensitive properties for electronic skin applications,' *Nature Nanotechnology*, vol. 7, pp. 825–832, Dec 2012.

[17] D.H. Kim, N. Lu, R. Ma, *et al.*, 'Epidermal electronics,' *Science*, vol. 333, pp. 838–843, Aug 12, 2011.

[18] J.R. Windmiller and J. Wang, 'Wearable electrochemical sensors and biosensors: A review,' *Electroanalysis*, vol. 25, pp. 29–46, Jan 2013.

[19] J.R. Windmiller, A.J. Bandodkar, G. Valdes-Ramirez, S. Parkhomovsky, A.G. Martinez, and J. Wang, 'Electrochemical sensing based on printable temporary transfer tattoos,' *Chemical Communications*, vol. 48, pp. 6794–6796, Jul 11, 2012.

[20] S. Bauer, 'Flexible electronics: Sophisticated skin,' *Nature Materials*, vol. 12, pp. 871–872, Oct 2013.

[21] W.H. Yeo, Y.S. Kim, J. Lee, *et al.*, 'Multifunctional epidermal electronics printed directly onto the skin,' *Advanced materials*, vol. 25, pp. 2773–2778, May 28, 2013.

[22] S. Xu, Y. Zhang, L. Jia, *et al.*, 'Soft microfluidic assemblies of sensors, circuits, and radios for the skin,' *Science*, vol. 344, pp. 70–74, Apr 4, 2014.

[23] D. Diamond, S. Coyle, S. Scarmagnani, and J. Hayes, 'Wireless sensor networks and chemo/biosensing,' *Chemical Reviews*, vol. 108, pp. 652–679, Feb 2008.

[24] D. Morris, S. Coyle, Y.Z. Wu, K.T. Lau, G. Wallace, and D. Diamond, 'Bio-sensing textile based patch with integrated optical detection system for sweat monitoring,' *Sensors and Actuators B-Chemical*, vol. 139, pp. 231–236, May 20, 2009.

[25] K.J. Fraser, V.F. Curto, S. Coyle, *et al.*, 'Wearable electrochemical sensors for monitoring performance athletes,' *Proceedings of the SPIE 8118, Organic Semiconductors in Sensors and Bioelectronics IV*; San Diego, USA, 2011. pp. 81180C–81180C-12.

[26] W. Boucsein, *Electrodermal Activity*, New York: Springer; 1992.

[27] A. Sano and R.W. Picard, 'Toward a taxonomy of autonomic sleep patterns with electrodermal activity,' *Proceedings of the IEEE Engineering in Medicine and Biology Society Annual Conference*, New York, NY, August 30–September 3, 2011. vol. 2011, pp. 777–780.

[28] R.R. Fletcher, K. Dobson, M.S. Goodwin, *et al.*, 'iCalm: wearable sensor and network architecture for wirelessly communicating and logging autonomic activity,' *IEEE Transactions on Information Technology in Biomedicine*, vol. 14, pp. 215–223, Mar 2010.

[29] H. Eydgahi, 'Design and evaluation of iCalm: A novel, wrist-worn, low-power, low-cost, wireless physiological sensor module,' S. M., Dept. of Electrical Engineering and Computer Science, Massachusetts Institute of Technology, Boston, 2008. Accessed October 2015. http://hdl.handle.net/1721.1/44690.

[30] M.Z. Poh, N.C. Swenson, and R.W. Picard, 'A wearable sensor for unobtrusive, long-term assessment of electrodermal activity,' *IEEE Transactions on Bio-Medical Engineering*, vol. 57, pp. 1243–1252, May 2010.

[31] O. Chetelat, R. Gentsch, J. Krauss, and J. Luprano, 'Getting rid of the wires and connectors in physiological monitoring,' *2008 30th Annual International Conference of the IEEE Engineering in Medicine and Biology Society*, New York, NY, August 20–25, 2008, vols. 1–8, pp. 1278–1282.

[32] M. Rapin, M. Proenca, F. Braun, *et al.*, 'Cooperative dry-electrode sensors for multi-lead biopotential and bioimpedance monitoring,' *Physiological Measurement*, vol. 36, pp. 767–783, Apr 2015.

[33] S.K. Vashist, 'Non-invasive glucose monitoring technology in diabetes management: A review,' *Analytica Chimica Acta*, vol. 750, pp. 16–27, Oct 31, 2012.

[34] J.J. Zhang, J.E. McCreanor, P. Cullinan, *et al.*, 'Health effects of real-world exposure to diesel exhaust in persons with asthma,' *Research Report*, pp. 5–109; discussion 111–123, Feb 2009.

[35] H.S. Zahran, C. Bailey, and P. Garbe, 'Vital signs: Asthma prevalence, disease characteristics, and self-management education: United States, 2001–2009,' *MMWR. Morbidity and Mortality Weekly Report*, vol. 60, pp. 547–552, May 6, 2011.

[36] M.C. McAlpine, H. Ahmad, D.W. Wang, and J.R. Heath, 'Highly ordered nanowire arrays on plastic substrates for ultrasensitive flexible chemical sensors,' *Nature Materials*, vol. 6, pp. 379–384, May 2007.

[37] F. Tsow, E. Forzani, A. Rai, *et al.*, 'A wearable and wireless sensor system for real-time monitoring of toxic environmental volatile organic compounds,' *IEEE Sensors Journal*, vol. 9, pp. 1734–1740, Dec 2009.

[38] Y. Hou, L.T. Song, P. Zhu, *et al.*, 'Single-cell exome sequencing and monoclonal evolution of a JAK2-negative myeloproliferative neoplasm,' *Cell*, vol. 148, pp. 873–885, Mar 2, 2012.

[39] H.T. Li, X.Y. Mu, Y.N. Yang, and A.J. Mason, 'Low power multimode electrochemical gas sensor array system for wearable health and safety monitoring,' *IEEE Sensors Journal*, vol. 14, pp. 3391–3399, Oct 2014.

[40] M.C. Buzzeo, O.V. Klymenko, J.D. Wadhawan, C. Hardacre, K.R. Seddon, and R.G. Compton, 'Kinetic analysis of the reaction between electro-generated superoxide and carbon dioxide in the room temperature ionic liquids 1-ethyl-3-methylimidazolium bis(trifluoromethylsulfonyl)imide and hexyltriethylammonium bis(trifluoromethylsulfonyl)imide,' *Journal of Physical Chemistry B*, vol. 108, pp. 3947–3954, Mar 25, 2004.

[41] R. Wang, T. Okajima, F. Kitamura, and T. Ohsaka, 'A novel amperometric O-2 gas sensor based on supported room-temperature ionic liquid porous polyethylene membrane-coated electrodes,' *Electroanalysis*, vol. 16, pp. 66–72, Jan 2004.

[42] C.N. Scanaill, S. Carew, P. Barralon, N. Noury, D. Lyons, and G.M. Lyons, 'A review of approaches to mobility telemonitoring of the elderly in their living environment,' *Annals of Biomedical Engineering*, vol. 34, pp. 547–563, Apr 2006.

[43] S. Helal, W. Mann, H. El-Zabadani, J. King, Y. Kaddoura, and E. Jansen, 'The Gator Tech Smart House: A programmable pervasive space,' *Computer*, vol. 38, pp. 50–60, Mar 2005.

[44] B. Logan, J. Healey, M. Philipose, E.M. Tapia, and S. Intille, 'A long-term evaluation of sensing modalities for activity recognition,' *UbiComp 2007: Ubiquitous Computing, Proceedings*, Berlin, Germany, September 21–24, 2007, vol. 4717, pp. 483–500.

[45] M. Rantz, M. Aud, G.L. Alexander, *et al.*, 'TigerPlace: An innovative educational and research environment,' in *AAAI Fall Symposium: AI in Eldercare: New Solutions to Old Problems*, Palo Alto, California, November 7–9, 2008, pp. 84–91.

[46] G. Singla, D.J. Cook, and M. Schmitter-Edgecombe, 'Recognizing independent and joint activities among multiple residents in smart environments,' *Journal of Ambient Intelligence and Humanized Computing*, vol. 1, pp. 57–63, Mar 2010.

[47] Q. Ni, A.B. Garcia Hernando, and I.P. de la Cruz, 'The elderly's independent living in smart homes: A characterization of activities and sensing infrastructure survey to facilitate services development,' *Sensors*, vol. 15, pp. 11312–11362, 2015.

[48] P. Rashidi and A. Mihailidis, 'A survey on ambient-assisted living tools for older adults,' *IEEE Journal of Biomedical and Health Informatics*, vol. 17, pp. 579–590, May 2013.

[49] K.K. Peetoom, M.A. Lexis, M. Joore, C.D. Dirksen, and L.P. De Witte, 'Literature review on monitoring technologies and their outcomes in independently living elderly people,' *Disability and Rehabilitation. Assistive Technology*, vol. 10, pp. 271–294, Jul 2015.

[50] M.R. Alam, M.B. Reaz, and M.A.M. Ali, 'A review of smart homes – past, present, and future,' *IEEE Transactions on Systems Man and Cybernetics Part C – Applications and Reviews*, vol. 42, pp. 1190–1203, Nov 2012.

[51] D. Ding, R.A. Cooper, P.F. Pasquina, and L. Fici-Pasquina, 'Sensor technology for smart homes,' *Maturitas*, vol. 69, pp. 131–136, Jun 2011.

[52] L. Cheng, C. Eng, L.Z. Nieman, A.S. Kapadia, and X.L. Du, 'Trends in colorectal cancer incidence by anatomic site and disease stage in the United States from 1976 to 2005,' *American Journal of Clinical Oncology*, vol. 34, pp. 573–580, Dec 2011.

[53] V. Vimarlund and S. Wass, 'Big data, smart homes and ambient assisted living,' *Yearbook of Medical Informatics*, vol. 9, pp. 143–149, 2014.

[54] H. Liu, H. Darabi, P. Banerjee, and J. Liu, 'Survey of wireless indoor positioning techniques and systems,' *IEEE Transactions on Systems Man and Cybernetics Part C – Applications and Reviews*, vol. 37, pp. 1067–1080, Nov 2007.

[55] R. Harper (ed.), *Inside the Smart Home*, Bristol: Springer Science & Business Media; 2003.

[56] L.C. De Silva, C. Morikawa, and I.M. Petra, 'State of the art of smart homes,' *Engineering Applications of Artificial Intelligence*, vol. 25, pp. 1313–1321, 2012.

[57] M. Brezovan and C. Badica, 'A review on vision surveillance techniques in smart home environments,' *Control Systems and Computer Science (CSCS), 2013 19th International Conference*, New York, NY, May 29–31, 2013, pp. 471–478.

[58] B. Ni, G. Wang, and P. Moulin, 'Rgbd-hudaact: A color-depth video database for human daily activity recognition,' in A. Fossati, J. Gall, X. Ren, K. Konolige (eds.) *Consumer Depth Cameras for Computer Vision*, (London, UK: Springer, 2013), pp. 193–208.

[59] L.C. De Silva and B. Darussalam, 'Audiovisual sensing of human movements for home-care and security in a smart environment,' *International Journal on Smart Sensing and Intelligent Systems*, vol. 1, pp. 220–245, 2008.

[60] V. Vaidehi, K. Ganapathy, K. Mohan, A. Aldrin, and K. Nirmal, 'Video based automatic fall detection in indoor environment,' *Recent Trends in Information Technology (ICRTIT), 2011 International Conference*, New York, NY, June 3–5, 2011, pp. 1016–1020.

[61] S.N. Robinovitch, F. Feldman, Y. Yang, *et al.*, 'Video capture of the circumstances of falls in elderly people residing in long-term care: An observational study,' *The Lancet*, vol. 381, pp. 47–54, 2013.

[62] A. Yorozu and M. Takahashi, 'Gait measurement system for the elderly using laser range sensor,' *Applied Mechanics and Materials*, vols. 490–491, pp. 1629–1635, 2014.

[63] S. Nishiguchi, M. Yamada, K. Uemura, *et al.*, 'A novel infrared laser device that measures multilateral parameters of stepping performance for assessment of all risk in elderly individuals,' *Aging Clinical and Experimental Research*, vol. 25, pp. 311–316, 2013/06/01 2013.

[64] T.C. Davies, T. Vinumon, L. Taylor, and J. Parsons, 'Let's Kinect to increase balance and coordination of older people: Pilot testing of a balloon catching game,' *International Journal of Virtual Worlds and Human Computer Interaction*, vol. 2, pp. 37–46, 2014.

[65] E. Stone, M. Skubic, M. Rantz, C. Abbott, and S. Miller, 'Average in-home gait speed: Investigation of a new metric for mobility and fall risk assessment of elders,' *Gait & Posture*, vol. 41, pp. 57–62, 2015.

[66] A. Ejupi, Y.J. Gschwind, T. Valenzuela, S.R. Lord, and K. Delbaere, 'A Kinect and inertial sensor-based system for the self-assessment of fall risk: A home-based study in older people,' *Human–Computer Interaction*, vol. 31, pp. 261–293, 2015.

[67] J. Yun and S.S. Lee, 'Human movement detection and identification using pyroelectric infrared sensors,' *Sensors*, vol. 14, pp. 8057–8081, 2014.

[68] J. Yun and S.S. Lee, 'Human movement detection and identification using pyroelectric infrared sensors,' *Sensors*, vol. 14, pp. 8057–8081, 2014.

[69] F. Ahmad, R.M. Narayanan, and D. Schreurs, 'Application of radar to remote patient monitoring and eldercare,' *Radar, Sonar & Navigation, IET*, vol. 9, pp. 115–115, 2015.

[70] Y.Z. Wang, Q.H. Liu, and A.E. Fathy, 'CW and pulse-Doppler radar processing based on FPGA for human sensing applications,' *IEEE Transactions on Geoscience and Remote Sensing*, vol. 51, pp. 3097–3107, May 2013.

[71] O. Postolache, P. Girão, E. Pinheiro, *et al.*, 'Multi-usage of microwave Doppler radar in pervasive healthcare systems for elderly,' *Instrumentation and Measurement Technology Conference (I2MTC), 2011 IEEE*, New York, NY, May 10–12, 2011, pp. 1–5.

[72] J. Li, L.B. Liu, Z.F. Zeng, and F.S. Liu, 'Advanced signal processing for vital sign extraction with applications in UWB radar detection of trapped victims in complex environments,' *IEEE Journal of Selected Topics in Applied Earth Observations and Remote Sensing*, vol. 7, pp. 783–791, Mar 2014.

[73] C.E. Phillips, J. Keller, M. Popescu, *et al.*, 'Radar walk detection in the apartments of elderly,' *Engineering in Medicine and Biology Society (EMBC), 2012 Annual International Conference of the IEEE*, New York, NY, August 28–September 1, 2012, pp. 5863–5866.

[74] M.G. Amin, F. Ahmad, Y.D. Zhang, and B. Boashash, 'Micro-Doppler characteristics of elderly gait patterns with walking aids,' *SPIE Defense+ Security*, vol. 9461, pp. 94611A–94611A-6, 2015.

[75] M. Mercuri, D. Schreurs, and P. Leroux, 'SFCW microwave radar for in-door fall detection,' *Biomedical Wireless Technologies, Networks, and Sensing Systems (BioWireleSS), 2012 IEEE Topical Conference*, New York, NY, January 15–18, 2012, pp. 53–56.

[76] S. Suzuki, T. Matsui, M. Kagawa, T. Asao, and K. Kotani, 'An approach to a non-contact vital sign monitoring using dual-frequency microwave radars for elderly care,' *Journal of Biomedical Science and Engineering*, vol. 6, p. 704, 2013.

[77] G. Sun, M. Kubota, M. Kagawa, N.Q. Vinh, A. Kurita, and T. Matsui, 'A screening method based on amplitude probability distribution analysis for detecting the disordered breathing using microwave radar respiration signals,' *Microwave Conference Proceedings (APMC), 2012 Asia-Pacific*, 2012, pp. 37–39.

[78] M. Pavel, T. Hayes, I. Tsay, *et al.*, 'Continuous assessment of gait velocity in Parkinson's disease from unobtrusive measurements,' *2007 3rd International IEEE/EMBS Conference on Neural Engineering, vols 1 and 2*, New York, NY, May 2–5, 2007, pp. 704–707.

[79] T. Frenken, O. Wilken, and A. Hein, 'Technical approaches to unobtrusive geriatric assessments in domestic environments,' *Proceedings of the 5th Workshop on Behaviour Monitoring and Interpretation, BMI*, 2010, pp. 63–74.

[80] N.A. Kerr, M.M. Yore, S.A. Ham, and W.H. Dietz, 'Increasing stair use in a worksite through environmental changes,' *American Journal of Health Promotion*, vol. 18, pp. 312–315, Mar–Apr 2004.

[81] N.M. Boers, D. Chodos, P. Gburzynski, *et al.*, 'The smart condo project: Services for independent living,' *Smart Healthcare Applications and Services*, 2011.

[82] C. Rocker and M. Ziefle, *E-health, assistive technologies and applications for assisted living: Challenges and solutions*, USA: Hershey, PA, Medical Information Science Reference (an imprint of IGI Global), 2011.

[83] T. Yamazaki, 'The ubiquitous home,' *International Journal of Smart Home*, vol. 1, pp. 17–22, 2007.

[84] J. Singh, U. Madhow, R. Kumar, S. Suri, and R. Cagley, 'Tracking multiple targets-using binary-proximity sensors,' *Proceedings of the Sixth International Symposium on Information Processing in Sensor Networks*, New York, NY, April 25–27, 2007, pp. 529–538.

[85] J. Singh, R. Kumar, U. Madhow, S. Suri, and R. Cagley, 'Multiple-target tracking with binary proximity sensors,' *ACM Transactions on Sensor Networks*, vol. 8, Aug 2011.

[86] V. Mehta and M. El Zarki, 'A bluetooth based sensor network for civil infrastructure health monitoring,' *Wireless Networks*, vol. 10, pp. 401–412, Jul 2004.

[87] R. Fukui, H. Ifuku, M. Watanabe, M. Shimosaka, and T. Sato, 'Easy-to-install system for daily walking ability assessment using a distance sensor array,' *Journal of Ambient Intelligence and Smart Environments*, vol. 7, pp. 375–387, 2015.

[88] M. Eldib, F. Deboeverie, W. Philips, and H. Aghajan, 'Sleep analysis for elderly care using a low-resolution visual sensor network,' in A.A. Salah,

B.J.A. Krose, D.J. Cook, (eds.) *Human Behavior Understanding*, Basel, Switzerland: Springer International Publishing, 2015, pp. 26–38.

[89] R.H. Wang, S.M. Gorski, P.J. Holliday, and G.R. Fernie, 'Evaluation of a contact sensor skirt for an anti-collision power wheelchair for older adult nursing home residents with dementia: Safety and mobility,' *Assistive Technology*, vol. 23, pp. 117–134, 2011.

[90] C. Lauterbach, A. Steinhage, and A. Techmer, 'A large-area sensor system underneath the floor for ambient assisted living applications,' in S.C. Mukhopadhyay, O.A. Postolache, *Pervasive and Mobile Sensing and Computing for Healthcare*, Berlin, Germany: Springer, 2013, pp. 69–87.

[91] L. Isella, M. Romano, A. Barrat, *et al.*, 'Close encounters in a pediatric ward: Measuring face-to-face proximity and mixing patterns with wearable sensors,' *PloS one*, vol. 6, Feb 28, 2011.

[92] M.Q. Quwaider, C.L. Daigle, S.K. Biswas, J.M. Siegford, and J.C. Swanson, 'Development of a wireless body-mounted sensor to monitor location and activity of laying hens in a non-cage housing system,' *Transactions of the Asabe*, vol. 53, pp. 1705–1713, Sep–Oct 2010.

[93] Q. Lin, D. Zhang, L. Chen, H. Ni, and X. Zhou, 'Managing elders' wandering behavior using sensors-based solutions: A survey,' *International Journal of Gerontology*, vol. 8, pp. 49–55, 2014.

[94] A. Fong and B. Fong, 'Indoor air quality control for asthma patients using smart home technology,' *Consumer Electronics (ISCE), 2011 IEEE 15th International Symposium*, New York, NY, June 14–17, 2011, pp. 18–19.

[95] P.D. Sloane, C.S. Williams, C.M. Mitchell, *et al.*, 'High-intensity environmental light in dementia: Effect on sleep and activity,' *Journal of the American Geriatrics Society*, vol. 55, pp. 1524–1533, 2007.

[96] A. Al-Fuqaha, M. Guizani, M. Mohammadi, M. Aledhari, and M. Ayyash, 'Internet of things: A survey on enabling technologies, protocols and applications,' *Communications Surveys & Tutorials, IEEE*, vol. 17, no. 4, pp. 2347–2376, Fourthquarter 2015.

[97] S.D.T. Kelly, N.K. Suryadevara, and S.C. Mukhopadhyay, 'Towards the implementation of IoT for environmental condition monitoring in homes,' *Sensors Journal, IEEE*, vol. 13, pp. 3846–3853, 2013.

[98] M. Cook, T. Myers, and J. Trevathan, 'A prototype home-based environmental monitoring system,' *International Journal of Smart Home*, vol. 7, pp. 393–407, 2013.

[99] M. Sidheswaran, '*Performance Evaluation of Real Time Formaldehyde Monitors: PTR-MS and Interscan 4160-500B Portable Monitor*,' Lawrence Berkeley National Laboratory: Lawrence Berkeley National Laboratory. LBNL Paper LBNL-6357E. Retrieved from: http://escholarship.org/uc/item/78s3r63, 2014.

[100] H.S. Ahn, B.K. Jung, J.R. Park, and J.C. Joo, 'Graphene oxide based selective VOCs sensor for indoor air quality monitoring,' *Materials Science Forum*, vol. 804, pp. 39–42, 2015.

[101] M. Sleiman, H. Destaillats, and L.A. Gundel, 'Solid-phase supported profluorescent nitroxide probe for the determination of aerosol-borne reactive oxygen species,' *Talanta*, vol. 116, pp. 1033–1039, Nov 15, 2013.

[102] J.M. Genkinger, L. Stigter, W. Jedrychowski, *et al.*, 'Prenatal polycyclic aromatic hydrocarbon (PAH) exposure, antioxidant levels and behavioral development of children ages 6–9,' *Environmental Research*, vol. 140, pp. 136–144, 2015.

[103] P. Chongthanaphisut, T. Seesaard, and T. Kerdcharoen, 'Monitoring of microbial canned food spoilage and contamination based on e-nose for smart home,' *Electrical Engineering/Electronics, Computer, Telecommunications and Information Technology (ECTI-CON), 2015 12th International Conference*, New York, NY, June 24–27, 2015, pp. 1–5.

[104] U. Tisch, I. Schlesinger, R. Ionescu, *et al.*, 'Detection of Alzheimer's and Parkinson's disease from exhaled breath using nanomaterial-based sensors,' *Nanomedicine*, vol. 8, pp. 43–56, 2013.

[105] N. Christodoulides, R.D.L.G. Ii, G.W. Simmons, *et al.*, 'Application of programmable bio-nano-chip system for the quantitative detection of drugs of abuse in oral fluids,' *Drug and Alcohol Dependence*, 1 August 2015, vol. 153, pp. 306–313.

[106] D.M. Holstius, A. Pillarisetti, K.R. Smith, and E. Seto, 'Field calibrations of a low-cost aerosol sensor at a regulatory monitoring site in California,' *Atmospheric Measurement Techniques*, vol. 7, pp. 1121–1131, 2014.

[107] S. Steinle, S. Reis, and C.E. Sabel, 'Quantifying human exposure to air pollution—moving from static monitoring to spatio-temporally resolved personal exposure assessment,' *Science of the Total Environment*, vol. 443, pp. 184–193, 2013.

[108] B.J. Apelberg, L.M. Hepp, E. Avila-Tang, *et al.*, 'Environmental monitoring of secondhand smoke exposure,' *Tobacco control*, vol. 22, pp. 147–155, 2013.

[109] N. Doesken and H. Reges, 'The value of the citizen weather observer,' *Weatherwise: The power, the Beauty, the Excitement*, vol. 63, pp. 30–37, 2010.

[110] X. Li, 'A low-cost wireless sensor network system using raspberry pi and arduino for environmental monitoring applications,' University of North Texas, 2014.

[111] M.C. Barco, 'Analysis, improvement, and development of new firmware for the smart citizen kit ambient board,' (Master's Thesis). Retrieved from: http://mcolom.perso.math.cnrs.fr/download/TFG_UOC.pdf, 2015.

[112] Attour, A. Thierry Burger-Helmchen, I. Capdevila, and M.I. Zarlenga, 'Smart city or smart citizens? The Barcelona case,' *Journal of Strategy and Management*, vol. 8, pp. 266–282, 2015.

[113] Welcome to Aclima. (2016). Retrieved from: https://aclima.io/about https://aclima.io/about

[114] L.M. Chen, C. Nugent, M. Mulvenna, D. Finlay, and X. Hong, 'Semantic smart homes: Towards knowledge rich assisted living environments,' *Intelligent Patient Management*, vol. 189, pp. 279–296, 2009.

[115] M.D. Hu, Z.Q. Wei, Y.J. Li, and X.D. Xing, 'Ontology-based device description model in smart home,' *Advances in Applied Science and Industrial Technology*, Pts 1 and 2, vol. 798–799, pp. 468–471, 2013.

[116] E. Kim and J. Choi, 'An ontology-based context model in a smart home,' *Computational Science and Its Applications – ICCCSA 2006*, Pt 4, vol. 3983, pp. 11–20, 2006.

[117] K. Wongpatikaseree, M. Ikeda, M. Buranarach, T. Supnithi, A.O. Lim, and Y. Tan, 'Activity recognition using context-aware infrastructure ontology in smart home domain,' *2012 Seventh International Conference on Knowledge, Information and Creativity Support Systems (KICSS 2012)*, New York, NY, November 8–10, 2012, pp. 50–57.

[118] M. Sohn, S. Jeong, and H.J. Lee, 'Self-evolved ontology-based service personalization framework for disabled users in smart home environment,' *2013 Seventh International Conference on Innovative Mobile and Internet Services in Ubiquitous Computing (IMIS 2013)*, New York, NY, July 3–5, 2013, pp. 238–244.

[119] R. Bogue, 'Towards the trillion sensors market,' *Sensor Review*, vol. 34, pp. 137–142, 2014.

[120] U. N. D. O. Economic, *World population ageing, 1950–2050*: New York: United Nations; 2002.

[121] C. F. D. Control and Prevention, 'The Merck Company Foundation: The State of Aging and Health in America 2007,' *The Merck Company Foundation*, Whitehouse Station, New Jersey.

[122] G. Demiris and H. Thompson, 'Smart homes and ambient assisted living applications: from data to knowledge-empowering or overwhelming older adults?,' *IMIA Yearbook*, vol. 6, pp. 51–57, 2011.

[123] J. Bryzek, 'Emergence of a $ Trillion MEMS sensor market,' *Proceed ings of Sensor Conference, Santa Clara*, Mar 2012: http://www.sensors con.org/English/Archives/201203/Presentations/Janusz_Bryzek_SensorsCon 2012.pdf.

Chapter 4

Computer vision for active and assisted living

Rainer Planinc[1], *Alexandros Andre Chaaraoui*[2], *Martin Kampel*[1], *and Francisco Florez-Revuelta*[3]

4.1 Introduction

The field of computer vision has been growing steadily and attracting the interest of both researchers and the industry. Especially in the last decade, enormous advances have been made with regard to automated and reliable recognition of image or video content such as face, object and motion recognition [1–3] and gesture and activity recognition [4, 5]. Additionally, recent advances in 3D scene acquisition such as the Microsoft Kinect depth sensor represent a huge leap forward, enabling 3D modelling and body pose estimation in real time with low cost and mostly simple setup solutions. Such advances are very relevant to the field of active and assisted living (AAL). Traditionally binary sensors have been employed to provide the infrastructure of a smart home upon which services can then be provided to assist people and provide comfort, safety and eHealth services, among others. However, binary sensors reach their limits when complex scenarios have to be taken into account that require a broader knowledge of what is happening and what a person is doing. One or more visual sensors can provide very detailed information about the state of the environment and its inhabitants when combined with the aforementioned pattern recognition and machine learning techniques. For this reason, computer vision is becoming more and more popular for assisted living solutions.

In this chapter, we are going to review how cameras are employed in AAL and what the current state of the art is related to applications and recognition techniques. We will distinguish between traditional RGB cameras and depth sensors (or multi-modal approaches, where both techniques are combined), whose recent great impact in the field deserves an individual overview. With the goal of introducing professionals of other AAL fields to computer vision in general and specifically to its application to assisted living, we will review the image processing pipeline, from different camera types that can be installed in care centres and people's homes to

[1]Vienna University of Technology, Computer Vision Lab, Favoritenstr. 9/183-2A-1040 Vienna, Austria
[2]Google, Inc., Zurich, Switzerland
[3]Faculty of Science, Engineering and Computing, Penrhyn Road, Kingston upon Thames, Surrey KT1 2EE, UK

recent advances in image and video feature extraction and classification. For depth sensors, specific applications, feature estimation techniques and the most successful data representations are reviewed to detail how these differentiate from the traditional RGB approaches.

The remainder of this chapter is structured as follows: Section 4.2 reviews the most common applications of computer vision in AAL and related projects and studies, and provides then an overview of the different stages of a traditional image processing pipeline including insights about its application to AAL. Section 4.3 introduces depth sensors and their main advantages that have made them so popular, it then continues with the data representations that are used among the state of the art for skeletal human motion analysis. Finally, Section 4.4 confronts the observed progress and advantages with the concerns of continuous monitoring and private spaces and related limitations, and concludes this chapter.

4.2 Using RGB cameras

Even though the idea of using cameras to monitor older or impaired people easily raises privacy concerns, computer vision has been considered widely for AAL [6, 7]. This is due to the multiple types of AAL scenarios in which the use of cameras would still be acceptable, such as in public facilities, i.e. nursing centres and hospitals, and during specific activities or events, such as tele-rehabilitation or safety assessment. Since image and video can provide very rich data about a person's activity, research has also been carried out to enhance monitoring systems with security [8] and privacy protection techniques [9, 10]. In this sense, cameras can provide rich sensor data for human monitoring, not only complementing systems with networks of binary sensors, but potentially replacing them in the near future.

4.2.1 Applications

In this section, we will briefly go through the main applications that video cameras have enabled in AAL scenarios. These applications go from event detection to person–environment interaction, support to people with cognitive impairment, affective computing and assistive robots. However, the following applications stand out among the state of the art.

Human behaviour analysis From basic motion tracking [11], through human action and activity recognition [12–14] to long-term behaviour analysis [7], these fields have been studied extensively for AAL applications. The potential recognition of activities of daily living (ADLs), which can lead to monitoring habits and routines related to a person's health, as well as to abnormal behaviour detection, which is of special interest for early detection of mental impairment has been of great interest. In this sense, performing an ADL, such as a kitchen task, can serve as a functional measure of cognitive health [15]. Recently, there is also an increasing interest in the use of wearable cameras for recognising ADLs [16–18].

Fall detection Over a decade of work can be found on using RGB cameras for fall detection. Early work from Nait-Charif and McKenna relies on tracking and ellipse modelling techniques to detect falls as unusual activities [19]. A very similar work can be found in [20], and multi-camera networks have been employed in [21] and [22] among multiple others [23].

Tele-rehabilitation Therapies based on rehabilitation exercises or gaming can benefit from visual monitoring allowing to apply semi-automated evaluations of the performed tasks. For instance, exergames have been developed for stroke rehabilitation [24] or to rehabilitate dynamic postural control for people with Parkinson's disease [25].

Gait analysis The uniqueness of human gait has traditionally led to its application to human identification [26]. Nonetheless, human gait is also a valuable indicator of the mobility and health of a person. Interestingly, due to the complex mental coordination process involved, physical frailty can also be associated with an increased risk of cognitive impairment [27]. Automatic visual gait analysis has also been employed for fall prevention by assessing the risk [28, 29].

Physiological monitoring Image processing techniques have recently been developed to measure some physiological variables without direct contact with the user, e.g. heart rate [30] and respiratory motion [31]. Monitoring over time the semeiotic face signs have also been used to assess cardio-metabolic risks [32].

4.2.2 Image processing stages

In order to take advantage of image- and video-based data from one or multiple RGB cameras, the image streams have to be analysed. For this purpose, different pattern recognition and machine learning techniques are commonly applied depending on the targeted application and the level of temporal and semantic complexity of the event that has to be detected. For this purpose, the video stream is processed through a pipeline of processing stages that allow one to apply computer vision techniques from person identification to activity recognition. In this chapter, these have been divided based on the objective of each processing stage, namely image acquisition, image pre-processing, and feature extraction and classification.

In the following, each of these processing stages will be described focusing on how these are applied to AAL scenarios and the related works that can be found among the state of the art.

4.2.2.1 Image acquisition

Nowadays, a variety of video cameras can be found for monitoring and surveillance purposes. Cameras can be divided based on their intended place of installation such as outdoors or indoors, their mechanical capacities such as bullet-type or pan-tilt-zoom cameras, or based on their in-built features such as motion detection or night vision, among others. Specifically, in AAL scenarios, mostly traditional indoor bullet-type cameras have been used, along with omnidirectional cameras. The latter have the advantage of having an increased field of view by means of a fish-eye lens. This allows to cover, for instance, a complete room as shown in Figure 4.1 from a centric

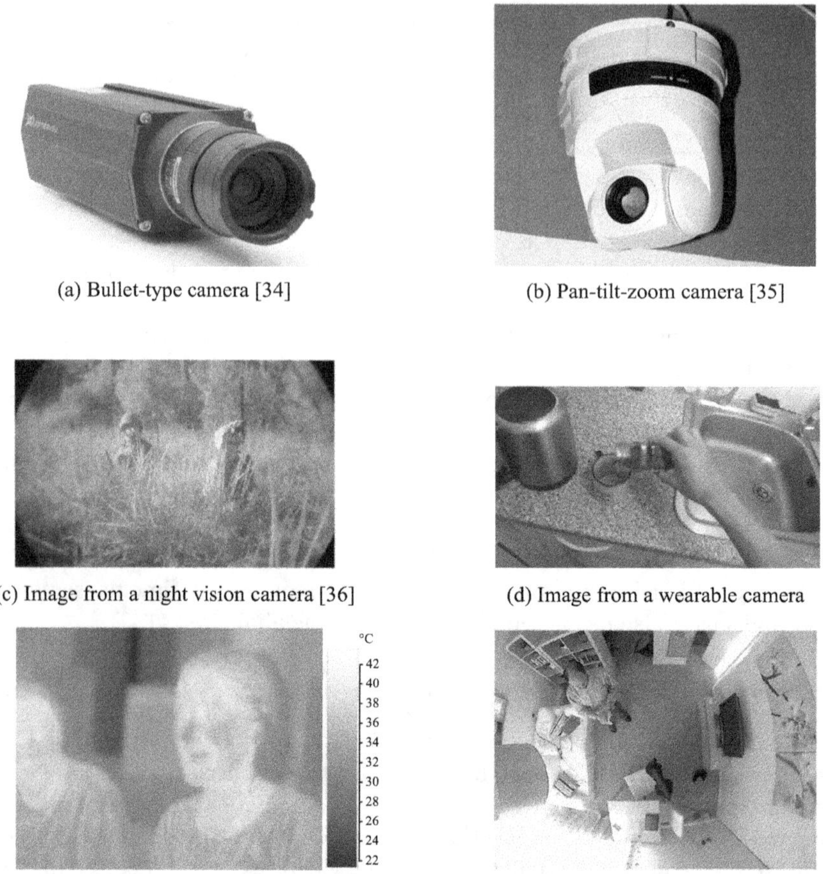

(a) Bullet-type camera [34]

(b) Pan-tilt-zoom camera [35]

(c) Image from a night vision camera [36]

(d) Image from a wearable camera

(e) Image from a thermal camera [37]

(f) Image from an omnidirectional camera [38]

Figure 4.1 These figures show respectively different types of cameras and images

view point of the ceiling, if its height and the camera's field of view are sufficient. Omnidirectional cameras have been proposed for example in [33] for a home-care robotic systems. However, capturing naturally occurring activities is challenging due to the inherently limited field of view of fixed cameras, the occlusions created by a cluttered environment, and the difficulty of keeping all relevant body parts visible, mainly hands, as the torso and head may create occlusions. This is the reason why wearable cameras, such as GoPro® or Google Glass™, are beginning to be employed in assisted living applications.

Besides RGB cameras, several other image capturing technologies have also been employed for assisted living scenarios. Depth cameras, based either on time-of-flight (TOF) or structured light, have been very popular recently. Computer vision methods can take advantage of depth data enabling, for example, 3D scene understanding and markerless human body pose estimation. This has led to a significant amount of research effort and results in the state of the art. For this reason, depth sensors are considered separately in Section 4.3. Thermal cameras, which acquire the infrared radiation of the scene, also facilitate person segmentation and pose estimation.

Although for video surveillance the traditional CCTV is still employed in most cases, in AAL scenarios these have been replaced for internet protocol (IP) cameras, where the image transition occurs over local area networks, which are typically used in smart homes also for other purposes such as binary sensor networks and internet-based services. A central point of processing, either inside the building or remotely, receives the camera streams for their storage and analysis. Additionally, cameras can provide features such as on-camera recording, and some basic image analysis, as the aforementioned motion detection, which can trigger the recording if desired.

Using networks of multiple cameras leads to additional constraints, since multi-view calibration and multi-camera geometry have to be taken into account. The work from Aghajan and Cavallaro [38] analyses these topics, along with distributed camera networks, multi-camera topologies and optimal camera placement. However, in AAL scenarios, like smart homes or care centres, other environmental sensors are also employed, which typically rely on a middleware, i.e. the interplatform service-oriented software that integrates sensor and actuator protocols of different manufacturers [39]. As a consequence, the system architecture will also constrain how a multi-camera network can be deployed and where the image streams can be analysed.

In [40], several recent assistive smart home projects are reviewed and it can be observed that RGB cameras are used widely in AAL for applications as activity recognition and fall detection.

4.2.2.2 Image pre-processing

Since for the applications mentioned in Section 4.2.1 the main interest is focused on the recorded people, the part of the image that contains the human silhouette, i.e. the region of interest (ROI), has to be extracted. Blob detection techniques make it possible to identify these ROI based on colour, brightness, shape or texture. In this stage, sensor-specific image pre-processing methods can be applied to filter noise, increase contrast or enhance colours. In order to separate the ROI from the rest of the image, most frequently motion segmentation techniques are applied, which rely on the fact that the people in the image are in motion whereas the background is rather static. Image segmentation techniques as codebook representation [41], Gaussian mixture learning [42], and *GrabCut* [43] are frequently used among the state of the art. However, alternative approaches can be found too. For example, in Reference [44] silhouettes are obtained based on contour saliency combining both colour and thermal images.

After the foreground pixels have been identified, blob detection techniques group neighbouring pixels based on different criteria such as connectivity, colour, shape and width and height ratios, and identify the image regions that should be considered as a single object (namely a *blob*). Once a region of interest is obtained, the containing pixels have to be described and normalised in a suitable manner in order to apply pattern recognition techniques or learn and classify them with machine learning algorithms. Additionally, a dimensionality reduction is usually desirable, since the increasing

(a) Image enhancement based on contrast correction [45]

(b) Pedestrian detection [46]

(c) Person segmentation obtained from Figure 4.1(f)

(d) Human silhouettes corresponding to different activities.

Figure 4.2 Result examples of image pre-processing methods are shown respectively for noise reduction, blob detection, background segmentation and silhouette extraction techniques

spatial and temporal resolution of video data would otherwise make real-time methods infeasible. Figure 4.2 shows examples of different pre-processing techniques that are typically performed before the feature extraction and recognition stages can be initiated.

For example, in [47] a view-invariant fall detection system is developed relying on view-invariant human pose estimation. The video stream provided by a monocular camera is first downsampled (or upsampled if necessary) to 15 fps to ensure stable real-time execution and then converted to 8-bit greyscale images. As part of the pre-processing, foreground extraction is performed based on the work of the W^4 system [48] using a non-parametric background model that learns greyscale levels and variations on a pixel-by-pixel basis assuming an empty background. The foreground is then detected based on the deviation of the learned model. An erosion filter is employed to delete noise, and blobs are obtained based on connectivity and a minimum size. Additionally, a temporal segmentation based on motion energy boundaries is performed to segment the continuous stream in individual sequences that can be analysed in isolation. This processing then allows to continue with pose modelling and recognition.

4.2.2.3 Feature extraction and recognition

Once the necessary pre-processing stages have been executed, image representations based either on the whole image or on the detected ROIs are generated in order to obtain the characteristic information that defines the event to be detected. These are the so-called visual features. Image and video features can be distinguished as dense features, which represent the data with a global (also known as *holistic*) descriptor, or sparse features, which use a set of local representations of the ROI or even of the whole image.

A very popular feature for human detection are histograms of oriented gradients (HOGs). Dalal and Triggs [49] proposed to evaluate normalised local histograms of image gradient orientations in a dense grid on a gamma and colour normalised image. In [50], the authors combined this approach with a similar feature for oriented optical flow (histogram of oriented flow –HOF–) in order to capture both shape and motion, leading to the state-of-the-art standard for human detection. In [51], this method is used in addition to holistic features extracted from raw trajectory cues for recognition of ADL of healthy subjects and people with dementia using the URADL dataset [52].

Another well-established image representation are motion history images (MHI) [53], where in this case both shape and motion are captured in a single bidimensional feature. First, background segmentation is applied to a sequence of images, which are typically downsampled in both size and frame rate. The segmented foreground of each image of the sequence is then combined by assigning to each coordinate of the feature vector a value that represents the recentness of the motion in that pixel. These values can then be mapped to greyscale intensities, where pixels with more recent motion appear brighter. This allows to encode the temporal evolution of the motion as well as its spatial location. This feature is used for example in [20] to detect different types of falls and activities.

SIFT [54], SURF [55], and other local descriptors are widely used to detect and characterise objects and persons. SIFT is a gradient-based algorithm that detects and describes local features in images; it is invariant to image translation, scaling, and rotation. These are usually clustered into different classes, named visual words, building a codebook. Then an image can be characterised with a bag of words [56] (BoW), a vector counting the occurrence or frequency of each visual word.

Since in the aforementioned pre-processing stage can include silhouette extraction, it is also common to build a holistic descriptor of the individual's silhouette. Seeing that the shape of the silhouette is defined by its boundary, contour representations can lead to very summarised and descriptive features. In Reference [58], such a representation is proposed for the recognition of human actions. The feature is built using the distances between the contour points and the centroid of the silhouette in order to obtain location invariance. The vector of distances is then downsampled to a fixed size and normalised to unit sum to obtain also scale invariance. This feature has been used successfully in [57] combining also multiple view points by means of feature concatenation (see Figure 4.3). It has been further improved in [59], where

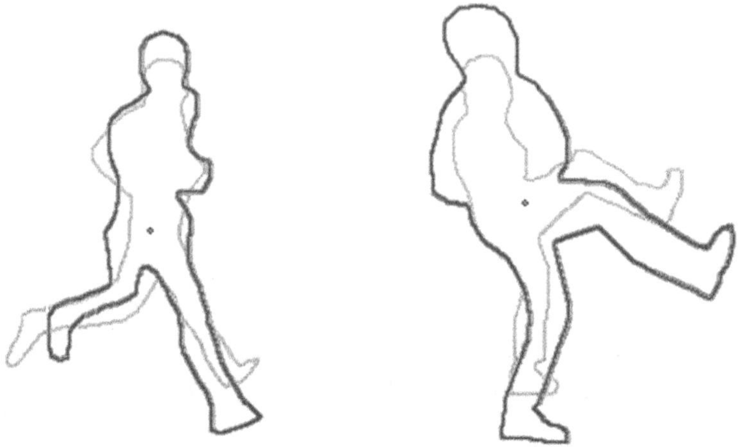

Figure 4.3 Centred silhouettes from two viewpoints (respectively in light and dark grey) used for contour-based multi-view pose representation in [57]

the silhouette is divided into radial sectors, and the statistical range of distances to the contour is used as a characteristic value, leading to further dimensionality reduction and reduced noise sensitivity. Finally, the work is applied to a visual monitoring system that enables multi-view recognition of human actions to provide care and safety services such as detection of home accidents and telerehabilitation [10].

Sparse features such as key points have also been used extensively, as in [52], where a proposal is made based on velocity histories of tracked key points. The obtained key points are tracked with a KLT tracker and their velocity histories are computed. The features are also augmented with additional data, including the absolute initial and final positions of the key points and the relative positions with respect to the position of an unambiguously detected face if present. The local appearance information is encoded relying on horizontal and vertical gradients and PCA-SIFT [60], and also colour information of the same area is encoded based on PCA-reduced colour histograms. Both texture and colour information are not used directly, instead a codebook is obtained using K-means clustering and the nearest-neighbour element is assigned to represent each key point.

These features are then classified with different recognition techniques based on learning and data analysis methods, such as the aforementioned BoW technique, or others. These can be classified into two categories, data-driven approaches, which learn generative or discriminative models from user data and infer the corresponding class, or knowledge-driven approaches, which take advantage of domain-specific knowledge and rules. Analysing these is out of scope in this chapter; however, it is covered extensively in the present book and we refer the reader to consult the corresponding chapters.

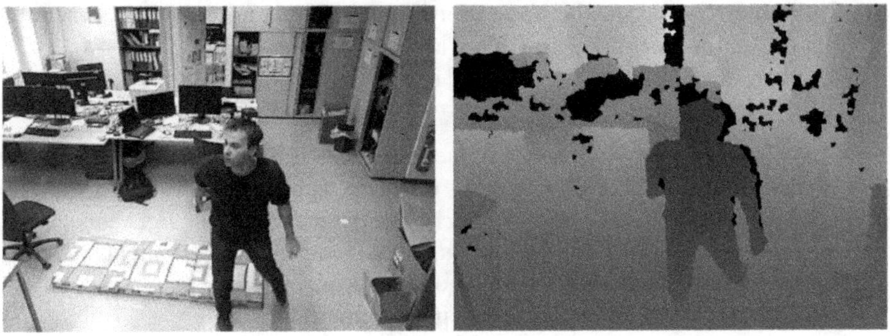

Figure 4.4 RGB camera image and its corresponding depth image

4.3 Using depth sensors

Advances in the development of 3D sensors motivate their use instead of cameras or wearable sensors, since they provide advantages like privacy protection and improved robustness when it comes to behaviour modelling, gesture recognition or activity recognition. In contrast to RGB-based analysis, depth-based approaches do not process RGB colour images, but depth or range images measuring distances from objects to the sensor. Figure 4.4 shows an RGB camera image together with its corresponding depth image: depth images do not visualise the scene with colours, but the grey level indicates the distance of the objects and its surroundings to the sensor. The darker the colour, the closer the object is to the sensor. On the other hand, brighter colours are used for objects at a higher distance. Black holes within the depth image indicate reflecting or absorbing areas where no valid depth measurement is available and are caused due to the functionality of the sensor. In contrast to RGB-based approaches, only the silhouette is detected and thus no conclusions whether the person is wearing clothes or in an emotional state can be obtained since neither the clothes, nor the face are visible. Hence, the appearance of the person is fully protected. However, this is the case only if processing is solely based on depth images. RGB-D-based approaches combine depth information together with appearance information in order to obtain and combine more details. Although these approaches reduce the privacy of elderly, colour information can be taken into consideration, thus allowing to perform a more in depth analysis.

The use of depth sensors became popular due to the introduction of the Microsoft Kinect in 2010 as an add-on for the Xbox console. It was the first low-cost 3D sensor and thus received a lot of attention from the research community [61]. The first version of the Kinect consists of an RGB camera, an infrared projector as well as an infrared camera. The functionality of the Kinect is based on structured light imaging, where the projector emits a pre-defined infrared light pattern to the scene [62, 63]. Due to the spatial arrangement of the pattern and its varying sizes, as well as distortions depending on the distance to the camera, the depth camera captures the light pattern and an on-board chip calculates a depth map. In contrast to the first versions of the

Kinect, the functionality of the Kinect for Windows v2 is not based on structured light, but on ToF in order to achieve more accurate results.[1]

The main advantages of depth-based sensors, especially within the context of AAL, can be summarised as follows:

- *No additional light source needed:* due to the use of infrared light, sensors also work during the night (e.g., when falls of elderly people occur).
- *Sensor is robust to changing lighting conditions:* switching the lights on and off does not affect the results of the depth images. However, direct sunlight interferes with the projected infrared pattern and thus, no depth value can be calculated. This restricts the use of the sensors to indoor environments only.
- *No calibrated camera setup is needed:* in contrast to the use of a calibrated multiple camera setup in order to calculate a 3D reconstruction, no calibration is needed.
- *Standard algorithms can be applied to depth information:* standard algorithms from computer vision (e.g. foreground/background segmentation, tracking) can be applied to depth data directly.
- *Protection of privacy:* if only depth information is processed, privacy is protected since the appearance of the person is not recognised in depth images. However, if a combined analysis of RGB and depth images is performed, privacy is not protected.

Typical applications of RGB-D approaches within the context of AAL are:

- Fall detection
- Rehabilitation
- Serious gaming
- Pose analysis
- Gesture-based interfaces
- Robotics

Behaviour modelling or activity recognition of humans depends on the correct detection of humans and their pose in the scene [64]. Person detection and tracking is a complex problem due to the non-rigid nature of persons, complex motion, and occlusions, among others [65]. Kinect and other depth sensors allow for a partial reconstruction of the scene geometry, which facilitates the problem. This section lists a selection of recent methods designed for or applicable to Kinect sensors, categorised based on the data representation they operate on (skeletal, depth maps, point clouds, or plan-view maps).

4.3.1 Skeletal

Human pose estimation is proposed by Shotton *et al.* [66], allowing to extract body parts and skeleton joints based on depth images. Figure 4.5 depicts the pipeline for the skeleton joint calculation [67]: foreground detection is performed at the beginning

[1] http://blogs.microsoft.com/blog/2013/10/02/collaboration-expertise-produce-enhanced-sensing-in-xbox-one/, accessed 09-September-2015.

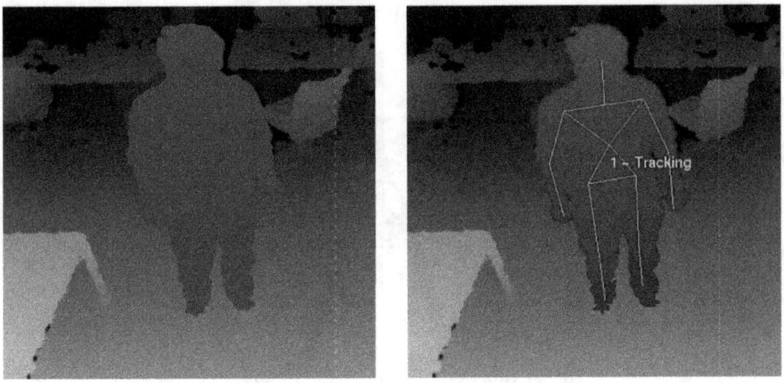

Figure 4.5 Skeleton tracking

in order to separate humans from the background. The use of a randomised forest allows to label different body parts in the depth image. Clustering of pixels hypothesise body joint positions, which are refined using a skeleton model exploiting temporal and kinematic constraints.

The skeleton tracking algorithm is optimised for people facing the sensor, either standing or sitting in front of the sensor. Hence, it can fail, if the person is not directly facing the sensor. The number and type of detected joints depends on the SDK to be used – two SDK are mainly used together with the Kinect [63]: the official SDK provided by Microsoft, being able to track 20 respectively 25 skeleton joints (Kinect v1 and Kinect v2) and the OpenNI[2] provided by PrimeSense, allowing to track 15 different skeleton joints [63]. PrimeSense developed the first sensor of the Kinect for Microsoft and introduced, in cooperation with Asus, their own 3D sensor[3]: Asus Xtion pro, offering almost the same hardware specification as the Microsoft Kinect, but built in a smaller case. In contrast to the Microsoft Kinect, the Asus Xtion pro does not contain an RGB camera, but only a depth sensor and thus allows to fully respect the privacy within the context of AAL, since it is technically not possible to obtain an RGB image from this sensor.

Figure 4.6 depicts an overview of the estimated skeleton joints for both versions of the Kinect. In the first version, 20 skeleton joints are estimated whereas in the second version of the Kinect, 25 skeleton joints are estimated. Five skeleton joints were added and hands are modelled in more detail in order to allow a more accurate gesture control.

4.3.2 Depth maps

Methods operating on depth maps frequently utilise histogram-based features and supervised learning for person detection [68–70]. Conceptually, these methods are

[2]http://www.openni.org, accessed 10-April-2014.
[3]https://www.asus.com/de/Multimedia/Xtion_PRO/, accessed 04-August-2015.

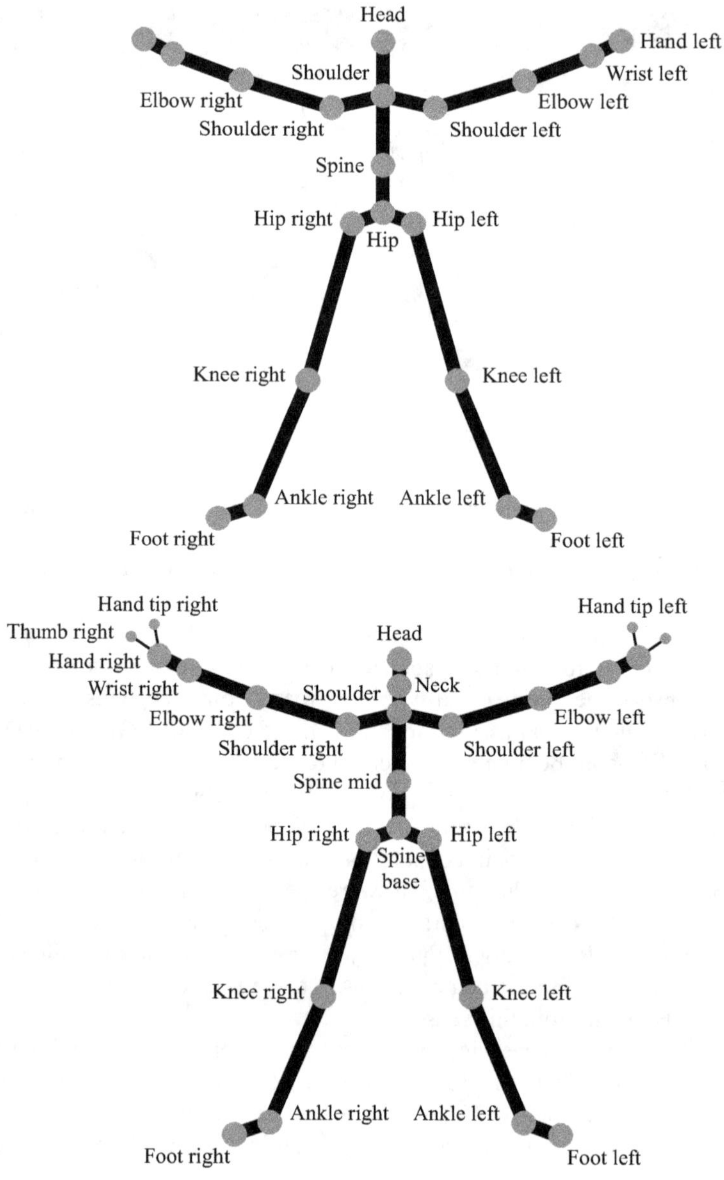

Figure 4.6 Skeleton joints of Kinect v1 (top) and Kinect v2 (bottom)

similar to seminal work by Dalal and Triggs [49] that introduced histograms of oriented gradients as powerful features for person detection in greyscale images. These features model the local appearance of objects by means of gradient distribution and, when consolidated to larger blocks, constitute powerful feature descriptors. These descriptors are then classified as (not) representing a person using a support vector machine. Spinello and Arras [68] as well as Wu *et al.* [70] proposed similar descriptors for depth maps, which model the local distribution depth gradient orientations.

4.3.3 Point clouds

Another approach is to first reproject depth map pixels to world coordinates and to operate on the resulting point cloud. Kelly *et al.* [71] follow this approach, clustering the point cloud using an iterative top-down approach based on 3D proximity tests. Thresholds are computed dynamically from the observed maximum height and the golden ratio, which allows for estimating the proportion of persons based on their height. Subsequently the obtained clusters are analysed with respect to under- and over-segmentation via ellipse fitting on shoulder and head height, respectively. This method is able to cope with significant occlusions but is prohibitively slow (less than 1 fps including tracking).

Hegger *et al.* [72] compute surface normals from subsampled point clouds and cluster points using an efficient top-down method that results in small clusters of adjacent points.

4.3.4 Plan-view maps

A reason why methods operating on point clouds are slow is the large number of points (up to 307200) and the fact that clustering is a complex task. Hegger *et al.* [72] alleviate this problem by subsampling the point cloud. This is accomplished by discretising the continuous space into cubic cells, which significantly reduces the number of points at the expense of resolution. A similar yet more extreme approach to data reduction is to utilise the so-called plan-view maps, two-dimensional representations of the scene as viewed from the top and under orthographic projection [73].

4.3.5 Accuracy

The resolution of a measuring device describes the smallest details it is able to resolve. With regard to depth sensors it is desirable to distinguish between depth resolution and spatial resolution. In this text, the term depth resolution denotes the smallest difference in distance the sensor can distinguish, while spatial resolution refers to the minimum size of reliably detectable objects.

The depth resolution of the sensor restricts detectable scene changes. If the depth resolution is too low, fallen persons or parts thereof may not be distinguishable from the floor. Kinect sensors can distinguish between two object distances only if their difference is large enough. This is because the sensor can derive disparities with only limited accuracy. In fact, the sensor distinguishes between 1024 distinct disparities and thus distances [74], while the measuring range spans approximately 10 m.

Figure 4.7 depicts a person lying flat on the floor at a distance of approximately 6 m from the sensor. Distances are represented by greyscale level, where darker represents closer distances. Due to the limited depth resolution, limbs are only partially distinguishable from the floor. This is illustrated by the fact that greyscale levels in

Figure 4.7 Illustration of the effect of limited depth resolution on fall detection

regions of arms or legs do not always differ from those of the floor beneath them (left side of the vertical line). The minimum size objects must have in order to be reliably (i.e. continuously) detectable from a certain distance depends on the spatial resolution of the sensor. According to a datasheet by the developer of the depth sensing technology of the Kinect, the spatial resolution is 3.4 mm at an object distance of 2 m.

The precision describes the variability between multiple measurements of the same object under stable conditions. With regard to the Kinect, the precision indicates the closeness of repeated distance measurements. This criteria is particularly important with respect to background subtraction algorithms, as their performance depends on data stability. Knowledge of the sensor precision allows for estimating the performance of such algorithms and aids in proper configuration.

Smisek *et al.* [61] evaluated the accuracy of the Kinect sensor (first version), by analysing its depth resolution. The depth resolution of the Kinect is within the range between 0.65 mm at a distance of 0.5 m and up to 685 mm at a distance of 15.7 m. These results indicate that the use of the Kinect for indoor environments is feasible, although the accuracy decreases with higher distance. Results show that the depth resolution within a range up to 10 m is below 300 mm. Moreover, the performance of the Kinect is compared to the performance of a stereo-camera (two Nikon D60 SLR) as well as a ToF (SwissRanger SR-4000) system. Smisek *et al.* [61] showed that the Kinect performs similar to a stereo system with medium resolution and outperforms the ToF system in terms of accuracy and costs. Stoyanov *et al.* [75] evaluated the Kinect and two ToF sensors based on ground truth data obtained by a laser sensor. For short distances, the Kinect slightly outperformed both ToF sensors and performed similarly to the accuracy of the laser sensor. However, no sensor achieved the accuracy of the laser sensor on longer distances.

Ditta [76] compared the skeleton tracking of the Kinect with a marker-based system from Vicon and showed that the errors of the Kinect are approximately 5 mm within a range of 1–3 m in comparison to the Vicon system. Galna *et al.* [25] evaluated the use of the Kinect for the detection of movement symptoms for people with Parkinson's disease and compared their results with a Vicon system. Normal actions (standing, walking and reaching) are combined with actions from the Unified Parkinson's disease Scale and include, amongst others, hand clasping and finger tapping. The timing of movement is measured accurately as well as extensive movements are detected accurately. Only when monitoring fine movement, the Kinect is not able to obtain accurate results and thus is outperformed by the Vicon system. Overall, Galna *et al.* [25] conclude that the Kinect can accurately detect most movements related to Parkinson's Disease. The accuracy of the Kinect within exergames is evaluated and compared to a Vicon system by van Diest *et al.* [77]. The outcome of their evaluation shows that the Kinect accurately detects movements of the trunk, but does not detect the movement of hands and feet accurately, resulting in a difference of up to 30% in comparison to the Vicon system. The reasons for the lower accuracy of the Kinect is the reduced resolution (640 × 480) in comparison to the Vicon system (4704 × 3456) and the low and irregular sampling frequency [77].

Plantard *et al.* [78] propose a framework to simulate 500 000 poses at a workplace in combination with different positions of the Kinect, in order to perform an automatic large-scale evaluation of the accuracy of the Kinect. The results show that the accuracy depends on the specific pose as well as the position of the Kinect. Although most results are accurate (e.g. error of the shoulder position is 2.5 cm), positions with partial occlusions results in the failure of the skeleton tracking algorithm.

Moreover, the performance of the skeleton tracking system during six different exercises is evaluated by Obdrzalek *et al.* [79]. Again, a marker-based tracking system provides ground truth data and it is shown that Kinect has a great potential. However, since exercises are performed either sitting or while touching a chair, the skeleton tracking algorithm fails when body parts are occluded or a chair is presented. Although problems with the skeleton tracker are reported, Obdrzalek *et al.* [79] state that within a more controlled environment, tracking results are better. They conclude that during general postures a variability of about 10 cm can be observed in comparison to the marker-based tracking system. Nevertheless, these results show that the use of a Kinect is feasible, but depends on the application and context. Due to its low costs in comparison to other 3D sensors and its accuracy, the Kinect is used in computer vision for achieving different and diverse tasks: approaches using the Kinect for object tracking and recognition, human activity analysis, hand gesture analysis, and 3D mapping of indoor environments are just few of them [63].

4.4 Conclusion

Over the course of the last years, vision-based solutions gained of interest within the field of AAL. However, since AAL solutions are used in private spaces (e.g. bedrooms, bathrooms), privacy issues need to be considered. The use of cameras in private spaces without any privacy protection mechanisms yields in low acceptance by the end user. Hence, privacy needs to be protected either by design, or by explicitly taking appropriate measures in order to prevent subjects from carrying out camera-avoiding techniques to sabotage the monitoring [80]. Although using depth sensors results in 3D data where the person cannot be identified, the location of the sensor provides information about the person itself. Especially when being used in homes for older adults, the connection between the depth image (where the person cannot be identified) and the room number (where the system is placed) allows to implicitly identify the person on the depth image. Again, appropriate protection need to be considered in order to ensure that only authorised personal has access to this personal data. Moreover, when using data from different sources, the aggregation of data might result in private information, violating the privacy of the user. Especially if a person is being tracked over multiple cameras/sensors, his or her whole trajectory is available and thus information about the behaviour of the person is gathered.

Although vision-based systems need to be designed carefully in order to protect the privacy, their flexibility and adaptability is one of their biggest advantages. The application of AAL systems can be extended easily, since only software needs to be updated in order to provide additional features and no installation of additional sensors is needed. Thus allows to install vision-based systems according to the specific needs of the current context and the possibility to adopt the system to changing needs later on. Moreover, using vision-based systems results in a big amount of information to be processed, allowing to analyse complex scenarios and offering possibilities for various applications within the field of AAL.

References

[1] S. G. Kong, J. Heo, B. R. Abidi, J. Paik, and M. A. Abidi, "Recent advances in visual and infrared face recognition—a review," *Computer Vision and Image Understanding*, vol. 97, no. 1, pp. 103–135, 2005. [Online]. Available: http://www.sciencedirect.com/science/article/pii/S1077314204000451

[2] P. F. Felzenszwalb and D. P. Huttenlocher, "Pictorial structures for object recognition," *International Journal of Computer Vision*, vol. 61, no. 1, pp. 55–79. [Online]. Available: http://dx.doi.org/10.1023/B:VISI.0000042934.15159.49

[3] R. Poppe, "Vision-based human motion analysis: An overview," *Computer Vision and Image Understanding*, vol. 108, no. 1–2, pp. 4–18, 2007, Special Issue on Vision for Huma–Computer Interaction. [Online]. Available: http://www.sciencedirect.com/science/article/pii/S1077314206002293

[4] S. Mitra and T. Acharya, "Gesture recognition: A survey," *IEEE Transactions on Systems, Man, and Cybernetics, Part C: Applications and Reviews*, vol. 37, no. 3, pp. 311–324, May 2007.

[5] T. B. Moeslund, A. Hilton, and V. Krüger, "A survey of advances in vision-based human motion capture and analysis," *Computer Vision and Image Understanding*, vol. 104, no. 2–3, pp. 90–126, 2006, special Issue on Modeling People: Vision-based understanding of a person's shape, appearance, movement and behaviour. [Online]. Available: http://www.sciencedirect.com/science/article/pii/S1077314206001263

[6] F. Cardinaux, D. Bhowmik, C. Abhayaratne, and M. S. Hawley, "Video based technology for ambient assisted living: A review of the literature," *Journal of*

Ambient Intelligent Smart Environments, vol. 3, no. 3, pp. 253–269, Aug. 2011. [Online]. Available: http://dl.acm.org/citation.cfm?id=2010465.2010468

[7] A. A. Chaaraoui, P. Climent-Pérez, and F. Flórez-Revuelta, "A review on vision techniques applied to human behaviour analysis for ambient-assisted living," *Expert Systems with Applications*, vol. 39, no. 12, pp. 10 873–10 888, 2012. [Online]. Available: http://www.sciencedirect.com/science/article/pii/S0957417412004757

[8] T. Winkler and B. Rinner, "Security and privacy protection in visual sensor networks: A survey," *ACM Computing Surveys*, vol. 47, no. 1, pp. 2:1–2:42, May 2014. [Online]. Available: http://doi.acm.org/10.1145/2545883

[9] A. Bamis, D. Lymberopoulos, T. Teixeira, and A. Savvides, "The behaviorscope framework for enabling ambient assisted living," *Personal and Ubiquitous Computing*, vol. 14, no. 6, pp. 473–487, 2010. [Online]. Available: http://dx.doi.org/10.1007/s00779-010-0282-z

[10] A. A. Chaaraoui, J. R. Padilla-López, F. J. Ferrández-Pastor, M. Nieto-Hidalgo, and F. Flórez-Revuelta, "A vision-based system for intelligent monitoring: human behaviour analysis and privacy by context," *Sensors*, vol. 14, no. 5, pp. 8895–8925, 2014.

[11] J. Aggarwal and Q. Cai, "Human motion analysis: A review," *Computer Vision and Image Understanding*, vol. 73, no. 3, pp. 428–440, 1999. [Online]. Available: http://www.sciencedirect.com/science/article/pii/S1077314298907445

[12] T. B. Moeslund and E. Granum, "A survey of computer vision-based human motion capture," *Computer Vision and Image Understanding*, vol. 81, no. 3, pp. 231–268, 2001. [Online]. Available: http://www.sciencedirect.com/science/article/pii/S107731420090897X

[13] R. Poppe, "A survey on vision-based human action recognition," *Image and Vision Computing*, vol. 28, no. 6, pp. 976–990, 2010. [Online]. Available: http://www.sciencedirect.com/science/article/pii/S0262885609002704

[14] J. Aggarwal and M. Ryoo, "Human activity analysis: A review," *ACM Comput. Surv.*, vol. 43, no. 3, pp. 16:1–16:43, Apr. 2011. [Online]. Available: http://doi.acm.org/10.1145/1922649.1922653

[15] B. Kröse, T. van Oosterhout, and T. van Kasteren, "Activity monitoring systems in health care," in *Computer Analysis of Human Behavior*, A. A. Salah and T. Gevers, Eds. London: Springer, 2011, pp. 325–346. [Online]. Available: http://dx.doi.org/10.1007/978-0-85729-994-9_12

[16] S. Karaman, J. Benois-Pineau, V. Dovgalecs, R. Mégret, J. Pinquier, R. André-Obrecht, Y. Gaëstel, and J.-F. Dartigues, "Hierarchical hidden markov model in detecting activities of daily living in wearable videos for studies of dementia," *Multimedia Tools and Applications*, vol. 69, no. 3, pp. 743–771, 2014.

[17] H. Pirsiavash and D. Ramanan, "Detecting activities of daily living in first-person camera views," in *IEEE Conference on Computer Vision and Pattern Recognition (CVPR)*. Providence, USA: IEEE, 2012, pp. 2847–2854.

[18] T.-H.-C. Nguyen, J.-C. Nebel, and F. Florez-Revuelta, "Recognition of activities of daily living with egocentric vision: A review," *Sensors*, vol. 16, no. 1, p. 72, 2016. [Online]. Available: http://www.mdpi.com/1424-8220/16/1/72

[19] H. Nait-Charif and S. McKenna, "Activity summarisation and fall detection in a supportive home environment," in *Proceedings of the 17th International Conference on Pattern Recognition, 2004. ICPR 2004*, vol. 4, Aug 2004, pp. 323–326.

[20] C. Rougier, J. Meunier, A. St-Arnaud, and J. Rousseau, "Fall detection from human shape and motion history using video surveillance," in *21st International Conference on Advanced Information Networking and Applications Workshops, 2007, AINAW '07*, vol. 2, May 2007, pp. 875–880.

[21] R. Cucchiara, A. Prati, and R. Vezzani, "A multi-camera vision system for fall detection and alarm generation," *Expert Systems*, vol. 24, no. 5, pp. 334–345, 2007.

[22] H. Aghajan, J. C. Augusto, C. Wu, P. McCullagh, and J.-A. Walkden, "Distributed vision-based accident management for assisted living," in *Pervasive Computing for Quality of Life Enhancement*. Springer, Berlin, Heidelberg, 2007, pp. 196–205.

[23] M. Mubashir, L. Shao, and L. Seed, "A survey on fall detection: Principles and approaches," *Neurocomputing*, vol. 100, pp. 144–152, 2013, Special Issue: Behaviours in video. [Online]. Available: http://www.sciencedirect.com/science/article/pii/S0925231212003153

[24] D. Webster and O. Celik, "Systematic review of Kinect applications in elderly care and stroke rehabilitation," *Journal of NeuroEngineering and Rehabilitation*, vol. 11, no. 1, p. 108, 2014. [Online]. Available: http://www.jneuroengrehab.com/content/11/1/108

[25] B. Galna, G. Barry, D. Jackson, D. Mhiripiri, P. Olivier, and L. Rochester, "Accuracy of the Microsoft Kinect sensor for measuring movement in people with parkinson's disease," *Gait & Posture*, vol. 39, no. 4, pp. 1062–1068, 2014. [Online]. Available: http://www.sciencedirect.com/science/article/pii/S0966636214000241

[26] L. Wang, T. Tan, H. Ning, and W. Hu, "Silhouette analysis-based gait recognition for human identification," *IEEE Transactions on Pattern Analysis and Machine Intelligence*, vol. 25, no. 12, pp. 1505–1518, Dec 2003.

[27] V. J. Verlinden, J. N. van der Geest, A. Hofman, and M. A. Ikram, "Cognition and gait show a distinct pattern of association in the general population," *Alzheimer's & Dementia*, vol. 10, no. 3, pp. 328–335, 2014. [Online]. Available: http://www.sciencedirect.com/science/article/pii/S1552526013001362

[28] D. Lim, C. Kim, H. Jung, D. Jung, and K. J. Chun, "Use of the Microsoft Kinect system to characterize balance ability during balance training," *Clinical Interventions in Aging*, vol. 10, p. 1077, 2015.

[29] A. Dubois and F. Charpillet, "A gait analysis method based on a depth camera for fall prevention," in *Engineering in Medicine and Biology Society (EMBC), 2014 36th Annual International Conference of the IEEE*, Aug 2014, pp. 4515–4518.

[30] H.-Y. Wu, M. Rubinstein, E. Shih, J. Guttag, F. Durand, and W. T. Freeman, "Eulerian video magnification for revealing subtle changes in the world," *ACM*

Transactions on Graphics (Proceedings SIGGRAPH 2012), vol. 31, no. 4, 2012.

[31] F. Tahavori, M. Alnowami, and K. Wells, "Marker-less respiratory motion modeling using the Microsoft Kinect for Windows," pp. 90 360K-90 360K-10, 2014. [Online]. Available: http://dx.doi.org/10.1117/12.2043569

[32] S. Colantonio, G. Coppini, D. Germanese, D. Giorgi, M. Magrini, P. Marraccini, M. Martinelli, M. A. Morales, M. A. Pascali, G. Raccichini *et al.*, "A smart mirror to promote a healthy lifestyle," *Biosystems Engineering*, vol. 138, pp. 33–43, 2015.

[33] M.-L. Wang, C.-C. Huang, and H.-Y. Lin, "An intelligent surveillance system based on an omnidirectional vision sensor," in *2006 IEEE Conference on Cybernetics and Intelligent Systems*, June 2006, pp. 1–6.

[34] "A photo of the lumenera Li045C intelligent IP camera with on-camera video analytics, released in 2006," By Lumihaychuk [CC BY-SA 3.0 (http://creativecommons.org/licenses/by-sa/3.0)], via Wikimedia Commons, last accessed 2015-09-26.

[35] "Axis model 214 pan-tilt-zoom IP camera," By Kkmurray (Own work) [CC BY-SA 3.0 (http://creativecommons.org/licenses/by-sa/3.0)], via Wikimedia Commons, last accessed 2015-09-26.

[36] "Two american soldiers pictured during the 2003 Iraq war seen through an image intensifier," By w:en:User:AlexPlank (w:en:Image:Nightvision) [Public domain], via Wikimedia Commons, last accessed 2015-09-26.

[37] "A false color image of two people taken in long-wavelength infrared (body-temperature thermal) light," By Cody.pope [CC BY-SA 3.0 (http://creativecommons.org/licenses/by-sa/3.0)], via Wikimedia Commons, last accessed 2016-05-03.

[38] H. Aghajan and A. Cavallaro, *Multi-camera Networks: Principles and Applications*. Academic Press, Burlington, USA, 2009.

[39] H. Nakashima, H. Aghajan, and J. C. Augusto, *Handbook of Ambient Intelligence and Smart Environments*. Springer Science & Business Media, New York, USA, 2009.

[40] P. Rashidi and A. Mihailidis, "A survey on ambient-assisted living tools for older adults," *IEEE Journal of Biomedical and Health Informatics*, vol. 17, no. 3, pp. 579–590, May 2013.

[41] K. Kim, T. H. Chalidabhongse, D. Harwood, and L. Davis, "Real-time foreground–background segmentation using codebook model," *Real-Time Imaging*, vol. 11, no. 3, pp. 172–185, 2005, Special Issue on Video Object Processing. [Online]. Available: http://www.sciencedirect.com/science/article/pii/S1077201405000057

[42] D.-S. Lee, "Effective gaussian mixture learning for video background subtraction," *IEEE Transactions on Pattern Analysis and Machine Intelligence*, vol. 27, no. 5, pp. 827–832, May 2005.

[43] C. Rother, V. Kolmogorov, and A. Blake, ""grabcut": Interactive foreground extraction using iterated graph cuts," *ACM Transactions on Graphics*, vol. 23,

no. 3, pp. 309–314, Aug. 2004. [Online]. Available: http://doi.acm.org/10.1145/1015706.1015720

[44] J. Davis and V. Sharma, "Fusion-based background-subtraction using contour saliency," in *IEEE Computer Society Conference on Computer Vision and Pattern Recognition – CVPR Workshops, 2005*, June 2005, pp. 11–11.

[45] "Example of photo editing: Contrast correction. left side of the image is unmodified, right side has been touched with gimp," By Toniht at en.wikipedia [Public domain], from Wikimedia Commons, last accessed 2015-09-26.

[46] "Pedestrian detection example," By Milwaukee (WIS) N 5th St "Tree, Rain, Wind" Pedestrian 1.jpg: vincent desjardins derivative work: Indif [CC BY 2.0 (http://creativecommons.org/licenses/by/2.0)], via Wikimedia Commons, last accessed 2015-09-26.

[47] Z. Htike, S. Egerton, and K. Y. Chow, "A monocular view-invariant fall detection system for the elderly in assisted home environments," in *2011 7th International Conference on Intelligent Environments (IE)*, July 2011, pp. 40–46.

[48] I. Haritaoglu, D. Harwood, and L. Davis, "W4: real-time surveillance of people and their activities," *IEEE Transactions on Pattern Analysis and Machine Intelligence*, vol. 22, no. 8, pp. 809–830, Aug 2000.

[49] N. Dalal and B. Triggs, "Histograms of oriented gradients for human detection," in *IEEE Computer Society Conference on Computer Vision and Pattern Recognition, CVPR 2005*, vol. 1, June 2005, pp. 886–893 vol. 1.

[50] N. Dalal, B. Triggs, and C. Schmid, "Human detection using oriented histograms of flow and appearance," in *Computer Vision–ECCV 2006*, Lecture Notes in Computer Science, A. Leonardis, H. Bischof, and A. Pinz, Eds. Springer, Berlin, Heidelberg, 2006, pp. 428–441.

[51] K. Avgerinakis, A. Briassouli, and I. Kompatsiaris, "Activity detection and recognition of daily living events," in *Proceedings of the 1st ACM International Workshop on Multimedia Indexing and Information Retrieval for Healthcare*, MIIRH '13. New York, NY, USA: ACM, 2013, pp. 3–10. [Online]. Available: http://doi.acm.org/10.1145/2505323.2505327

[52] R. Messing, C. Pal, and H. Kautz, "Activity recognition using the velocity histories of tracked keypoints," in *2009 IEEE 12th International Conference on Computer Vision*, Sept 2009, pp. 104–111.

[53] A. F. Bobick and J. W. Davis, "The recognition of human movement using temporal templates," *IEEE Transactions on Pattern Analysis and Machine Intelligence*, vol. 23, no. 3, pp. 257–267, 2001.

[54] M. Brown and D. G. Lowe, "Invariant features from interest point groups," in *British Machine Vision Conference (BMVC)*, Cardiff, UK, 2002.

[55] H. Bay, A. Ess, T. Tuytelaars, and L. Van Gool, "Speeded-up robust features (surf)," *Computer Vision and Image Understanding*, vol. 110, no. 3, pp. 346–359, 2008.

[56] J. Niebles, H. Wang, and L. Fei-Fei, "Unsupervised learning of human action categories using spatial-temporal words," *International Journal of Computer Vision*, vol. 79, no. 3, pp. 299–318, 2008, cited By 712.

[Online]. Available: http://www.scopus.com/inward/record.url?eid=2-s2.0-45049084813&partnerID=40&md5=501fd1c88bae3a8f2aabe7deafe5cb72

[57] A. A. Chaaraoui, P. Climent-Pérez, and F. Flórez-Revuelta, "Silhouette-based human action recognition using sequences of key poses," *Pattern Recognition Letters*, vol. 34, no. 15, pp. 1799–1807, 2013, smart Approaches for Human Action Recognition. [Online]. Available: http://www.sciencedirect.com/science/article/pii/S0167865513000342

[58] Y. Dedeoğlu, B. U. Töreyin, U. Güdükbay, and A. E. Çetin, "Silhouette-based method for object classification and human action recognition in video," in *Computer Vision in Human–Computer Interaction*. Springer, Berlin, Heidelberg, pp. 64–77.

[59] A. A. Chaaraoui and F. Flórez-Revuelta, "A low-dimensional radial silhouette-based feature for fast human action recognition fusing multiple views," *International Scholarly Research Notices*, vol. 2014, Article ID 547069, 2014.

[60] Y. Ke and R. Sukthankar, "Pca-sift: a more distinctive representation for local image descriptors," in *Proceedings of the 2004 IEEE Computer Society Conference on Computer Vision and Pattern Recognition, CVPR 2004*, vol. 2, June 2004, pp. II–506–II–513.

[61] J. Smisek, M. Jancosek, and T. Pajdla, "3D with Kinect," in *Proceedings of the International Conference on Computer Vision Workshops (ICCV Workshops)*. Barcelona, Spain: IEEE, Nov. 2011, pp. 1154–1160. [Online]. Available: http://ieeexplore.ieee.org/lpdocs/epic03/wrapper.htm?arnumber=6130380

[62] D. Fofi, T. Sliwa, and Y. Voisin, "A comparative survey on invisible structured light," in *Electronic Imaging 2004*, J. R. Price and F. Meriaudeau, Eds. SPIE, May 2004, Bellingham, USA, pp. 90–98. [Online]. Available: http://proceedings. spiedigitallibrary.org/proceeding.aspx?articleid=837538

[63] J. Han, L. Shao, D. Xu, and J. Shotton, "Enhanced computer vision with Microsoft Kinect sensor: a review." *IEEE Transactions on Cybernetics*, vol. 43, no. 5, pp. 1318–1334, Oct. 2013. [Online]. Available: http://www.ncbi.nlm.nih.gov/pubmed/23807480

[64] Planinc R., "Behavior Modeling based on Depth Data in the Context of Ambient Assisted Living", PhD Thesis, Vienna University of Technology, 2015.

[65] A. Yilmaz, O. Javed, and M. Shah, "Object Tracking: A Survey," *ACM Computing Surveys*, vol. 38, no. 4, pp. 1–45, 2006.

[66] J. Shotton, A. Fitzgibbon, M. Cook, T. Sharp, M. Finocchio, R. Moore, A. Kipman, and A. Blake, "Real-time human pose recognition in parts from single depth images," in *Proceedings of the Conference on Computer Vision and Pattern Recognition (CVPR)*. Colorado Springs, USA: IEEE, 2011, pp. 1297–1304.

[67] P. Kohli and J. Shotton, *Consumer Depth Cameras for Computer Vision*, A. Fossati, J. Gall, H. Grabner, X. Ren, and K. Konolige, Eds. London: Springer London, 2013. [Online]. Available: http://www.springerlink.com/index/10.1007/978-1-4471-4640-7 http://link.springer.com/10.1007/978-1-4471-4640-7

[68] L. Spinello and K. Arras, "People detection in RGB-D data," in *Proeedings of the International Conference on Intelligent Robots and Systems*, 2011, pp. 3838–3843.

[69] S. Ikemura and H. Fujiyoshi, "Real-time human detection using relational depth similarity features," in *Proceedings of the Asian Conference on Computer Vision*, 2011, pp. 25–38.

[70] S. Wu, S. Yu, and W. Chen, "An attempt to pedestrian detection in depth images," in *Proceedings of the Chinese Conference on Intelligent Visual Surveillance*, 2011, pp. 97–100.

[71] P. Kelly, N. E. O'Connor, and A. F. Smeaton, "Robust pedestrian detection and tracking in crowded scenes," *Image and Vision Computing*, vol. 27, no. 10, pp. 1445–1458, 2009.

[72] F. Hegger, N. Hochgeschwender, G. Kraetzschmar, and P. Ploeger, "People Detection in 3D Point Clouds Using Local Surface Normals," in *Robot Soccer World Cup 2012*, Lecture Notes in Computer Science. Springer, Berlin, Heidelberg, 2013, vol. 7500, pp. 154–165.

[73] D. Beymer, "Person counting using stereo," in *Proceedings of the Workshop on Human Motion*, Los Alamitos, USA, 2000, pp. 127–133.

[74] K. Khoshelham and S. Elberink, "Accuracy and resolution of Kinect depth data for indoor mapping applications," *Sensors*, vol. 12, no. 2, pp. 1437–1454, 2012.

[75] T. Stoyanov, A. Louloudi, H. Andreasson, and A. J. Lilienthal, "Comparative evaluation of range sensor accuracy in indoor environments," in *Proceedings of the European Conference on Mobile Robots (ECMR)*, Örebro, Sweden, 2011, pp. 19–24. [Online]. Available: http://www.aass.oru.se/Research/Learning/publications/2011/Stoyanov_etal_2011-ECMR11-Comparative_Evaluation_of_Range_Sensor_Accuracy_in_Indoor_Environments.pdf\nhttp://oru.diva-portal.org/smash/record.jsf?pid=diva2:540987

[76] T. Dutta, "Evaluation of the Kinect sensor for 3-D kinematic measurement in the workplace," *Applied Ergonomics*, vol. 43, no. 4, pp. 645–649, Jul. 2012. [Online]. Available: http://www.sciencedirect.com/science/article/pii/S0003687011001529

[77] M. van Diest, J. Stegenga, H. J. Wörtche, K. Postema, G. J. Verkerke, and C. J. Lamoth, "Suitability of Kinect for measuring whole body movement patterns during exergaming," *Journal of Biomechanics*, vol. 47, no. 12, pp. 2925–2932, 2014. [Online]. Available: http://www.sciencedirect.com/science/article/pii/S0021929014003984

[78] P. Plantard, E. Auvinet, A.-S. Pierres, and F. Multon, "Pose estimation with a Kinect for ergonomic studies: evaluation of the accuracy using a virtual Mannequin," *Sensors*, vol. 15, pp. 1785–1803, 2015. [Online]. Available: http://www.mdpi.com/1424-8220/15/1/1785/

[79] S. Obdrzalek, G. Kurillo, F. Ofli, R. Bajcsy, E. Seto, H. Jimison, and M. Pavel, "Accuracy and robustness of Kinect pose estimation in the context of coaching of elderly population," in *Proceedings of the Annual International Conference*

of the IEEE Engineering in Medicine and Biology Society (EMBS). San Diego, USA, IEEE, Aug. 2012, pp. 1188–1193.

[80] K. Caine, S. Šabanović, and M. Carter, "The effect of monitoring by cameras and robots on the privacy enhancing behaviors of older adults," in *2012 7th ACM/IEEE International Conference on Human–Robot Interaction (HRI)*, March 2012, pp. 343–350.

Chapter 5
A data fusion approach for identifying lifestyle patterns in elderly care

Mohamed Eldib[1], *Tongda Zhang*[2], *Francis Deboeverie*[1], *Wilfried Philips*[1], *and Hamid Aghajan*[1,2]

5.1 Abstract

The motion sensor technology and the network of cameras have been explored separately as attractive solutions for building in-home monitoring systems for elderly care. Each of these technologies offers advantages over others while suffering from certain limitations. Motion detectors usually offer a privacy preserving solution, but do not yield granular information about the user's activities. Cameras, on the other hand, offer access to details of activities of daily life, but are regarded with caution in terms of coping with user privacy concerns. In this chapter, we provide an informative and a highly updated review of sensor fusion approaches. Then, we introduce an in-home monitoring system for elderly care, which is based on information collected from a network of PIR motion detection sensors and low-resolution cameras with 30×30 pixel arrays. The data fusion method we used for that system is based on different activity features. From the PIR sensors, the level of the daily occupation and the active level of the occupant are extracted. From the low-resolution cameras, other features such as activeness, sleep duration, visitors, and TV activity are extracted. Generally speaking, the proposed monitoring system fuses the heterogeneous sensor information to identify the occupant's lifestyle pattern. The system employs K-means clustering algorithm to group the observed days into different lifestyle patterns such as restful isolated days, restful social days, busy isolated days, and busy social days. To evaluate our system, experiments were conducted on six months of real-data. The results show promising performance to identify a certain lifestyle pattern or changes which occur over time.

5.2 Introduction

Our society is characterised by a steadily growing proportion of elderly people. The majority of these elders have a strong desire to age independently in their own home

[1]Ghent University, TELIN-IPI-IMINDS, Sint-Pietersnieuwstraat 41, Ghent, Belgium
[2]Ambient Intelligence Research Lab, David Packard Building, Stanford, CA 94305–9515, USA

environment. However, with age the physical limitations increase and the social environment changes, which often leads to decreasing activity levels, need for assistance with daily life tasks and social isolation. A study shows that in the UK, more than 80% of people aged between 65 and 74 years do not meet the recommended level of physical activity [1]. In addition, with age, the need for help with Activity of Daily Livings (ADLs) increases due to physical, cognitive, and perceptual decline of the elder people. Many family caregivers feel overburdened taking care of an elderly family member while also having a career and being a parent [2].

Ageing-in-place presents itself as a promising solution for health care systems. Recently, ageing-in-place has gained a lot of attention in recent years due to the fact that many senior citizens prefer to age and maintain their independence as long as possible in their own homes [3], rather than in a residential care, because of their preferences regarding emotional and physical associations, memories and comfort. Ageing-in-place promotes the well-being of elder people without scarifying the quality of life in a familiar environment and maintains the valuable social networks. The success of ageing-in-place depends on Ambient Assisted Living (AAL) tools which have witnessed tremendous improvements in the last few years. AAL tools provide supervision and assistance with ADL to prevent, cure, and improve wellness and health conditions of seniors.

Over the past decade, the sensor technology has been through an enormous improvement in terms of low-power usage, low-cost, high-capacity, and small size, which greatly increase the potential of supporting healthy and active ageing-in-place [4, 5]. There are different research projects aimed at creating multi-sensor environments that use a variety of sensor technologies to enable independent living. In these projects, sensor fusion techniques have been extensively employed with the aim of fusing and aggregating data from different sensors. The purpose of using sensor fusion in multi-sensor environments is to have a decreased detection error probability and an increased reliability by using data from multiple distributed sources.

Complementary information from multiple sensors allows features in the environment to be perceived that are difficult or impossible to perceive using just the information from each individual sensor operating separately. For example, in a smart home setting multiple sensors such as motion, temperature, and pressure sensors collect complementary data about cooking activity. Motion sensors can provide data about a human presence in the kitchen area, while temperature sensors provide clues to whether the stove is on and pressure or vibration sensors can indicate whether any kitchen objects are being used. While these three sensor classes may independently be weak at characterising the cooking activity, fusing them together leads to a stronger model. Information fusion is worthwhile for reducing uncertainty. Data collected from sensors can be noisy and incomplete. As a result, activity learning methods that directly utilise raw sensor data are also prone to yield erroneous results. However, when different sensors and methods produce varying levels of errors and each method is reasonably accurate independently, a combination of multiple experts should reduce overall classification error and as a consequence yield more accurate outputs.

Human activity recognition in elderly homes generally uses two main sensor categories: (1) wearable sensors and (2) ambient sensors. Various wearable sensors, such as electrocardiogram (ECG), glucometer, electromyography (EMG), pulse oximeter, and e-textile sensors, could be attached to the subject of interest to monitor his vital signs such as the heartbeat, respiration, blood pressure, glucose level, and muscle activity. However, wearable sensors are still facing a few disadvantages such as limited battery life, high cost, easily forgotten to be worn by the user, and the rigid requirement to be worn on specific body parts to provide reliable measurements. In the second category, ambient sensors are installed in the home environment by mounting them on the wall or the ceiling, embedded in furniture and appliances. Passive Infrared Motion (PIR) sensors, Kinect sensors, visual sensors, and Radio Frequency Identification (RFID) are the most popular among researchers. It is important to find the right combination of ambient sensors to install in the smart home, taking into account the challenges that come with the choice of each sensor. There are other different kinds of ambient sensors such as pressure sensors to measure pressure on mat or chair, smart tiles to measure pressure on floor, and magnetic switches to detect door/cabinet opening/closing.

In this chapter, we will first explore some of the multi-sensor environment projects. Then, we will summarise the state-of-the-art sensor fusion approaches for assisted living. Finally, we will present a multi-sensor fusion approach to identify different lifestyle patterns of an elder occupant in our smart home testbed.

5.3 Multi-sensor environment projects

As a result of the recent advancement in sensor technologies and the growing application demands, a substantial number of smart home projects and initiatives have been undertaken. CogWatch [6] is an EU project that provides a cognitive rehabilitation system to monitor patients with apraxia and action disorganisation syndrome while staying in their own homes. CogWatch used Kinect and a set of sensorised tools. Kinect allows the system to track the user hand movements and behaviour when executing the task by providing the hands' position. While, the sensorised tools are designed to provide data that can be used to recognise or analyse interactions with everyday objects, such as a mug, a jug, or a kettle. In the fusion module, the measurements from the sensors are synchronised and transformed into a feature vector to identify human activities.

The GatorTech [7] smart house project at the University of Florida integrates several existing smart devices such as mailboxes, driving simulators, ultrasonic location tracking, smart front doors, smart mirrors, smart bathrooms, smart blinds, smart displays, smart floors, and smartwaves. To systematically integrate the various devices, appliances, and sensors, a sensor platform is introduced to represent any attached object in a pervasive space simply as a Java program, more specifically, as an OSGi service bundle. HERMES[1] is an EU project which deploys intelligent audio and visual processing to alleviate the elderly cognitive decline. In the audio processing system, several attributes are extracted such as speaker identification and emotion detection.

[1] http://www.fp7-hermes.eu/

In the visual processing system, face detection, person tracking, and face identification are extracted. Finally, in the low-level information fusion component, the contextual cues from both audio and visual processing systems are combined with a view to improving the recognition accuracy obtained from uni-modal recognition systems.

The Aware Home [8] project at Georgia Tech uses multiple integrated sensors such as ultrasonic sensors, smart floor sensors, and video. By using tracking and sensing technologies, the Aware Home provides assistance to elder people to help them to find the most common lost objects such as keys, wallets, glasses, and remote controls. More specifically, a radio-frequency (RF) tag is installed in each object. Then, via LCD touch panels placed throughout the house, the elderly could interact with the system for the possible assistance. Finally, the system helps the elderly to look for the lost object using audio cues (e.g., "your glasses are in the bedroom"). For example, a person is identified as having an object using RF and the person is last seen at the entrance using smart floor. Then, the system concludes that the person has left with the object.

GER'HOME [9] project uses a multi-sensors analysis for everyday elderly activity monitoring. A video monitoring platform fed by a network of cameras and contact sensors are installed. The vision component is used to detect and track people and to recognise predefined video events. For instance, a typical video event is a person sitting down on a chair. While the sensor component is used to recognise predefined contact events. In the multi-modal composite component, the video and contact event recognition aims at recognising complex temporal events that combine both video and contact events, and those events are based on a priori knowledge. As an example, meal preparation entails at least motion in the kitchen and use of cabinets where food, plates and/or utensils are stored. Depending on the type of meal it may involve the use of appliances, such as stove, oven, or microwave oven.

The UbiMon project [10] proposed a framework to collect information from wearable and implantable sensors to a local processing unit such as: PDA or smart phone through a body network, before transmitting to a remote database.

5.4 Sensor fusion approaches

In this section, we describe some sensor fusion approaches that can integrate sensory data from multiple sensors to provide more reliable and accurate information. There are several potential advantages of multi-sensor fusion approaches such as reliability and complementarity. The overall uncertainty can be decreased with the integration or the fusion of the redundant information and thus serve to increase the accuracy with which the information is perceived by the system. Typically, fusion of data/information can be carried out on three levels of abstraction closely connected with the flow of the activity learning process: data, feature, and classifier fusion.

5.4.1 Data fusion level

Multi-sensor data fusion refers to the stage in the integration process where there is an actual combination of different sources of sensory information into one representational format. Given a group of sensors, a weighted average of the redundant sensor data provides a simple approach to fuse the events generated by different sensors. Kalman filter is more often the most popular approach as it results in estimates for the fused data that are optimal in a statistical sense. Besides that, the Kalman filter does not require heavy computational expense. Kalman filtering assumes that the sensors can be modelled as a linear system and sensor noise can be modelled as a Gaussian distribution. The algorithm operates in two steps. In the first step, referred to as the prediction step, the Kalman filter uses the actual sensor values as observation sequences, in order to estimate the current sensor values along with their uncertainties. In the second step, the fused estimate is obtained by the weighted average of the sensor values, with more weights given to the values with higher certainty. Kalman filtering can be performed in real time using the current sensor measurements and the previously calculated sensor states with their uncertainty values.

Durrant-Whyte and Henderson [11] provided a resource on techniques for fusing raw sensor data and Kalman filtering for data fusion. Medjahed *et al.* [12] proposed an automatic in-home healthcare monitoring system for distress situations detection. The system uses a set of sensors such as physiological sensors (e.g., cardiac frequency), microphones, infrared sensors, debit sensors, and state-change sensors. A data fusion approach based on fuzzy logic with a set of rules directed by medical recommendations, is used to fuse the various sensor outputs.

5.4.2 Feature fusion level

Feature fusion refers to the process of combining features extracted from multiple sensors. The naive technique for fusing features from multiple sensors is to concatenate the features extracted from each sensor into a single, longer feature vector. This is an effective technique when the multiple sensors and features extracted from the raw sensor data are independent of each other. When the features extracted from different features are not independent, then dimensionality reduction or feature selection techniques can be used to fuse the different features. Fleury *et al.* [13] presented a feature fusion approach to detect loss of autonomy. For this purpose, they used location (presence infrared) sensors, microphones for sound and speech recognition, wearable sensor constituted of accelerometers and magnetometers, which indicate walking periods and postural transfers, door contact on appliances (fridge, cupboard, and dresser), and finally, temperature and hygrometry sensors. The features were then put together to create a feature vector significantly representative of an activity. Finally, SVM classifier is employed to detect seven ADLs.

Libal *et al.* [14] utilised audio-visual sensors to automatically recognise ADLs. The cameras provide human 3D location information as visual features, and the microphones provide monolithic acoustic segments as audio features. At the feature fusion level, the audio-visual features are concatenated. Then, various GMM classifiers are used to detect the ADLs. An ontology-based approach [15] was used to model the human status context such as body posture and the environment context such as the semantic information about the scene. The data provided by a set of cameras for person detection, and wearable accelerometer sensors attached to objects of daily living for environment events triggering such as TV remote control or doors use are employed to build a human status context model and environment context model. Then, a rule-based approach is used for processing and integrating both model types at activity decision level. Demongeot *et al.* [16] described a health smart home framework to ensure the medical security and quality of life for patients who need home-based medical monitoring. In order to follow the patient activity, the system uses generic sensors such as infrared sensors coupled with an accelerometer located on the body of the patient, and with magnetic contacts on the room doors to detect a fall or an abnormally long or frequent presence in the toilet. Also, the system measures the respiratory rhythm, the heart rate, and the pulse amplitude using vegetative integrity sensors. Finally, specific disorder sensors are used to detect distress syndrome or asthma. The classical factorial analysis is used for the feature fusion.

5.4.3 Classifier fusion level

At the third level of sensor fusion, various individual classification models are first trained using data from each sensor independently. In the second step, the individual classifier outputs are combined to find the actual output value. This type of fusion is also referred to as decision fusion or mixture of experts, and has been studied with a number of proposed approaches. There are two common approaches for fusing output from the classifiers. In the first approach, the goal is to find a single best classifier or a group of best classifiers whose outputs are then considered for making the final prediction. In the second approach, a group of methods perform simple fusion functions. Such functions operate directly on the outputs of the individual classifiers to determine the best combination to make the final prediction.

One popular approach of classifier fusion level is the Dynamic Classifier Selection (DCS) [17]. In DCS, the single best classifier is extracted that is most likely to output the correct classification label for a particular data point (such as a sensor event sequence). DCS divides samples from sensor training data and selects the best classifier for each division. Given a data point, we first determine the most likely division to which it belongs and generate its label using only the output of the best classifier for the corresponding division. The idea of using division-based DCS is to estimate each classifier's accuracy in a local region of the feature space and generate a final output from the most locally accurate classifier. Different metrics can be used for dividing the training samples. These include dividing samples into a subgroup for which there is agreement between the classifier output and the ground truth and a subgroup for which there is no such agreement and groupings based on the features of the input samples. DCS approaches strongly depend on the availability of sensor event training data and hence can be excessively influenced by noisy data samples. Furthermore, determining only the locally best classifier will result in loss of additional information that could have been gained from other classifiers for making a more accurate prediction.

Simple fusion functions are the most obvious choice when constructing a multiple classifier system. Common fusion functions are listed below:

- Simple voting strategy: Voting strategies can be applied to various classifier system assuming that each classifier gives a single class label as output. Voting strategies include unanimous voting in which all classifiers agree on the same class label, simple majority in which more than half the number of classifiers agree on the same class label, and majority voting in which the class label is a label that is most often predicted by the classifiers.

- Weighted majority voting: If the classifiers in the ensemble do not have identical performance, then classifiers that are more accurate are assigned higher weights when making the final prediction. The accuracy of the classifiers can be used to determine the weights.
- Highest average classification probability: When the classifiers also output posterior classification probabilities, a simple strategy is to pick the class label with the highest average probability of classification.

Another approach to classifier fusion is to formulate it as a supervised learning problem. This approach simply treats the outputs of various classifiers as input to a second-level classifier, and uses classical supervised learning techniques, such as Hidden Markov Model (HMM), Support Vector Machine (SVM), Conditional Random Field (CRM), decision trees, and Gaussian Mixture Model (GMM) to generate the final classification output. The selection of the second-level classifier relies on the distribution of its input, which in this case are probabilities that are skewed towards 0 and 1. As a result, algorithms that assume the data follows a normal distribution are not likely to fare well. Classifier fusion is useful when activity information is gathered from a redundant set of noisy sensors and data from each sensor alone is sufficient to model the activity. The meta-classifier, a classifier that integrates the output of various classifiers, exploits the redundancy intrinsic in the sensor network. An example of this situation is a setting where there are various accelerometers gathering movement information while an individual performs an activity. Accelerometers can be selected according to their contribution to classification accuracy as assessed during system training. Bayesian methods are also applicable for classifier fusion when classifiers output posterior probabilities.

In [18], a classifier fusion approach that combines video cameras and environmental sensors (e.g., electrical sensor) is proposed to recognise ADLs. The authors used prior knowledge about the activities and the environment to model the ADLs in a supervised way. A multi-sensor surveillance system using heterogeneous sensors data (video and accelerometer recordings) to identify activities during a medical clinical protocol is proposed in [19]. The system contains vision and accelerometer components. The vision component is used to detect moving objects and classify them according to a set of objects of interest (e.g., person). The accelerometer component analyses person's activity in relation to the activity models previously defined by an expert using an ontology language. Then, the multi-sensor fusion herein presented is performed at the decision level to integrate the vision-based features and the accelerometer-based features to identify the body posture of the person in situations where camera cannot give accurate information about the people detection on the scene due to occlusion.

Chen *et al.* [20] presented a system for detecting human interaction events in a nursing home using multi-modal technology. Two major sensors, video cameras and microphones, are used for detecting human activities. A dynamic Bayesian network is applied to model the temporal and semantic relationships. Luo *et al.* [21] described the three levels of multi-sensor fusion and discussed techniques for each level. Velik [22] introduced a powerful activity recognition model based on brain-inspired multi-modal data mining methods. Khaleghi *et al.* [23] provide a comprehensive survey of the state-of-the-art data fusion methods. Ruta and Gabry [24] provided an overview of methods for combining classifier outputs for multi-sensor fusion. Kuncheva [25] presented a thorough discussion on the theoretical underpinnings of classifier fusion strategies. Gao *et al.* [26] and Zappi *et al.* [27] employed a multi-sensor fusion for combining data from multiple accelerometers for recognising human locomotion. Hong *et al.* [28] discussed a methodology for fusing information gathered from uncertain smart home sensor readings using Dempster-Shafer theory.

5.5 Overview of the service flat setup

The system proposed in this chapter is based on algorithms developed in a fundamental research project "SONOPA: Social Networks for Older adults to Promote an Active Life".[2] In Sonopa, the aim is to combine a social network with activity recognition in a smart home environment to stimulate and support activities and daily life tasks. Sonopa suggests suitable activities and social connections to the senior citizen automatically, proactively, and at the optimal time, while providing a simple bridge to the social network of the senior citizen. Sonopa achieves this by analysing both physical and online activities of senior citizen users in their smart homes.

The in-home monitoring system is deployed in a house of two floors in the UK. The system is equipped with low-resolution visual sensors [29], PIR sensors, and two computers to record the data. Tables 5.1 and 5.2 summarise the different capabilities and properties of three ambient sensors: PIR, Kinect, and visual sensors. We chose not to use high-resolution cameras like Kinect sensor because cameras can raise people's privacy concerns. In this case, post-processing algorithms are required to solve privacy issues [30]. Figure 5.1 shows the layout of the house. The participant is

Table 5.1 Comparison between the different capabilities of PIR, visual, and Kinect sensor. H, M, and L stand for High, Medium, and Low values, respectively

Technology	Location detection	Presence detection	Shape detection	Tracking
PIR sensors	L	M	Not possible	Single-person
Visual sensors	H	H	L	Multi-person
Kinect sensors	H	H	H	Multi-person

[2] http://www.sonopa.eu/

Table 5.2 Comparison between the different properties of PIR, visual, and Kinect sensors. H, M, and L stand for High, Medium, and Low values, respectively

Properties	**PIR sensors**	**Visual sensors**	**Kinect sensors**
Network density	H	M	M
Resolution	Single pixel (on/off)	30 × 30 pixels	IR depth sensor: 640 × 480 pixels Color camera: 1280 × 1024 pixels
Space occupancy	L	L	H
Cost	L	M	H
Privacy concern	L	L	H
Operation (Lighting)	No	Yes	IR depth sensor: No Color camera: Yes
Applicability	Indoors	Indoors	Indoors/Outdoors
Battery life	H	M	L

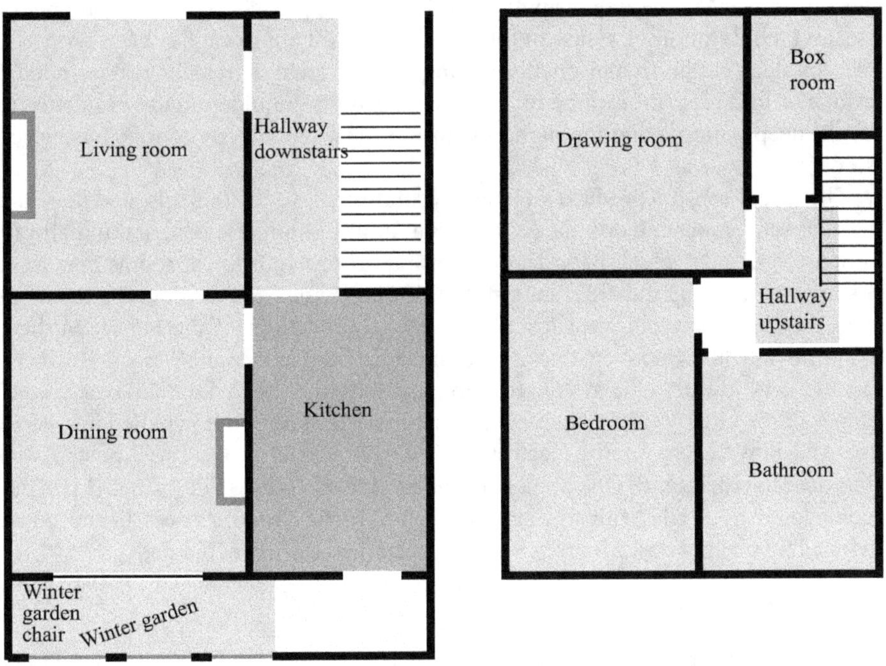

Figure 5.1 Ground floor (left) and first floor (right) of the house in the UK

an 81-year-old lady. She is living entirely autonomously. The system started running in August 2014 until March 2015. The ground truth is collected from diaries where the senior citizen wrote down her activities such as sleep time, wake-up time, visitors, dinner time, and out-of-home time. The diaries are verified by an informal caregiver (e.g., family member).

5.5.1 Low-resolution visual sensor

The use of visual sensors is suitable for ADLs because of the preservation of privacy. In [31], a network of low-resolution visual sensors with an image size of 30 × 30 pixels has been installed in a service flat of a senior citizen living alone. The authors proposed a novel measure to find similar patterns of behaviour between each pair of days from the user's detected positions, based on heatmaps and Earth mover's distance (EMD). Then, an exemplar-based approach is used to identify sleeping, eating, sitting activities, and walking patterns. In a laboratory setup of 5 low-resolution visual sensors [32, 33], users's locations and mobility statistics were obtained from a robust people tracker based on recursive maximum likelihood principles. Xingzhe *et al.* [34] analysed the behaviour patterns of an elderly person using statistical features extracted from the senior citizen tracks such as the time of getting up and going to bed, the walking distance over a day, and the number of tracks detected at a specific area. Then, the statistical features are clustered using a RANdom SAmple Consensus (RANSAC)-principle method to detect the behaviour patterns. Eldib *et al.* [35] measured the socialisation level of a senior citizen using low-resolution visual sensors by detecting visits via an HMM. Then, they applied a rule-based approach to infer the number of visits and the duration of visits. In [36], a video-based approach has been proposed to detect sleep duration and quality among older adults. The authors developed a network of low-resolution cameras that are placed on the older adult's residence to analyse sleep patterns and the nightly bathroom visits indirectly to recognise sleep disorders.

The visual sensors produce an image resolution of 30 × 30 pixels, and there are 10 visual sensors installed in the living room, dining room, kitchen, and in hallway, in order to get as much activity data as possible from the different rooms that serve the test person during the different times of the day. Figure 5.2 displays the locations where the visual sensors were installed in the ground floor of the house. In these images privacy is maintained, thus it is for instance hard to recognise faces. However, those are very useful in the system to recognise activities, behaviour and behavioural changes of the elderly. The visual sensors are prototypes and thus installed in a wired setup and powered by mains electricity. The visual sensors are fixed on the wall using double-sided tape. One visual sensor produced a video file of 6 GB per day accompanied by a file with timestamps. In total, the visual sensors in the house produced 60 GB per day. Monthly visits are arranged to collect the data.

5.5.2 PIR sensors

The PIR sensors are installed in all rooms of the test house, both on the ground floor and the first floor. There are 14 battery powered ZigBee PIR sensors, 1 ZigBee gateway and 1 Zigbee router. The PIR sensors produce very little data compared to the visual sensors. The data is stored locally in a Postgres database in one of the two computers. The sensors are glued to the ceiling railings and the furniture using double-sided tape (unobtrusively to the furniture in order not to spoil the paint and the wallpaper). When necessary, small flexible metal wedges were used to get the right angle. Some rooms have two sensors installed for better coverage of the areas. However, we believe that

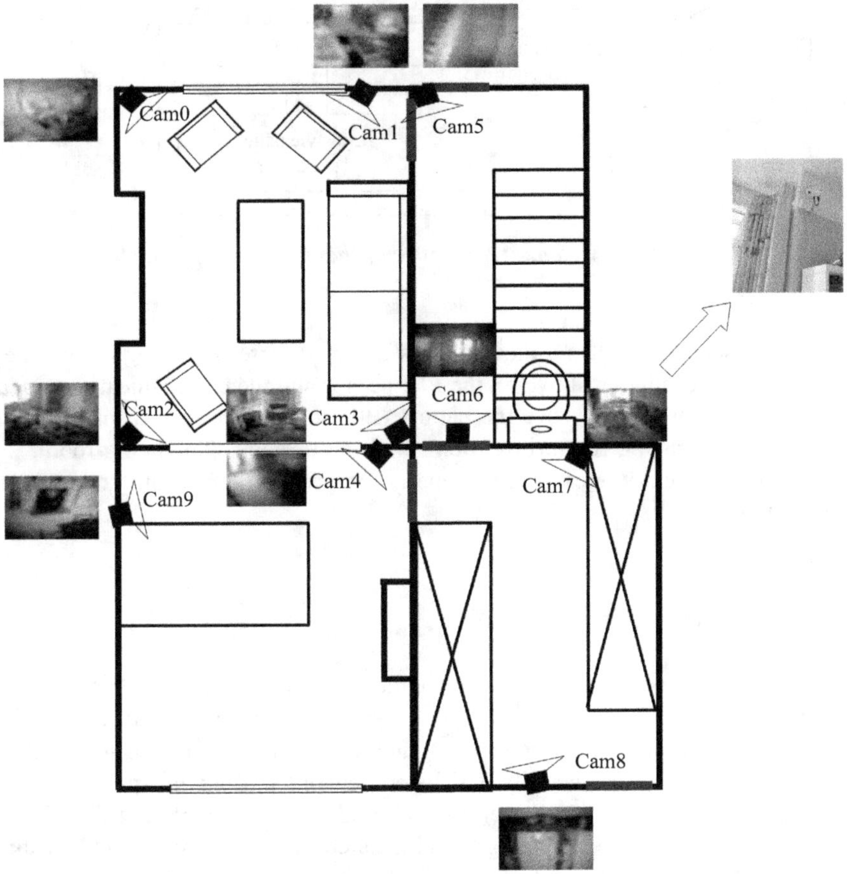

Figure 5.2 Room layout of the ground floor and the visual sensor positions

a future setup can be done with fewer sensors, taking into consideration that in rooms with multiple sensors such as the living room, both sensors provide very similar event patterns.

Despite the popularity of PIR sensors, they are known to have the following problems: (1) highly bursty output which limits PIR systems to single-person scenarios; (2) self-triggering due to sudden changes in environmental conditions such as heating, ventilating, and air conditioning; (3) PIR sensors cannot sense people who are standing still [37]. In comparison, visual sensors and the Kinect sensors are capable of detecting the person even when he is standing still, because people tend to move parts of the upper body (head, shoulders, and hands), which could be easily detected by foreground detection algorithms [38, 33].

The sensors are connected to a ZigBee gateway (Zigbee to Ethernet), which is in the utilities closet, either directly or through a Zigbee Router (Zigbee to Zigbee) forming a multi-hop network. The router is used to cover the sensors upstairs. The

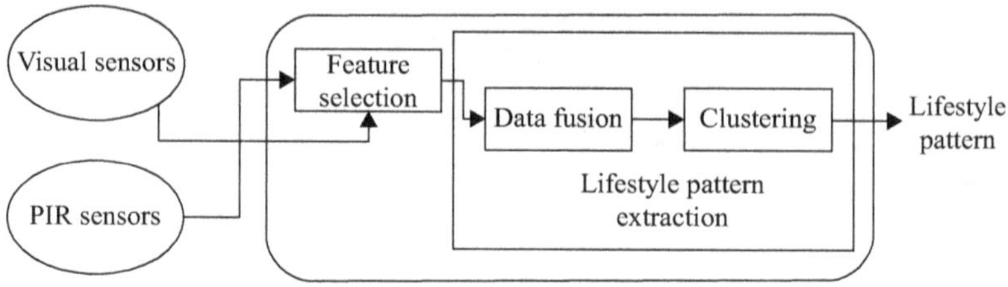

Figure 5.3 System architecture of the proposed approach

sensors only send data when a movement is detected. After reporting a movement, they sleep for two minutes in which the PIR sensor and the communication module are switched off in order to save the battery. PIR sensors are located in: bedroom, bathroom, 2 × drawing room, box room, hallway upstairs, 2 × living-room, 2 × dining room, kitchen, winter garden (general), winter garden (reading corner), and entrance hall.

5.6 System overview

Our proposed system contains several modules: feature selection, data fusion, and clustering. In the feature selection module, several features are selected and extracted from the visual sensors and PIR sensors such as sleep duration, visitors, activity level in each room, and TV activity. The heterogeneous sensors information is fused in the data fusion module. In the clustering module, the features are divided into two groups: (1) features representing social/isolated lifestyle pattern and (2) features representing busy/restful lifestyle pattern. Then, the K-means clustering algorithm is used to cluster each group of features to the corresponding lifestyle pattern. Finally, a certain day is categorised to one of predefined lifestyle patterns. Figure 5.3 shows an overview of the proposed approach.

5.6.1 Feature selection

From the low-resolution visual sensors, we extracted several features such as sleep duration, visitors, TV activity, and activeness. In our previous study to detect sleep duration in older adults [36], a network of low-resolution visual sensors is placed on the older adult's residence in Belgium to detect his wake-up time, his sleep time, and whether he wakes up in the middle of the night using a rule-based approach based on the motion patterns. We follow the same approach here to extract the sleep duration feature. The number of visits and the duration of visits features are extracted using an HMM and a rule-based approach as in [35]. TV status feature is detected by computing the intensity values of the TV's region of interest. If the intensity proceeds a threshold, then the TV status is on. Otherwise, the TV status is off. Finally, the activeness (activity level) feature is computed for each room using similarity heatmaps

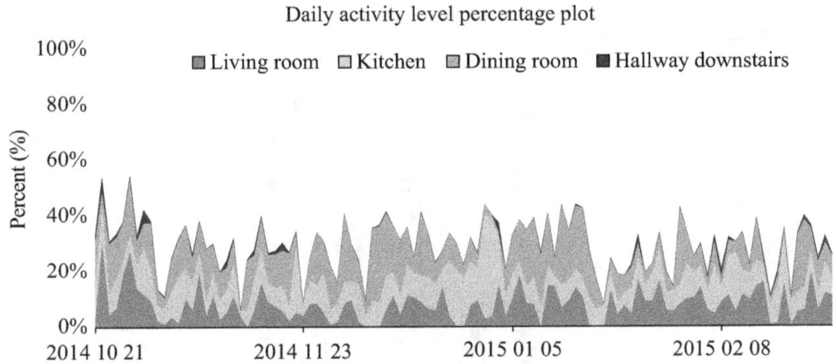

Figure 5.4 Visual sensor: daily activity level percentage

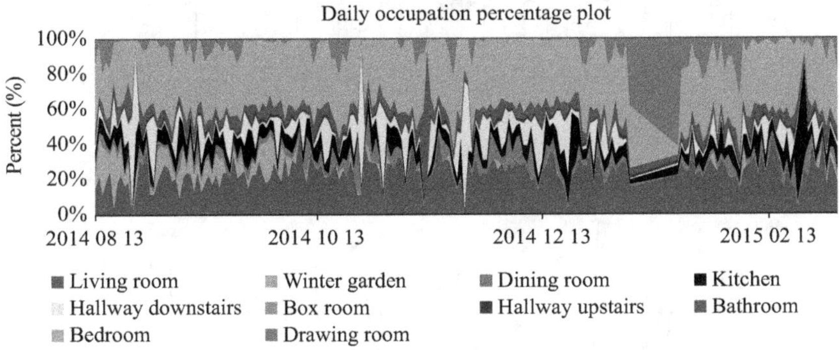

Figure 5.5 PIR sensor: daily occupation percentage

and exemplar-based approach [31, 39]. Figure 5.4 provides the visualisation of the activity level percentage of the occupant.

From the PIR sensors, we extracted two major feature categories: activeness level and room occupation. More specifically, the activeness level measures how active the occupant is during a certain period of time by considering the frequency of PIR sensors' activation; the room occupation, on the other hand, quantifies how the occupant spends time in different rooms in the house by combining PIR sensors' location information with the activation events we observed. To provide a better perspective of the occupant's actual activity, we apply different time scales on both the activeness level and the room occupation features, such as daily measure and hourly measure. For example, Figures 5.5 and 5.6 provide the visualisations of the daily occupation percentage and the activeness level of the occupant, respectively.

The visual sensor and PIR sensor features are of incomparable units. For instance, the TV status feature is represented in seconds while the activeness feature is represented in terms of the number of the sensor transitions. The TV status feature has a variance that is orders of magnitude larger than the activeness feature. In this situation, leaving variances unequal is equivalent to putting more weight on variables

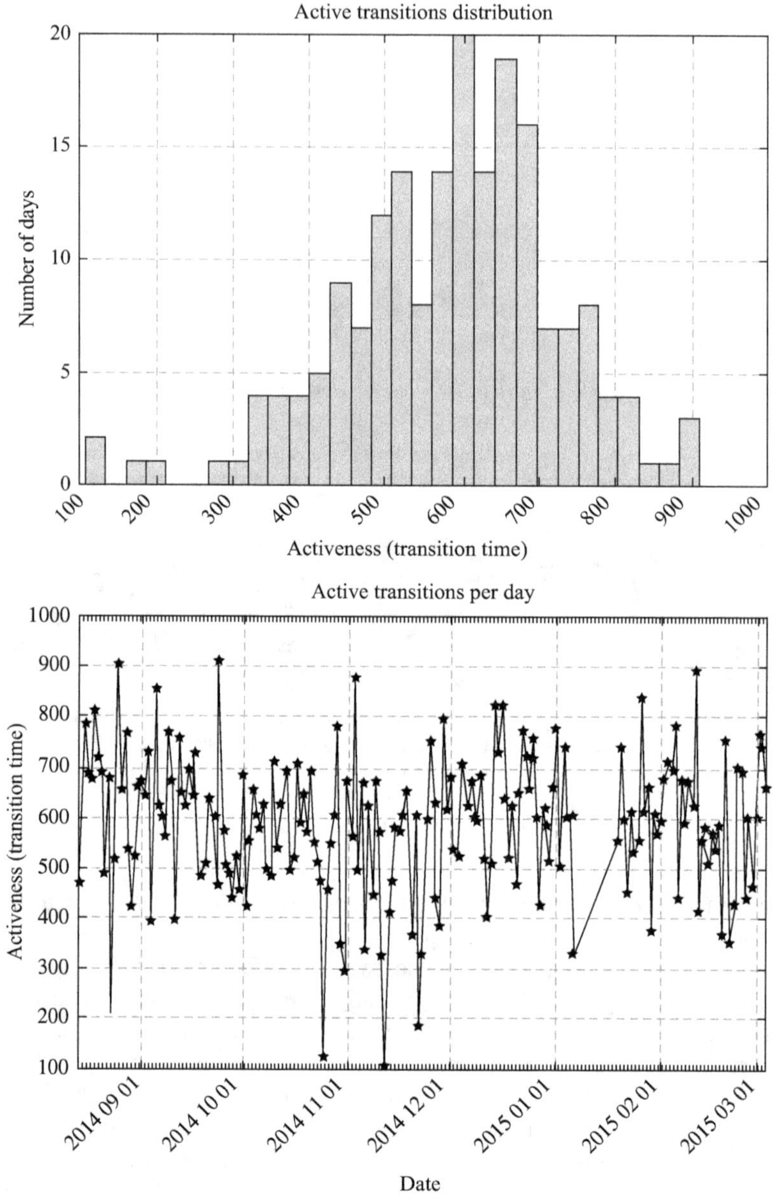

Figure 5.6 PIR sensor: daily activeness level

with smaller variance. Since K-means clustering is used later, the clusters will tend to be separated along variables with greater variance. For instance, the TV status feature will have disproportionately more importance in its computation with respect to the activeness feature. We apply normalisation to give all features an equal weight.

Table 5.3 Combination of lifestyle patterns

$C^i \oplus C^j$	**Lifestyle pattern**
$C_1 \oplus C_3$	Busy social
$C_1 \oplus C_4$	Busy isolated
$C_2 \oplus C_3$	Restful social
$C_2 \oplus C_4$	Restful isolated

5.6.2 Lifestyle pattern extraction

We aim at identifying certain lifestyle patterns or changes which occur over time. Having extracted the visual sensor and PIR sensor features. We propose a simple approach for feature fusion. Let $X_d = (x_1 \ldots x_m)$ be the visual sensor feature vector of day d. Let $Y_d = (y_1 \ldots y_n)$ be the PIR sensor feature vector of day d. Then, we concatenate the features extracted from each sensor into a single, longer feature vector $Z_d = X_d \oplus Y_d$. Let Z_d^i be the set of the concatenated features to represent busy and restful lifestyle patterns. Let Z_d^j be the set of the concatenated features to represent social and isolated lifestyle patterns. In Z_d^i, the selected features are the sleep duration, TV activity, activity level from the visual sensors in each room, daily activeness summary and the daily occupation time distribution among different locations from PIR sensors. In Z_d^j, the number of visits and the duration of visits features are selected.

In the clustering step, given a sequence $Z_D^i = (Z_{d_1}^i \ldots Z_{d_l}^i)$ representing busy and restful lifestyle pattern features in the dataset D. The K-means clustering algorithm is used to group the days in Z_D^i into two clusters. Let $C^i \in (C_1, C_2)$ be the possible cluster output of Z_D^i, where C_1 represents days with busy lifestyle pattern and C_2 represents days with restful lifestyle pattern. Similarly, the K-means clustering is used to group the days in the given sequence $Z_D^j = (Z_{d_1}^j \ldots Z_{d_l}^j)$ representing social and isolated lifestyle patterns features into two clusters. Let $C^j \in (C_3, C_4)$ be the cluster output of Z_D^j, where C_3 represents days with social lifestyle pattern and C_4 represents days with isolated lifestyle pattern. Then, the lifestyle patterns from each cluster are fused by simply concatenating the results of C^i and C^j. Finally, a certain day is categorised to one of the following lifestyle patterns: restful isolated day, restful social day, busy isolated day, and busy social day. Table 5.3 shows the possible combination of the lifestyle patterns.

5.7 Experiments

To evaluate the performance of our proposed approach, we collected six months of real-data recordings using a network of 10 low-resolution visual sensors producing an image of 30 × 30 pixels at a frame rate of 120 fps and 14 PIR sensors. Video capturing is time-synchronised. The data of each day corresponds to a 24-h period starting from midnight. The senior citizen has not given instructions about the way

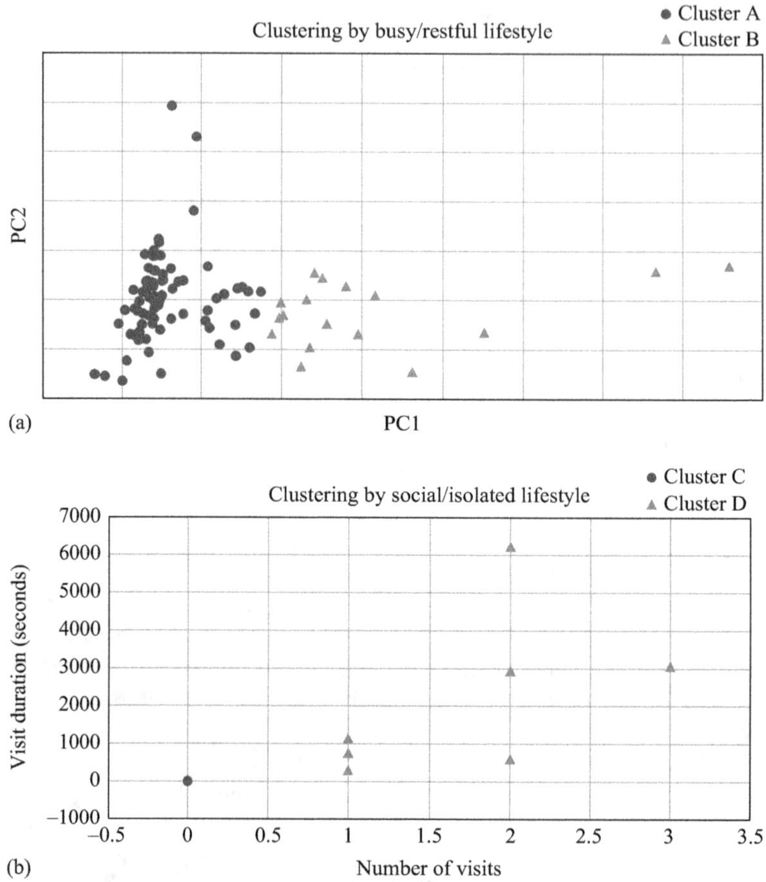

Figure 5.7 Scatter plot of different lifestyle patterns: (a) busy/restful lifestyle pattern; (b) social/isolated lifestyle pattern

and the time of performing the activities. The number of the fused features to identify busy and restful lifestyle patterns is 19 features, while the number of the fused features to identify social and isolated lifestyle patterns is 3 features.

In order to identify which cluster belongs to which lifestyle pattern, we selected indicative features to interpret each cluster. We perform clustering analysis by selecting the activity level and the daily occupation of the living room to identify a busy or a restful lifestyle pattern. The activity level is represented by a value between 0 and 1, a high activity level is represented by a small value, while a low activity level is represented by a high value. To visualise the data and the clusters, the Principle Component Analysis (PCA) is used to reduce the fused features into two components. Figure 5.7(a) provides the visualisation of the projected data and the clusters. The average activity level and the daily occupation of the living room for cluster A are 0.087 (low activeness) and 5837 s, and for cluster B are 0.044 (high activeness) and 2156 s, respectively.

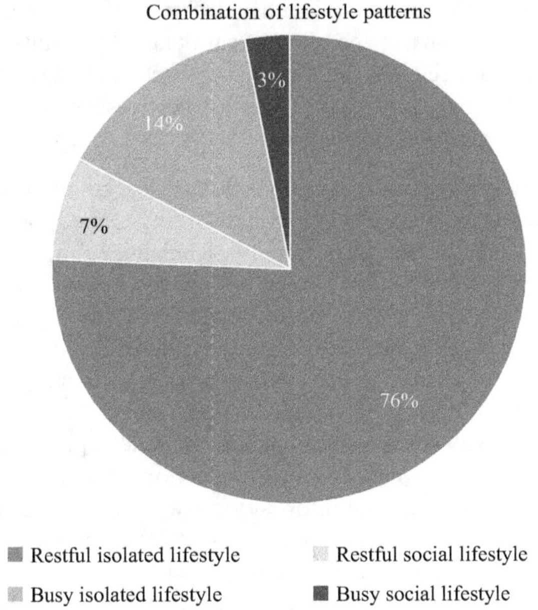

Figure 5.8 Analysis of lifestyle patterns combination

From the cluster analysis, the occupant has a low activity level and spent more time in the living room for cluster A than cluster B. This shows that cluster A represents the restful lifestyle pattern and cluster B represents the busy lifestyle pattern.

Similarly, we select the number of visits and the duration of visits to identify the social and the isolated lifestyle patterns. Figure 5.7(b) provides the visualisation of the social/isolated data points and their corresponding clusters. The average number of visits and the duration of visits for cluster C are 0 visits and 0 s, and for cluster D are 2 visits and 1594 s, respectively. Given the information about each cluster, cluster C represents the isolated lifestyle pattern and cluster D represents the social life style pattern.

In the last set of experiments, we aim at categorising each day to one of the possible combination of the lifestyle patterns described in Table 5.3. Figure 5.8 shows the percentage of each of the lifestyle patterns combination. The occupant tends to have more restful days with no visitors (76% of the days). On the other hand, the occupant has more visitors in the restful days (7% of the days) than the busy days (3% of the days).

5.8 Conclusion and future challenges

In this selective chapter, we looked at several fusion approaches. In the context of smart home projects, we proposed an in-home monitoring system for elderly care using a network of low-resolution cameras and PIR sensors. First, we extracted features from the different sensors. Then, we introduced a data fusion approach to integrate the heterogeneous sensor information to identify different lifestyle patterns. Our fusion approach was evaluated on six months of real-data. The results showed different lifestyle patterns of the occupant. There are still some challenges that sensor fusion approaches can tackle next as well as ideas for how researchers can build a more reliable multi-sensor environment systems:

- *Sensor fusion with other sources of information*: Recent improvements in pervasive computing, sensors, and wireless networking make it possible to tap these sources of information to learn models of behaviour in a non-intrusive way. However, other sources of information are available as well to help to describe behaviours. Dictionaries provide definitions of behaviours. The internet in general and the web pages specifically are literary sources that mention behaviours or reference them by name in the context of a larger plot. The social networks contain a lot of information on human actions, interactions, and locations on a regular basis. Activity learning approaches can benefit greatly from information sources outside sensor events to learn models of behaviour and to relate outer sources of knowledge with the information that is automatically captured from sensors. By taking advantage of the power of these rich data sources, we can not only learn better models and recognise activities but we can also note how strong is the correlation between the discovered-sensed actions and activities with people's understanding of human behaviour.
- *Multi-sensor datasets*: Unfortunately, the proposed sensor fusion approaches are evaluated on separate datasets which make comparisons impossible. There is a need for standard multi-sensor benchmark datasets and made available to the research community, which would facilitate comparing alternative fusion algorithms and learning techniques. For obvious ethical reasons, it might not be possible to collect a dataset with actual activities of older people but the simulated activities should be made as realistic as possible by consulting specialist clinicians.
- *Multi-level sensor fusion*: Single level sensor fusion shorten the capacity and robustness of in-home monitoring systems, due to the weaknesses in uncertainty, missing observation, and incompleteness of a single sensor. Therefore, there is an obvious need to combine and fuse multi-sensor data to gain more high robustness and flexibility and the multi-level sensor fusion is needed in advanced in-home monitoring systems. For example, a four fusion level for decision making from fusion levels of the time-varying data can be designed. In the low-level fusion, methods can integrate the multi-sensor data. While in the medium level fusion, methods can integrate data and feature to obtain fused feature or decision. Finally, in the high-level fusion, methods can integrate feature and decision to obtain the final decision.

Acknowledgements

This research has been performed in the context of the European AAL project "Sonopa". This research has been financed by the agency for Innovation by Science and Technology (IWT), the Belgian National Fund for Scientific Research (FWO Flanders), and iMinds.

References

[1] Department of Health, UK., "At least five a week: Evidence on the impact of physical activity and its relationship to health," Tech. Rep., 2004.

[2] R. Schulz and P. R. Sherwood, "Physical and mental health effects of family caregiving," *American Journal of Nursing*, vol. 108, p. 23–27, 2008.

[3] G. Doblhammer and U. Ziegler, "Future elderly living conditions in Europe: Demographic insights," in *Gender, Health and Ageing*, G. M. Backes, V. Lasch, and K. Reimann, Eds. VS Verlag für Sozialwissenschaften, 2006, pp. 267–292.

[4] D. Ding, R. A. Cooper, P. F. Pasquina, and L. Fici-Pasquina, "Sensor technology for smart homes," *Maturitas*, vol. 69, no. 2, pp. 131–136, 2011.

[5] H. Alemdar and C. Ersoy, "Wireless sensor networks for healthcare: A survey," *Computer Networks*, vol. 54, no. 15, pp. 2688–2710, 2010.

[6] M. Pastorino, A. Fioravanti, M. T. Arredondo, J. M. Cogollor, J. Rojo, M. Ferre, and et al., "Preliminary evaluation of a personal healthcare system prototype for cognitive erehabilitation in a living assistance domain," *Sensors*, vol. 14, no. 6, pp. 10 213–10 233, 2014.

[7] S. Helal, W. Mann, H. El-Zabadani, J. King, Y. Kaddoura, and E. Jansen, "The gator tech smart house: a programmable pervasive space," *Computer*, vol. 38, no. 3, pp. 50–60, March 2005.

[8] G. D. Abowd and E. D. Mynatt, *Designing for the Human Experience in Smart Environments*. John Wiley & Sons, Inc., 2005, pp. 151–174.

[9] N. Zouba, F. Bremond, M. Thonnat, and V. T. Vu, "Multi-sensors analysis for everyday elderly activity monitoring," in *Proceedings of SETIT 2007*, Tunis, Tunisia, March 2007.

[10] J. W. P. Ng, B. P. L. Lo, O. Wells, M. Sloman, N. Peters, A. Darzi, and et al., "Ubiquitous Monitoring Environment for Wearable and Implantable Sensors (UbiMon)," in *6th International Workshop on Ubiquitous Computing*, 2004.

[11] H. Durrant-Whyte and T. Henderson, "Multisensor data fusion," in *Springer Handbook of Robotics*, B. Siciliano and O. Khatib, Eds. Springer, Berlin, Heidelberg, 2008, pp. 585–610.

[12] H. Medjahed, D. Istrate, J. Boudy, J.-L. Baldinger, and B. Dorizzi, "A pervasive multi-sensor data fusion for smart home healthcare monitoring," in *Fuzzy Systems (FUZZ), 2011 IEEE International Conference on*, June 2011, pp. 1466–1473.

[13] A. Fleury, M. Vacher, and N. Noury, "Svm-based multimodal classification of activities of daily living in health smart homes: Sensors, algorithms, and first experimental results," *IEEE Transactions on Information Technology in Biomedicine*, vol. 14, no. 2, pp. 274–283, March 2010.

[14] V. Libal, B. Ramabhadran, N. Mana, F. Pianesi, P. Chippendale, O. Lanz, and G. Potamianos, "Multimodal classification of activities of daily living inside smart homes," in *Distributed Computing, Artificial Intelligence, Bioinformatics, Soft Computing, and Ambient Assisted Living*, Lecture Notes in Computer Science, S. Omatu, M. Rocha, J. Bravo, F. Fernández, E. Corchado, A. Bustillo, and J. Corchado, Eds. Springer, Berlin, Heidelberg, 2009, vol. 5518, pp. 687–694.

[15] Y. Cao, L. Tao, and G. Xu, "An event-driven context model in elderly health monitoring," in *Ubiquitous, Autonomic and Trusted Computing, 2009. UIC-ATC '09. Symposia and Workshops on*, July 2009, pp. 120–124.

[16] J. Demongeot, G. Virone, F. Duchêne, G. Benchetrit, T. Hervé, N. Noury, and V. Rialle, "Multi-sensors acquisition, data fusion, knowledge mining and alarm triggering in health smart homes for elderly people," *Comptes Rendus Biologies*, vol. 325, no. 6, pp. 673–682, 2002, longevite et vieillissement.

[17] G. Giacinto and F. Roli, "Methods for dynamic classifier selection," in *1999 Proceedings of the International Conference on Image Analysis and Processing*, 1999, pp. 659–664.

[18] N. Zouba, F. Brémond, and M. Thonnat, "Multisensor Fusion for Monitoring Elderly Activities at Home," in *6th IEEE International Conference on Advanced Video and Signal Based Surveillance AVSS09*, Genoa, Italy, Sep. 2009.

[19] C. F. Crispim-Junior, F. Bremond, and V. Joumier, "A Multi-Sensor Approach for Activity Recognition in Older Patients," in *The Second International Conference on Ambient Computing, Applications, Services and Technologies – AMBIENT 2012*, IARIA. Barcelona, Spain: XPS/ThinkMindTM Digital Library, Sep. 2012, in press. [Online]. Available: https://hal.inria.fr/hal-00726184

[20] D. Chen, R. Malkin, and J. Yang, "Multimodal detection of human interaction events in a nursing home environment," in *Proceedings of the 6th International Conference on Multimodal Interfaces*, ICMI '04. New York, NY, USA: ACM, 2004, pp. 82–89. [Online]. Available: http://doi.acm.org/10.1145/1027933.1027949

[21] R. Luo, C.-C. Yih, and K. L. Su, "Multisensor fusion and integration: approaches, applications, and future research directions," *IEEE Sensors Journal*, vol. 2, no. 2, pp. 107–119, Apr 2002.

[22] R. Velik, "A brain-inspired multimodal data mining approach for human activity recognition in elderly homes," *Journal of Ambient Intelligence and Smart Environments*, vol. 6, no. 4, pp. 447–468, 2014.

[23] B. Khaleghi, A. Khamis, F. O. Karray, and S. N. Razavi, "Multisensor data fusion: A review of the state-of-the-art," *Information Fusion*, vol. 14, no. 1, pp. 28–44, 2013.

[24] D. Ruta and B. Gabrys, "An overview of classifier fusion methods," *Journal of Computing and Information Systems*, vol. 7, no. 1, pp. 1–10, 2000.

[25] L. I. Kuncheva, "A theoretical study on six classifier fusion strategies," *IEEE Transactions on Pattern Analysis and Machine Intelligence*, vol. 24, no. 2, pp. 281–286, Feb. 2002.

[26] L. Gao, A. Bourke, and J. Nelson, "A system for activity recognition using multi-sensor fusion," in *Engineering in Medicine and Biology Society, EMBC, 2011 Annual International Conference of the IEEE*, Aug 2011, pp. 7869–7872.

[27] P. Zappi, C. Lombriser, T. Stiefmeier, E. Farella, D. Roggen, L. Benini, and G. Tröster, "Activity recognition from on-body sensors: Accuracy-power trade-off by dynamic sensor selection," in *Wireless Sensor Networks*, Lecture Notes in Computer Science, R. Verdone, Ed. Springer, Berlin, Heidelberg, 2008, vol. 4913, pp. 17–33.

[28] X. Hong, C. Nugent, M. Mulvenna, S. McClean, B. Scotney, and S. Devlin, "Evidential fusion of sensor data for activity recognition in smart homes," *Pervasive and Mobile Computing*, vol. 5, no. 3, pp. 236–252, 2009, Pervasive Health and Wellness Management.

[29] M. Camilli and R. Kleihorst, "Demo: Mouse sensor networks, the smart camera," in *Distributed Smart Cameras (ICDSC), 2011 Fifth ACM/IEEE International Conference on*, Aug 2011, pp. 1–3.

[30] M. Ziefle, C. Rocker, and A. Holzinger, "Medical technology in smart homes: Exploring the user's perspective on privacy, intimacy and trust," in *2011 IEEE 35th Annual Computer Software and Applications Conference Workshops (COMPSACW)*, July 2011, pp. 410–415.

[31] M. Eldib, B. B. Nyan, F. Deboeverie, X. Xie, H. Aghajan, and W. Philips, "Behavior analysis for aging-in-place using similarity heatmaps," in *8th ACM/IEEE International Conference on Distributed Smart Cameras, Proceedings*. ACM/IEEE, 2014, p. 6.

[32] B. B. Nyan, F. Deboeverie, M. Eldib, J. Guan, X. Xie, J. Niño Castañeda, and et al., "Human mobility monitoring in very low resolution visual sensor network," *Sensors*, vol. 14, no. 11, pp. 20 800–20 824, 2014.

[33] M. Eldib, B. B. Nyan, F. Deboeverie, J. Nino, J. Guan, V. de Velde, and et al., "A low resolution multi-camera system for person tracking," in *2014 IEEE International Conference on Image Processing (ICIP)*, Oct 2014, pp. 378–382.

[34] X. Xie, F. Deboeverie, M. Eldib, W. Philips, and H. Aghajan, "Phd forum: Analyzing behaviors patterns of the elderly from low-precision trajectories," in *Proceedings of the International Conference on Distributed Smart Cameras*, ICDSC '14. New York, NY, USA: ACM, 2014, pp. 47:1–47:2.

[35] M. Eldib, F. Deboeverie, D. Van Haerenborgh, W. Philips, and H. Aghajan, "Detection of visitors in elderly care using a low-resolution visual sensor network," in *Proceedings of the 9th ACM International Conference on Distributed Smart Cameras*, ACM, 2015, p. 6.

[36] M. Eldib, F. Deboeverie, W. Philips, and H. Aghajan, "Sleep analysis for elderly care using a low-resolution visual sensor network," in *Human Behavior Understanding*, Lecture Notes in Computer Science, A. A. Salah, B. J. Kröse, and D. J. Cook, Eds. Springer International Publishing, 2015, vol. 9277, pp. 26–38. [Online]. Available: http://dx.doi.org/10.1007/978-3-319-24195-1_3

[37] T. Teixeira, G. Dublon, and A. Savvides, "A survey of human-sensing: Methods for detecting presence, count, location, track, and identity," *ENALAB technical report*, 2010.

[38] F. Cardinaux, D. Bhowmik, C. Abhayaratne, and M. S. Hawley, "Video based technology for ambient assisted living: A review of the literature," *Journal of Ambient Intelligent Smart Environments*, vol. 3, no. 3, pp. 253–269, Aug. 2011.

[39] F. Deboeverie, J. Hanca, R. Kleihorst, A. Munteanu, and W. Philips, "A low-cost visual sensor network for elderly care," *SPIE NEWSROOM*, p. 4, Dec. 2014.

Chapter 6
Towards interoperable enhanced living environments

Susanna Spinsante[1], Ennio Gambi[1], Laura Montanini[1], Laura Raffaeli[1], Lambros Lambrinos[2], Virginie Felizardo[3], Nuno Pombo[3], and Nuno Garcia[3]

6.1 Introduction

In the world of the Internet of Things (IoT), objects create a pervasive and interconnected system, using multiple communication technologies [1]. A *thing* can be seen as any man-made or natural object provided with the ability to transmit data over a wired but, in most cases, a wireless network connection. A *thing* in the IoT paradigm, can be, for example, an armchair equipped with force sensors, able to send user's behavioural data, or a plant which, relying on humidity and brightness sensors, can notify the user its need to be watered or exposed to light.

Fitness [2], health and emergency wearable devices [3], smart appliances [4], smart buildings, and so on are emerging as main IoT application domains, but each proposed solution features its own architecture and data models. Therefore, the integration of different objects using different technologies and media, is one of the most difficult challenges to face. As asserted by authors in [5], IoT may be defined as a "dynamic global network infrastructure with self configuring capabilities based on standard and interoperable communication protocols, where physical and virtual things have identities, physical attributes, and virtual personalities, use intelligent interfaces, and are seamlessly integrated in the information network". Thus, interoperability in IoT infrastructures represents a nodal issue. The ambitious goal of interconnecting any possible object cannot be achieved without addressing several basic challenges, among which the following ones may be mentioned [6]:

- mechanisms to facilitate and enable devices cooperation and discovery;
- solutions to reduce the number of gateways;
- personalization from a user-centric perspective.

[1]Dipartimento di Ingegneria dell'Informazione, Università Politecnica delle Marche, Italy
[2]Dept. of Communication and Internet Studies, Cyprus University of Technology, Cyprus
[3]Instituto de Telecomunicações and Universidade da Beira Interior, Portugal

Despite several solutions have been proposed to address these problems, a standard has not been defined yet and the situation is very fragmented [7], often preventing an effective go-to-market process for industries and SMEs operating in this field [8].

In this chapter, the solutions adopted in Smart Homes (SHs) and Enhanced Living Environments (ELEs), to address the issues related to interoperability among devices, systems, and platforms, are analysed and discussed. Further, the chapter also presents interoperable standards focused on either the network connectivity, or the data format regarding its exchange, storage, and security. These topics are relevant for Active and Assisted Living (AAL) contexts, because they are the cornerstone to enable both effective connections among different devices, and data management in terms of its acquisition, storage, and availability.

The chapter is organized in two main parts, each one dealing with the aforementioned topics, and a final section of common concluding remarks by means of which the main open issues, identified approaches, and future perspectives are presented.

6.2 Smart homes, living environments, and the need for interoperability

A smart home can be described as an environment characterized by heterogeneous devices, supplied by different manufacturers (e.g. home appliances, sensors, actuators, mobile phones, wearable devices), supporting various technologies (e.g. wireless, power-line communication), and protocols (e.g. ZigBee SEP, ZigBee HA, Z-Wave) [9, 10]. In order to fully exploit the advantages of such an architecture, interoperability is one of the most critical issues to overcome [11, 12]. The diversity of protocols and technologies makes the seamless integration between devices and services difficult, and prevents the full implementation of smart home environments.

Different strategies can be applied, with the purpose of interconnecting all kinds of devices, by means of different communication technologies in the same environment. Although solutions relying on the use of a middleware platform for integration are being accepted, the capability to use a single connectivity protocol in a smart home would truly decrease the complexity caused by the integration of different protocols, and would be far more efficient [9]. Looking at the current trends in technology development, an integration approach based on middleware as a common abstraction layer, provides the faster and more flexible answer for the interoperability problem among heterogeneous devices already co-existing within the home environment. Often, these heterogeneous systems are developed in isolation and consist of various operating systems, different programming platforms, and different tiers of services. Many middleware-based projects have been developed in recent years; some of them have become gradually more consolidated and diffused. Among them, OpenIoT [13], a project co-funded by the Seventh Framework Programme (FP7) of the European Union, aims to provide a means for the creation and management of cloud environments or local servers employing IoT resources, regardless of the way they are employed. This is facilitated through an open source middleware platform enabling the communication between Internet connected devices (such as sensors, actuators, and intelligent devices), and semantic web services. Interoperability between connected objects, sharing of information captured by the sensors, and privacy protection are the main challenges to face. To promote the spreading of this platform, the needs and expectations of SMEs operating in the IoT field have been analysed, to propose several open source solutions that can be easily adapted to specific IoT business domains. For the smart living domain, the focus is on citizens, especially elderly, offering outdoor and indoor services that use open data to simplify the elderly everyday life. In a long-term evolution, the emergence of a single standard should be envisioned, including the features of all the others, possibly open, supported and observed by industry and academy, to make the heterogeneous systems talk each other and interoperate in a federated manner regardless of operating platform [14]. Nowadays, to cope with numerous non-interoperable standards, developers typically implement integration platforms. An integration platform is basically a combination of hardware and software components, able to convert messages from different formats to others, and to solve the interoperability problem by unifying numerous different protocols and data representations under one agreed format [15]. In practice, the platform acts as a central server for managing communications among devices. Beyond the communication, interoperability is realized by some kind of logic that analyses the available information, and makes decisions to execute commands. The devices implementing the proper logic can be referred to as *control systems*.

Some proposals try to solve the problem of communications among heterogeneous devices by exploiting low-level solutions, based on gateways or hubs. In [16], a novel configurable smart IoT gateway is proposed: it is a bridge which connects a traditional network to sensor networks, and can be customized for different applications. In order to do that, authors adapt the PCI-E standard interface to realize unified external interfaces: users can develop appropriate user cards according to different applications. The use of a protocol to translate the different sensor data into a uniform format provides scalability and flexibility.

In [17], the authors state that, before proceeding with higher level solutions, it is necessary to resolve low-level problems, and, particularly, to allow different hubs to communicate with each other in an interoperable way. In fact, they believe that a "premature standardization could risk stifling innovation". In order to avoid this, they suggest tools enabling hub developers to address the interoperability problem. However, in [18], the authors do not agree with such a position, claiming that solutions based on gateways and, therefore, also on hubs, are not true answers to the issue. They propose an alternative architecture which relies on Resource Name Services (RNS), intended primarily for smart homes, by defining the participating parties and the relationships between them, and designing the specific process of the services based on this architecture. Some key points emerge from the analysis of the state-of-the-art literature. They have been identified also in [19], as what is needed to achieve more interoperable and synergetic AAL solutions:

- inter-organizational collaboration;
- user-centred studies;

- increased standardization efforts;
- focus on open systems.

User-centred design is a common requirement mentioned in several projects. According to [20], for example, the development of technology based on a human centred design is a fundamental key to meet the needs, and improve the quality of life of elderly people.

The lack of interoperability in smart home systems can be considered at different levels. According to [21], in fact, cooperation among devices requires interoperability at four layers: protocol layer, logic layer, semantic layer, and intelligence layer. The solution presented in [21] promotes interoperability in a smart home environment based on the logic and semantic layers. The system focuses on service-oriented communication and web services, specifically a high level rule-based (with XML-based rules) control system is proposed, for building automation systems, together with the adoption of standard guidelines for the interoperability at the lower level. Moreover, in an ideal fully integrated smart home, a *single unified smart ubiquitous application* is necessary to control all the operations. This idea leads from the IoT concept to the Web of Things (WoT) [22, 23]. As discussed in [24], while the IoT focuses on device interconnection at the network layer, the WoT can be seen as an approach to interoperability at the application layer. The same paper proposes a Web-based framework to facilitate the development of applications for the smart home. In fact, the presence of incompatible standards and protocols creates difficulties in the integration of appliances from different vendors, and the development of even simple applications requires advanced programming skills, and a big amount of time, thus making the delivery of new interoperable products into the market a non-trivial task.

6.3 Projects and frameworks overview

In this section some projects are reviewed, dealing with ELEs and SHs environments. Projects that evolved over the time are considered, eventually extended after their formal duration to exploit the results of previously concluded work.

6.3.1 UniversAAL

UniversAAL, standing for *UNIVERsal open platform and reference Specification for Ambient Assisted Living*, is the only project funded under the FP7 Call 4, with the mission of studying previous AAL projects and consolidating the state of the art toward an open cross-application platform [20]. The open platform and reference specification aim to make it technically and economically feasible to design and deploy innovative AAL solutions. The UniversAAL platform is built upon three main pillars:

1. the runtime support, a set of software components that can be installed on different types of hardware used in an AAL environment. It provides core services for the execution of AAL applications;

2. the developer support, a set of tools and guidelines to develop services in accordance to a common reference architecture;
3. the community support, built around the concept of *uStore*, a shop where end users (elderly and their care providers) can find complete AAL services, including all necessary software, hardware and human resources, published by service providers.

UniversAAL promotes an AAL Open Association (AALOA), with the role to further support and continue the work initiated by the project even after its completion. The association tackles the provision of a shared open framework for developers, service providers, research institutions, and end-user representatives to discuss, design, develop, evaluate, and standardize a common service platform in the field of AAL. With respect to interoperability, the platform enables applications, components, and services the following functionalities [25]:

- applications may run on distributed heterogeneous systems, such as mobile phones, general-purpose computing and interaction devices, share data and functionalities;
- components, such as sensors, actuators and other equipment, can be dynamically plugged and played;
- application-level services can be added based on existing capabilities.

The whole platform was tested in some of the available living labs provided by the project partners. Feedback from these experiments is being used to improve the platform, and there are also plans for further extension and maintenance based on the open source model. The UniversAAL platform is used in several on-going projects, such as ReAAL and OpenAAL. Moreover, the project uSmAAL shows a recent implementation of AAL services based on the universAAL platform.

6.3.1.1 ReAAL

The project ReAAL, started in 2013, is currently ongoing, and it aims to promote standards, guidelines and open platforms for interoperable solutions in the domain of active and independent living. ReAAL aims to validate the role of common open platforms in putting interoperability standards in place; to measure the related socio-economic impact and consider the ethical, legal, quality of life, and user experience impacts; and to analyse the effectiveness of pilot projects and derive replication guidelines, from which publishing findings and recommendations. ReAAL builds on several elements, such as: (i) an existing open platform, exactly UniversAAL; (ii) existing pilots, in order to benefit from their infrastructure that will help to reduce the distribution costs, the experienced staff already gone through the deployment process, and the experienced end users, that might be able to challenge ReAAL with more targeted requirements; (iii) existing applications and services, that can be utilized in the independent-living scenarios of ReAAL.

6.3.1.2 OpenAAL

OpenAAL, the open source middleware for ambient-assisted living [26], is the outcome of a joint open source initiative for the development of a flexible and powerful middleware for AAL scenarios. It was initiated during the SOPRANO Integrated project co-funded under the European FP6: its objective was to develop a next generation of SHs based on ambient intelligence, for a more independent life of older people in their familiar environment [27]. The experiences gained from the first initiative led to the development of the new version of OpenAAL (v. 2.0), based on UniversAAL.

6.3.1.3 uSmAAL

Based on the UniversAAL framework, a case study called uSmAAL, has been implemented by Stengler *et al.*, in order to demonstrate the reliability of the open platforms in the real world [28]. The aim of the study is, therefore, to minimize the gap between the requirements of the real world use cases and the standard solutions available so far. The uSmAAL system is able to read data from various sensors and combine them to detect relevant situation; to use functionalities provided by different devices and organize them; and to obtain a proper behaviour in response to the detected situation. The uSmAAL system relies on the UniversAAL platform functionalities, enriching the interoperable open framework through a modular architecture, and by using additional components, called Exporter. The insertion of these components allows to add and remove further hardware devices, regardless of the technology used: it represents a key interoperability element in uSmAAL. The project has been extensively tested, both through functional and stress tests, in order to demonstrate its effective reliability and stability, in a real life environment.

6.3.2 *DOMOINSTANT*

DOMO.IN.STA.NT (DOmotic INteroperability STAndard NeTwork) is a universal solution for the smart home proposed by the Laboratory of Domotic at ISTI-CNR in Pisa (Italy). It allows to create innovative services accessible by all types of users through different interfaces designed for PCs, mobile devices, TV, or remote control. All the home automation devices such as sensors and actuators, even adopting different standards of communication, become interoperable thanks to *DomoNET*, a Java open source software delivered under GPL license [29]. DomoNET offers a middleware infrastructure providing a high level of abstraction, necessary to bring together different home automation protocols such as X10, KNX, UPnP, Jini, and others, as shown in Figure 6.1. The core of the framework is represented by *domoNetWS*, a Web Server that manages the cooperation between the nodes of the system and builds a kind of unique virtual environment composed by various devices of the home automation system. Gateways manage the communication between the different types of systems and the overall system.

In DOMO.IN.STA.NT, interoperability is achieved by means of the semantic abstraction of heterogeneous systems: at the initialization stage, the abstract representation of each device is created. In particular, two elements are provided: the description of the features and functionalities of the device, and the standardization of messages describing an event, a command or a response. This way, all the devices are represented in a similar manner, regardless of the underlying technology. The DomoNet project is extended in [30], aiming to develop a software able to

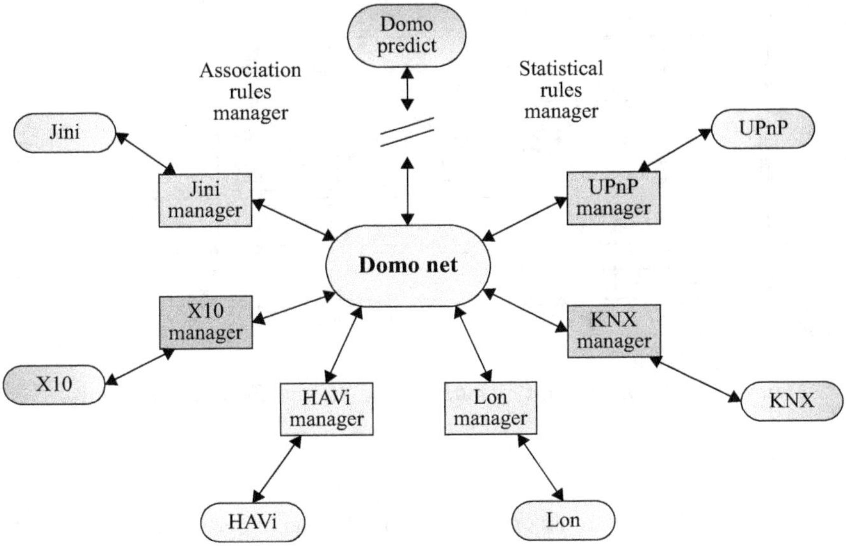

Figure 6.1 DomoNet framework architecture

directly interface the domotic devices, according to the IoT principles. The extension of DomoNet consists of adding IPv6 capabilities to the domotic system. Each device is associated to an IPv6 address, and thanks to this, it is linked to all the other devices within the network. Calling its IPv6 address, it is possible to access the Web interface of a device in order to obtain information on its status and capabilities.

6.3.3 AllJoyn

AllJoyn is a collaborative project of the AllSeen Alliance, aimed at the Internet of Everything (IoE). The AllJoyn framework provides open source software API and a set of services for interoperability between connected devices and software applications belonging to different manufacturers. A Smart Spaces Working Group within the AllSeen Alliance deals with creating specifications for SHs and Smart Buildings. In this context, many deliverables have been released, such as a proposal for the smart home service framework [31], and specifications for the high-level interface design [32], in 2014.

In a general view, the AllJoyn framework allows to discover proximal devices and applications, to adapt the framework to specific devices, to transport data between different devices, and to interoperate between diverse operating systems. The provided Software Development Kit (SDK) is open, and runs on popular platforms such as Linux, Android, iOS and Windows. The project includes blocks and fundamental services to implement the discovery of nearby devices, the connection, the safety, and dynamic management of the network.

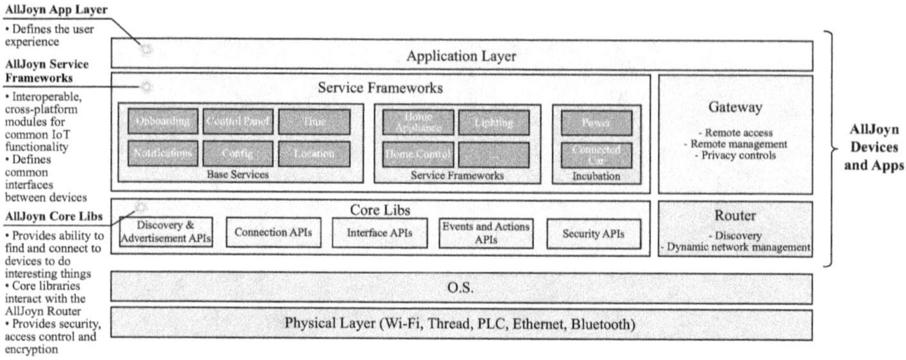

Figure 6.2 High-level architecture of the AllJoyn Software Framework

As shown in Figure 6.2, the project tries to solve the interconnection and interoperability problems at a high level, through a range of services contained in the AllJoyn Service Frameworks which support:

- onboarding functionality: it allows the devices to be connected together, through intermediaries (smartphones, tablets, smart TV, etc.);
- notification functionality: it provides support APIs, in order to enable the devices to receive or transmit information in different formats (image, audio, video, text, etc.);
- control panel functionality: it lets the smartphone or tablet control a connected device, through the user interface, without the need of knowing the manufacturer or the type;
- device information and configuration functionality: it allows an IoT product to transmit information (such as type, vendor, series number), and the user to assign a name and a password to the device;
- audio functionality: it foresees the audio stream transmission from a device to another.

From a lower level point of view, the AllJoyn network architecture is composed by two types of nodes: Routing Nodes (RN) and Leaf Nodes (LN). The LN can connect only to the RN, while the RN can connect to other RN; therefore AllJoyn can be thought of as a mesh network of star networks.

Although AllJoyn is one of the most promising frameworks for the IoT system development, several shortcomings were observed [33] which make the AllJoyn system unsuitable for the management of complex smart environments. Communication between devices connected to different subnets is not supported; furthermore, AllJoyn lacks capability to store and manage large amounts of data. To compensate for these shortcomings, Villari *et al.* [33] proposed an improved software solution, called AllJoyn Lambda. It exploits the Lambda architectural model, integrating the AllJoyn system with MongoDB (a very efficient distributed NoSQL database), and

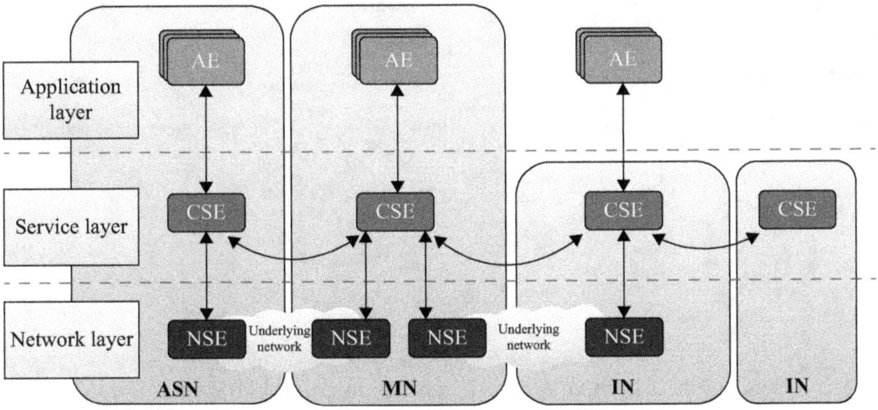

Figure 6.3 oneM2M architecture

Apache Strom (a distributed system for real time processing of data). The authors demonstrate the feasibility and effectiveness of the proposed solution by analysing a SH case study.

6.3.4 OneM2M

OneM2M [34] is a global organization aiming at establishing a scalable and interoperable standard for device communications and services used in Machine-to-Machine (M2M) and IoT applications. Standardization is needed in order to deliver the scalability and flexibility the market requires to maximize the potential of M2M and IoT. The objective is to standardize interfaces across networks, applications, and devices, so they are applicable to the entire ecosystem.

Members of the oneM2M Global Initiative cooperate in the production of globally applicable, access-independent M2M service layer specifications related to M2M solutions. This common service layer is intended to be embedded within various hardware and software components to ensure M2M devices can communicate on a global scale. The oneM2M architecture sketched in Figure 6.3 includes functional entities called *nodes*, known as application dedicated node (ADN), application service node (ASN), middle node (MN), and infrastructure node (IN). At the lower network layer, Network Service Entities (NSEs) allow communications with the underlying IP network, whereas, at the upper application layer, Application Entities (AEs) enable logical functions for end-to-end M2M applications. Within a single node, communications between NSE and AE take place by exploiting the Common Service Entity (CSE). A CSE is a logical entity that is instantiated in an M2M node and comprises a set of service functions called Common Services Functions (CSFs). The information within the system are represented by a resource-based model, in which each piece of information is stored as a resource. Each resource belongs to a default type that determines its semantics.

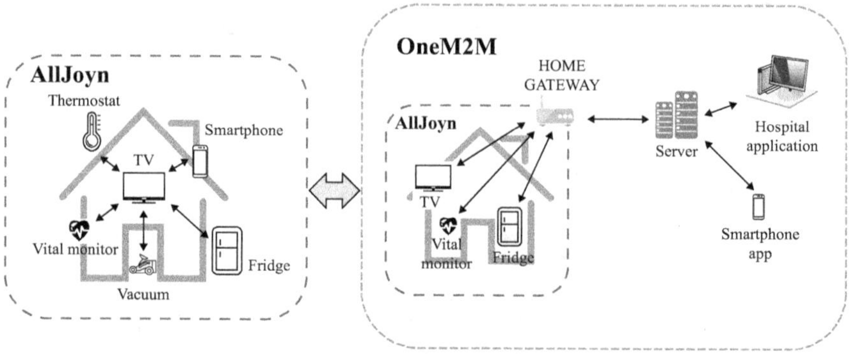

Figure 6.4 Example of interaction among oneM2M and AllJoin

Table 6.1 AllJoin and oneM2M comparison

	AllJoin	OneM2M
Network architecture	Peer-to-peer in LAN	Server-to-client in WAN
API style	RPC (RMI) API	Resource-based API
Discovery style	Proactive discovery	Passive discovery

OneM2M may co-exist and interoperate with AllJoin, as shown in a sample scenario in Figure 6.4. AllJoyn deals with interconnecting the various objects in the home over a LAN peer-to-peer network; oneM2M, thanks to its WAN client-server architecture, allows the device control and access to information from remote platforms, such as web interfaces and smartphone apps. At the same time, however, oneM2M and AllJoin present some differences, as highlighted in Table 6.1.

6.3.5 openHAB

OpenHAB [35] is a vendor and technology agnostic open source automation software, which provides an integration platform for all the home technological devices. It enables communication among devices regardless of any protocol or vendor specification. OpenHAB is open source, and maintained by a big community of developers instead of a single company. OpenHAB is a java-based software, which can run on every device supporting a Java Virtual Machine (JVM), such as Windows, Linux, MacOS but also Raspberry Pi and Beaglebone Black. APIs are provided for the integration in other systems.

A key concept in openHAB is the *item* that can be considered as a capability: openHAB does not need to know if an item is related to a physical device, or to a virtual source, like a Web service. All features offered by openHAB use the *item* abstraction, which means that there is no reference to device-specific parameters (like IP addresses, IDs) in automation rules, User Interface (UI) definition, and so on. This makes it extremely easy to replace one technology by another, without any changes in rules and UIs.

OpenHAB has a powerful rule engine included. Rules are written in a java-like language, and can be triggered by different types of events, like changes of a certain device status, time or system events. Apart from defining a flexible logic, openHAB rules provide a set of different actions which can be undertaken by the rule, like sending commands and updates to devices, sending tweets or e-mails, and so on. OpenHAB comes with a generic textual configuration for its UIs, the so-called Sitemap. The sitemap is a tree structure of widgets, which define the different pages of a UI and their content. The Sitemap is supposed to be a suitable UI model for different kinds of UIs, so that the user does not have to configure each of them, in case multiple UIs are set up. Item UI providers offer a dynamic way to configure the UI. An item UI provider can, for example, define what widget should be used, and can dynamically define icons and labels (which might depend on the current state of the item) for the widgets.

6.3.5.1 Architecture

A very relevant aspect, appreciated by the openHAB community, is its modular architecture. It is very easy to add new features (like the integration with another system through a binding), moreover it is possible to add and remove such features at runtime. The openHAB project is split into two main parts: the openHAB *runtime*, running on the server and truly executing operations; and the openHAB *designer*, a configuration tool for the openHAB runtime. It comes with user-friendly editors to configure runtime, to define UIs, and to implement rules.

The UI may be implemented as native Android/iOS applications, and as a HTML/JS solution, that can be accessed by means of a Web browser. The openHAB runtime is a set of OSGi (Open Service Gateway initiative) bundles deployed on an OSGi framework (Equinox). It is therefore a pure Java solution and needs a JVM to run. The OSGi enables to build applications with a highly modular architecture (composed of bundles), providing a service-oriented programming.

OpenHAB supports several home automation technologies and systems: each device, technology or platform integrated in it, is supported by a specific bundle. There are also bundles providing integration and communication with social networks, instant messaging, cloud IoT platforms, and much more. Some examples are: KNX, EnOcean, Asterisk, MQTT, Insteon, Xively, and more. The scheme given in Figure 6.5 is an overview over the main bundles and how they depend on each other.

OpenHAB has two different internal communication channels: an asynchronous event bus, and a stateful repository, which can be queried. The event bus is the basic service of openHAB. All bundles that do not require stateful behaviour should use it to inform other bundles about events, and to be updated by other bundles on external events. There are mainly two types of events: commands which trigger an action or a state change of some item/device, and status updates which inform about a status change of some item/device (often as a response to a command). It is important to notice that openHAB is not meant to reside on actual hardware devices.

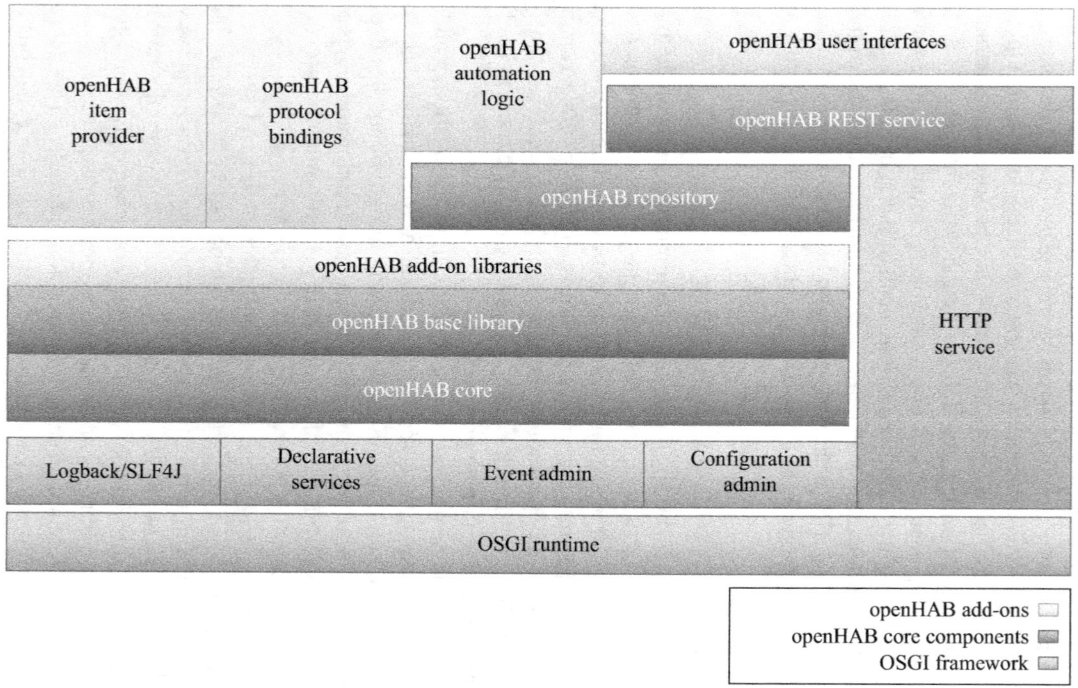

Figure 6.5 openHAB architecture

Instead, it rather serves as an integration hub between such devices, and as a mediator between different protocols supported between devices. Therefore, in a typical installation, there will be usually a single instance of openHAB running on some central server.

OpenHAB offers the so-called Item Repository, which is connected to the event bus and keeps track of the current status of all items. For example, a UI needs to display the current state of items at the time the user accesses the system. Also the automation logic execution engine needs always to be informed about the current states. The Item Repository avoids each bundle to cache states themselves for their internal use; it also makes sure that the state information is synchronized and up-to-date for all those bundles, providing the possibility to persist states in the file system or in a database, so that they are even maintained throughout a system restart.

6.3.6 Thread

Designed for consumers and devices in and around the home, Thread [36] is an IPv6 networking protocol built on open standards for low-power 802.15.4 mesh networks that can easily and securely connect hundreds of devices to each other and directly to the cloud. The non-profit Thread Group is focused on making Thread the foundation for the IoT in the home, by educating product developers and consumers on the unique features and benefits of Thread and ensuring a great user experience through rigorous, meaningful product certification. Thread is backed by industry-leading companies including ARM, Big Ass Fans, Freescale Semiconductor, Nest Labs, OSRAM, Qualcomm, Samsung Electronics, Silicon Labs, Somfy, Tyco, and Yale Security. Since opening its membership in October 2014, Thread has quickly grown to more than 220 members.

The Thread Group announced on July, 2015 the release of Thread, the new IP-based wireless networking protocol designed for low-power connected products in the home. Product developers who are members of the Thread Group can access the technical specifications and documentation to build Thread-compliant products.

Following the release of the Thread 1.0 technical specification, the Thread Group launched in November 2015 the first phase of its product certification program. The program provides rigorous testing for products that use Thread to help ensure that they work together effortlessly and securely right out of the box. More than 30 products and components from Thread Group members have been submitted for certification, and Thread certified software stacks from ARM, Freescale and Silicon Labs will be available soon. Thread products will be tested to validate device behaviour for commissioning, networking functionality, security, and operation in Thread's network, and may bear the "BUILT ON THREAD" or "THREAD CERTIFIED COMPONENT" logo to help consumers and product developers identify them on the market. Unlike other industry standards, where implementers charge royalties or produce something proprietary, Thread's royalty-free licensing terms are designed to aid rapid adoption and reduce confusion by creating a royalty-free pledge between members, so that the entire industry can rapidly adopt Thread whilst minimizing costs.

6.4 Network and data interoperability

This section presents an overview of standards that focus on areas such as network connectivity and the exchange, storage, and security of data. The overview of these topics is relevant for AAL contexts, because they are the cornerstone to enable both effective connections among different devices, and data management in terms of its acquisition, store, and availability.

6.4.1 Java for intelligent network (Jini)

Jini technology consists of an infrastructure and programming model which address the basic issues of how clients and services connect and discover each in order to build an *ad hoc* service network [37]. A service is an entity that can be used by a program, or another service. Computation, storage, software filtering, hardware devices are such examples of services.

Based on Java language, Jini uses all the object-oriented features so as to provide an API for remote invocation of objects. As depicted in Figure 6.6, the typical workflow is composed by the following stages: (1) Discover stage: the network service discovers available lookup services (LUS); (2) Join stage: the network sends the service proxy to LUS; (3) Discover stage: the network client discovers available LUS; (4) Lookup stage: the network client sends a request to LUS to find for desired services; (5) Receive stage: the LUS sends the registered service proxy to network client; (6) Use stage: the network interacts directly with the network service using the service proxy.

Each service provides an object with service attributes and communication rules, called proxy, to move around the network. Each client needs to discover and join (register) prior to use an available service on the network. Then, a service proxy is sent to all lookup services on the network. The lookup service is equivalent to a directory or index of available services and represents the central bootstrapping mechanism for the system providing a point of contact between the system and its users. Thus, when a service is required, its proxy is sent to the requesting client. Afterward, the proxy manages all the communications between the client and the service provider.

However, to preserve the network resilience and flexibility, when a service joins the network, it is only available for a limited leased time which may be negotiated before it expires. When a lease expires its lookup is removed automatically. In addition, all the communication between services are accomplished using Java Remote Method Invocation (RMI). The RMI provides mechanisms to find, activate, and garbage collect object groups, and offering multicast, replication, basic security and confidentiality. Thus, the RMI allows not only data to be passed from object to object around the network but also full objects, including code. Much of Jini simplicity system is enabled by this ability to move code around the network in a form that is encapsulated as an object.

6.4.2 Universal plug and play

Universal plug and play (UPnP) is a protocol for automatic discovery and device interoperability on a peer-to-peer network [38]. This network is divided into devices,

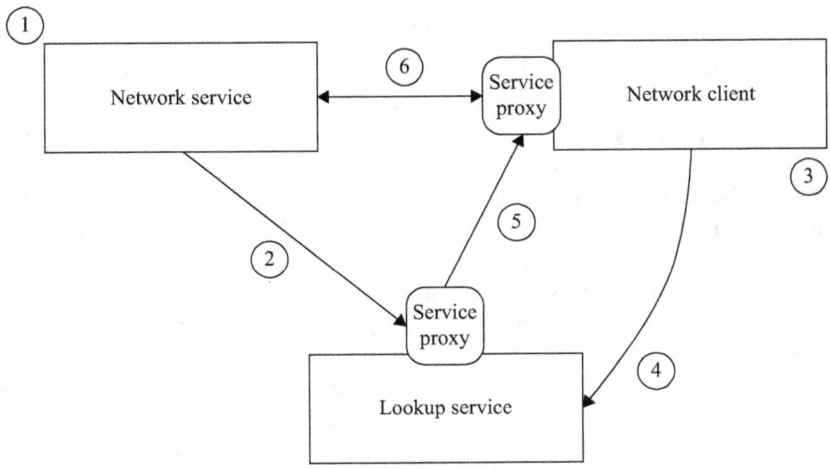

Figure 6.6 The Jini workflow

or service providers, and controls points. On the one hand, either a smartphone, or a laptop is considered as a control point. On the contrary, AAL systems are recognized as devices. Likewise Jini, UPnP can support *ad hoc* service and devices network, and its specification is divided into several stages, namely: addressing, discovery, and description specify automatic integration of devices and services; control, eventing, and presentation specify how to use them. Moreover, this protocol enables the UPnP template language (UTL) based on an XML schema for describing services. In summary, UPnP enables a device to dynamically join a network, obtain an IP address, convey its capabilities and learn about the presence and capabilities of other devices [39]. In addition, devices can automatically communicate with each other directly without any additional configuration.

6.4.3 Home electronic system

The Home electronic system (HES) standard provides a multi-application context in order to offer a free configuration of the interoperations among devices in a system. This standard aims to define the interface between the different devices and the network. It may interconnect the power network, cable TV, and/or telephone, as an example for security, and video control purposes.

6.4.4 Home audio/video interoperability

The (Home audio/video interoperability) HAVi specification was defined by several consumer electronics companies aiming at providing a standard for interoperability. This standard provides a lightweight configuration of network addresses, or device drivers. Each connected device automatically announces its capabilities and presence to every other devices in the network.

6.4.5 Medical standards, interoperability, and challenges

Equally important to assure the interconnection of all devices in an ELE scenario is to guarantee that data that is collected and processed is formatted to standards that allow other systems, including e-health systems, to further read and process these data.

To ensure interoperability throughout the different architectural models and also among stakeholders in the e-health and ELE contexts, the use of standards must be enforced. The standardization of health data with Health Level Seven (HL7) [40] and Digital Imaging Communications in Medicine (DICOM) [41] (among others) allows the interoperability between several e-health systems as these standards are among the most important [42, 43] concerning this purpose. HL7 is really a set of standards ranging from primary and foundational standards to education and awareness, encompassing seven categories. As HL7 deals with all aspects of communicating information within a health information system, this is relevant for ELE informations systems too as, sooner or later, these two types of systems will have to exchange information. DICOM is older than HL7, having been published initially in 1993. As stated in its official website, DICOM "(. . .) is the international standard for medical images and related information (ISO 12052) (. . .)" [41]. DICOM is a standard that defines protocols for exchange of medical images that can have some associated information.

When collecting, storing and transmitting sensible medical or health information, systems and stakeholders must abide to security rules, in order to prevent data corruption, data mishandling, data impersonation or data loss. Data privacy and data security require sophisticated, consent-based access control models and permissions regimes [44, 45]. The eight fields that concern this area are well identified (e.g. in the Accenture 2010 white paper), as follows: Interoperability, Data Integrity, Access Control, Security, Data Handling, Data Quality, Consent, and Compliance. These areas are often cited as challenges that systems must meet in order to make information usable in a pervasive manner.

Making the bridge from the e-health systems to the ELE systems, we can draw an almost immediate correspondence between the eight e-health challenges regarding information and the challenges that must be met for the ELE information systems. This is the case, first and foremost, with Interoperability. Heterogeneous as ELE systems are, assuring the interoperability of systems through transparent data exchange, supported in standard formats, is one of the most critical requirements [19]. Nevertheless, it can still be observed that full semantic interoperability is not common.

Following the requirement for interoperability, the challenges of security, in a wider sense, need to be addressed. These challenges are present at the machine and communication services that support the data communication, often referred to as the security challenge, but are also present from a human perspective and from a data perspective. From a human perspective, the focus is set on the areas of data access control and data handling. Data access control means that there are roles assigned to users, and that each role is capable to perform a defined set of operations on a defined set of data. These operations include the common Create, Read, Update, and Delete, but data handling extends this to the right to copy, right to archive, right to classify, and finally, to traceability of operations performed over such data. Such activities must also be encompassed within a program that includes user training, and data loss/mishandling liability assessment. From a data perspective, the security issues are met when data integrity and data quality are maintained. This means that the integrity of the structure of complex data records is taken care of in all steps of the data processing and transmission processes. Moreover, throughout these processes, data is still useful and meaningful, i.e. data quality is preserved.

Finally, two additional issues for assuring that full interoperability can be achieved are consent and compliance. Consent means, of course, that the user or users are fully aware, understand and agree to the use of the information the systems collect, and how it is processed. This may be an extremely difficult task to complete, as many times the users are undisclosed persons who wander in a shopping mall, or use some public or private facility. Governments often solve these problems by issuing very strict data privacy laws, or, on the opposite side, abstaining from completely legislate on these matters, leaving the users to decide, possibly in courts of law, whether the information was used in respect of its inalienable rights, or not. When governments choose to legislate, systems must comply to the legislation, and this is called compliance. Yet, compliance does not have to be confined to laws, but must also consider regulations, best known practices and very often, plain old traditional wisdom. The key goal the systems' manufacturers seek is to deploy the system, maximizing its usability and outputs, but limiting to a maximum potential liabilities.

6.5 Conclusion

The vast interest in the IoT has led to the development of many solutions in multiple sectors. In this chapter we concentrated on AAL and Smart Homes which are two domains that are closely interrelated. Of particular interest is the interoperability between the solutions deployed which as expected rely on different software and hardware platforms. Our attention has focused on consolidated proposals, evolved over the time, which are a base for other projects. In particular, we presented three major projects and their evolution: UniversAAL, Domoinstant and AllJoyn.

The UniversAAL project, whose evolution is called ReAAL, is one of the interoperable solutions in the AAL field currently most widespread. There are many projects that have developed over this platform, among them OpenAAL and uSmAAL.

A solution more focused on the Smart Home concepts is, instead, Domoinstant, offering a Java middleware infrastructure, allowing various communication protocols to interact. The project has evolved, thanks to the addition of the IPv6 capabilities, enabling the web interfaces to interact directly with the home automation system.

Finally, the AllJoyn project has been presented. Although not specifically oriented to the smart home, it provides specific proposals in this area. This framework provides software API and a set of services oriented to the interoperability between heterogeneous connected devices and different software applications. A case study has been cited to demonstrate the effectiveness of this solution in the Smart Home context. However, it integrates further software tools, such as Mongo DB and Storm, in order to make up for some deficiencies identified by the authors, in the AllJoyn

Table 6.2 A brief summary of the projects described in Section 6.3

Project	Field	Key points
UniversAAL	AAL	Creation of an open platform and reference specification. Enable heterogeneous devices to be plugged and played in the same system, add application-level services
ReAAL	AAL	Based on UniversAAL. Put standards in place, analyse the effectiveness of pilot projects, consider quality of life and user experience impacts
OpenAAL	AAL	Open source middleware exploiting ambient intelligence to support independent living of elderly at home. Second version based on UniversAAL
uSmAAL	AAL	Insertion of additional components to improve the UniversAAL framework. Case study to test the reliability and stability in the real world
DOMO.IN.STA.NT	Smart home	Semantic abstraction, all the home automation devices are represented in the same manner
DomoNet Extension	Smart home	Assignment of IPv6 capabilities to the domotic devices enables a direct interaction by means of web interfaces
AllJoyn	Smart home	Enable communication among connected devices and software applications, both through a framework providing high-level services and a low-level communication protocol
AllJoyn Lambda	Smart home	Improve AllJoyn through a software solution that enables communication and management of devices in large-scale smart environments

system. Table 6.2 shows a brief summary of the objectives and characteristics of the various projects previously described.

The projects and solutions presented in this chapter are some of the most affirming in the Smart Home and AAL context, although, as already stated, a standing solution for interoperability has not yet been defined. However, these projects are the starting point for improvements and further developments; this indicates that we are on track for the adoption of an effective standard in such fields.

Acknowledgements

This work was partially co-funded in the framework of the Italian National Technology Cluster project "Shared Interoperable Home Ecosystems for a Green, ComfortabLe and Safe Living" (SHELL), project code: CTN01_00128_111357.

References

[1] G. Kortuem, F. Kawsar, D. Fitton, and V. Sundramoorthy, "Smart objects as building blocks for the internet of things," *IEEE Internet Computing*, vol. 14, no. 1, pp. 44–51, January 2010.

[2] J. Wei, "How wearables intersect with the cloud and the internet of things: Considerations for the developers of wearables," *IEEE Consumer Electronics Magazine*, vol. 3, no. 3, pp. 53–56, July 2014.

[3] B. Lo, H. Ip, and G.-Z. Yang, "Transforming health care: Body sensor networks, wearables, and the internet of things," *IEEE Pulse*, vol. 7, no. 1, pp. 4–8, January 2016.

[4] A. Yachir, Y. Amirat, A. Chibani, and N. Badache, "Event-aware framework for dynamic services discovery and selection in the context of ambient intelligence and internet of things," *IEEE Transactions on Automation Science and Engineering*, vol. 13, no. 1, pp. 85–102, January 2016.

[5] O. Vermesan, P. Friess, P. Giullemin, S. Gusmeroli, H. Sundmaeker, A. Bassi, et al., "Internet of things strategic research roadmap," in *Internet of Things: Global Technological and Societal Trends*. River Publishers, Aalborg, Denmark, 2011, pp. 9–52.

[6] M. Vega-Barbas, D. Casado-Mansilla, M. Valero, D. Lopez-de Ipina, J. Bravo, and F. Florez, "Smart spaces and smart objects interoperability architecture (s3oia)," in *2012 Sixth International Conference on Innovative Mobile and Internet Services in Ubiquitous Computing (IMIS)*, July 2012, pp. 725–730.

[7] M. Spring, "The future of standardization: Are we destined to repeat history?" *Computer*, vol. 49, no. 1, pp. 99–101, January 2016.

[8] C. Perera, C. Liu, and S. Jayawardena, "The emerging internet of things marketplace from an industrial perspective: A survey," *IEEE Transactions on Emerging Topics in Computing*, vol. 3, no. 4, pp. 585–598, December 2015.

[9] H. J. O. Albuquerque and G. S. de Aquino Junior, "A proxy-based solution for interoperability of smart home protocols," in *2014 Eighth International Conference on Complex, Intelligent and Software Intensive Systems*. Institute of Electrical & Electronics Engineers (IEEE), July 2014. [Online]. Available: http://dx.doi.org/10.1109/CISIS.2014.96

[10] J. Juarez, J. Rodriguez-Mondejar, and R. Garcia-Castro, "An ontology-driven communication architecture for spontaneous interoperability in home automation systems," in *Emerging Technology and Factory Automation (ETFA), 2014 IEEE*, September 2014, pp. 1–4.

[11] W. Chen, S.-Y. Kuo, and H.-C. Chao, "Service integration with UPnP agent for an ubiquitous home environment," *Information Systems Frontiers*, vol. 11, no. 5, pp. 483–490, 2008. [Online]. Available: http://dx.doi.org/10.1007/s10796-008-9122-3

[12] J. Mitsugi, S. Yonemura, H. Hada, and T. Inaba, "Bridging upnp and zigbee with coap: Protocol and its performance evaluation," in *Proceedings of the Workshop on Internet of Things and Service Platforms*, ser. IoTSP '11. New York, NY, USA: ACM, 2011, pp. 1:1–1:8. [Online]. Available: http://doi.acm.org/10.1145/2079353.2079354

[13] (2013) Openiot project. [Online]. Available: http://www.openiot.eu/

[14] T. Perumal, A. R. Ramli, C. Y. Leong, K. Samsudin, and S. Mansor, *Web-Based Information Technologies and Distributed Systems*. Paris: Atlantis Press, 2010, Interoperability Among Heterogeneous Systems in Smart Home Environment,

pp. 141–157. [Online]. Available: http://dx.doi.org/10.2991/978-94-91216-32-9_7

[15] H. Jarvinen, A. Litvinov, and P. Vuorimaa, "Integration platform for home and building automation systems," in *Consumer Communications and Networking Conference (CCNC), 2011 IEEE*, January 2011, pp. 292–296.

[16] S. Guoqiang, C. Yanming, Z. Chao, and Z. Yanxu, "Design and implementation of a smart iot gateway," in *Green Computing and Communications (GreenCom), 2013 IEEE and Internet of Things (iThings/CPSCom), IEEE International Conference on and IEEE Cyber, Physical and Social Computing*, August 2013, pp. 720–723.

[17] M. Blackstock and R. Lea, *IoT interoperability: a hub-based approach*. IEEE, Cambridge, MA, vol. 10, 2014, pp. 79–84.

[18] C. Yang, B. Yuan, Y. Tian, Z. Feng, and W. Mao, "A smart home architecture based on resource name service," in *2014 IEEE 17th International Conference on Computational Science and Engineering (CSE)*, Dec 2014, pp. 1915–1920.

[19] M. Memon, S. R. Wagner, C. F. Pedersen, F. H. A. Beevi, and F. O. Hansen, "Ambient assisted living healthcare frameworks, platforms, standards, and quality attributes," *Sensors*, vol. 14, no. 3, p. 4312, 2014. [Online]. Available: http://www.mdpi.com/1424-8220/14/3/4312

[20] M. Tazari, F. Furfari, F. Valero, S. Hanke, O. Hoftberger, D. Kehagias, M. Mosmondor, R. Wichert, and P. Wolf, "The universaal reference model for aal." in *Handbook of Ambient Assisted Living*. IOS Press, 2012, pp. 610–625.

[21] H. Jarvinen and P. Vuorimaa, "Interoperability for web services based smart home control systems," in *Proceedings of the 10th International Conference on Web Information Systems and Technologies*, 2014, pp. 93–103.

[22] S. Ray, Y. Jin, and A. Raychowdhury, "The changing computing paradigm with internet of things: A tutorial introduction," *IEEE Design Test*, vol. PP, no. 99, pp. 1–1, 2016.

[23] V. Pendyala, S. Shim, and C. Bussler, "The web that extends beyond the world," *Computer*, vol. 48, no. 5, pp. 18–25, May 2015.

[24] A. Kamilaris and A. Pitsillides, "Towards interoperable and sustainable smart homes," in *IST-Africa Conference and Exhibition (IST-Africa), 2013*, May 2013, pp. 1–11.

[25] (2013) Reaal project. [Online]. Available: http://www.cip-reaal.eu/about/the-platform/

[26] (2012) Openaal project. [Online]. Available: http://www.openaal.org

[27] M. Klein, A. Schmidt, and R. Lauer, "Ontology-centred design of an ambient middleware for assisted living: The case of soprano," in *Towards Ambient Intelligence: Methods for Cooperating Ensembles in Ubiquitous Environments (AIM-CU), 30th Annual German Conference on Artificial Intelligence 2007*, 2007.

[28] J. Stengler, G. Gaikward, and H. B. Hmida, "Towards the deployment of open platform aal services in real life-advantages and lessons learned – uSmAAL: A case study for implementing intelligent AAL services in real life based on the open platform universAAL," in *Proceedings of the 1st International Conference on Information and Communication Technologies for Ageing Well and e-Health*, 2015, pp. 67–74.

[29] V. Miori, D. Russo, and C. Concordia, "Meeting people's needs in a fully interoperable domotic environment," *Sensors*, vol. 12, no. 12, pp. 6802–6824, May 2012. [Online]. Available: http://dx.doi.org/10.3390/s120606802

[30] V. Miori and D. Russo, "Domotic evolution towards the IoT," in *2014 28th International Conference on Advanced Information Networking and Applications Workshops*. Institute of Electrical & Electronics Engineers (IEEE), May 2014. [Online]. Available: http://dx.doi.org/10.1109/WAINA.2014.128

[31] (2014) Smart home service framework proposal. [Online]. Available: https://wiki.allseenalliance.org/_media/tsc/technical_steering_committee/allseen_smart_home_service_framework_proposal_2014-06-23.pdf

[32] (2014) Smart home service interface design. [Online]. Available: https://wiki.allseenalliance.org/_media/tsc/technical_steering_committee/allseen_smart_home_service_framework_high_level_interface_design_2014-07-03.pdf

[33] M. Villari, A. Celesti, M. Fazio, and A. Puliafito, "Alljoyn lambda: An architecture for the management of smart environments in iot," in *Smart Computing Workshops (SMARTCOMP Workshops), 2014 International Conference on*, Nov 2014, pp. 9–14.

[34] OneM2M, "White paper: The interoperability enabler for the entire m2m and iot ecosystem," http://www.onem2m.org/images/files/oneM2M-whitepaper-January-2015.pdf, January 2015.

[35] (2015) openhab. [Online]. Available: http://www.openhab.org/

[36] (2015) Thread group. [Online]. Available: http://www.threadgroup.org

[37] P. Hasselmeyer and M. Voß, "Monitoring component interaction in jini federations."

[38] A. K. Eidsvik, R. Karlsen, G. Blair, and P. Grace, "Interfacing remote transaction services using {UPnP}," *Journal of Computer and System Sciences*, vol. 74, no. 2, pp. 158–169, 2008, web and Mobile Information Systems 2006. [Online]. Available: http://www.sciencedirect.com/science/article/pii/S0022000007000475

[39] T. Zahariadis and K. Pramataris, "Multimedia home networks: standards and interfaces," *Computer Standards & Interfaces*, vol. 24, no. 5, pp. 425–435, 2002. [Online]. Available: http://www.sciencedirect.com/science/article/pii/S0920548902000685

[40] (2007) Hl7 reference information model. [Online]. Available: http://www.hl7.org/implement/standards/

[41] (2007) Digital imaging communications in medicine. [Online]. Available: http://dicom.nema.org/

[42] R. L. Arenson, K. P. Andriole, D. E. Avrin, and R. G. Gould, "Computers in imaging and health care: now and in the future," *Journal of digital imaging*, vol. 13, no. 4, pp. 145–156, 2000.

[43] T. J. Eggebraaten, J. W. Tenner, and J. C. Dubbels, "A health-care data model based on the hl7 reference information model," *IBM Systems Journal*, vol. 46, no. 1, pp. 5–18, 2007.

[44] (2010) Accenture: Information governance, the foundation for effective e-health (white-paper). [Online]. Available: http://www.cas-ag.eu/nl-en/Pages/insight-information-governance-effective-ehealth-summary.aspx

[45] E. AbuKhousa, N. Mohamed, and J. Al-Jaroodi, "e-health cloud: opportunities and challenges," *Future Internet*, vol. 4, no. 3, pp. 621–645, 2012.

Chapter 7
Reasoning systems for AAL

Gorka Azkune[1], *David Ausín*[1], *and Diego López-de-Ipiña*[1]

7.1 Introduction

The ability to reason about the environment, resources, and inhabitants is key to provide a proper assistance. Humans interact with their environments in order to perform certain activities, which change the state of the environment. Hence, reasoning systems in AAL have to be based on human activities and interactions with the environment. In this chapter, reasoning systems for human activities and environments will be analysed.

Generally speaking, there are three functional layers in AAL reasoning systems which cope with human activities:

1. *Activity monitoring layer*: This layer is composed by the sensors deployed in the home environment and the low-level processing routines applied to the raw sensor data. It is very important to decide which parameters of human–environment interactions have to be monitored and what kinds of sensors are the most suitable ones to monitor those parameters. Once the sensors have been deployed and their information is grabbed, low-level processing routines may be applied to get meaningful information.
2. *Activity modelling layer*: It builds on top of the sensor information obtained in the previous layer. The aim of this layer is to obtain the activity models which describe and encapsulate the meaningful information about activities, such as the location, duration, or usage of objects. There are two main approaches for activity modelling: data- and knowledge-driven approaches.
3. *Activity inference layer*: Using the information obtained by sensors and the activity models defined in the previous layer, activity inference attempts at inferring the ongoing activities. The reasoning techniques applied for such an inference process strongly depend on the activity modelling approach.

[1]DeustoTech – Deusto Institute of Technology, University of Deusto, DeustoTech-INTERNET, Faculty of Engineering, 4th Floor, D-8, Avda. Universidades 24, 48007 Bilbao, Bizkaia, Spain

In consequence, this chapter is structured as four sections. Section 7.2 introduces the most representative activity monitoring approaches of the literature. In Section 7.3, the two main approaches for activity modelling will be analysed. Due to the tight relationship between modelling and inference, both topics will be covered in the same section. Finally, Section 7.4 shows an example of a promising approach for reasoning in AAL, which can be classified inside the knowledge-driven approaches. More concretely, Web Ontology Language (OWL) modelling and reasoning will be analysed in more detail.

7.2 Activity monitoring

The idea behind activity monitoring is to deploy sensors in human-populated environments in order to capture environmental parameters, human–environment interactions and other human actions, which provide relevant information about the activities carried out by those humans. In that sense, researchers have to carefully plan which parameters should be captured for their purposes and select and deploy the most suitable sensors in consequence.

Although there are several different types of sensors which can be used for activity monitoring, two categories are usually distinguished: vision-based and sensor-based activity monitoring approaches. Vision-based monitoring systems are classified apart due to their special nature regarding information processing techniques. It can be said that vision-based activity recognition is a discipline by itself, where the sensor choice heavily determines the techniques to be used in the next layers (Section 7.2.1). On the other hand, sensor-based activity monitoring covers all the other sensor infrastructures except vision systems. That is why there is a great heterogeneity in this category, ranging from wearable sensors to dense sensors (Section 7.2.2).

7.2.1 Vision-based activity monitoring

Vision-based activity monitoring is the use of visual sensing facilities such as video cameras to monitor human behaviour and environmental changes. The generated sensor data are video sequences or digitized visual data. The approaches in this category exploit computer vision techniques, including feature extraction, structural modelling, person and object recognition, movement segmentation, action extraction, and movement tracking to analyse visual observations for pattern recognition.

A very good survey has been provided by Poppe in [1]. The author focuses specially on the recognition of human movement at various levels of abstraction. More concretely, he follows a hierarchy where primitive actions, actions and activities are distinguished. An action primitive is an atomic movement that can be described at the limb level. An action consists of action primitives and describes a, possibly cyclic, whole-body movement. Finally, activities contain a number of subsequent actions, and give an interpretation of the movement that is being performed. For example, “left leg forward” is an action primitive, whereas “running” is an action. “Jumping hurdles” is an activity that contains starting, jumping, and running actions. The survey is very focused on human movement and does not treat vision-based approaches for context modelling and recognition. Turaga *et al.* [2] present another survey which is more focused on the higher-level recognition of human activities.

Vision-based approaches have been mainly applied to surveillance and security scenarios, rather than to AAL. However, there are several interesting research work in the AAL domain. A good example is provided by Hoey *et al.* [3], who present a real-time vision-based system to assist a person with dementia wash their hands. Another example can be found in the work performed by Brdiczka *et al.* [4]. Their paper addresses the detection of activities of individuals in a smart home environment ("walking", "standing", "sitting", "interacting with table", or "sleeping"). The developed system is based on a robust video tracker that creates and tracks targets using a wide-angle camera.

One of the principal obstacles for vision-based systems to be applied in AAL scenarios is the privacy problem. As noted by Chen *et al.* [5] cameras are generally perceived as recording devices and many people are reluctant to install them in their day-to-day environments. This problem has boosted the adoption of other monitoring approaches, mainly sensor-based monitoring systems which will be presented in the next section.

7.2.2 Sensor-based activity monitoring

The second category of activity monitoring is known as sensor-based activity monitoring and it is based on the use of emerging sensor network technologies. As a consequence of the latest developments in electronics, a plethora of miniaturised sensors have appeared in the market, such as contact sensors, RFID, accelerometers, audio or motion detectors. Due to their nature, those sensors are very suitable to install in AAL environments and they allow monitoring many interesting features of human–environment interactions.

The generated sensor data from sensor-based monitoring are mainly time series of state changes and/or various parameter values that are usually processed through data fusion, probabilistic or statistical analysis methods and formal knowledge technologies for posterior activity recognition. Even though there are different sensors that can be used in this category, sensor-based activity monitoring can be further classified into two categories: (i) wearable sensor-based activity monitoring (Section 7.2.2.1), where sensors are attached to an actor under observation, and (ii) dense sensing-based activity monitoring (Section 7.2.2.2), where sensors are attached to objects that constitute the activity environment. Wearable sensors, where modern smartphones are included, often use inertial measurement units and RFID tags to gather an actor's behavioural information. This approach is effective for recognising physical movements such as physical exercises. In contrast, dense sensing infers activities by monitoring human–object interactions through the usage of multiple multi-modal miniaturised sensors.

7.2.2.1 Wearable sensors

Wearable sensors refer to those sensors that can be worn by humans. Thus, wearable sensors are located in the human body, taking advantage of some of the complements humans tend to wear, such as clothes, eyeglasses, belts, shoes, wristwatches, and, recently, mobile devices. They can be used to collect information such as body position and movement and/or physiological parameters, e.g. heart rate, skin temperature, or arterial pressure. Researchers have found that different types of sensor information are effective for classifying different types of activities, but usually those activities are related to physical activities.

Wearable sensor-based activity monitoring suffers from limitations, which can be classified into three categories:

1. *Operational limitations*: Most wearable sensors need to run continuously and be operated hands-free. This may have difficulties in real-world application scenarios.
2. *Practical limitations*: These include the acceptability or willingness to use wearable sensors and the viability and ability to wear them.
3. *Technical limitations*: These include the size, ease of use, battery life, and effectiveness of the approach in real-world scenarios.

A way of overcoming such problems is to make use of existing gadgets that have already been carried out on a daily basis like smartphones as intelligent sensors for activity monitoring, recognition, and assistance. Modern smartphones have multiple features which make them ideal candidates for activity monitoring: (i) they have integrated sensors such as inertial measurement units, GPS, compass, and cameras, (ii) they have powerful processing units and memory, (iii) they offer high connectivity options, and (iv) they provide tools to interact with users (tactile screen, speakers). The adoption of smartphones by the scientific community has been in place for a while [6, 7] and is expected to gain large-scale uptake given the latest development and affordability of such palm-held electronic devices.

Obviously, wearable sensors are not suitable for monitoring activities that involve complex physical motions and/or multiple interactions with the environment. In some cases, sensor observations from wearable sensors alone are not sufficient to differentiate activities involving simple physical movements (e.g., making tea and making coffee). As a result, dense sensing-based activity monitoring has emerged, which is described below.

7.2.2.2 Dense sensing

Dense sensing-based activity monitoring is based on attaching sensors to objects that a user manipulates while performing activities. Hence activity monitoring is carried out by detecting user–object interactions. The approach is based on real-world observations that activities are characterised by the objects that are manipulated during their performance. A simple indication of an object being used can often provide powerful clues about the activity being undertaken. As such it is assumed that activities can be recognised from sensor data that monitors human interactions with objects in the environment. By dense sensing, the way and scale with which sensors are used is characterised. While using dense sensing a large number of sensors, normally low-cost low-power and miniaturised, are deployed in a range of objects or locations within an environment for the purpose of monitoring movement and behaviour. Following the dense sensing paradigm, sensors are moved from human bodies to human-populated environments.

The dense sensing paradigm has been widely used for creating ambient intelligent applications such as smart homes, because sensors are embedded within environments. The pre-eminence of dense sensing in AAL, via the smart home paradigm, can be seen in [8, 9]. Sensors in a smart home can monitor an inhabitant's movements and environmental events so that activities can be inferred based on the sensor observations, thus providing just-in-time context-aware assistance. For instance, a switch sensor in the bed can strongly suggest sleeping or relaxing, and pressure mat sensors can be used for tracking the movement and position of people within the environment.

Since the introduction of the idea in early 2000s [10, 11], extensive research has been undertaken to investigate the applicability of the approach in terms of sensor types, modalities and applications. For example, Tapia *et al.* [12] use environmental state-change sensors to collect information about interaction with objects and recognise activities that are of interest to medical professionals such as toileting, bathing, and grooming. Wilson *et al.* [13] use four kinds of anonymous and binary sensors, motion detectors, break-beam sensors, pressure mats, and contact switches for simultaneous tracking and activity recognition. Wren *et al.* [14] employ networks of passive infrared motion sensors to detect presence and movement of heat sources. With this captured data they can recognise low-level activities such as walking, loitering, and turning, as well as mid-level activities such as visiting and meeting.

Dense sensing activity monitoring has also some drawbacks. One of the most obvious ones is the need to install a lot of sensors in human populated environments, which may involve considerable infrastructure changes (install pressure mats or movement sensors), or the problem of adding sensors to any new object introduced in the environment (attach contact sensors to glasses bought in the supermarket). The practical feasibility of the approach is probably linked to introducing sensors in the production steps of objects and building phases of the environments. Another drawback is the simple information provided by the sensors, which cannot be enough if very detailed activities have to be recognised. A contact sensor may indicate an interaction between a human and an object, but it cannot provide any information about how the object is being used by the human.

7.3 Activity modelling and inference

Due to the tight relationship between modelling and inference, both layers will be exposed together in this chapter. Two main approaches will be distinguished for modelling and inference, namely the data-driven approaches (Section 7.3.1) and knowledge-driven approaches (Section 7.3.2). Recently, activity recognition systems which combine the principles and techniques from both approaches have appeared (Section 7.3.3).

Data-driven approaches aim at learning activity models directly from data, using machine learning and data mining techniques. On the other hand, knowledge-driven approaches use knowledge engineering and management tools to build activity models from expert knowledge rather than from data. As such, both data- and knowledge-driven approaches are suitable for different reasoning systems. Data-driven approaches are suitable for probabilistic and statistical inference and reasoning, whereas knowledge-driven approaches allow symbolic and logic reasoning.

7.3.1 Data-driven approaches

In terms of the techniques and methods used for activity modelling and recognition, abstracting from specific activity monitoring systems, data-driven activity modelling is one of the most successful approaches. The main idea of these approaches is to obtain activity models from user-generated data and use probabilistic and statistical inference to recognise activities. Data-driven activity modelling can be generally classified into two main categories: generative and discriminative.

The generative approach attempts to build a complete description of the input or data space, usually with a probabilistic model such as a Bayesian Network. A popular generative model is the Naïve Bayes classifier, which has been used with promising results for activity recognition [4, 15]. However, the Hidden Markov Model (HMM) is probably the most popular generative approach that includes temporal information, which is very important for activity recognition. They have been applied to vision-based activity recognition (see [16, 17] for early examples). But also to other monitoring approaches (e.g. pedestrian motions [18], office activities [19], smart homes [13]).

In contrast, the discriminative approach only models the mapping from inputs (data) to outputs (activity labels), focusing only on solving the classification problem rather than the representation problem. Typical discriminative approaches include the Nearest Neighbour [10], decision trees [20], Support Vector Machines [21, 22], and meta-level classifiers [23].

There are many approaches in the literature which cannot be clearly classified into discriminative or generative categories, but rather use a combination of both. The Independent Lifestyle Assistant (ILSA) is an example, as it uses a combination of heuristic rules and statistical models of sequential patterns of sensor firings and time intervals to help a person with planning and scheduling [24]. PEAT (the Planning and Execution Assistant and Trainer) is a cognitive assistant that runs on a mobile device, and helps compensate for executive functional impairment. PEAT uses reactive planning to adjust a user's schedule based on their current activities. Activity recognition in PEAT is based on what the user is doing, and on data from sensors on the mobile device. These are fed into an HMM, the outputs of which are combined with the reactive planning engine [25].

Other work has investigated how activities can be modelled with a combination of discriminative and generative approaches [26], how common sense models of everyday activities can be built automatically using data mining techniques [27], and how human activities can be analysed through the recognition of object use, rather than the recognition of human behaviour [28]. This latter work uses DBNs to model various activities around the home, and a variety of radio frequency identification (RFID) tags to bootstrap the learning process. Some authors have attempted to compare discriminative and generative models [10, 23], generally finding the discriminative models yield lower error rates on unseen data, but are less interpretable. Gu *et al.* use the notion of emerging patterns to look for frequent sensor sequences that can be associated with each activity as an aid for recognition [29]. Omar *et al.* present a comparative study of a variety of classification methods for analysing multi-modal sensor data from a smart walker [30].

One of the biggest problems of data-driven approaches, regardless their category, is the need of large-scale labelled databases. This problem arises because activity recognition is posed as a supervised learning problem. However, due to the required effort and time for annotating activity databases, supervised methods do not scale up well in practice. Annotating activity data imposes a burden on annotators and users and often introduces a source of error in the process. Rashidi and Cook show how to overcome the problem of depending on labelled activity databases in [31]. They use a non-labelled database, where they extract activity clusters using unsupervised learning techniques. More concretely, they develop a new activity modelling method named COM, which stands for Continuous, varied Order, Multi Threshold. COM not only discovers discontinuous patterns and their variations, but is able to better handle real-life data by dealing with different frequencies/sensor problem. All patterns discovered by COM are then used as inputs for a hierarchical agglomerative clustering algorithm. The clustering algorithm aims at grouping those sensor patterns which are similar in terms of their structure. An activity structure is defined as the start time, duration and region or location. Those clusters are finally used to train a boosted HMM, which is shown to be able to recognise discovered activities.

Rashidi and Cook bring a new paradigm to data-driven activity recognition in their work, showing the real possibility of using an unsupervised model and avoid database annotation problems. However, their approach still suffers some other traditional problems of data-driven approaches. Firstly, even though they do not use labelled databases, they do not overcome the "cold start" problem, since they have to collect data in order to discover activities and train HMMs. Besides, as other data-driven approaches, they have many difficulties when generalising what they have discovered and learnt from previous users, since every user has his/her own ways of performing activities. The price to pay for avoiding labelled databases is that the approach cannot set activity granularity and that discovered activities do not have any semantic meaning. The first problem refers to the impossibility of deciding at design phase what kind of activities will be recognised. For example, if a user usually washes dishes after preparing meal, COM will discover this sequence as a typical activity. However, the discovered activity is actually a sequence of two lower-grain activities. The second problem refers to the fact that discovered activities are only sequences of sensors with their location and time information, but without any semantic meaning. A human expert is needed afterwards to interpret discovered activities and label them with semantic tags. Notice that this a posteriori labelling step may be difficult since activity granularity is not set.

The other big problem of data-driven approaches is related to activity model generalisation. As activity models are learnt from specific users' behavioural data, the acquired models are personal. Those personal models cannot be generalised and applied to other users. To overcome this problem, some researchers have already worked on transfer learning techniques. A complete survey on transfer learning techniques for activity recognition can be found in [32]. The idea behind transfer learning is to take advantage of the learnt models whenever a new training process is launched in a new environment and/or for a new user. This way, the new training phase can be boosted, transferring the knowledge acquired in a previous scenario. Information about sensors, specific activities, or locations can be very similar and can provide interesting hints to train new models with few data.

Transfer learning scenarios usually distinguish between the source domain and the target domain. The first domain is composed by the environment, sensors, activities and person where initial activity models have been learnt. The second domain refers to the environment, sensors, activities, and person where the new models have to be learnt. Notice that there can be big differences between both domains. For instance, sensors may be completely different, some new activities may appear in the target domain or environments may vary slightly. Depending on those differences, learning techniques vary. However, the most influential parameters to define the learning techniques are related to the availability of labelled data in both domains. Following the classification established in [32], depending on the availability of labelled data, the next scenarios are distinguished:

- Informed supervised transfer learning: when labelled data is available for both target and source domains. Those scenarios are solved using inductive learning techniques.
- Uninformed supervised transfer learning: labelled data is only available in the source domain, so transductive learning approaches are applied.
- Informed unsupervised transfer learning: in this case, labelled data is available only in the target domain. Inductive learning techniques are used for those kinds of scenarios.
- Uninformed unsupervised transfer learning: there is no labelled data available, neither in the target nor in the source domain. Unsupervised learning techniques have to be used to work in those scenarios.

Transfer learning can mitigate the problem of model generalisation, and thus, reusability, but it cannot solve those problems completely. Notice that whenever a new domain is approached, even using transfer learning techniques, a new training phase has to be carried out. The merit of transfer learning is to use previously learnt models to enhance and accelerate new training processes.

7.3.2 Knowledge-driven approaches

Knowledge-driven activity modelling and recognition intends to make use of rich domain knowledge and heuristics for activity modelling and pattern recognition. The rationale is to use various methods, in particular, knowledge engineering methodologies and techniques, to acquire domain knowledge. The captured knowledge can then be encoded in various reusable knowledge structures, including activity models for holding heuristics and prior knowledge in performing activities, and context models for holding relationships between activities, objects and temporal and spatial contexts. Compared to data-driven activity modelling that learns models from large-scale datasets and recognises activities through data intensive processing methods, knowledge-driven activity modelling avoids a number of problems, including the requirement for large amounts of observation data, the inflexibility that arises when each activity model needs to be computationally learnt, and the lack of reusability that results when a person's activity model is different from another's.

Knowledge structures can be modelled and represented in different forms, such as schemas, rules, or networks. This will decide the way and the extent to which knowledge is used for further processing such as activity recognition, prediction and assistance. In terms of the manner in which domain knowledge is captured, represented and used, knowledge-driven approaches to activity modelling and recognition can be roughly classified into two main categories: mining-based and logic-based approaches.

The objective of a mining-based approach is to create activity models by mining existing activity knowledge from publicly available sources such as how-tos (e.g. those at *ehow.com* or *wikihow.com*), recipes (e.g. from *epicurious.com*), training manuals, experimental protocols, and facility/device user manuals. Mining-based activity modelling was initially investigated by researchers from Intel Research [28]. Representative examples of mining the web to generate probabilistic activity models can be found in [33, 34].

A logic-based approach adopts the idea of representing activities as knowledge models that can be formally specified using various logical formalisms, thus establishing an equivalence between activity and knowledge modelling. Using standard knowledge engineering tools, activity recognition, and other advanced application features such as prediction and explanation can be mapped to knowledge-based inference such as deduction, induction, and abduction.

The literature shows many logical modelling methods and reasoning algorithms in terms of logical theories and representation formalisms. Some researchers treat the problem of recognising activities as the problem of recognising plans, where given the actions of a person, the final goal has to be inferred [35, 36]. Some other researchers use the well-known Event Calculus formalism introduced by Shanahan [37]. For example, Chen *et al.* [38] proposed an EC-based framework in which sensor activations are modelled as events and object states as properties, and where activity recognition is mapped to abductive reasoning tasks.

Finally, ontology-based approaches have emerged as a result of the developments carried out in the Semantic Web community. Compared with data-driven and mining-based approaches, ontology-based approaches offer several interesting features: firstly, ontological activity models can capture and encode rich domain knowledge and heuristics in a machine understandable and processable way. Secondly, Description Logic-based descriptive reasoning along a time line can support incremental progressive activity recognition and assistance as an activity unfolds. Thirdly, it provides an easy and flexible way to capture a user's activity preferences and styles, thus facilitating personalised assistance. Finally, the unified modelling, representation, and reasoning for activity modelling, recognition, and assistance makes it natural and straightforward to support the integration and interoperability between contextual information and activity recognition. Due to the recent promising research contributions in this area such as [33, 39, 40], ontology-based modelling and reasoning approaches are further analysed in Section 7.4.

7.3.3 Hybrid approaches

Upon analysing carefully the advantages and disadvantages of knowledge- and data-driven approaches, it can be seen that they are complementary at some extent. For instance, while knowledge-driven approaches are weak in handling uncertainty and temporal information, data-driven ones are good at it. On the other hand, while data-driven approaches cannot build reusable generic activity models, knowledge-driven approaches can. This complementarity has recently raised the need to research on hybrid approaches to activity modelling. The main idea of hybrid approaches is to fuse data- and knowledge-driven approaches in order to solve the problems of both approaches in a single system.

Tran *et al.* [41] and Healaoui *et al.* [42] use Markov Logic Networks (MLN) to model temporal relations between actions and activities. The first one uses MLNs to naturally integrate common sense reasoning with uncertain analyses produced by computer vision algorithms for object detection, tracking, and movement recognition. The second work is focused on providing activity models for interleaved and concurrent executions using qualitative and quantitative temporal relationships. With a similar objective, Steinhauer *et al.* [43] combine HMMs with Allen logic, providing another hybrid approach example. All those approaches encode and use temporal knowledge and rely on automatically extracting these temporal patterns from datasets.

Another hybrid approach example is provided by Riboni *et al.* [44]. Their approach combines statistical inference techniques with ontological activity modelling and recognition. Statistical inferencing is performed based on raw data retrieved from body-worn sensors (e.g., accelerometers) to predict the most probable activities. Then, symbolic reasoning is applied to refine the results of statistical inferencing by selecting the set of possible activities performed by a user based on his/her current context. By decoupling the use of context information, statistical inferencing becomes more manageable in terms of necessary training data, while symbolic reasoning can more effectively select candidate activities taking into account context-dependent ontological relationships.

Chen *et al.* [45] present an ontology-based hybrid approach to activity modelling. They combine knowledge-based activities with specially developed learning algorithms to overcome the "cold start" problem, model reusability and incompleteness. The distinguishable feature of the approach from existing approaches is that activity modelling is not a one–off effort, instead, a multi-phase iterative process that interleaves knowledge-based model specifications and data-driven model learning. This approach is a first step to solve the static nature of activity models in knowledge-driven systems. However, the approach has a limitation: it considers that seed activity models contain all the actions performed by any user to perform an activity and hence, it is limited to learn only descriptive properties. This means that if different users perform the same activity by means of different sets of actions, the system will not be able to learn those distinct actions.

The problem of learning personalised actions is addressed by Azkune *et al.* in [46]. They define generic and incomplete knowledge-based activity models, where only the necessary actions to perform an activity are included. Once a user has been monitored during a period of time, they use data-driven learning techniques to capture the personalised actions performed by different users, thus creating a dynamic activity modelling approach which obtains generic and personalised models in the same process.

7.4 Ontologies in AAL

Ontologies are a good solution to model complex domains, thanks to their expressivity and formal semantics. Moreover, they have become one of the most popular methods to model the context [47]. In AAL, the context is any information that can be utilised to define the state of an entity [48]. Entities comprise any person, place, or object that is considered important for the iteration between the user and the application, including themselves. On the other hand, ontologies have emerged as a new technique for activity recognition, especially in the field of activity recognition based on objects [49].

The first version of the OWL ontology language was published in 2002. Since then, several revisions have appeared, including a new version, OWL 2, which was released in 2009. OWL 2 was born to overcome the shortcomings present in OWL 1 [50]:

- Expressivity limitations due to the lack of necessary constructs for complex domains.
- Syntax issues, such as difficulty to be parsed correctly.
- Absence of metamodelling in OWL 1 Lite and OWL 1 DL ontologies. Metamodelling is possible only in OWL 1 Full.
- An inefficient ontology import system.
- OWL 1 DL forbids the usage of annotations properties in axioms.
- Several problems related to OWL 1 semantics.
- The implementation of OWL 1 Full may not be possible.
- OWL 1 Lite computational complexity is ExpTime, while OWL 1 DL is NExpTime-complete.
- Determine if an ontology is OWL 1 Lite, OWL 1 DL or OWL 1 full is not an easy task.

Based on the previous facts, it is recommended to employ OWL 2 instead of OWL 1 when it is possible. So, we focus on OWL 2, which is the newest version.

There are several syntaxes that can be employed to write an OWL 2 ontology. However, RDF/XML syntax support is mandatory for any conformant OWL 2 tool. Other syntaxes are OWL/XML which is an alternative to RDF/XML; Manchester syntax is oriented to be more human readable; Turtle which is an alternative to RDF; and the functional-style syntax whose aim is to define the language structure.

The meaning of OWL 2 ontologies is assigned by the direct semantics [51] and the RDF-based semantics [52]. The former is very similar to the semantics of the description logic $\mathcal{SROIQ}$. However, OWL includes some features, such as keys, that are not included in $\mathcal{SROIQ}$. On the other hand, ontologies must meet certain conditions in order to be translated into it. These ontologies are called OWL 2 DL ontologies. In contrast, the RDF-Based Semantics not only is fully compatible with the RDF Semantics [53], but it also can interpret any OWL 2 ontology. Any ontology that can only be interpreted under the RDF-based semantics is member of OWL 2 Full.

One of the drawbacks of OWL is that the more expressive an ontology is, the more complex the reasoning task is. This disadvantage should be taken into account by ontology engineers in order to provide the performance required by each AAL scenario. Fortunately, OWL 2 offers the OWL 2 profiles, which are sub-languages of OWL 2 created to offer a better performance in exchange for losing some expressivity. OWL 2 defines three profiles OWL 2 EL, OWL QL, and OWL RL. Their features are described in [54]. Any OWL 2 DL ontology can be interpreted under RDF-semantics. In contrast, under the direct semantics it is only possible to interpret OWL 2 DL ontologies.

Before modelling a domain in OWL, we should know that OWL ontologies follow the open world assumption. This means that if a fact is not explicitly stated, it is impossible to determine its truthfulness, in contrast to the closed world assumption, in which every fact which is not explicitly defined as true is false.

A complete explanation of the OWL 2 language structure can be found in the W3C recommendation: OWL 2 Web Ontology Language Structural Specification and Functional-Style Syntax (Second Edition) [55].

7.4.1 Tools

One of the main advantages of OWL ontology language for modelling domains is that there are hundreds of tools available. Some of them are follows:

- Protégé[1] is an ontology editor that offers an intuitive user interface to define concepts and relationships between the individuals of the domain.
- TopBraid Composer[2] is semantic web modelling tool that supports SPARQL queries.
- Vitro[3] is a web-based ontology editor.
- Pellet[4] is an OWL 2 java reasoner.
- HermiT[5] is another OWL 2 java reasoner.
- Fact++[6] is a C++ DL reasoner that provides a partial OWL 2 DL support.
- OWLlink API[7] is an implementation of the OWLlink protocol. This protocol offers a simple way to divide a semantic web application into a client and a server. The

[1]http://protege.stanford.edu/
[2]http://www.topquadrant.com/tools/modeling-topbraid-composer-standard-edition/
[3]http://vitro.mannlib.cornell.edu/
[4]https://github.com/complexible/pellet
[5]http://hermit-reasoner.com/
[6]https://code.google.com/p/factplusplus/
[7]http://owllink-owlapi.sourceforge.net/

server performs the reasoning tasks and the client queries the server through http requests.

- Jena[8] is a semantic web framework that allows developers to handle ontologies swimmingly. It also provides several reasoners, but no one offers a full OWL support. Besides, it contains a SPARQL query engine.
- OWL API[9] is another semantic web framework to develop semantic web applications.
- Derivo SPARQL DL API[10] is an extension of the SPARQL DL language that offers a SPARQL DL query engine.

7.4.2 Beyond OWL

As in other domains, uncertainty is also present in intelligent environments [56] and affects the decision-making process. This task requires context information in order to respond to the users' needs. Data in intelligent environments are provided by several sensors and services in real time. But, unfortunately, these sensors can fail, run out of battery, or be forgotten by the user, in the case of wearable devices. On the other hand, the services can also be inaccessible due to network connectivity problems or technical difficulties on the remote server. Nonetheless, that unavailable information may be essential to answer correctly user's requirements.

In these cases in which the information of the domain comes with quantitative uncertainty or vagueness, ontology languages are less suitable [57]. Uncertainty is usually considered as the different aspects of the imperfect knowledge such as vagueness or incompleteness. In addition, the uncertainty reasoning is defined as the collection of methods to model and reason with knowledge in which Boolean truth values are unknown, unknowable or inapplicable [58]. Other authors [59, 57] consider that there are enough differences to distinguish between uncertainty and vague knowledge. According to them, uncertainty knowledge is comprised by statements that are either true or false, but we are not certain about them due to our lack of knowledge. In contrast, vagueness knowledge is composed of statements that are true to certain degree due to vague notions.

Taking into account the last definition, we can differentiate between uncertainty and vagueness approaches. In the former one, there are probabilistic uncertainty and possibilistic approaches. Some of the most well-known probabilistic approaches are Pronto reasoner [60], a probabilistic extension of $\mathcal{SROIQ}$; BayesOWL [61], which employs a set of rules to convert an ontology into a Bayesian Network; and PR-OWL [62], which is based on multi-entity Bayesian networks (MEBNs). From possibilistic approaches, PossDL reasoner [63], which extends Pellet for uncertainty reasoning and inconsistency handling, stands out. On the other hand, in vagueness approaches we can find works such as fuzzyDL [64], which is an expressive fuzzy Description Logic reasoner; and Almeida's work [65], which defines a method to model the vagueness in intelligent environments, to name a few. Despite the multiple available approaches,

[8]https://jena.apache.org/index.html
[9]http://owlapi.sourceforge.net/
[10]http://www.derivo.de/en/resources/sparql-dl-api.html

none has become a standard and new research results in the area of uncertainty appear every year.

7.5 Conclusions

Human activity recognition is a key enabler for Ambient Assisted Living and provides a paradigmatic example of how reasoning capacities can be used in such scenarios. In order to detect and recognise human activities, first of all, humans have to be monitored using sensors (Section 7.2). The information grabbed by those sensors feed the modelling and inference layers, where the reasoning takes place. In this chapter, the most used inference and reasoning techniques have been introduced.

Reasoning systems for AAL have been divided into three different categories, namely, the data-driven approach 7.3.1, the knowledge-driven approach 7.3.2, and the hybrid approach 7.3.3. The most representative examples of those approaches have been presented, describing the advantages and disadvantages. Due to the emergence of the Semantic Web and its application to AAL scenarios, Section 7.4 has been devoted to describe the features of OWL and associated semantic reasoners, which can be classified as a knowledge-driven approach.

Future research on reasoning systems for AAL will try to address the current problems, such as the trade-off between generic and personalised activity models, building dynamic models that evolve in time and overcoming the dependency to labelled data. One of the most promising research in the short-term may come from hybrid approaches, which fuse the data- and knowledge-driven ideas. However, a new trend in machine learning has recently emerged: deep learning [66]. Deep neural networks have proven to be very well suited for sensor fusion, hierarchical representation learning and transfer learning. Besides, deep neural networks are also outperforming other approaches for natural language processing. Those features make deep learning a very promising approach for future activity recognition systems.

Acknowledgement

This research work has taken place within FRASEware project (TIN2013-47152-C3-3-R) granted by Spanish Ministry of Economy and Competitiveness (MINECO).

References

[1] Poppe R. A survey on vision-based human action recognition. Image and Vision Computing. 2010;28(2):976–990.

[2] Turaga P, Chellappa R, Subrahmanian VS, Udrea O. Machine recognition of human activities: a survey. IEEE Transactions on Circuits and Systems for Video Technology. 2008 Nov;18(11):1473–1488.

[3] Hoey J, Poupart P, von Bertoldi A, Craig T, Boutilier C, Mihailidis A. Automated handwashing assistance for persons with dementia using video and a

partially observable Markov decision process. Computer Vision and Image Understanding. 2010 May;114(5):503–519.

[4] Brdiczka O, Reignier P, Crowley J. Detecting individual activities from video in a smart home. 11th International Conference in Knowledge-Based Intelligent Information and Engineering Systems. KES 2007, XVII Italian Workshop on Neural Networks, Vietri sul Mare, Italy, September 12–14, 2007. Proceedings, Part I. Springer Berlin Heidelberg; 2007. p. 363–370.

[5] Chen L, Hoey J, Nugent C, Cook D, Yu Z. Sensor-based activity recognition. IEEE Transactions on Systems, Man, and Cybernetics-Part C. 2012;42(6): 790–808.

[6] Gellersen H, Schmidt A, Beigl M. Multi-sensor context-awareness in mobile devices and smart artifacts. Mobile Networks and Applications. 2002;7(5):341–351.

[7] Schmidt A, Laerhoven KV. How to build smart appliances? IEEE Personal Communications. 2001;8(4):66–71.

[8] Chan M, Estève D, Escriba C, Campo E. A review of smart homes—Present state and future challenges. Computer Methods and Programs in Biomedicine. 2008;91(1):55–81.

[9] Nugent C, Mulvenna M. Experiences in the development of a smart lab. International Journal of Biomedical Engineering and Technology. 2009;2(4):319–331.

[10] Bao L, Intille S. Activity recognition from user-annotated acceleration data. In: Science LNiC, editor. Pervasive Computing. Vienna, Austria: Springer Berlin Heidelberg; 2004. p. 1–17.

[11] Patterson D, Liao L, Fox D, Kautz H. Inferring high-level behavior from low-level sensors. 5th International Conference in Ubiquitous Computing, UbiComp 2003, Seattle, WA, USA, October 12–15, 2003. Proceedings. Springer Berlin Heidelberg; 2003. p. 73–89.

[12] Tapia E, Intille S, Larson K. Activity recognition in the home using simple and ubiquitous sensors. Pervasive Computing. 2004;3001:158–175.

[13] Wilson D, Atkeson C. Simultaneous tracking and activity recognition (STAR) using many anonymous, binary sensors. Third International Conference, PERVASIVE 2005, Munich, Germany, May 8–13, 2005. Proceedings. Springer Berlin Heidelberg; 2005. p. 62–79.

[14] Wren C, Tapia E. Toward scalable activity recognition for sensor networks. Second International Workshop in Location-and Context-awareness, LoCA 2006, Dublin, Ireland, May 10–11, 2006. Proceedings. Springer Berlin Heidelberg; 2006. p. 168–185.

[15] Cook D, Schmitter-Edgecombe M. Assessing the quality of activities in a smart environment. Methods of Information in Medicine. 2009;48(5):480.

[16] Galata A, Johnson N, Hogg D. Learning structured behaviour models using variable length Markov models. In: IEEE International Workshop on Modelling People, Kerkyra, Greece. IEEE; 1999. p. 95–102.

[17] Starner T, Pentland A. Visual recognition of American sign language using hidden Markov models, in Proc. Intl. Workshop on Automatic Face and Gesture Recognition, Zurich, Switzerland, June 1995.

[18] Brand M, Oliver N, Pentland A. Coupled hidden Markov models for complex action recognition. In: Proceedings of Computer Vision and Pattern Recognition. IEEE; 1997. p. 994–999.

[19] Oliver N, Garg A, Horvitz E. Layered representations for learning and inferring office activity from multiple sensory channels. Computer Vision and Image Understanding. 2004;96(2):163–180.

[20] Maurer U, Rowe A. Location and activity recognition using eWatch: a wearable sensor platform. In: Ambient Intelligence in Everyday Life. Springer Berlin Heidelberg; 2006. p. 86–102.

[21] Brdiczka O. Learning situation models in a smart home. IEEE Transactions on Systems, Man, and Cybernetics, Part B: Cybernetics. 2009;39(1):56–63.

[22] Chen D, Yang J, Wactlar H. A study of detecting social interaction with sensors in a nursing home environment. In: Computer Vision in Human–Computer Interaction. Springer Berlin Heidelberg; 2005. p. 199–210.

[23] Ravi N, Dandekar N, Mysore P, Littman M. Activity recognition from accelerometer data. In: Proceedings of the Seventeenth Innovative Applications of Artificial Intelligence Conference, AAAI; 2005, Menlo Park, California. p. (5) 1541–1546.

[24] Guralnik V, Haigh K. Learning models of human behaviour with sequential patterns. In: Proceedings of the AAAI-02 workshop "Automation as Caregiver"; 2002. p. 24–30.

[25] Modayil J, Levinson R, Harman C. Integrating sensing and cueing for more effective activity reminders. In: AAAI Fall Symposium: AI in Eldercare: New Solutions to Old Problems; 2008. p. 60–66.

[26] Lester J, Choudhury T, Kern N. A hybrid discriminative/generative approach for modeling human activities. In: IJCAI; 2005. p. (5)766–772.

[27] Pentney W, Philipose M, Bilmes J. Structure learning on large scale common sense statistical models of human state. Proceedings of AAAI, Chicago, Illinois, July 13–17, 2008. p. 1389–1395.

[28] Wu J, Osuntogun A. A scalable approach to activity recognition based on object use. In: IEEE 11th International Conference on Computer Vision, 2007. ICCV 2007. IEEE; 2007. p. 1–8.

[29] Gu T, Wu Z, Tao X, Pung H, Lu J. epSICAR: An emerging patterns based approach to sequential, interleaved and concurrent activity recognition. In: IEEE International Conference on Pervasive Computing and Communications (PerCom), Galveston, Texas, 2009. p. 1–9.

[30] Omar F, Sinn M, Truszkowski J. Comparative analysis of probabilistic models for activity recognition with an instrumented walker. In: Proceedings of the 26th Conference on Uncertainty in Artificial Intelligence; 2010. p. 392–400.

[31] Rashidi P, Cook D. COM: A method for mining and monitoring human activity patterns in home-based health monitoring systems. ACM Transactions on Intelligent Systems and Technology. 2013;4(4):64.

[32] Cook D, Feuz K, Krishnan N. Transfer learning for activity recognition: a survey. Knowledge and Information Systems. 2013;36(3):537–556.

[33] Tapia EM, Choudhury T, Philipose M. Building reliable activity models using hierarchical shrinkage and mined ontology. In: Proceedings of PERVASIVE; 2006. p. 17–32.

[34] Palmes P, Pung H, Gu T, Xue W, Chen S. Object relevance weight pattern mining for activity recognition and segmentation. Pervasive and Mobile Computing. 2010;6(1):43–57.

[35] Wobcke W. Two logical theories of plan recognition. Journal of Logic and Computation. 2002;12(3):371–412.

[36] Bouchard B, Giroux S, Bouzouane A. A smart home agent for plan recognition of cognitively-impaired patients. Journal of Computers. 2006;1(5):53–62.

[37] Shanahan M. Solving the Frame Problem: A Mathematical Investigation of the Common Sense Law of Inertia. MIT Press, Cambridge, MA, 1997.

[38] Chen L, Nugent C. A logical framework for behaviour reasoning and assistance in a smart home. International Journal of Assistive Robotics and Mechatronics. 2008;9(4):20–34.

[39] Chen L, Nugent C, Wang H. A knowledge-driven approach to activity recognition in smart homes. IEEE Transactions on Knowledge and Data Engineering. 2012;24(6):961–974.

[40] Riboni D, Bettini C. OWL 2 modeling and reasoning with complex human activities. Pervasive and Mobile Computing. 2011;7(3):379–395.

[41] Tran S, Davis L. Event modeling and recognition using Markov logic networks. In: Forsyth, David A; Torr, Philip H S; Zisserman A, editors. Computer Vision–ECCV 2008. Marseille: Springer Berlin Heidelberg; 2008. p. 610–623.

[42] Helaoui R, Niepert M, Stuckenschmidt H. Recognizing interleaved and concurrent activities using qualitative and quantitative temporal relationships. Pervasive and Mobile Computing. 2011;7(6):660–670.

[43] Steinhauer H, Chua S. Utilising temporal information in behaviour recognition. In: AAAI Spring Symposium; 2010. p. 54–59.

[44] Riboni D, Bettini C. COSAR: hybrid reasoning for context-aware activity recognition. Personal and Ubiquitous Computing. 2011;15(3):271–289.

[45] Chen L, Nugent C, Okeyo G. An ontology-based hybrid approach to activity modeling for smart homes. IEEE Transactions on Human–Machine Systems. 2014;44(1):92–105.

[46] Azkune G, Almeida A, López-de-Ipiña D, Chen L. Extending knowledge-driven activity models through data-driven learning techniques. Expert Systems with Applications. 2015 Apr;42(6):3115–3128.

[47] Strang T, Linnhoff-Popien C. A context modeling survey. In: Workshop on Advanced Context Modelling, Reasoning and Management, UbiComp 2004 – The Sixth International Conference on Ubiquitous Computing, Nottingham/England; 2004.

[48] Dey AK. Understanding and using context. Personal Ubiquitous Computing. 2001 Jan;5(1):4–7. Available from: http://dx.doi.org/10.1007/s007790170019.

[49] Chen L, Nugent C. Ontology–based activity recognition in intelligent pervasive environments. International Journal of Web Information Systems. 2009;5(4):410–430.

[50] Grau BC, Horrocks I, Motik B, Parsia B, Patel-Schneider P, Sattler U. OWL 2: The next step for OWL. Web Semantics: Science, Services and Agents on the World Wide Web. 2008;6(4):309–322. Semantic Web Challenge 2006/2007.

[51] Grau BC, Patel-Schneider P, Motik B. OWL 2 Web Ontology Language Direct Semantics (second edition). W3C; 2012. Http://www.w3.org/TR/2012/REC-owl2-direct-semantics-20121211/.

[52] Schneider M. OWL 2 Web Ontology Language RDF-Based Semantics (second edition). W3C; 2012. Http://www.w3.org/TR/2012/REC-owl2-rdf-based-semantics-20121211/.

[53] Hayes P. RDF Semantics. W3C; 2004. Http://www.w3.org/TR/2004/REC-rdf-mt-20040210/.

[54] Motik B, Grau BC, Horrocks I, Fokoue A, Wu Z. OWL 2 Web Ontology Language Profiles (second edition). W3C; 2012. Http://www.w3.org/TR/2012/REC-owl2-profiles-20121211/.

[55] Patel-Schneider P, Parsia B, Motik B. OWL 2 Web Ontology Language Structural Specification and Functional-Style Syntax (second edition). W3C; 2012. Http://www.w3.org/TR/2012/REC-owl2-syntax-20121211/.

[56] Ramos C, Augusto JC, Shapiro D. Ambient intelligence – The next step for artificial intelligence. IEEE Intelligent Systems. 2008;23(2):15–18.

[57] Lukasiewicz T, Straccia U. Managing uncertainty and vagueness in description logics for the semantic web. Web Semantics: Science, Services and Agents on the World Wide Web. 2008;6(4):291–308.

[58] W3C. Uncertainty Reasoning for the World Wide Web. W3C; 2008. Http://www.w3.org/2005/Incubator/urw3/XGR-urw3/.

[59] Baader F, Küsters R, Wolter F. The Description Logic Handbook. New York, NY, USA: Cambridge University Press; 2003. p. 219–261. Available from: http://dl.acm.org/citation.cfm?id=885746.885753.

[60] Klinov P, Parsia B. Pronto: a practical probabilistic description logic reasoner. In: International Workshop on Uncertainty in Description Logics; 2010.

[61] Ding Z. BayesOWL: A Probabilistic Framework for Uncertainty in Semantic Web. PhD thesis, University of Maryland, Baltimore, 2005.

[62] Costa PCG. Bayesian semantics for the Semantic Web. George Mason University; 2005.

[63] Qi G, Ji Q, Pan JZ, Du J. PossDL – a possibilistic DL reasoner for uncertainty reasoning and inconsistency handling. 7th Extended Semantic Web Conference, ESWC 2010, Heraklion, Crete, Greece, May 30 – June 3, 2010, Proceedings, Part II. Springer; 2010. p. 416–420.

[64] Bobillo F, Straccia U. fuzzyDL: An expressive fuzzy description logic reasoner. In: FUZZ-IEEE; 2008. p. 923–930.

[65] Almeida A, López-de-Ipiña D. Assessing ambiguity of context data in intelligent environments: towards a more reliable context managing system. Sensors. 2012;12(4):4934. Available from: http://www.mdpi.com/1424-8220/12/4/4934.

[66] Bengio Y, Courville A, Vincent P. Representation Learning: A Review and New Perspectives. Arxiv. 2012 June. Available from: http://arxiv.org/abs/1206.5538.

Chapter 8
Person–environment interaction

Praminda Caleb-Solly[1]

This chapter presents an overview of human–computer interaction and design issues for person–environment interaction that are applied in Ambient Assisted Living (AAL), with consideration of the accessibility needs of people with age-related physical and cognitive impairments when designing interfaces. A review of the recent technologies that have been used to facilitate interaction via different modalities is presented, and the advantages and disadvantages they offer in relation to different physical and cognitive impairments. In addition, this review also considers the impact of stress, attention, social context, user experience and cognitive load on the usability of these technologies within specific AAL scenarios.

This chapter also presents some key interaction models and considers how the recent AAL technologies designed for specific application areas can be assessed against these to ensure effectiveness and efficiency of user interaction.

This chapter concludes with a set of best practice guidelines for designing person–environment interactions in the AAL context.

8.1 Introduction

In 1996 Mark Weiser envisaged a future where he conjectured that we would "dwell with computers" that would be integrated into different objects, clothes, cups, light switches [1]. These would subtly provide us with up-to-date information, such as whether a garment or cup needed washing, lights needed turning off to preserve energy when not required etc. His use of the word "dwell" was purposeful, as he perceived our relationship to be more than just interacting with this technology, but that of co-existing with it. He had earlier coined the term "*ubicomp*" or "*ubiquitous computing*" [2] and this concept has morphed into the paradigms of "pervasive computing" and "ambient intelligence" that have become well established today. The notion is one of seamless blending of technology into one's environment to the extent that it becomes a natural part of life, providing contextually relevant information in an intuitive manner, and intelligently adapting the environment to personalised needs.

[1]Bristol Robotics Laboratory, Department of Engineering Design and Mathematics, Faculty of Environment and Technology, University of the West of England, Frenchay Campus, Bristol BS16 1QY, UK

Smart environments, comprising integrated sensors, displays and actuators, particularly those embedded into physical objects and the surrounding fabric of the built environment, have the potential to offer unobtrusive sensing and control. Ambient and contextual information that can be gleaned at a glance can not only help to reduce cognitive load associated with traditional interactive systems, but also help in making us more efficient and effective by providing timely and pertinent information to support decision making.

However, by the very nature of their design, ambient systems can present a challenging dilemma for accessible interaction. Interaction in an ambient context is often facilitated through a range of modalities such as speech, gesture, touch, vision, affective and physiological response monitoring, and EEG analysis for brain–computer interaction. These interaction modalities can lack adequate visibility and affordances, and require new forms of interaction vocabularies and protocols, that might be unfamiliar to users, thereby reducing their usability. These issues can be compounded by the fact that the progression of ageing-related sensory impairments, such as the deterioration of vision, hearing, speech, haptic perception, reduced muscle strength and balance, result in diminished ability of the person to interact effectively via these modalities. In addition, ageing-related impairments can progress at varying levels of severity and affect people in different ways due to other co-morbidities. This can mean that a once effective mode of interaction gradually starts to become unsuitable, with the progressive decline of the person's condition rendering the system ineffective, without a clearly detectable point of failure. Ensuring effective interaction therefore requires a clear understanding of the impact of ageing-related impairments, and how these might be ameliorated by improving the design of the system. Additionally, having an understanding of the nature of the impairments could enable the designer to incorporate ability for the system itself to track, and thus respond to changes by adapting the interfaces.

The next section explores different aspects of person–environment interaction with AAL technologies. Following this, an overview of ageing-related effects on the body that designers need to consider when developing AAL technology interfaces is provided. Successful interaction is predicated on ensuring ease of use, as well as user comfort, both ergonomic and psychological. Human factors research and inclusive design methodologies have provided a wealth of guidelines and heuristics in regards to designing systems that are intuitive and easy to use. These are discussed later in the chapter in the context of specific AAL system applications and functionalities.

8.2 Interaction within an ambient assisted living environment

Context awareness underlies successful interaction within an AAL environment. Poslad [3] defines three main types of external environment context-awareness which are: physical environment context, relating to physical aspects of the environment such as location, light levels, temperature, time of day; human or user context, relating to the identity of the user, their preferences, task requirements, social contexts and activities, their experience and prior knowledge; and the Information and Communication

Technology (ICT) context, which relates to the particular components and services comprising the systems that are available.

Context awareness enables a system to dynamically adapt its behaviour, either based on a set of pre-set rules or conditions, or in an intelligent manner, based on learning algorithms that use context data for modelling behaviour. This enables an ambient assistive system to function proactively and in a non-deterministic manner, resulting in an environment that is sensitive, adaptive and responsive to user needs [4]. Enhanced user awareness can result in improved user experience by also taking into consideration the user's affective or emotional state, and can include recognising user intention, enabling improved system responsiveness by sensing or anticipating user action and goals.

The nature and scope of an interaction is determined by the specific AAL task or scenario and the purpose of the interaction. In their survey of AAL tools for older adults, Rashidi and Mihailidis [5] identify six overarching AAL application areas: (1) Cognitive Orthotics, which include daily reminders and planners, medication reminders, navigators and wandering prevention tools; (2) Continuous Health Monitoring, which include vital signs monitoring, sleep monitoring and Activities of Daily Living (ADL) monitoring; (3) Therapy, which includes tele-health and tele-rehabilitation; (4) Emergency Detection, which includes fall, hazard and medical emergency detection; (5) Emotional Well-being, which includes social connectedness and facilitating communication; and 6) Persuasive applications, whose aim is to promote well-being and medical compliance.

The objectives of the interactions in these application areas is to inform, assist, guide or provide support for decision making, for example to help people make healthy lifestyle choices by encouraging and motivating them to adhere to behaviours that promote good health. As the interaction in these contexts is purposeful, it is also important to consider the characteristics of the information provided. Such information needs to be made easily interpretable, meaningful, engaging, actionable and personalised, not only for the specific end-users, but also for other stakeholders who might be using this information to augment the support they are providing the end-users.

There are a growing number of examples that illustrate the use of creative techniques being devised to try and ensure that the objectives of the interaction are achieved, particularly for persuasive AAL applications. An early example is the use of gamification, as presented in a study by Klasnja *et al.* [6] who trialled the use of a mobile system for encouraging physical activity. Their UbiFit system consisted of an interactive application used for journaling and reviewing daily activities and tracking progress, and a *glanceable* display that used an abstract, stylistic representation of the physical activities depicted using a garden metaphor. The garden bloomed with different types of flowers that represented different types of activities, and the successful achievement of user goals were represented by the appearance of butterflies. In addition to employing an engaging form of information visualisation, Klasnja *et al.*'s research illustrated some key lessons from a design perspective for interaction. These included the importance of supporting persistent cognitive activation of a person's health goals where goals can be activated through environmental cues that provided a constant reminder; encouraging an extensive range of healthy behaviours, giving credit for a diverse range of activities that the person might engage in, and hence adapting to and acknowledging individual behaviours and allowing for customisation; focusing on long-term patterns of activity that could help people make informed decisions and choices that were reflective of their overall habits and support them in seeing trends in their behaviour rather than one–off events and learn to moderate their behaviour better; and facilitating but not depending on social support, as people preferred to share positive achievements rather than lapses in activity, which could hamper motivation and introduce social friction.

In order for an AAL system to provide information that is meaningful to the person, in the way described by Klasnja *et al.*, it has to be made relevant to a person's own operational parameters or frame of reference. For instance – something that informs the person that they need to engage in more activity has to be calibrated to their own thresholds of activity levels. While a person could indicate these threshold levels themselves, these values need to be represented in terms of the measurement sensors being used in the system and calibrated in order to be valid. Additionally most people will not be aware of their overall patterns of behaviour in relation to the aspects that can be measured and this would be something that the system would need to elicit, using parameters provided by an expert. This implies a period of learning, where the system is placed in situ, monitoring and measuring over time in order to calculate baseline metrics. Also, such a system has to be adaptable and sensitive to other correlated aspects – such as, if the person has been unwell or if they were particularly active the previous day. Setting up of operational parameters, authorising data access, specifying rules for monitoring the relevant behaviours and identifying baseline patterns, would still require appropriate system configuration interfaces to enable fine-tuning for personalisation of these systems. The interaction with these interfaces also needs to be considered as part of the ambient system design.

The availability of appropriately labelled data to generate reliable baseline behavioural information that can be used to build interactions that are meaningful and contextually relevant is highly problematic. Fiorini *et al.* [7] explored the effectiveness of using a "busyness" measure for identifying patterns in unlabelled sensor data to develop a personalised model of user behaviour that recognised a specific activity. Such models are vital for enabling effective person–environment interaction that is based on recognising and responding to changes in specific activities that are usually carried out in a specific environment. So for instance, in the case study example considered by Fiorini *et al.*, where the specific activity being considered was making hot drinks in the kitchen, an AAL-based person–environment interaction that could be triggered might be a light that could glow in the kitchen cupboard to remind the person to make a hot drink when they entered the kitchen. This would be triggered if the system detected a deviation from their usual pattern for hot-drink making for the day.

8.2.1 Interaction modalities

The range of modalities for interaction continues to grow with the development of novel and more reliable sensing technologies and designers' innovation in using them more effectively. The AALIANCE2 project, which defines a roadmap for the next generation of European Ambient Assisted Living innovation, produced a report listing emerging interface technologies [8]. These are categorised as spatial, sensorial, natural language and multimodal interfaces.

In our everyday communications, gesture is a vital and often integral non-verbal communication component accompanying speech. When studying the semantics of gestures used as part of communication, it is found that gesture can take on a number of forms. McNeill [9] considers different types of gestures as part of a continuum which extends from gesticulation, where gestures accompanies speech providing emphasis, imagery and metaphorical representation, to sign language, where speech becomes redundant as the gestures are structured as part of a vocabulary and have specific meaning. As such, gestures can be continuous movements through space, discrete or specific hand poses or pointing movements that need knowledge of the environment to enable their interpretation.

Spatial interfaces for gestural interaction include 3D movement trackers which incorporate accelerometers and optical sensors such as the Nintendo Wii remote [10] or data gloves [11]; eye-trackers that register eye movement in real time to control a pointer by tracking the user's gaze pattern [12]; gesture-based interaction where hand movements are used without touching the display, such as using a Microsoft Kinect [13] or Leap motion device [14], or identifying deictic gestures through machine vision using cameras, or sensing symbolic gestures using a Myo armband [15] that analyses electromyographic (EMG) signals when specific hand poses are made.

Interfaces for capturing gestures also encompass tactile or touch screens where a user interacts using a finger or pointing device. Multi-touch screens, tables and smart boards where the user can perform gestures with their fingers (such as flicking, pinching or drawing shapes) in addition to direct interaction, enable enhancing the scope of the interaction. Pressure sensitive surfaces also incorporate a 3D element to 2D touch gestures, such as the Apple iPhone 6S [16], which also provides haptic feedback.

Sensorial interfaces are defined as digital augmentation of physical objects through sensory perception and include augmented reality, stereoscopic vision, binaural sound which is a technique that enable people with visual impairment to find the source of a sound in a 3D space [17], tactile feedback whereby haptic displays can utilise a combination of the skin's perception of touch, temperature and pressure, and including force and positional feedback by using gaming-type interaction devices [18]. The use of scent-based interfaces is also emerging as a possible interaction modality [19].

Natural language interfaces use voice recognition and map phonemes captured into words from a known vocabulary set, and voice synthesis for text to speech conversion. More sophisticated systems can also incorporate prosody to improve the tonal characteristics of speech generated. Handwriting recognition is also categorised under natural language understanding.

Multimodal interfaces [20] can process two or more interaction modalities, providing flexibility from the user's perspective in preferred choice of interaction. This is particularly useful when there is redundancy in the system and the user can choose which modality to use for the interaction. Alternatively, the multimodality might be complementary or augment other forms of interaction, which could result in more robust interaction where the user can focus their attention on other tasks or information. For people with accessibility needs, multimodal interaction provides a particular advantage, especially when there is redundancy, as information that is unclear or missed due to problems with one interaction channel, can be compensated by the information provided via another.

8.2.2 Spatial relevance of person–environment interaction

Our sensory systems are exposed to constant streams of information. In order to keep from becoming overwhelmed, and the fact that we have limited cognitive capacity to consciously process all this information simultaneously, we only process information through selective attention [21]. Also, habituation can occur as a result of repeated exposure to continuous stimuli, especially where the information is not perceived to be salient enough to focus our attention [22], and particularly when it repeated in the same way or format [23]. A careful balance therefore needs to be achieved; too much change and movement and the information could be seen as distracting or even irritating; too little, and it might get diffused into the background and be ignored. Therefore in the case of information from an AAL system, it is particularly important for designers to consider how to prevent habituation and keep information salient, meaningful and engaging.

As part of their project of developing ubiquitous working environments and considering ambient information displays, Streitz *et al.* [24] postulated three concentric "zones of interaction" based on the proximity of people to an information display. The proximity was determined by sensors which could be adapted to the surrounding spatial conditions and enabled Streitz *et al.* to introduce "distance dependent semantics" which helped define the type of information that would be shown and the nature of the interaction that would be made available. The "Ambient Zone" is the area furthest away from the information display and outside the range of the display's sensors where people experience general information, the next closest is defined as the "Notification Zone," when particular person-specific information is displayed, and the "Interaction Zone" when the person is very close to the display and is able to interact with it, physically or via a wireless device. This concept provides a useful framework for person–environment interaction in the AAL context. An individual's accessibility needs could be built into this framework whereby not only the information content is modified to make it pertinent to how the person can be expected to utilise the information, but also how the information is presented in terms of its design and format.

8.2.3 Interaction initiative

The notion of "zones" highlights another key aspect of person–environment interaction within an ambient intelligent system, which is the nature of the interaction – implicit or explicit. Schmidt [25] described implicit interaction as being based on the system being able to understand a person's behaviour or activity in a given situation without the need for direct or explicit interaction with a system interface. In order to perceive and interpret contextual behaviour, appropriate sensing and learning systems to develop the underlying models for facilitating implicit interaction are required. Blasco *et al.* [26] describe a modular structure of such a system developed as part of a smart kitchen for AAL. A range of user activity is sensed as implicit information, such as opening and closing of cupboards, presence in the kitchen, when lights are left on etc. and this is used in conjunction with user profile data, as well as explicit input via the user interfaces. The use of implicit interactions help to develop a more contextually sensitive and responsive system – such as a host who refills your glass with water when she sees that it is empty without you explicitly having to ask. However, this example illustrates one of the difficulties of accurate interpretation based on just one variable, your glass being empty, and using default assumptions, that you were previously drinking water and this is what you want again. An experienced host might be more vigilant of their guests' needs, they might notice them glancing at their empty glass, and then follow their gaze to the juice on the table, and use this as cue to perhaps offer them juice instead of water. This scenario could be considered in relation to assistive robots, which are also seen as part of AAL system ecologies and where learning preferences and baseline patterns of behaviour supports autonomous operation. Ju and Leifer [27] put forward an implicit interaction framework that helps designers consider where and how the use of the implicit interactions could help make systems more usable and useful. Their framework identifies two core components of interactions, (1) who takes the initiative – the user or the system, and (2) the level of attentional demand required of the user. Ambient agents are seen to be more proactive in terms of initiative, operating in the background and providing assistance as and when needed.

8.3 Interaction models

With potentially multiple sources of information from ambient assistive technologies in a person's environment that they could interact with, both in an implicit and explicit manner, effective management of the cognitive load on users and providing a good user experience is crucial. Surie *et al.* [28] introduce the concept of *egocentric interaction*, describing an interaction-modelling framework that is person-centric as opposed to being device-centric. Their framework augments the classical user-centred approach in human–computer interaction (HCI) by emphasising the following: (a) *situatedness* of the person with regard to interactions and activities, (b) attention to the complete local environment, stressing that proximity of the person is fundamental in determining the focus of the interaction, (c) the changeability of the environment and person–environment relationship and (d) that equal attention is paid to virtual and physical objects, circumstances and people in the interaction. By considering these aspects, Surie *et al.* define a useful infrastructure for an ambient ecology that acknowledges the complexity of a system comprising multiple and parallel ambient intelligent applications, physical and virtual smart objects and a person at the heart of it all. As part of their interaction paradigm, they consider people as mobile agents in a dynamically changing environment, populated by physical and virtual objects. Incorporating key system elements such as an object manager, activity recogniser and situation monitor, communicating with an overall interaction manager, provides the scope for a co-ordinated schema to facilitate and customise the interaction and hence the user experience.

In an early study of user-centred system design, Norman [29] defined a seven-stage model of action, where the objective is to provide designers with a structure that can help them analyse interaction tasks. This model assumes that the initiative originates from the user when they first establish a goal and intention to act in order to achieve their goal, which is expressed in psychological terms. The system that they are interacting with however has physical states, mechanisms and variables, and these have to map onto how the user conceptualises the action sequence in order to carry out the task. Norman describes discrepancies that might arise between the user's psychological intentions and system's physical variables as a *gulf* that must be bridged – the *Gulf of Execution*. The system's input interface and modalities need to be designed so that the user understands what actions can be performed to achieve their goal. Well-designed affordances (physical, perceptual, cognitive) [30] can provide the user with clues on what to do, which relate to the user's experience of interaction conventions and can help to bridge this gulf. Once the user has carried out an action, they need to interpret the system state and evaluate whether the outcome of their action matches their original goal. Norman calls this the *Gulf of Evaluation* that needs to be bridged – where the output from the system interface has to be perceptually processed, interpreted and then evaluated by the user to compare it with their goal. There is also a potential for a temporal delay in the change of the system state, which can also interfere with their process of evaluation. Good design principles of visibility, mapping, clear conceptual models and continuous feedback can help to bridge the *Gulf of Evaluation*.

These two models provide a useful framework for identifying, and then analysing the different elements comprising an interaction to define the most appropriate design options or support structured evaluation of existing interaction schemas.

8.4 Understanding the ageing process and considerations for interaction

The accumulation of cellular and molecular damage over a period of time is widely considered to be the general cause of ageing [31]. Ageing studies have shown that individual genetic, environmental and medical factors can result in the ageing process impacting on a range of physical and cognitive functions in the body in different ways [32]. Ageing-related sensory impairment includes a decline in vision, hearing, touch and psychomotor coordination [33]. Accurate and efficient recognition, processing and interpretation of information determine successful interaction and communication, so changes in sensory and perception abilities are particularly crucial to consider.

8.4.1 Vision

The effects of ageing on vision include reduced ability to read small print, as well as problems with focus for distance vision. While these can be corrected with prescription glasses, it should be noted that people might not always be wearing their glasses, particularly when they are not actively engaging in a task where they deem it necessary to be doing so. Margrain [33] citing Evans *et al.* (2002) states that one in five over the age of 75 in Britain has impaired vision and "age-related macular degeneration" which affects photoreceptors in the central retina accounts for half of all blind registrations in the UK. Within an AAL environment, where information might be provided via displays that are integrated into the environment, consideration would need to be given to screen sizes and design of visual content that allows for reconfiguration of the graphical elements. In addition, contrast sensitivity, colour perception, sensitivity to light, as well as the ability of the eyes to resolve temporal events, declines with age. This has significance for displays that are in poorly lit rooms, which might require the use of additional sensors for person-detection to change the display brightness and contrast levels accordingly. While an automated adjustment process is desirable, providing features for manual control and customisation should be mandatory.

8.4.2 Hearing

One of the most prevalent chronic health conditions is hearing loss and Margrain [33] states that according to epidemiological evidence, approximately 17% of the UK population age 61–80 are unable to hear sounds that equate to a moderate whisper and 35% of those over 50 experience less substantial degrees of hearing loss. The cause of hearing loss can be either conductive, sensorineural or both. Acute hearing loss can jeopardise ability for independent living by affecting people's ability to hear alarms, sirens and doorbells, and ambient assisted technology could offer support by compensating for that loss by utilising an alternative interaction modality. However, for people with mild to moderate hearing loss, which impacts ability for communication, this could have severe implications for systems utilising speech-based interaction, which is often preferred as it feels more natural by providing a more intuitive interaction experience. People suffering hearing loss report difficulty in understanding speech where their difficulties relate to confusion or misinterpretation between phonetically similar words, particularly with loss of ability to hear speech in the higher frequency range, which certain consonants that are critical to recognising speech occupy. Accuracy of understanding speech is also affected by other factors such as loudness, speech rate, availability of contextual information and background noise [34]. While hearing aids offer improvements, hearing aid ownership and usage is relatively low. The reasons for this given by people include the high cost of aids, a belief that they did not need a hearing aid and a fear of stigmatisation associated with using them [35]. Speech-based interaction can also pose a barrier where speech synthesis does not include prosody to enhance semantics [36] and where the cognitive, physical and emotional cost of conversation repair (asking for the system to repeat the information) is perceived to be too high [37] resulting in a frustrating experience.

Results from a study by Lin *et al.* [38] suggest that hearing loss is associated with accelerated cognitive decline and incident cognitive impairment in older adults. In the context of providing more cognitive stimulation, ambient assistive systems could provide a useful function in supporting people with a focus on stimulating interaction through other modalities.

8.4.3 Touch

The loss of touch sensation can reflect age-related decrease in sensory innervation and neuronal processing decline, with the decline in vibro-tactile sensation varying with frequency, with sensitivity being most affected at 100 Hz [33]. This might have implication for the design of vibro-tactile stimulation patterns in wearables that might be part of an AAL solution, taking into account the fact that haptic perception and recognition can be reduced as a result of reduced cutaneous sensitivity [39].

8.4.4 Cognition

Cognition relates to activities such as thinking, reasoning and problem solving and decision making. Fisk *et al.* [40] provide a set of cognitive constructs that are fundamental to the process of interaction. These include working, semantic, prospective and procedural memory, attention, spatial cognition and language comprehension. Memory, in particular, plays a crucial role in successful interaction, which each component playing a specific role in comprehending, retrieving and learning information. Designers need to consider the function of each type of memory construct to address the issues that might occur due to impairments and provide corresponding augmentations and support. These include reducing cognitive load, not requiring people to remember too much during the task, keeping each task step short and providing clear and consistent feedback and signposting for the next steps. People have a limited capacity to process information, and through selective attention only certain information is processed in more depth. As such, there are particular implications for multimodal interface design that process two of more user input modes or provide information via more than one modality. Oviatt as in Jacko *et al.* [41] considers the design of multimodal systems and suggests that there should be complementarity of content, rather than redundancy. This will ensure that age-related decline in ability to attend and respond, will not be compounded by competing sources of attention that could potentially confuse and add to the cognitive load of user.

8.4.5 Physical ability

While progression of ageing-related impairments can cause communication barriers, due to as deterioration of vision, hearing and speech, a key aspect to consider in designing enabling technologies is physical frailty. Increased frailty from muscle weakness and osteoporosis, and tendency for rapid muscular fatigue, are often the result of poor physical health or impact of long-term chronic conditions such as arthritis or diabetes that can result in a tendency for an inactive lifestyle further adding to the decline. Additionally, impaired visual acuity, proprioception and vestibular function, as well as a slowing of reaction time, contribute to increased instability [42]. AAL systems that use gesture-based interaction could prove particularly problematic for this group of people, with risk of physical discomfort [43]. Touch screen, gesture-based interaction and even speech-based interaction, assume a certain level of physical ability and endurance. For some frail older people, even short periods of interactions with a speech-based interaction system can be quite physically fatiguing, as general speech recognition technology requires clear enunciation, articulation, and a certain level of alertness to follow synthetically generated speech without the non-verbal cues that people rely heavily on for interpretation of semantics.

8.4.6 Summary

The effects of ageing are individual and change in a variable manner over time, taking place at different rates often with a range of other co-morbidities existing simultaneously. It is therefore unhelpful and potentially detrimental to consider "older adults" as a homogenous group, but rather develop a clear empathetic and scientific understanding of the different physiological and psychological changes associated with ageing. This also suggests that part of developing an effective interaction system for people with declining sensory and perception ability, is to imbue the system with the ability to detect decline and adapt the interface accordingly.

It should also be noted that the medicalisation of ageing and old age results in negative societal images, perspectives, as well as low personal expectations. As discussed by Tullee (p. 2) [44], this "discourse of decline" is used to "legitimate the social and economic marginalisation experienced by older people." The International Classification of Functioning, Disability and Health [45] provides a more holistic model, placing importance on functioning and level of impairment, and encompassing the physiological impairment (body functions and structure) and its impact on activities and participation, while recognising that environmental and personal factors (contextual factors) also influence functioning and disability.

8.5 Person–environment interaction design guidelines

This section identifies best-practice guidelines to be considered when designing for person–environment interaction. This is by no means an exhaustive list, and as the research in this area grows it is expected that more specific recommendations based on comprehensive evaluation studies will continue to emerge.

8.5.1 Understanding context of use

Due to the breadth of the types of information that needs to be communicated between the system and user, and the accessibility issues resulting from co-occurring age-related impairments, it is vital that the starting point for designing any interactive AAL system is gaining an in-depth understanding of who is going to use the system and how it will be integrated into their everyday life. New design concepts and innovative ways of deploying the existing technologies will lead to systems being appropriated in a variety of forms that can only emerge through longer term use and understanding of the technology. Therefore, early and continuous user-centred studies based on an iterative prototyping design methodology are crucial to understanding and addressing barriers, constraints and opportunities for making the interaction meaningful.

The various elements of the context of use include the physical environment within which the interaction will take place as well as the characteristics of the person, or people, involved in the interaction. There will also be temporal dimensions to both these aspects. The physical environment might have different noise, light and occupancy levels for instance, that could vary throughout the day or on different days of the week or year. Also physical and psychological characteristics of the person could vary, such as their fatigue and concentration levels, based on the time of day or activities with which they were engaged. It is gaining an understanding of these more subtle aspects and variations that often provide the level of information needed for successful interaction. Holtzblatt [46] defined Contextual Design as a step-by-step process for collecting field data using techniques such as contextual inquiry, interpretation sessions for creating a shared understanding of the gathered data and developing work models that reveal the structure of the activities that people engage in. Work models as described by Holtzblatt are diagrams that represent aspects such as sequences of steps comprising activities, people's responsibilities and communication and coordination with others carried out as part of the activities that they engage in, cultural and social values influencing their behaviours, the physical layout of their environment and the constraints that it might impose, as well as artefacts that are used. All of this information need to be consolidated to create a coherent representation of the users and their needs and activities. Typical outcomes are affinity diagrams, which help to capture and bring together insights and issues across all users, helping the design team to identify common themes while maintaining individual variations, as well as generating personas that encapsulate core user characteristics and experiences, with a view to keeping the design scope focussed.

Additionally, methods such as storyboarding [47] can help to illustrate and highlight key elements of the user's interaction as situated in the environment, depicting the physical and/or virtual embodiment of the system and the user's interactions with, and reaction to, the system elements. These methods can be invaluable for the design team, providing unique insights for making decisions about suitable modalities for interaction and the scope of the interaction.

8.5.2 Usability guidelines and principles

Successful Person–Environment Interaction will have many dimensions. These include, the usability of the interaction interface(s) or technologies, overall user experience and whether the interaction is achieving the desired effect in the context of the AAL scenario. It is also important to consider how these elements change over time, given changes in the person's conditions and corresponding AAL needs.

The field of HCI provides a wealth of research on usability heuristics for interaction design. Heuristics are usually derived from empirical research in particular contexts and while re-using these requires careful consideration of generalisability to other contexts, they provide valuable insights for designers. Well-established sets of general usability heuristics and principles have been proposed by usability experts such as Nielsen [48], Norman [49] and Shneiderman [50].

Key design principles put forward by Norman include providing a good conceptual model, where it is clear to the user as to how the system is responding to their explicit or implicit inputs. People establish a mental model of how the system operates from using visible aspects of the system's features, which include affordances that provide discernable clues for how interaction elements should be manipulated, constraints that only allow user actions that are correct and accepted, and natural mapping which takes advantage of physical analogies and cultural standards. When these aspects are not intuitive, then the ability of the user to learn how to use the system is negatively impacted. This is particularly noticeable in person–environment interactions where more unfamiliar or novel forms of interaction are employed. For example, in a study on the intuitive learnability of touch gestures for technology-naïve older adults, Mihajlov *et al.* [51] found that a more complex gesture, which involved a specific rotation movement using two fingers, was harder to learn for the older participants in their study. Gesture-based interaction often lacks visible signifiers that clearly indicate or provide clues as to the appropriate gesture for a given action. Visibility of available control actions and system status, and corresponding meaningful feedback, are vital elements to enable usable interaction. Nielsen identified the need for employing consistency and standards for interaction control elements as crucial determinants of good usability.

Shneiderman *et al.* [52] explore the use of innovative approaches for interactive information visualisation that enable presentation of multivariate data in formats which prompt people to actively explore the data to elicit meaningful relationships between elements and make comparisons to support decision making. Carenazzi *et al.* [53] identify inclusive design principles for environment to human, and human to environment, communication. These include making the information *perceivable*, *understandable* and *actable*, for all sensory abilities and ambient conditions. It is vital for interaction designers to work alongside healthcare professionals to understand the salient parameters and aspects of the data that can be used to promote better health awareness and health behaviours in people, and also identify what elements will provide valid signifiers of progress to provide encouragement and engagement over time.

8.5.3 User experience

Oli and Benyon [54] defined a useful user-centred design (UCD) reporting template as a means of capturing and formalising outputs from a UCD process that helps to articulate key contextual elements of the user interface and interaction. They also include a section on stakeholder interests and expectations, as well as goals to be achieved. These are critical aspects of the user experience that need to be considered as they dictate long-term emotional engagement with, and acceptance of the technology [55]. Eliciting and understanding these is facilitated by the use of creative methods, such as cultural probes [56] which when used alongside other requirements elicitation methods, can help designers gain deeper insights into people's real-world routines and aspirations. This can provide valuable information to ensure that individual preferences can be considered that enable richer user experiences.

In addition, elements such as the aesthetics of the technology and the pleasure and satisfaction that people get from using it, are equally important attributes that need to be considered. Carenazzi *et al.* [53] also highlight the importance of comfort and pleasure in the interaction experience. Donald Norman [57] in his book on emotional design discusses the importance of making things attractive to promote a better user experience. He argues that attractive things make people feel good, which results in a positive emotional state, where we are more open to explore and solve problem creatively, and hence not become overwhelmed when we are not able to find a solution immediately. He links this to the three level of human information processing, the automatic prewired layer – the visceral level; the processes that control everyday behaviour – the behavioural level; and the contemplative part of the brain – the reflective level. At the visceral level, certain conditions associated with warm comfortable places, bright-saturated hues, harmonious sounds, and smooth and sensuous feelings and shapes, produce an automatic positive emotional affect. At the behaviour level it is important to attend to functionality, understandability, usability and physical feel, which results in something being intuitively easy to use. Norman postulates that at the reflective level, design is about culture and the meaning of a product or its use, which evokes or corresponds to personally identifiable values or meaning. If we find something corresponds with our own core values and motivations, then we are happy to continue to use it or even recommend it to others. As all these aspects are bound by our individual experiences and backgrounds, the only way designers can truly understand and appropriately respond is by engaging in a UCD process.

8.5.4 Privacy, trust and data security

For AAL technologies, privacy concerns can often underpin people's willingness to use these systems, particularly if facilitating person–environment interaction requires the collection and mining of large amounts of daily activity data. Person–environment Interaction systems often use video-based information to enable person and activity identification in an AAL context. In their study of 165 participants' views of video-based system for monitoring health data in the home, Ziefle *et al.* [58] reported insecurity about the trustfulness of these systems and fear of illegal data transfer and access, and personal data getting altered or lost due to technical problems.

Transparency of what data is being collected, and who has access to it, should be clearly communicated to the user, including easily accessible options to change settings relating to these. If allowing the user to access and change system settings impacts on its performance, then this could compromise the effectiveness of the system and ensuring continued safe and ethical operation might need to be assessed, with alternative options being made available. Padilla-Lopez *et al.* [59] propose a level-based visualisation scheme to work with video-based intelligent monitoring systems, using a number of visualisation modes which focus on protecting identity and appearance.

Memon *et al.* [60] conducted a comprehensive review of the AAL field with a focus on frameworks, platforms, standards and quality attributes. They found that amongst other factors, data accuracy, cost, security and privacy were the major challenges for AAL systems which result in unacceptable installation and deployment complexity, security threats and poor user experience. They identified a number of standards such as IEEE 11073, HL7, ISO-13485 and IEC 60601 for exchanging data; however they found that researchers working in silos and using closed and proprietary protocols for data exchange are compromising the deployment of these systems in real-world scenarios. In order to enhance user acceptance of AAL systems, there needs to be more effort in employing and refining existing standards. At the same time effort also has to be put in developing the interfaces that enable users to understand and be confident about the use and security of their data in order to engender trust in the systems.

8.5.5 Evaluating effectiveness of person–environment interaction in AAL systems

Clear definition of system requirements and understanding of context of use at the outset of developing the technology results in the identification of outcome measures to evaluate the effectiveness of the system and the user's interaction with it. Additionally, an iterative prototyping process, which includes a continuous process of user-based evaluation, also helps to refine and understand the scope of what "effectiveness" means from the user's perspective.

The overarching focus of AAL systems is to facilitate independent active living, health and wellness. The concept of wellness itself is highly complex; Strout *et al.* [61] consider W. Hettler's "Six dimensions of Wellness Model" which is useful for providing high-level criteria. These dimensions include occupational, social, intellectual, physical, emotional and spiritual wellness. Of these, physical wellness, which is defined as "commitment to self-care through regular participation in physical activity, healthy eating and appropriate health care utilisation" and emotional wellness, defined as "ability to acknowledge personal responsibility for life decisions and their outcomes with emotional stability and positivity" could be useful markers when gauging effectiveness from a user's perspective.

Due to the complexity of integrating a prototype system into a real-world environment, it is not always possible for users to fully experience the system in order to able to provide information for ecologically valid evaluation studies. As stated earlier, the nature of what constitutes effectiveness is closely linked to the AAL application area, and the benefits and impact of the interventions are not always realised for weeks, months and even years. This makes it very difficult to define specific, measurable and realistic criteria for evaluating against system objectives that go beyond general usability criteria.

In a bid to identify a comprehensive approach to evaluation that addresses the complexities of evaluating effectiveness, it is useful to draw on complementary fields of research that provide additional perspectives. Shea *et al.* [62] review universal design evaluation methods and identify four categories of Universal Design evaluations that include (1) checklist evaluations, (2) value-driven evaluations, (3) holistic evaluations and (4) invisible evaluations. Checklist evaluations, in relation to evaluating effectiveness of person–environment interaction in AAL systems, would include drawing on specific aspects that determine usability of the interaction technology and interfaces, when considered in relation to the person's particular impairments. Usability evaluation methods include Heuristic Evaluation [48] and methods such as Cognitive Walkthroughs [63]. Value-driven evaluations focus on more user experience goals and aspects relating to Hettler's wellness dimensions. These can be assessed via obtaining feedback from the user in a number of ways, including questionnaires, interviews or workshops. Additionally, other approaches such as getting the user to keep activity journals or logs could be used, as well as using measures that are indicative of physical wellness. Holistic evaluations, measure both usability and user experience attributes, and also incorporate methods relating to observational studies that evaluate interaction in-situ, using a cross-section of normative sources and feedback from users, as well as professional domain experts. Invisible designer evaluations represent a reflective and self-evaluative approach adopted by the designer, which is part of their professional practice, often emerging from incorporating multiple perspectives that have been elicited via co-design and participatory design processes.

8.6 Conclusion

Designing AAL technology to prolong and improve a healthy lifespan requires understanding the impact the ageing process has on a person's ability to interact with the technology. This involves working closely with end-users as part of a UCD process that is situated in the environments in which the technology will be used. It also requires designers to work with stakeholders from healthcare to understand what interventions work best and what information people being supported, and those supporting them, need to access. Effective interaction also extends to understanding cultural and socio-economic contexts. As highlighted by Virone *et al.* [64], we therefore also need to be mindful of ethical aspects of information sharing and the consequent redefining of family and social relationships, roles and responsibilities. Developing person–environment interaction strategies that guide and support people to make improvements need a nuanced approach, not only taking into account individual impairments, but also providing the appropriate information in a way that will be both engaging and persuasive, and becomes a natural and integral part of their everyday experience. As we seek to empower people to take charge of their own health, responding to information that they might never have had access to before, we also have to be aware of the responsibility and potential stress and worry living with an AAL system could create.

So while it is important for a designer of assistive technology to respond to specific accessibility needs to ensure there are no barriers to its use, it is vital to also focus on people's ability and aspirations, rather than just disabilities and constraints, as understanding people's desires and motivations can lead to more positive, innovative and creative design solutions.

References

[1] Weiser M. Open house. Interactive Telecommunications Program of New York University, Review 2, 1996.

[2] Weiser M. Ubiquitous computing. Computer. *IEEE Computer Society*. 1993; 26(10):71–72.

[3] Poslad S. *Ubiquitous computing: smart devices, environments and interactions*. John Wiley & Sons, 2009.

[4] Sadri F. Ambient intelligence: a survey. *ACM Computing Surveys (CSUR)*. 1 Oct 2011;43(4):36.

[5] Rashidi P, Mihailidis A. A survey on ambient-assisted living tools for older adults. *Biomedical and Health Informatics, IEEE Journal of*. 2013 May; 17(3):579–90.

[6] Klasnja P, Consolvo S, McDonald D W, Landay J A, Pratt W, Using mobile & personal sensing technologies to support health behavior change in everyday life: lessons learned. In *AMIA Annual Symposium Proceedings*, American Medical Informatics Association, 2009, 338.

[7] Fiorini L, Caleb-Solly P, Tsanaka A, Cavallo F, Dario P, Melhuish C, The efficacy of "busyness" as a measure for behaviour pattern analysis using unlabelled sensor data – a case study. In *IET International Conference on Technologies for Active and Assisted Living – TechAAL, UK*, 2015.

[8] AALIANCE2 project report. ALLIANCE2 Roadmap, http://www.aaliance2.eu/sites/default/files/files_list/AA2_D2.3_Roadmap_rev4.9_20130422_0.pdf, 2013, public deliverable.

[9] McNeill D. *Gesture and thought*. University of Chicago Press, USA, 2008.

[10] Nintendo Wii Remote, https://www.nintendo.co.uk/Wii/Accessories/Accessories-Wii-Nintendo-UK-626430.html, date accessed 12 Dec 2015.

[11] Premaratne P. *Historical development of hand gesture recognition. Human computer interaction using hand gestures*. Springer Singapore, 2014, 5–29.

[12] Porta M. A study on text entry methods based on eye gestures. *Journal of Assistive Technologies*. 2015;9(1):48–67.

[13] Grgurić, A., Mošmondor, M., Kušek, M., Stocklöw, C. and Salvi, D. Introducing gesture interaction in the Ambient Assisted Living platform universal. In *Telecommunications (ConTEL) 12th International Conference on IEEE*, 2013.

[14] López G, Quesada L, Guerrero LA. A gesture-based interaction approach for manipulating augmented objects using leap motion. In *Ambient Assisted Living. ICT-based Solutions in Real Life Situations,* Springer International Publishing, Switzerland, 1 Dec 2015, 231–243.

[15] Boyali A, Naohisa H, Osamu M. Spectral collaborative representation based classification by circulants and its application to hand gesture and posture recognition from electromyography signals. In *Proceedings of the International Conference on Image Processing, Computer Vision, and Pattern Recognition (IPCV)*, (WorldComp), 2015.

[16] Apple iPhone 6s http://www.apple.com/uk/iphone-6s/3d-touch/, date accessed: 1 Dec 2015.

[17] Carrasco E, Loyo E, Otaegui O, *et al. Autonomous navigation based on binaural guidance for people with visual impairment. Assistive technology: from research to practice*. IOS Press, Netherlands, 2013.
[18] El Saddik A, Orozco M, Eid M, Cha J. *Haptics technologies: bringing touch to multimedia*. Springer Science & Business Media, 15 Sep 2011.
[19] Tanikawa T, Hirose M. Interaction of olfaction with vision or other senses using olfactory display. *Human Olfactory Displays and Interfaces: Odor Sensing and Presentation*. 2013:86.
[20] Dumas B, Lalanne D, Oviatt S. Multimodal interfaces: a survey of principles, models and frameworks.In *Human machine interaction*, Springer Berlin Heidelberg, 2009, 3–26.
[21] Friedenberg J. Visual attention and consciousness. *Psychology Press*, NY, 2012.
[22] Horn G, Hinde RA. Short-term changes in neural activity and behaviour: a conference sponsored by King's college research centre. Cambridge: CUP Archive, 1970.
[23] Kim S, Wogalter MS. Habituation, dishabituation, and recovery effects in visual warnings. In *Proceedings of the Human Factors and Ergonomics Society Annual Meeting*, Sage Publications, 1 Oct 2009;53(20):1612–1616.
[24] Streitz NA, Röcker C, Prante T, Stenzel R, van Alphen D. Situated interaction with ambient information: facilitating awareness and communication in ubiquitous work environments. In *Tenth International Conference on Human-Computer Interaction (HCI International 2003)*, 22 Jun 2003.
[25] Schmidt, A. Implicit human computer interaction through context. *Personal Technologies*. 2000;4(2–3):191–199.
[26] Blasco, R., Marco, Á., Casas, R., Cirujano, D. and Picking, R. A smart kitchen for ambient assisted living. *Sensors*. 2014;14(1):1629–1653.
[27] Ju W, Leifer L. The design of implicit interactions: making interactive systems less obnoxious. *Design Issues*. 25 Jun 2008;24(3):72–84.
[28] Surie, D., Janlert, L.E., Pederson, T. and Roy, D. Egocentric interaction as a tool for designing ambient ecologies—the case of the easy ADL ecology. *Pervasive and Mobile Computing*. 2012;8(4):597–613.
[29] Norman DA. Cognitive engineering. User centered system design: New perspectives on human-computer interaction, 1986; Chapter 3: 31–61, CRC Press.
[30] Still JD, Dark VJ. Cognitively describing and designing affordances. *Design Studies*. 2013;34(3):285–301.
[31] López-Otín, C., Blasco, M.A., Partridge, L., Serrano, M. and Kroemer, G. The hallmarks of aging. *Cell*. 2013;153(6):1194–1217.
[32] Schneider E, ed. *Handbook of the biology of aging*. Elsevier, USA, 2012.
[33] Margrain TH. *Sensory impairment. The Cambridge Handbook of age and ageing*. Malcolm L. Johnson, ed. Cambridge: Cambridge University Press, 2005.
[34] Gordon-Salant S. Hearing loss and aging: new research findings and clinical implications. *Journal of Rehabilitation Research and Development*. 2005;42(4):9.

[35] Meyer C, Hickson L. What factors influence help-seeking for hearing impairment and hearing aid adoption in older adults? *International Journal of Audiology*. 1 Feb 2012;51(2):66–74.

[36] Campbell Li Y. Expressivity in interactive speech synthesis; some paralinguistic and nonlinguistic issues of speech prosody for conversational dialogue systems, in *Speech prosody in speech synthesis: modeling and generation of prosody for high quality and flexible speech synthesis*. Springer Berlin Heidelberg, 2015, 97–107.

[37] Carson AJ. What brings you here today? The role of self-assessment in help-seeking for age-related hearing loss. *Journal of Aging Studies*. 2005;19(2): 185–200.

[38] Lin, F.R., Yaffe, K., Xia, J., *et al*. Hearing loss and cognitive decline in older adults. *JAMA Internal Medicine*. 2013;173(4):293–299.

[39] Schleber F. *Aging and the senses. Handbook of mental health and aging*, Vol. 251, Academic Press, Inc, London, 2013.

[40] Fisk, A.D., Rogers, W.A., Charness, N., Czaja, S.J. and Sharit, J. *Designing for older adults: principles and creative human factors approaches*. CRC press, USA, 2009.

[41] Jacko JA., ed. *Human computer interaction handbook: fundamentals, evolving technologies, and emerging applications*. CRC press, Taylor & Francis Group, USA, 2012.

[42] Kenny RA. Mobility and falls, in Johnson M., ed. *The Cambridge Handbook of age and ageing*. Cambridge, UK: Cambridge University Press, 2005.

[43] Son, M., Jung, H., Yang, H. and Park, W. Evaluating physical discomfort associated with prolonged, repetitive use of gesture: a comparison of subjective rating protocols. In *Proceedings 19th Triennial Congress of the IEA*, Vol. 9, 2015.

[44] Tulle E. *Ageing, the body and social change: running in later life*. Basingstoke, UK: Palgrave Macmillan, 2008.

[45] World Health Organization. *International classification of functioning, disability and health: ICF*. World Health Organization, 2001.

[46] Holtzblatt K. Contextual Design, in Jacko, Julie A., ed. *Human Computer Interaction Handbook: Fundamentals, Evolving Technologies, and Emerging Applications*. CRC press, USA, 2012.

[47] Truong KN, Gillian RH, Gregory DA. Storyboarding: an empirical determination of best practices and effective guidelines. In *Proceedings of the 6th Conference on Designing Interactive Systems*. ACM, 2006.

[48] Nielsen J. Heuristic evaluation. *Usability Inspection Methods*. 1994;17(1): 25–62.

[49] Norman DA. *The design of everyday things: revised and expanded edition*. Basic books, New York, 2013.

[50] Shneiderman B, Plaisant C, Cohen M, Jacobs S. *Designing the user interface: strategies for effective human-computer interaction*, 5th Edition, Pearson, USA, 2013.

[51] Mihajlov M, Law EL, Springett M. Intuitive learnability of touch gestures for technology-naïve older adults. *Interacting with Computers*. 1 May 2015;27(3):344–356.

[52] Shneiderman B, Plaisant C, Hesse BW. Improving healthcare with interactive visualization. *Computer*. 2013 May;46(5):58–66.

[53] Catenazzi, N., De Luca, V., Sommaruga, L. and Botta, M. Guidelines to design inclusive ambient intelligence solutions for human activity sharing. In *Complex, Intelligent and Software Intensive Systems (CISIS), 2012 Sixth International Conference on. IEEE*, 2012.

[54] Mival O, Benyon D. *User experience (UX) design for medical personnel and patients. Requirements engineering for digital health*. Springer International Publishing, 2015, 117–131.

[55] Beringer, R., Sixsmith, A., Campo, M., Brown, J. and McCloskey, R. *The "acceptance" of ambient assisted living: developing an alternate methodology to this limited research lens. Toward useful services for elderly and people with disabilities*. Springer Berlin Heidelberg, 2011, 161–167.

[56] Caleb-Solly P, Flind A, Vargheese JP. Cameras as cultural probes in requirements gathering—exploring their potential in supporting the design of assistive technology. In *Computer-Based Medical Systems (CBMS), 2011 24th International Symposium,* 27 Jun 2011, 1–6.

[57] Norman DA. *Emotional design: Why we love (or hate) everyday things*. Basic books, New York, 2004.

[58] Ziefle M, Röcker C, Holzinger A. Medical technology in smart homes: exploring the user's perspective on privacy, intimacy and trust. In *Computer Software and Applications Conference Workshops (COMPSACW), 2011 IEEE 35th Annual IEEE*, 2011.

[59] Padilla-López JR, Chaaraoui AA, Gu F, Flórez-Revuelta F. Visual privacy by context: proposal and evaluation of a level-based visualisation scheme. *Sensors*. 4 Jun 2015;15(6):12959–12982.

[60] Memon, M., Wagner, S.R., Pedersen, C.F., Beevi, F.H.A. and Hansen, F.O. Ambient assisted living healthcare frameworks, platforms, standards, and quality attributes. *Sensors*. 2014;14(3):4312–4341.

[61] Strout KA, Howard EP. The six dimensions of wellness and cognition in aging adults. *Journal of Holistic Nursing*. 2012;30(3):195–204.

[62] O Shea EC, Pavia S, Dyer M, Craddock G, Murphy N. Measuring the design of empathetic buildings: a review of universal design evaluation methods. *Disability and Rehabilitation: Assistive Technology*. 2 Jan 2016;11(1):13–21.

[63] Mahatody T, Sagar M, Kolski C. State of the art on the cognitive walkthrough method, its variants and evolutions. *International Journal of Human–Computer Interaction*. 30 Jul 2010;26(8):741–785.

[64] Virone G, Sixsmith A. Toward information systems for ambient assisted living. In *The 6th International Conference of the International Society for Gerontechnology, Pisa, Italy*, 4 Jun 2008, 4–7.

Chapter 9
Data analytics for enabling connected health

Sanjeev Naguleswaran[1], *Kylie Wall*[1], *and Karen Grimmer*[2]

9.1 Introduction

Telehealth is an umbrella term used for health-related activities that occur remotely by employing information communication technologies. It embodies diagnosis, treatment, educational, caring and curative aspects of healthcare for holistic management [1]. The benefits of telehealth extend beyond saving costs and improving access to health services for individuals living in non-metropolitan areas. The benefits include the ability to augment face-to-face service delivery models with health information obtained from everyday living. The ability to obtain real-time data reflecting everyday living can be used to improve healthcare services by increasing the volume, variety and velocity of data to clinical management [2].

There is an increasing awareness of the need to collect and analyse multiple sources of data to demonstrate the health value and utility of telehealth interventions [3]. Patient outcomes, biological and functional markers of health status, health provider outcomes and cost have been reported in large-scale trials, with convincing evidence of positive effects of telehealth interventions for a range of conditions, using various health and satisfaction outcomes [4–6]. In addition to providing interventions for health conditions, telehealth technology can be applied to support Ambient Assisted Living, to enable vulnerable individuals (e.g. the elderly) live safely in their own homes for longer [7]. Technology to support ambient assisted living requires a more holistic approach, which broadens the information sourced for health conditions to meaningful information of daily living [8]. Individual's social, environmental, behavioural and emotional concerns need to be considered. Telehealth technology can be used to measure changing daily behaviours, loneliness, loss of appetite, compromised sleep, forgetfulness, reduced physical activity and loss of interest in activities previously pursued and enjoyed [9, 10]. Thus, new data items and advanced analytics will be needed for reporting holistic, Ambient Assisted Living information.

The main communication technologies used to capture daily activity data are 'external' and 'wearable' sensors. External sensors are commonly used for Ambient

[1]QSpectral Systems, Brisbane, Australia
[2]International Centre for Allied Health Evidence (iCAHE), University of South Australia, Adelaide, Australia

Assisted Living, particularly in Smart Homes [11]. Algorithms are applied to generate information about a persons' daily living by obtaining data from sensors, which can be attached to household furnishings, doors, walls or appliances (e.g. stove, refrigerator and faucets). Expertise to select appropriate sensors, process the raw sensor data and apply data analytics to generate valid and reliable health information is essential [12].

The alternate wearable sensors can be paired with wristbands, smart watches and clothing items (e.g. shoes). Sensors can also be 'worn' indirectly (e.g. carrying a smartphone). Real-time data are sourced from movement (e.g. GPS, accelerometers), environmental conditions (e.g. humidity, temperature) or physiological signals (e.g. heart rate, electrocardiogram). To illustrate, Dobkin [13] used pattern-recognition algorithms in a patient population to describe the type, quantity and quality of mobility-related activities in the community. He noted that remote access to continuous data concerning walking speed, duration and distance, gait asymmetry and smoothness of movements, exercise and skills practice can enable tailored therapies with performance feedback. Thus, activity data obtained from wearable sensors are not restricted by external, fixed sensors with a determinate range. Wearable sensors can encompass daily activity information that occurs outside the home.

Shared decision-making between individuals and healthcare providers is an important consideration for telehealth applications and acceptance [14, 15]. Engaging end-users ethically and respectfully requires substantial investment in the processes of shared decision-making, particularly as the success of telehealth requires acceptance and compliance with the technology (for instance to enhance safety, security, surveillance, quality of life and healthy behaviours). Brear [16] noted a decade ago of the challenges to be faced regarding the development of suitable data collection methodologies, as well as the multiple data items required to collect appropriate, comprehensive telehealth information. These concerns continue, as the need to collect appropriate, comprehensive and accurate data to evaluate the totality of effect of any new health technology is a challenge widely reported in the health implementation literature [17].

The recent advent of data analytics and context modelling provides solutions to aggregate the data from these disparate sensors in a meaningful manner. Ambient Assisted Living requires analytics on real-time streaming data. Constructing data analytics that is scalable, reliable, satisfies functional needs and considers latency requirements is necessary for this purpose. Applying streaming and learning algorithms to discover patterns will have more precise predictive power and timely alerts. Data can be obtained from multiple sources and aggregated to advance predictive, preventative and personalised health information for patients and healthcare providers.

This chapter aims to demonstrate that applying novel data analytics algorithms can improve the validity, reliability and predictability of human activity and healthcare information. Capturing real-time data and aggregating data attributes from multiple sources will also improve the connectivity between health technology and healthcare delivery systems.

9.1.1 Current situation

Personal alarms are a well-established form of monitoring in primary care that aims to provide safety assurances for vulnerable people living at home. They have been available for many years, offering a variety of supports that range in costs [18]. There are many limitations to personal alarms that are primarily associated with the alarms having to be fixed to individuals at all times. Thus, monitoring and safety is compromised during some daily routine activities, such as bathing and dressing where the monitoring device is often removed. Also, responsibility to activate the alarm is dependent on the individual wearing it. Thus, an alarm may not be able to be reached following a heavy fall or if an individual suffers an event that renders them unconscious. Furthermore, individuals with cognitive impairments may not activate their alarm when it is most needed.

The reliability of personal alarms has been challenged, due to the high incidence rate of false alarms. For example, the Ambulance Call Direct version in Australia has resulted in costly and unnecessary ambulance call outs [19]. Whilst the ambulance service generally (and generously) perceives these call outs as a community service to ‘check’ on the person who activated the alarm inadvertently, the cost has to be considered. The existing technology provides only partial, safety reassurance.

9.1.2 The future of healthcare delivery

Provision of efficient and effective services in the community, using holistic primary care approach that takes into account the needs, priorities and preferences of individuals is the way of the future in healthcare. The consequences of deconditioning as a result of tertiary hospital admissions, subsequent ongoing illness or unmanaged chronic diseases pose significant financial burdens to world economies. Tertiary hospital care and re-housing the elderly into residential care is significantly more expensive than living at home safely, with appropriate and reliable healthcare supports.

It is apparent that an autonomous, remote monitoring system is necessary where healthcare providers can monitor and engage remotely with individuals. The outcomes of current research efforts in this area will contribute significantly to satisfying the demand for better, more comprehensive and preventive healthcare. Affordable and effective healthcare will see a departure from centralised site-based healthcare services (provided in doctors’ offices, clinics, hospitals etc.), to ubiquitous, pervasive health monitoring in daily life situations to facilitate early intervention [20, 21]. Rapid technological advances have resulted in increasing computational performance in very small form factors (Figure 9.1). This has resulted in the possibility of on-body/around-body networks of miniscule wirelessly connected computing devices becoming a reality. The flexibility and potential for integration of sensor networks has led to a number of published studies [22–24].

There is a need for developing advanced analytics to effectively aggregate disparate sensor data in order to distinguish between normal activity and conditions and events that require intervention. The responsibility of activating alerts or interventions should be transferred from the user and rely on advanced technology to extract and

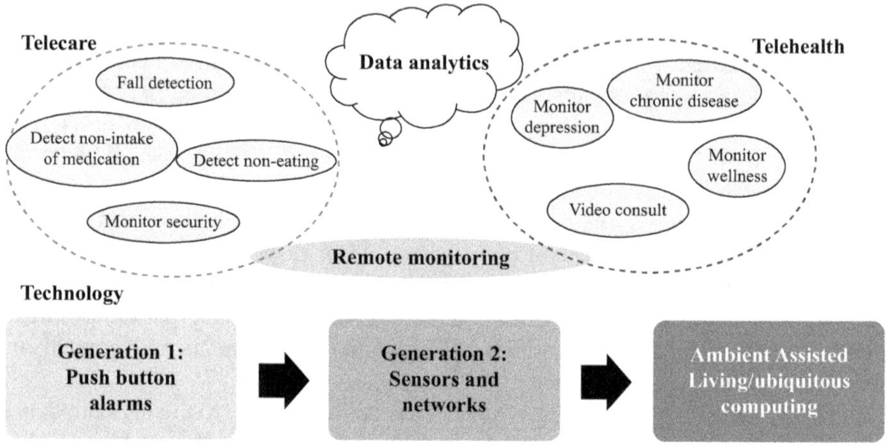

Figure 9.1 Remote monitoring enables healthcare professionals to monitor vulnerable individuals using various technological devices

report the necessary information. Telehealth should be designed to support Ambient Assisted Living, using a holistic management approach to living safely and independently [25]. This requires novel analytic algorithms that collate data from a variety of sources to provide more meaningful, reliable and comprehensive information for individuals and healthcare providers. In this chapter, we use data from typical daily living activities to demonstrate how novel analytics algorithms can improve on the current monitoring and alert capabilities. We present a decision support system based on the disparate data obtained from sensors. Data analytics algorithms are used to analyse the data in order to provide reliable insights and predictive capability. This enables appropriate alerts and flags to be raised to meet individual health requirements and permits the precision of computerised outcome measures. Additionally, the performance of the underlying sensor and communication network can be measured in different scenarios, enabling customised network configuration, which can be compared with evidence-based practice and key performance indicators to demonstrate reliability and robustness of the sensors.

9.2 Data analytics

To achieve reliable autonomous remote monitoring we need to effectively combine various sensors and associated technology. Therefore, the challenges are more complex than obtaining information from single sensor systems. This section describes activities in an exemplar dwelling scenario and how monitoring technology can be used to characterise those activities and by extension, risk situations. An extrapolation to more realistic scenarios, where the interplay of sensors on and around the body (e.g. fall detectors, personal alarms and blood pressure monitors), with other technology (e.g. flood detection and gas shut off solutions) together with the gateway communications and alert methods, can be incorporated within the methodology.

9.2.1 Machine learning

Machine Learning (ML) is a component Artificial Intelligence (AI) that focuses on learning from data. It primarily addresses the interface between learning, complexity and computation. Research in this area spans many theoretical topics as well as practical applications. ML concerns the extraction of insights from data, developing predictions to permit data-driven decision-making. Algorithms that constitute ML fall into two broad categories, supervised and unsupervised learning. Supervised learning is applicable when training data is available and typically most classification and regression algorithms fall within this category. Decision Trees, Neural Networks Support Vector Machines (SVM) and Bayesian networks can also be used under supervised learning. In the case of unsupervised learning clustering algorithms, such as k-means, clustering can be used to derive insights into the nature of the data. In practice, other categories, such as semi-supervised learning are encountered where the boundary between algorithm types is blurred, resulting in hybrid fit-for-purpose algorithms using any of the available approaches in an appropriate manner. ML theory is also closely connected to statistical inference where a basis function is applied to combine the features of a problem in order to infer the response for the subset of data used for training. Further details on ML can be found in References [26, 27].

The advent of more available computational resources has seen ML deployed in a number of areas including healthcare systems, where the data and computational power are harnessed to diagnosis and treatments. Decision trees are decision support tools based on a graph structure where the nodes are tests on an attribute of a problem. The attributes particular to sensor measurements could include chance of an event or measurement, measurement thresholds and the type of sensor. The decision tree learning can be used as a predictive model to map features or observations to a target value. In particular, a classification tree is defined as a decision tree where the target variable has a finite set of values (or classes). In contrast in a Regression Tree the target variable is continuous and real. Therefore, the resulting decision trees can be used to predict the value of a target variable based on several input variables [28].

A further enhancement to ML is the use of ensemble methods that take a number of simple models and combine them in a way to yield a final, overall picture. We present an ensemble technique based on Random Forest models for iteratively averaging multiple, deep decision trees, which are trained on different parts of the collected data, with the goal of reducing the variance. Each iteration creates a simple decision tree on randomly selected subsets of input variables and input data. The final result is the aggregation of all such trees [29].

9.3 Active daily living

To develop and evaluate methods for recognising data measured during the course of a sample period of activity, we used the Public Domain Dataset for Human Activity

Recognition Using Smartphones [30, 31]. This dataset was downloaded from the University of California, Irvine machine Learning Repository [32]. The experiments were carried out with a group of 30 volunteers aged from 19 to 48 years. Each person performed six activities (walking, climbing upstairs, sitting, standing and lying down) wearing a smartphone (Samsung Galaxy S II) on their waist. The experimenters captured 3-axial linear acceleration and 3-axial angular velocity at a constant rate of 50 Hz from the embedded accelerometers. The data were labelled through comparison with a video recording and randomly partitioned so that 70% of the participants generated the training data and 30% the test data. We first tested and customised the data pre-processing scripts accessed from the Matlab file exchange [33]. The raw accelerometer data is plotted in Figure 9.2 and it is immediately noted that lying down is already quite separated from the other activities. Walking and climbing stairs appear similar in all three accelerometer directions, while some discrimination between sitting and standing can be obtained by considering the data on different axes. It should be noted that this lack of immediate discriminatory capability arises when considering a single dataset from smartphone sensors. Combining data from other independent sources alleviates this problem.

9.3.1 Algorithm

ML methods that have been previously employed for recognition include Naive Bayes, SVMs, Threshold-based and Markov chains [34]. In particular, SVMs were used by the original experimenters [30] and also used in [35, 36]. In Reference 30 experiments were conducted using SVM binary classifiers, which are generalized to the multiclass case and using 10-fold Cross Validation as well as Gaussian kernels. The authors concluded that the use of ML techniques with the data acquired from smartphones in isolation provided an activity recognition performance that matched performance obtained by deploying a number of special purpose sensors.

After carrying out investigations where we tested the accuracy of various classifiers, we determined that ensembles of Decision Trees would provide a powerful method for identifying the states of Active Daily Living (ADL). Decision tree methods are invariant under scaling (i.e. more data inputs can be added over time and the attributes do not have to be normalised). In addition, various other transformations of inputs are robust to include irrelevant features, and produce transparent models for on-going analysis. For example, in building a reliable connected health model, data from multiple sensors, as well as historical patient records, need to be considered.

9.3.2 Feature engineering

The data obtained from the UCI dataset can be tabulated. The rows in a tabular dataset correspond to the measurement instances while columns correspond to attributes. These attributes can be used as features to solve the underlying prediction problem and they can also be manipulated to form more useful features that better represent the structure of the problem.

Feature Engineering is the process of transforming the data into a better representation of the problem so that the solution can be learnt. Knowledge or information

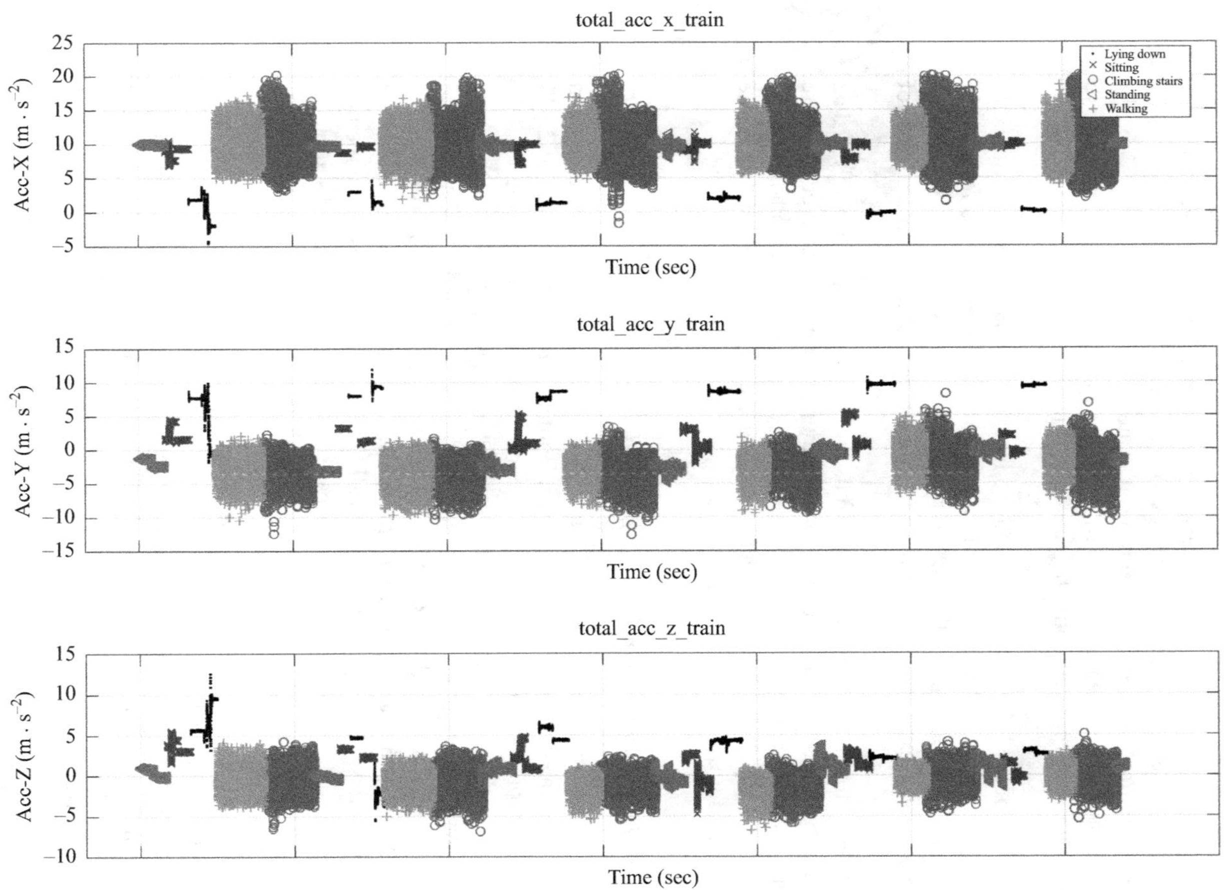

Figure 9.2 *Raw data showing each of the activities in a different shading and marking. The different activities are depicted in dark grey point for lying down, black cross for sitting, light black triangle for standing, light grey circle for climbing stairs and light grey cross for walking*

is introduced into the problem representation by this process. In general, the feature engineering process removes unnecessary and redundant attributes, creates new features and modifies feature types. These tasks can be accomplished through:

Binarisation – Boolean values through defining thresholds for numerical data
Discretisation – Convert continuous features to discrete features
Value transformation – Scaling and centring of values
Random Forest methods are not affected by scaling, and therefore, we do not have to consider this aspect due to our choice of algorithm.

In the case of the UCI dataset we used the fact that the data consisted of fixed-width sliding windows of 128 points and took the mean of these points for each of the measurements as one class of feature for each of the measurements. It was also possible to similarly extract the standard deviation of these points as another class of feature.

Principle Component Analysis (PCA) finds the underlying structure of the data by transforming or rotating the axes until the maximum variance is achieved or where the data are most spread-out. The new axes are the principle components. We used PCA on each of the sliding windows and obtained the mean of the first principle component axis for each of the measures.

9.3.3 Application of the ensemble method

The dataset with the new features was split into a training component for the algorithm to learn and a testing component for validation. Both components consisted of 2,947 data points with total accelerometer and gyroscope data from a smartphone.

The training data was parsed to the classification learner in Matlab™, where experiments with a number of classification methods were carried out. The Bagged trees ensemble method based on Random Forests was able to classify the labelled predicted data with 96.4% accuracy. This model was then used on the test dataset to predict the activities. The Bagged tree (also called bootstrapping) ensemble method is a model averaging method, where in the case of decision trees, an ensemble of a predefined number of trees is randomly selected for classification. A number of iterations are carried out with trees permuted in and out of the bag through random selection. The stability of the algorithm is improved while also reducing the variance without injecting too much bias through overfitting.

The confusion matrix showing the accuracy of prediction is shown in Figure 9.3. It can be seen from the figure that the model is 100% accurate at predicting the activity of lying down, but experiences confusion between the activities of walking and climbing stairs.

A further technique to measure the accuracy of the model using the Receiver Operational Characteristic (ROC) for the worst case walking activity, where the True Positive values are plotted against the False Positive values as given in Figure 9.4. The maximum area under the curve, of one, where the true-positive identification rises to one as fast as possible is the desired outcome, while a straight line through the origin

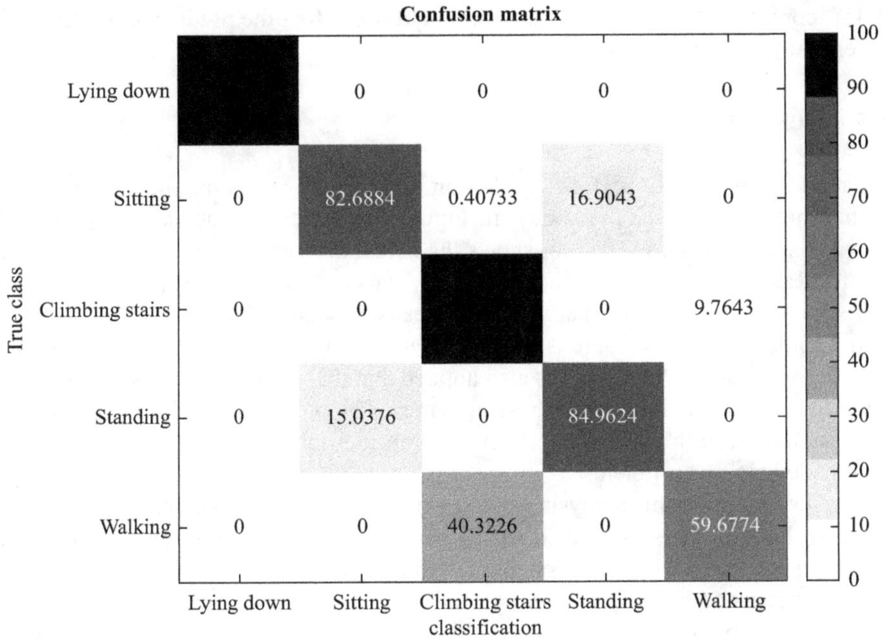

Figure 9.3 The confusion matrix depicting prediction accuracy of the model

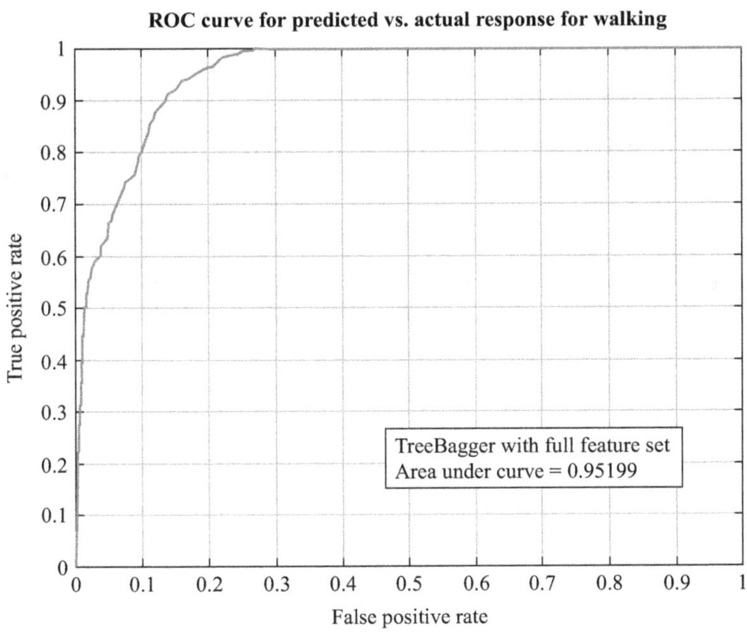

Figure 9.4 ROC curve showing that even in the worst case the area under the curve approaches unity

at 45°, corresponding to an area of 0.5, would mean that the predictive model is no better than randomly guessing an activity.

9.4 Results and analysis

The model was used to perform prediction tests on the test component of the data. A snapshot of the results is shown in Figure 9.5, where the actual and predicted activities are compared. The plot shows the actual activity as grey triangles and the predicted as dark grey circles. The lighter grey stars indicate accurate predictions. The more points covered by this latter marker means the better the prediction. As shown by the confusion matrix, there is some difficulty in discriminating between walking and climbing stairs. However, it is also noticed that this difficulty seems to exist only when there is a switch between these activities. The blending of additional external data, such as from a fixed motion detector, an inertial sensor or location detector, would alleviate this problem.

An extension of this analysis would be the use of anomaly detection methods to detect if the indicators associated with a particular activity for a given subject, are trending away from what has been learned to be normal. In the current example, the accelerometer and gyroscope readings may deviate from the normal measurements learned over a period of time. This would indicate a change in the subject's physical condition or with sufficient data fields an emergency event such as a fall or even predication of a condition about to occur requiring intervention.

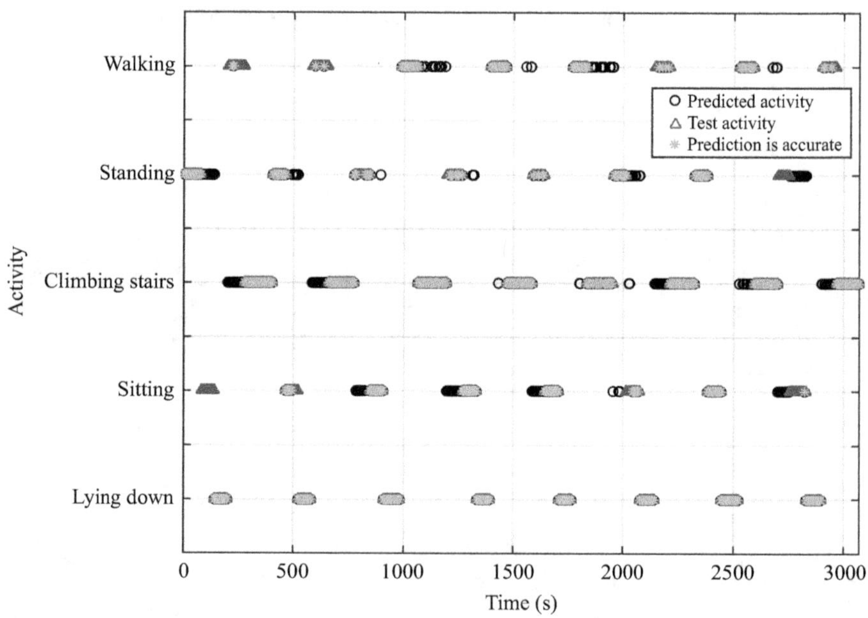

Figure 9.5 The plot shows the actual activity as triangles and the predicted as dark circles. The lighter starts indicate accurate predictions. The more points covered by this latter marker means the better the prediction

9.5 Conclusion

This chapter has shown that ML is a powerful method of identifying the activities of daily living. This has key implications for telehealth and Ambient Assisted Living in that the activities of a person at home can be monitored autonomously in order to provide a safe and secured environment.

Health-enabling technologies, and smart home technologies in particular, enable an individual to manage their own health, alleviating the effects of demographic change in society (such as the ageing population). In order to fully realise the health and wellness benefits of technologies that constitute Ambient Assisted Living, determining user requirements and developing a holistic approach to healthcare services should facilitate technology adoption. Therefore, devices need to function cooperatively rather than the single purpose monitoring regime of current devices. Health services need to be delivered using a broader spectrum of sources, ranging from traditional health providers to commercial providers.

The importance of being able to access disparate sources of data was highlighted in this example scenario by the fact that there was difficulty in accurate identification at the point of change between activities. As mentioned in the preceding discussion, other points of reference (or data dimensions) through acquisition of data from additional sensors would have alleviated the problem. In contrast to previously reported techniques, the method presented in this chapter based on Bagged decision trees, is able to seamlessly handle the addition of extra information and does not require extensive manipulation (such as scaling) of the features.

The potential application of data analytics by extending the methodology discussed in this chapter is immense. The incorporation of more ambient sensors into the dataset will provide extra data dimensions and context to identify other activities such as cooking, cleaning and gardening. The addition of anomaly detection techniques to the normal data for a particular person may improve safety as well as provide the potential for pre-emptive clinical intervention rather than reacting to an emergency event such as a fall. This would naturally reduce the demand on hospitals and other health infrastructure.

References

[1] Australian Department of Health and Aging Author Guide, http://www.health.gov.au/internet/main/publishing.nsf/Content/e-health-telehealth, accessed 12 September 2015.

[2] Vimarlund V, Wass S., Big Data, Smart Homes and Ambient Assisted Living. *Yearbook of Medical Informatics*. 2014, **9** (1), pp. 143–149.

[3] Dattakumar A., Gray K., Jury S., *et al.*, A Unified Approach for the Evaluation of Telehealth Implementations in Australia. Institute for a Broadband – Enabled Society, 2013, http://apo.org.au/node/35845, accessed 17 September 2015.

[4] Rosenberg C.N., Peele P., Keyser D., McAnallen S., Holder D., Results from a Patient-Centered Medical Home Pilot at UPMC Health Plan Hold

Lessons for Broader Adoption of the Model, *Health Affairs*, 2012, **31** (11), pp. 2423–2431.

[5] Thomason T.R., Hawkins S.Y., Perkins K.E., Hamilton E., Nelson B., Home Telehealth and Hospital Readmissions: A Retrospective OASIS-C Data Analysis. *Home Healthcare Now*, 2015, **33** (1), pp. 20–26.

[6] Steventon A., Bardsley M., Billings J., *et al.*, Effect of Telehealth on Use of Secondary Care and Mortality: Findings from the Whole System Demonstrator Cluster Randomised Trial, *British Medical Journal*, 2012, **344**, e3874.

[7] Morris M., Miller K., Santamaria N., *et al.*, Smart Technologies for Older People. Institute for a Broadband-Enabled Society, 2012, http://apo.org.au/node/29534, accessed 17 September 2015.

[8] Bastiaens H., Van Royen P., Pavlic D.R., Raposo V., Baker R., Older People's Preferences for Involvement in Their Own Care: A Qualitative Study in Primary Health Care in 11 European Countries. *Patient Education and Counseling*, 2007, **68** (1), pp. 33–42.

[9] Rockwood K., Mitnitski A., Limits to Deficit Accumulation in Elderly People. *Mechanisms of Ageing* and *Development*, 2006, **127** (5), pp. 494–496.

[10] Fried L.P., Ferrucci L., Darer J., Williamson J.D., Anderson G., Untangling the Concepts of Disability, Frailty, and Comorbidity: Implications for Improved Targeting and Care [review]. *Journal of Gerontology Series A: Biological Sciences* and *Medical Sciences*, 2004, **59** (3), pp. 255–M263.

[11] Demiris G, Skubic M, Rantz MJ, *et al.*, Facilitating Interdisciplinary Design Specification of 'Smart' Homes for Aging in Place. *Studies in Health Technology and Informatics*, 2006, **124**, p. 45.

[12] Ni Q., García Hernando A.B., de la Cruz I.P., The Elderly's Independent Living in Smart Homes: A Characterization of Activities and Sensing Infrastructure Survey to Facilitate Services Development. *Sensors*, 2015, **15** (5), pp. 11312–11362.

[13] Dobkin B.H. Wearable Motion Sensors to Continuously Measure Real-World Physical Activities. *Current Opinion in Neurology*, 2013, **26** (6), pp. 602–608.

[14] Wetzels R., Harmsen M., Van Weel C., Grol R., Michel W., Interventions for Improving Older Patients' Involvement in Primary Care Episodes. *Cochrane Database Systematic Reviews*, 2007.

[15] Schneider A., Körner T., Mehring M., Wensing M., Elwyn G., Szecsenyi J., Impact of Age, Health Locus of Control and Psychological Co-morbidity on Patients' Preferences for Shared Decision Making in General Practice. *Patient Education and Counseling*, 2006, **61** (2), pp. 292–298.

[16] Brear M., Evaluating Telemedicine: Lessons and Challenges. *Health Information Management Journal*, 2006, **35** (2), pp. 23–31.

[17] Hoomans T., Severens J.L., Economic Evaluation of Implementation Strategies in Health Care. *Implementation Science*, 2014, **9** (1), p. 168.

[18] My Aged Care [Internet] 2013 [updated 2013 July; cited 2013 May 20]. http://www.agedcareaustralia.gov.au/internet/agedcare/publishing.nsf/Content/Personal+and+medical+alarms. Accessed 15/05/2015.

[19] Woods M, Fitzgerald R, Macri S. Caring for Older Australians: Final Inquiry Report. Canberra 2011. ISSN 1447–1329; ISBN 978–1–74037–362–3.

[20] Bachmann C., Ashouei M., Low-Power Wireless Sensor Nodes for Ubiquitous Long-Term Biomedical Signal Monitoring. *IEEE Communications Magazine*. 2012, **50**, pp. 20–27.

[21] Wootton R., Realtime Telemedicine. *Journal of Telemedicne and Telecare*, 2006, **12** (7), pp. 328–336.

[22] Xiao Y., Takahashi D., Liu J., Deng H., Zhang J., Wireless Telemedicine and m-Health: Technologies, Applications and Research Issues. *International Journal of Sensor Networks*, 2011, **10** (4), pp. 202–236.

[23] Navarro K.F., Lawrence E., WSN Applications in Personal Healthcare Monitoring Systems: A Heterogeneous Framework. ETELEMED '10. *Proceedings of the 2nd International. Conference of eHealth, Telemedicine, and Social Medicine*, 10–16 February 2010. St Maarten, Netherlands, 2010. pp. 77–83.

[24] Cheng H., Zhuang W., Bluetooth-Enabled in-Home Patient Monitoring System: Early Detection of Alzheimer's Disease. *IEEE Wireless Communication*, 2010, **17** (1), pp. 74–79.

[25] Schulke A.M., Plischke H., Kohls N.B., Ambient Assistive Technologies (AAT): Socio-Technology as a Powerful Tool for Facing the Inevitable Sociodemographic Challenges. *Philosophy, Ethics, and Humanities in Medicine*, 2010, **5** (1), p. 8.

[26] Bishop C.M., *Pattern Recognition and Machine Learning*, New York: Springer, 2006.

[27] Hastie T., Tibshirani R., Friedman J. H., *The Elements of Statistical Learning*, New York: Springer, July 2001.

[28] Breiman L., Friedman J.H., Olshen R.A., Stone, C.J., *Classification and Regression Trees*, Monterey, CA: Wadsworth & Brooks/Cole Advanced Books & Software, 1984, ISBN 978–0–412–04841–8.

[29] Breiman L., Bagging Predictors. *Machine Learning*, 1996, **24**, pp. 123–140.

[30] Anguita D., Ghio A., Oneto L., Parra X., Reyes-Ortiz J., A Public Domain Dataset for Human Activity Recognition Using Smartphones. *21st European Symposium on Artificial Neural Networks, Computational Intelligence and Machine Learning, ESANN 2013*. Bruges, Belgium, 24–26 April 2013.

[31] Reyes-Ortiz J.-L., Oneto L., Samà A., Parra X., Anguita D., Transition-Aware Human Activity Recognition Using Smartphones, *Neurocomputing*, 2015, **171**, pp. 754–767.

[32] Machine Learning Repository, University of California, Irvine, http://archive.ics.uci.edu/ml/, accessed 15 May 2015.

[33] Prasanna S., Machine Learning Made Easy – Human Activity Learning Using Mobile Phone Data, 2015, http://www.mathworks.com/

matlabcentral/fileexchange/50232, MATLAB Central File Exchange, retrieved 18 May 2015.

[34] Mannini A., Sabatini A.M., Machine Learning Methods for Classifying Human Physical Activity from On-body Accelerometers. *Sensors*, 2010, **10** (2), pp. 1154–1175.

[35] Ravi N., Dandekar N., Mysore P., Littman M.L., Activity Recognition from Accelerometer Data. *Proceedings of the Seventeenth Conference on Innovative Applications of Artificial Intelligence*, IAAI, AAAI Press, 2005, pp. 1541–1546.

[36] Kwapisz J.R., Weiss G.M., Moore S.A., Activity Recognition Using Cell Phone Accelerometers. *SIGKDD Explorations Newsletter*, 2011, **12** (2), pp. 74–82.

Part 2

Applications for Active and Assisted Living

Chapter 10
Human gait analysis for frailty detection – quantitative techniques and procedures

Jesús Fontecha[1], *Iván González*[1], *Ramón Hervás*[1], *and José Bravo*[1]

In the last few years the amount of research regarding gait analysis has increased considerably. The emergence of new technologies makes it possible to accurately measure a wide amount of gait aspects, detecting multiple physical disabilities and syndromes. In this sense, frailty syndrome describes elderly people who are dependent on others to perform basic needs. Obviously, the study and detection of frailty considers a set of parameters from different domains mainly from a clinical viewpoint. However, the functional domain has been classically appreciated as the independence level of a person, and the gait activity as the main predictor of functional disorders. Thus, gait analysis can be used in the detection and diagnosis of frailty.

Development of mobile technologies and sensor devices facilitates the gait study, providing relevant information about the health condition of the person. Acquisition, segmentation, filtering, and depth analysis are the main stages in the treatment of the gait signal.

This chapter presents a detailed study of gait activity from a quantitative viewpoint, taking into account multiple sensors and devices, as well as several analysis techniques. We study the use of accelerometer-enabled devices, pressure sensors, and cameras to collect gait data, which provides details according to our experience about the pros and cons of using specific and general purpose devices. Also, the estimation of gait parameters and the application of classification algorithms in gait analysis are discussed.

We present three experimental systems that analyse gait parameters by using different devices and procedures and consider other relevant factors to support frailty detection. From the deployment of these systems, some results and conclusions are shown.

[1]MAmI Research Lab, Castilla-La Mancha University, Paseo Universidad, 4, 13071 Ciudad Real (Spain)

10.1 Introduction

Advances in the recognition of physical activities–specifically gait–are possible because of the growth of smart devices and sensors that facilitate this task. For example, the integration of accelerometers, gyroscopes and magnetometers into smartphones is now standard. This allows the gathering of movement data and provides other interesting capabilities such as processing, communication, and interaction with the user. There are particular sensors with communication capabilities that can work independently, having some differences with general devices like smartphones regarding battery, cost, size, and ease of placement. However, the combined use of several devices and sensors in an appropriate way would increase accuracy levels in the early stages of data acquisition. Gait is the main aspect in detection and monitoring of frailty syndrome in elderly people. As we will see throughout this chapter, the study of frailty involves multiple clinical domains in which function is the most important, but not the only. Frailty should be studied from an integral viewpoint, since it depends on clinical, functional, cognitive, and social fields. Although we focus on gait study, the rest of the domains should be taken into account in deeper analysis of frailty. A suitable gait analysis includes gait characterisation through the abstraction of a model of human movement, taking into account one or more of the following dimensions:

- *Spatio-temporal*: Derived parameters from gait events such as step/stride length, step/stride time, cadence.
- *Kinematics*: Related to the analysis of the angles described by the joints during gait.
- *Kinetics*: Studies human gait from the point of view of dynamics and external forces to which joints are exposed during the gait cycle (i.e., ground reaction force, inertia, gravity, etc.).
- *Muscle activity*: Electromyography.

Data acquisition from sensors is the first step for the gait study, and it directly depends on hardware components. Application of algorithms and data processing methods by means of software techniques and platforms provide us tools to facilitate the study and interpretation of the gait. Creation, identification, and analysis of gait patterns help experts personalise the study, being interested in making clinical decisions, and adopting next treatments.

Gait patterns extraction, as a part of gait analysis, is an important requirement for diagnosing and monitoring the clinical course of several disabilities or diseases to capture the inherent gait variability. This requires a controlled long-term gait monitoring to provide reliable analysis and characterisation of the human gait over time. Particularly, the inherent gait variability is a crucial aspect in the diagnosis of frailty syndrome and neurodegenerative diseases (such as *Parkinson* and *dementia*). Also, under the scope of gait analysis, several clinical applications require a real-time approach of gait monitoring (i.e., systems based on functional electrical stimulation to improve walking in people under rehabilitation therapy or to control the gait cycle of a neuroprosthesis based on gait phases detection).

In this chapter we focus on which mechanisms we have to detect and diagnose frailty syndrome based on gait analysis. We gain this information from the use of specific purpose sensors in next-gen devices, and study how these devices collect information to be processed by computer systems, improving detection, diagnosis and monitoring of elderly people with frailty.

Section 10.2 presents a clinical definition of frailty and the most important factors to detect this syndrome. In the next section the stages in gait analysis procedure are explained in detail, including data acquisition, segmentation and filtering, and data analysis. An overview of gait analysis mechanisms is presented in Section 10.4. In this case we differentiate between general and specific purpose devices. Section 10.5 details three experimental systems and study cases developed and deployed to analyse frailty in terms of gait, while also considering other factors. Finally, Section 10.6 discusses the conclusions of this chapter.

10.2 An overview of frailty syndrome: the importance of functional markers

Frailty is defined as a syndrome related to ageing, disability, and comorbidity. In elderly people, resistance and physiological reserves decrease, leading to cumulative wear and an increased risk of adverse health effects appearing as frailty signs. In the 1970s this concept was used to describe a group of elderly people who preserved their independence in a precarious way. Woodhouse [1] defined frail elderly people as those greater than 65 years of age who are dependent on other people to perform their basic needs, whereas Gillick [2] described frail older persons as old, debilitated individuals who cannot survive without help from others. More recently, Brocklehurst [3] defined frailty as the risk of losing the ability to live in the community. Certainly, there are many definitions of frailty, and the frailty state should be studied from multiples domains, as Rockwood [4] describes. Fried proposed a phenotype of frailty that sets out five criteria to determine whether a person is frail or not (three or more given criteria indicates the person is frail). This phenotype takes into account the most important symptoms and signs of a frail person [5]. These five criteria include:

- Involuntary weight loss (4.5 kg or more per year)
- General feeling of exhaustion
- Weakness (measured by grip strength)
- Slow gait speed (based on five metres of distance approx.)
- Low level of physical activity (less than 400 calories per week)

As we can see, three of these criteria refer to the functional domain (weakness, gait speed, and physical activity). Functional domain is the most important aspect in the study of frailty, but not the only one. Hamerman [6] described the difficulty of addressing the concept of frailty due to the large number of factors considered. In fact, detection and diagnosis of frailty must be studied in the following domains: clinical, functional, socio-economic, cognitive, and institutional. The application of tests and scales is widely used to assess many aspects of each domain. The scores achieved from these tests, the results of laboratory reports, and general information, among others, are analysed by physicians from a medical viewpoint. Social and psychological indicators are not usually considered because they do not have a direct relationship with the patient record. In [7], Fontecha presented the most common quantifiable clinical indicators regarding frailty, considering factors from these groups: anthropometric, nutritional, functional, cognitive, geriatric syndromes, and gait. The group of gait variables were acquired by using a smartphone device attached to the waist of the person during the completion of the Tinetti [8] test which assesses gait and balance.

The study of gait in elderly people helps detect frailty from a physical viewpoint, being one of the most important indicators in frailty detection and diagnosis as mentioned before. Traditionally, gait has been studied by means of observation applying different tests. In this sense, the Tinetti test is considered to be the most relevant and used by experts today. The aim of this test is to assess postural control and motor skills through two common and simple activities related to gait and balance.

- *Gait assessment*: From a questionnaire, the physician evaluates several aspects of gait. Thus, the elder must walk approximately eight metres at a normal pace while he/she being observed by their physician (see Figure 10.1). The maximum score of the test is 12 points, where each evaluated aspect can receive between zero

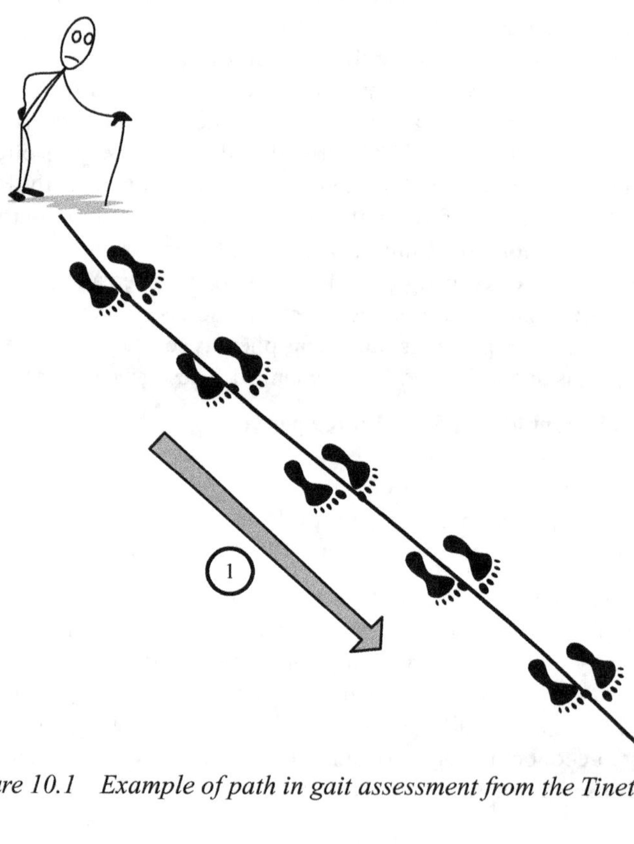

Figure 10.1 Example of path in gait assessment from the Tinetti test

and two points (a score of zero represents the most impairment, while two would represent independence of the patient). Some aspects to be evaluated include:
 - Initiation of gait
 - Step length and height
 - Step symmetry and continuity
 - Path
 - Trunk and walking stance
- *Balance assessment.* From another questionnaire, the physician can evaluate aspects related to the balance of the person. In this component, the maximum score is 16 points, and each aspect receives from zero to two points in the same way as the gait component. In this case, the elder sits on a chair and then gets up. The physician assess aspects such as: sitting balance, attempts to arise and standing balance among others.

Initially, the Tinetti test was focused on detecting the risk of falls; however, it can be used in the study of frailty. With ageing, gait becomes more rigid, stiff, and less flexible, slowing pace and step length [9]. In the next sections of the chapter we go further into the aspects of gait analysis because of the importance of this activity in frailty detection and diagnosis.

10.3 Gait analysis procedure

Gait is the most common physical activity in humans and can be recognised by using multiple devices or electromechanical components. Tri-axial accelerometer sensors are the usual mechanisms used to recognise any kind of physical activity, including gait. Currently, these elements can be embedded in many computational devices thanks to MEMS[1] technology, where sensors are becoming increasingly smaller and accurate.

Usually, accelerometers are attached to different parts of the human body to recognise certain parameters of physical activity, in our case gait. Moreover, there are other mechanisms and sensors to collect gait information, as well as multiple ways of filtering and analysing data, as explained in this section. Data acquisition, filtering, and analysis identify the most important stages in gait analysis procedure.

10.3.1 Data acquisition: sources and fundamentals

There are many quantitative ways to acquiring data from gait activity. We have identified the following three main groups of studies: those that employ different kinds of wearable sensors located on the human body, others that are based on computer vision systems, and ones that use pressure sensors placed on the floor. In the wearable approach, heterogeneous set ups and arrays of sensors of different natures and size can be attached to various parts of the human body (i.e., feet, ankles, knees, lower

[1]Microelectromechanical systems.

back at the level of vertebrae L2–L4, hips, trunk, etc.). These systems can include inertial sensors such as: tri-axial accelerometers, gyroscopes, and magnetometers; various types of pressure sensors, like force sensitive resistors (FSR) and piezoelectric pressure sensors; bend sensors to analyse flexion during gait; as well as electric field sensors to measure the foot clearance or EMG electrodes for electromyography, among others.

Tri-axial accelerometers have the ability to gather acceleration measures from the three axes with a specific time interval and accuracy level. Depending on the work or study to be performed, gathered data could be different; however, all of them should take into account the following characteristics:

- *Coordinate axis values*: Defined by the position and orientation of the sensor if it is attached to the human body. A different placement of the accelerometer could modify the orientation of coordinates and their values.
- *Frequency and period*: In a temporary domain, a configurable time interval can be set. This parameter corresponds to the time space between two consecutive values of each axis (resolution). The smaller this value, the greater the accuracy, but the computational cost will also be higher.
- *Timestamp*: This parameter allows us to know the date/time in which accelerometer values were collected.

Pressure sensors are also other components usually integrated in specific gait analysis devices (to get gait events and spatio-temporal parameters), and these are embedded into sensorised insoles or instrumented shoes [10–12]. This kind of sensor provides an estimation of the magnitude of the ground reaction force (GRF) under the foot in terms of voltage and capacity variation when pressure is exerted. There are several types of pressure sensors regarding their mode of operation, which has influence in the pressures range, sensitivity, and linearity. The most common are force sensitive resistors and piezoelectric sensors (see Figure 10.2).

Furthermore, there are highly elaborated wearable systems, as proposed by Bamberg *et al.* [13], where inertial and pressure sensors are complemented by the inclusion of other sensors, like bend sensors (also known as flexible goniometers), to gather data from flexion during the gait cycle and electric field proximity sensor (EFPS) or ultrasonic sensors to measure foot clearance and stride length.

A completely different data acquisition approach that has also been used for quantitative gait analysis through wearable devices is based on surface electromyography (SEMG). This technique allows measurement of the levels of muscle activity. When a muscle contracts, electrical activity generated as action potentials propagates along the muscle fibres; electrodes attached to the skin allow the capture of electrical activity and thus, muscle activity can be estimated. Electromyography has been used for activity monitoring, and particularly for gait cycle detection [14, 15] by generally acquiring the surface EMG from the Gastrocnemius and/or Soleus muscles of the leg.

Sensor features such as size, temperature of operation, and accuracy level depend on the hardware specification. These are established by the manufacturers and must be taken into account during the design and development of the wearable devices.

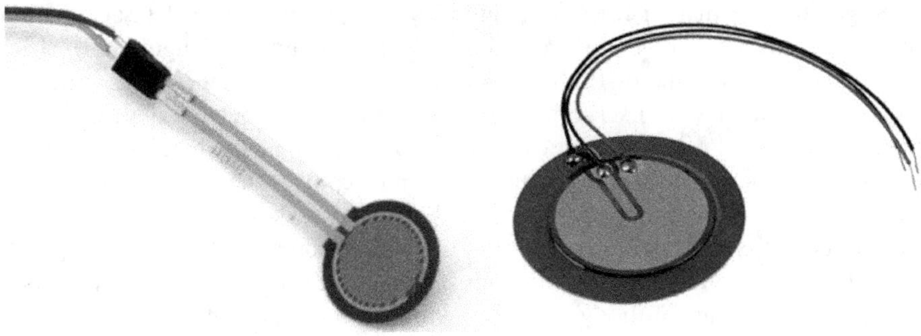

(a) (b)

Figure 10.2 *(a) Force sensitive resistor (Source: https://www.sparkfun.com/products/9375) and (b) Piezoelectric pressure sensor (Source: http://electronics.stackexchange.com/questions/18212/whats-the-thirdwire-on-a-piezo-buzzer). Images are CC BY-NC-SA 3.0)*

Based on our experience [16] and from research as proposed by Foerster [17], the use of a single accelerometer attached to the waist is an acceptable and sufficient option to collect movement data from gait exercise. Obviously, we can collect gait data in a more enriched form by using a combination of elements. For example, to achieve better distance measurement and thus estimate more accurate spatial gait parameters it is common to combine two or more inertial sensors (sensor fusion). One tri-axial accelerometer together with a single tri-axial gyroscope packaged into an inertial measurement unit (IMU) make up a baseline configuration to undertake this task. The gyroscope captures orientation through changes in angular velocity; for this reason, it has a tendency to drift over time due to the lack of a fixed frame of reference. The use of an additional accelerometer (usually combined with the signals from the gyroscope through a Kalman filter) minimises the propagating error and corrects the orientation estimation produced by the gyroscope. This is because the accelerometer captures changes in direction with respect to gravity, which is useful to orient the gyroscope to a more accurate angular displacement.

Likewise, computer vision-based systems have also been widely used for the assessment of biomedical parameters and human motion analysis [18], including gait data acquisition. Such systems can be classified into two approaches: marker and marker-less. The use of markers attached to the subject's body boosts accuracy and makes the measurements more robust during motion tracking; however, these systems are more invasive than those that are marker-less. In this context, it is worth mentioning that highly specialised systems used for gait data acquisition employing markers exist, such as the Vicon (Vicon Motion Systems Ltd., Oxford, UK) and the Optotrak (NDI Ltd., Ontario, Canada) motion capture systems, which are broadly used as the gold standard in numerous works about quantitative gait analysis. However, these systems are typically only available for clinical purposes due to their size and cost [19].

Regarding the marker-less video-based systems used for gait data acquisition, several methods are available: stereo camera systems (camera triangulation), structured light sensors, time-of-flight (ToF) cameras, laser scanners, and arrays of pyroelectric infrared (PIR) sensors. Regardless of the technique employed, the pursued objective is the scene depth measurement, also called range imaging or distance mapping. This consists of determining the point depth in the scene, allowing the construction of its three-dimensional representation. A common approach in gait data acquisition is to use background subtraction and some clustering techniques to extract human silhouettes from the depth image and compose a three-dimensional point cloud of the human body. This is useful for fitting a skeletal model [20] or for segmenting the low part of the body, projecting it on the ground and extracting footfalls [21] for further analysis.

Particularly, structured light sensors have become very popular in the last few years due to their acceptable resolution and low price. The mass sale of these sensors, as the one developed by PrimeSense (Kinect) and distributed by Microsoft as a game peripheral, has decisively contributed to this. Structured light consists of projecting a geometric light pattern onto the scene objects (see Figure 10.3). Three-dimensional information is built from the deformation of the projection of that pattern (captured by an infrared camera) with respect to the original non-deformed pattern.

Lastly, particular emphasis should be on the pressure sensors placed on the floor. There are two kinds of devices: instrumented walkways equipped with pressure sensors to measure gait parameters during short walking paths, and force platforms (also called force plates) designed to measure the forces and moments applied to its top surface as a subject stands, steps, or jumps on it. Both types of devices are regularly

Figure 10.3 Geometric infrared pattern projected onto the scene by a Kinect structured light sensor

Figure 10.4 *GAITRite walkway*

used in clinical studies looking at gait and balance. It is worth mentioning the commercial GAITRite walkway system (CIR Systems Inc., PA, USA) (see Figure 10.4) which provides accurate spatio-temporal parameters and is widely used as a ground truth in many gait analysis studies.

Apart from the sensorisation mechanisms used or if multiple approaches are combined into a single system, there is a growing interest in building gait data acquisition systems that can collect data in the course of a normal day of activity, avoiding the subjective viewpoint that exits in the clinical gait trials in a laboratory environment. Ideally, these systems have to work autonomously, transparently, and with slightly intrusive operability in the elders' environment.

10.3.2 Data segmentation and filtering

The data acquisition stage generates raw data from gait activity. Not all collected data are valid for analysis, and the original signals contain noise that should be removed. Thus, it is important to carry out the signal segmentation and filtering at this point.

Segmentation is the process in which the signal is divided into fragments with valuable data from the physical activity (gait in our case). Most of the time, the complete signal contains incorrect values that can be removed through segmentation. It is necessary to establish an appropriate segmentation criteria that depends on the aim of the analysis, determining which parts of the signal are important. Segmentation can be performed by direct action of the user or by means of a computational algorithm. Schlomer [22] uses the buttons of a device to decide the beginning and end of each signal fragment, while authors like Joselli [23] use mathematical equations to recognise the beginning and end of movement in each accelerometer signal fragment (considering x, y, and z coordinates).

Besides, specific segmentation algorithms or techniques can be implemented depending on the purpose of the work and could be based on time, sounds, etc. For instance, the GAITRite device requires manual segmentation, by direct action of the operator, every time a subject walks across it. Conversely, in computer vision systems, background subtraction techniques (based on simple thresholds or modelling the background using mixture of Gaussians) are usually used to detect movement and, thus, activity. In the vision-based system presented in Section 10.5.3 a more elaborated segmentation algorithm is implemented to identify straight gait paths over recorded trajectories.

Regardless of the segmentation method, there is another issue related to the segmented signal. Graphically, a fragment of the signal after the segmentation phase is similar to that shown in Figure 10.5.

The signal noise means a disturbance of the signal when it is transmitted, and this fact depends on way of transmission, environment, variation on gravity forces (in the case of inertial sensors), and other external factors [24]. Reduction of noise level is achieved by applying filters. These filters discriminate a given frequency or a frequency range (band) from an electrical signal that comes from devices such inertial sensors (e.g., accelerometers).

The most common filters in gait recognition are: Butterworth-pass filter, median filter, wavelet transforms, finite impulse response, and Kalman filters.

Concerning the acceleration signals, considering sensor location and filtering over the axis are crucial aspects before applying algorithms for the identification of

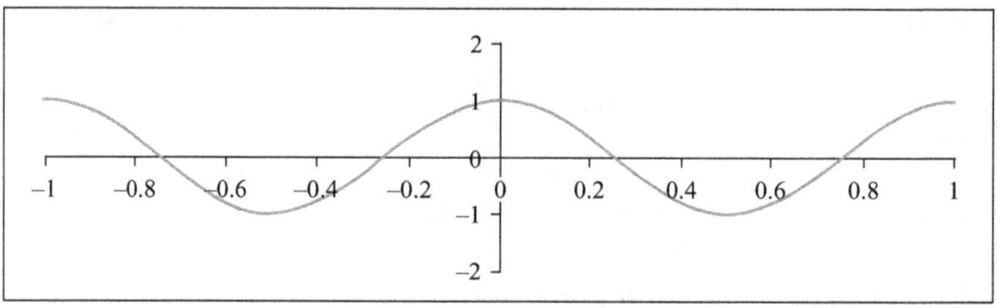

(a)

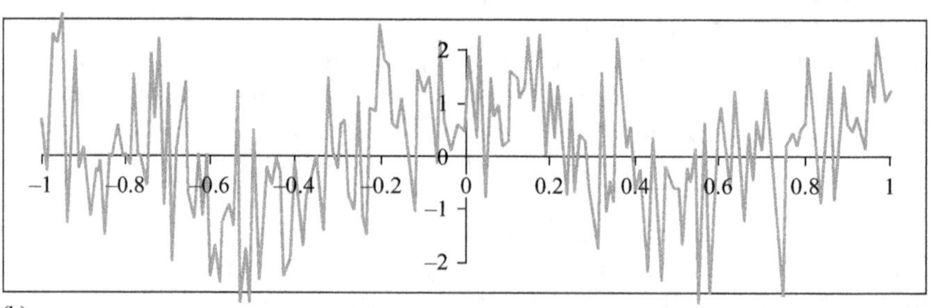

(b)

Figure 10.5 Graphical representation: (a) signal without noise, (b) signal with noise peaks related to temporal gait events like heel-strike (HS) and toe-off (TO). The filter type and its parameterisation directly affect the waveform and the peak detection criteria. For instance, in [25] antero-posterior acceleration at the trunk is filtered with a fourth-order Butterworth low-pass filter at 20 Hz, then a fourth-order 2-Hz Butterworth filter. HS is estimated as the time of peak forward acceleration preceding the change in sign from positive to negative of the 2-Hz filtered signal. In contrast, in [26] the vertical acceleration measured in the trunk was smoothed by integrating and then differentiating the acceleration signal using a Gaussian Continuous Wavelet transform (CWT). HS events were identified as the times of the minima of the smoothed signal.

Most of these segmentation and filtering methods are used in mobile systems, embedded into wearable sensors or mobile devices.

10.3.3 Data analysis: parameter estimation and identification of gait patterns

Once the signal has been pre-processed, its data is analysed in detail. Assessment of gait signal can be focused on frequency or time domains. When most of the analysis is performed in real time, it is more appropriate to work in the time domain. In this sense, there are many ways to assess data that come from external devices such as inertial sensors, cameras, and other smart devices. Based on data mining techniques, relevant data is collected, parameters are calculated, and patterns can be identified. The use of either procedure depends on the aim of the analysis.

When we are using accelerometer sensors and mobile devices, such as smartphones, for data processing, we calculate relevant features from gait activity through statistical measures. These statistical measures can be compared to measures from other signals. Some of these parameters are arithmetic mean, standard deviation, Pearson's coefficient, amplitude, etc. These can be easily calculated and used in classification tasks. Besides, computational processes allow the estimation of more variables, such as step/stride time, stance time, swing time, single/double support time, cadence, and velocity in gait movement. However, although velocity can be obtained by integrating the acceleration signals and velocity integration can also help estimate sensor position, this double integration accumulates big errors (integration drift) in the measurements, which makes it more difficult to use a simple tri-axial accelerometer for distance measurement and thereby the estimation of spatial parameters as step/stride length. Generally, reducing the integration drift is achieved assuming the condition of "zero velocity" when the foot is stationary on the ground [27]. In such a way, the integration is only performed at each step interval (between heel-strikes).

By means of a six-degrees-of-freedom IMU (tri-axial accelerometer + tri-axial gyroscope), the stationary phases when the foot is solid on the ground (identified as zero velocity instants) can be better demarcated as periods when the gyroscope output magnitude is small and the accelerometer output magnitude is close to gravitational (g). In this manner, these sensors can be combined to get more accurate orientation readings and distance estimations.

Many current works use classification, clustering, and pattern recognition to analyse gait activity. Also, a combination of these procedures is possible to perform a deeper and more complex analysis. In this case, it is appropriate to use more powerful devices to run algorithms and obtain reliable results, and data mining is presented as a suitable process for working with large amounts of data. The most used data mining techniques mainly include statistical models, decision trees, clustering, neural networks, and Bayesian networks. Obviously, each technique has multiple pros and cons; however, according to the purpose of the analysis we have to decide which one to use. This decision will take into account features such as performance, computational cost, accuracy, and communication capabilities. Relationships between these features are usually crucial in the system operation.

In the study case presented in Section 10.5.2 a probabilistic classifier helps differentiate several foot movement patterns with the goal of achieving walking forward discrimination. In this case, Fuzzy logic is used to demarcate gait phases. In [28], a probabilistic method is also used, although the goal is to automatically detect human strides from raw accelerometer signals. Concretely a Bayesian classifier is trained to identify characteristic peaks that correspond to gait events from others produced by shifts of the acceleration signals.

10.4 Overview of gait analysis systems and mechanisms

10.4.1 Specific purpose devices: sensors and tiny mechanisms

There are many sensors and small mechanisms to acquire data from environmental elements, even under particular conditions. Development of MEMS technology allows the creation of sensors that can be embedded into tiny devices. In Section 10.3.1, some of these sensors that acquire data from gait activity were described in detail.

Usually, specific purpose devices have the minimum requirements necessary to work, including the sensor, battery, wireless communication capabilities, and even small storage capacity. In case of gait recognition, accelerometer-based devices are widely used. An example of this is the WiTilt device from Sparkfun manufacturer,[2] a small device that incorporates a triple axis accelerometer sensor, a single axis gyroscope, a LiPo battery, and a Bluetooth module to communicate with other devices.

Moreover, sensors can be independent, so developers can integrate several modules or hardware components electronically. For example, pressure sensors can be embedded into insoles, as we mentioned in previous sections. Obviously, these independent sensors are connected to small motherboards or systems to satisfy data acquisition and analysis of gait activity.

One advantage of tiny sensors and small devices is that they can be attached to any body part (wearables) or environmental objects in a less intrusive way than other, bigger devices. In case of cameras, these allow users to wander without computational elements and sensors placed on the user's body; however, systems with cameras are more complex (e.g., more storage and processing needs) and they need to deal with security issues.

[2]https://www.sparkfun.com/

10.4.2 General purpose devices: smart and mobile devices

Since the last decade, the number of mobile and smart devices has increased considerably. Healthcare and assistive technologies demand the use of devices to control and monitor a wide range of activities. Obviously these devices have many more benefits regarding process capability, communication, human interaction, storage, and versatility. Some examples of these are: smartphones, smartwatches, tablets, and computers. In the case of physical activity monitoring and gait specifically, small and mobile devices are usually more appropriate and comfortable to carry. In this sense, smartphones are the most widely used general-purpose devices today. These devices satisfy many daily needs, and not just in the activity monitoring domain. However, there are many works and research studies that use smartphones as monitoring mechanisms as we have mentioned throughout this chapter.

Furthermore, general purpose devices can independently integrate complete gait analysis systems, as well as to work with other general and specific purpose devices. Depending on the context, some features to take into account include the number of devices, size, battery duration, ease of placement, etc. For example, in an integral solution, specific purpose devices can be used to gather gait data, while smartphones can be the mechanisms used in gait movement visualisation, which communicate with computers or servers by means of software components to perform analysis tasks.

In any case, focus groups of users should be taken into account when these systems are deployed. Developing less intrusive and more natural and intuitive solutions is as important as performing an exhaustive gait analysis.

Table 10.1 shows a comparison of relevant features between general and specific purpose devices.

Table 10.1 Features summary of general and specific purpose devices. A true tick indicates which device provides a better result on each feature

	Devices	
Features	**General purpose**	**Specific purpose**
Sensorised feature regarding gait movement (acceleration, pressure)		✓
Sensor accuracy	✓	✓
Processing capabilities	✓	
Storage	✓	
Communication	✓	✓
Communication range	✓	
Display and visualisation	✓	
Battery duration		✓
Device cost		✓
System cost	✓	
Size		✓
Development platform	✓	
Ease of placement		✓
Versatility	✓	

As we can see, general purpose devices have more advantages for use than unique devices. However, a combination of specific purpose devices would provide more complete analysis with better results, although some sensors are not embedded in general purpose devices (e.g., pressure sensors). Certainly, the use of general and/or specific purpose devices depends on the context, project costs, deployment environment, target users, and other factors also important in deciding what kind of device is more suitable.

10.5 Gait analysis as part of a comprehensive study of frailty: experimental applications and case studies

As mentioned in previous sections, gait is the most important aspect in frailty detection. We have developed three main systems to detect frailty in elderly people taking into account gait analysis and other relevant factors. In the first case, a mobile system for supporting physicians in frailty detection is described in detail. The second case presents an ambulatory system for gait monitoring by using pressure sensors embedded into insoles. Finally, the third system uses a structured light sensor to passively capture gait trajectories from the scene, identify straight paths, and monitor gait activity.

10.5.1 Frailty detection and diagnosis system by using accelerometer-enabled smartphones and clinical information

10.5.1.1 Description

We developed a system used in clinical environments to detect the frailty and pre-frailty of an elderly community. This system collects a complete set of relevant clinical parameters related to frailty, including factors from these domains: nutritional, anthropometric, cognitive, geriatric syndromes and functional. Parameters from the functional domain were gathered by means of an accelerometer-enabled smartphone attached to the patient's waist. Besides this, their own smartphone performed an estimation about the frailty condition of the patient taking into account the values of frailty factors of the rest of the elderly people in the community. The application of algorithms based on similarities allowed us to determine this fact [29].

Moreover, not all frailty parameters have the same degree of importance in frailty assessment. Geriatricians typically evaluate the importance of each variable depending on the context and particular case. According to most physicians, gait activity analysis is essential because it provides substantial information regarding the functional level of an individual. In particular, gait analysis using the tests introduced by Tinetti [8] provides a reliable measure of frailty (but in this system, we work with multiple clinical sources of frailty parameters, not only physical). From this activity several quantitative variables are calculated, including: standard deviation, amplitude, acceleration mean, and Pearson's coefficient of variation. Using

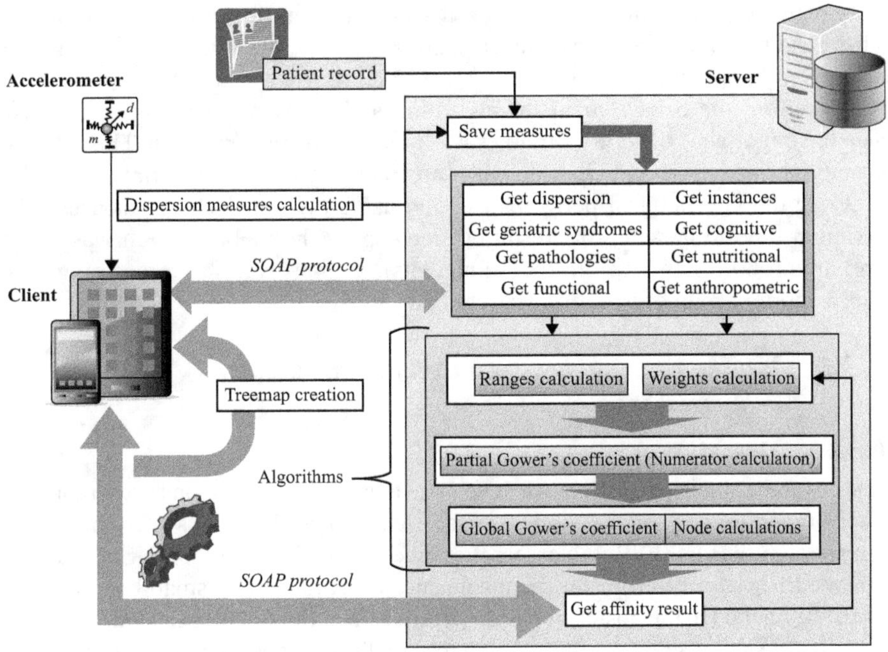

Figure 10.6 Schema of frailty detection system by using smartphones and clinical records

these variables and those from the other domains, the system applies a set of algorithms based on similarity coefficients to determine the health state of each evaluated person, and the treemaps approach to show the results in a mobile application. Figure 10.6 shows an overview of the schema, computational services, and data flows from the data acquisition of frailty variables (from patient record and accelerometer-enabled smartphone) to calculate results and present them on the smartphone screen as treemaps.

10.5.1.2 Results

We evaluated this system on a cluster of sixty instances from 20 elderly patients with an equal number of women and men (an instance consisted of a complete set of frailty variables associated with a given patient at a given moment). Men had an average age of 81.8 ± 4.74 while the women had an average age of 85.45 ± 3.22. All of the patients had adequate social abilities and were independent in their everyday activities, and none of them presented cognitive impairment. The developed mobile application was used during gait activity for collecting accelerometer data and calculating the statistical parameters from a specific patient. These parameters and those from nutritional, cognitive, and anthropometric domains of the same patient were used by the system services, and algorithms determined the similarity degree (percentage) between frailty patients in the community. The results aided the physicians in determining the frailty condition of the specific patient at a given moment, in relation to other patients in the elderly population.

Moreover, the behaviour of the algorithm and the final results were different when the importance of frailty variables was modified by the physician. This allowed an analysis focused on particular domains previously requested by geriatricians.

From the viewpoint of performance, and using the initial cluster of users, the maximum time to generate the resulting treemap on the mobile screen was 2.55 s (working on 60 instances from the 20 patients); we observed less processing time when a shorter group of variables was studied.

10.5.2 Gait monitoring system based on wireless sensorised insoles

10.5.2.1 Description

In this experimental application, we developed a gait phase detection system for continuous monitoring based on wireless sensorised insoles [30]. The system integrates four pressure sensors (FSRs) to assess the force exerted by each foot during walking with one tri-axial accelerometer per insole encapsulated inside a small box (see Figure 10.7) attached to the posterior part of the tibia, near the Achilles tendon. Raw data from the sensors is sent to a mobile application for gait analysis (see Figure 10.8), where gait sub phases are demarcated in real-time using a fuzzy rule-based inference algorithm (GPDA: Gait Phase Detection Algorithm).

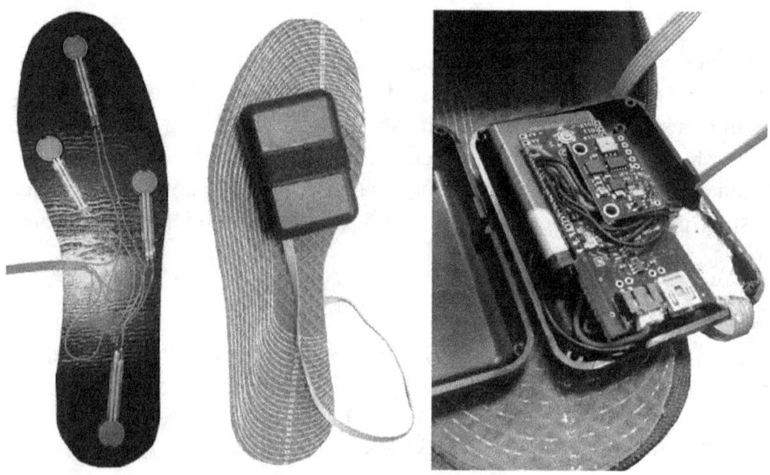

(a) (b) (c)

Figure 10.7 Insole hardware prototype: (a) arrangement of FSR sensors, (b) final prototype, (c) Arduino Fio microcontroller + 9DOF IMU (only accelerometer is being used) + LiPO batt. + Bluetooth module WLS125E1P

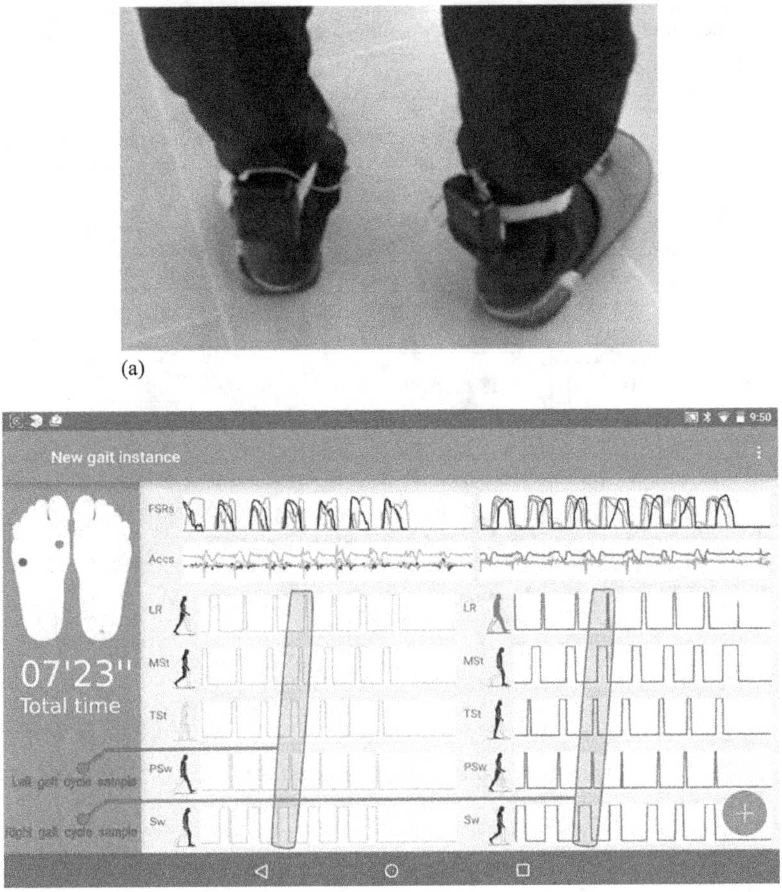

(a)

(b)

Figure 10.8 (a) insoles setup, (b) mobile app: gait instance recorder user interface

Additionally, to provide a solution that is insensitive to perturbations caused by non-walking activities, a probabilistic classifier (Gaussian naive Bayes (GNB) classifier) is employed to discriminate walking forward from other low-level activities, such as turning, walking backwards, lateral walking, etc. The main purpose of the walking forward discrimination is to provide a way to automatically segment gait's fragments over straight paths. This is because gait data acquisition for the estimation of spatio-temporal parameters in clinical settings generally only pays attention to straight paths since turns or abrupt changes in the gait trajectory distort the spatio-temporal measures (in our case the demarcation and duration of each stance sub phase and the swing phase).

Furthermore, an algorithm based on principal component analysis (PCA) is used for dimension reduction of the selected features which will feed the classifier. Its

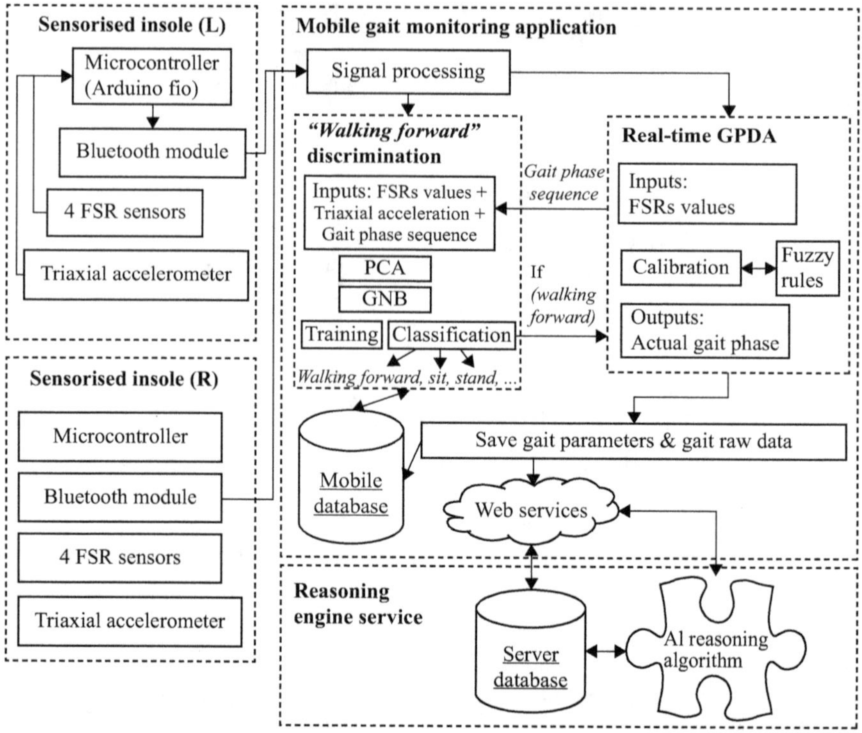

Figure 10.9 Overview of the insole system for continuous gait monitoring

use boosts the classifier's accuracy and reduces the time needed for foot movement pattern classification.

The combination of these proposed algorithms, together with the insole prototype, constitutes our first wearable approach towards a continuous gait assessment system, by means of the avoidance of non-walking influences.

An overview of the proposed system for continuous gait monitoring is shown in Figure 10.9. As can be observed, the estimated gait parameters (gait's sub phases timestamps) are saved in a mobile database that can be useful for local processing in applications with real-time requirements (i.e., to control the gait cycle of a neuroprosthesis or to provide instant feedback on gait shortcomings to people under rehabilitation, etc.) Alternatively, gait parameters can be sent via web services to a remote server where reasoning techniques can be applied, for example, to identify frailty patterns over long term monitoring and to analyse trends of entering into a frailty condition.

10.5.2.2 Results

To verify the GPDA the system was tested on five male subjects (average age of 33 ± 2). All presented a healthy gait pattern. The detected gait phases from the trials were compared with a theoretical model of human gait [31] generally accepted by the experts and used as a ground truth. For each gait trial, the average time of each gait phase was compared with the expected time for this particular phase in the reference system. As can be seen, the larger differences from the reference system were in the mid-stance phases (overestimated around 78 ± 41 ms) and in the pre-swing phases (underestimated around 59 ± 19 ms).

10.5.3 Computer vision system based on a structured light sensor

10.5.3.1 Description

We developed another completely different approach for quantitative gait analysis that consists of a computer vision system based on the utilisation of a low-cost, structured light sensor (the first version of the Kinect peripheral). In the developed application the main purpose was to achieve passive in-home gait monitoring in enabling the estimation of the heel-strike events and derived spatio-temporal parameters, specifically: step/stride time, step/stride length, gait velocity, cadence, path length, and ambulation time.

One of the advantages of using a structured light sensor (see Section 10.3.1) was the versatility provided in terms of the sensor's location. That means the Kinect might be located in line with the walking path, orthogonal to it, etc., so that the gait could be captured by means of the depth image and the three-dimensional representation of the human body in the scene. For instance, the Kinect could be located at a corner of the room's ceiling in such a way that a potential walking area could be monitored without occlusions.

The system was able to acquire gait measurements passively, over the course of the day (avoiding the subjective viewpoint that exists during clinical gait trials). The idea was to get the subject's point cloud (PCL) from the scene using background subtraction techniques; after that, lower limbs were segmented, cutting off 40 cm from the ground. Therefore, the floor plane equation should be first estimated using an implementation of a Random Sample Consensus (RANSAC) algorithm. Centroids were computed from the lower limb point clouds in each frame and then the whole centroid series was filtered through a Kalman filter in order to estimate the trajectory followed by the subject during the gait trial (see Figure 10.10a).

As gait trajectory could include turns or abrupt changes of direction, a two-stage analysis was performed to extract only straight paths, which are more useful for gait event extraction (see Figure 10.10b). The first stage of analysis consisted of computing how much, on average, the tangent to the trajectory changed. For a perfect line, this would be zero (as tangent is constant). Using this index as a measure of straightness, a first set of candidates for straight paths was obtained. The second stage of analysis defined the definitive set of straight paths fitting a line (using linear least squares regression) over each candidate path. The Root-Mean-Square Deviation index (RMSD) was computed as a straightness indicator; if this index was below a fixed threshold, for a specific candidate path, it was labelled as a definitive straight path.

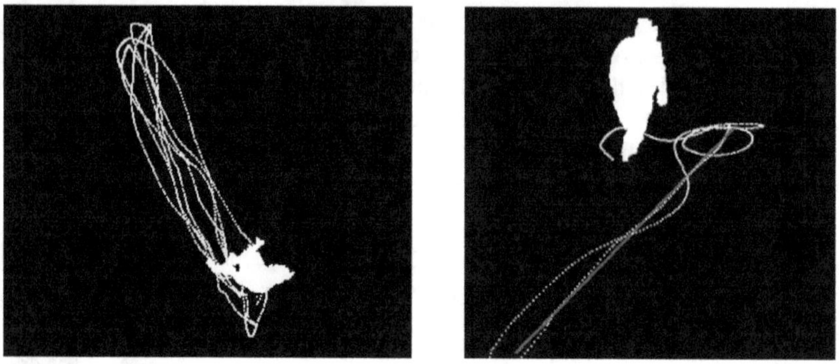

(a) (b)

Figure 10.10 (a) filtered trajectory during a gait trial in a home environment, (b) one straight path (solid line) identified in a trajectory (dotted line)

After identifying straight paths, lower limb point clouds were projected onto the floor plane (footfalls), normalised, and aligned to each concerned path line. The Pearson correlation coefficient was computed for each of these projections. Therefore, a signal or correlation time series was composed for each identified straight path. These signals were a little noisy and also needed filtering. Heel-strike events were located at the maximum and minimum peaks of the filtered signals; an algorithm for automatic detection of those peaks was developed.

An overview of the vision system is shown in Figure 10.11.

10.5.3.2 Results

A visual inspection of several gait tests at a low frame rate (2–3 Hz) confirmed the accuracy detection of heel-strikes. Nonetheless, an experimental protocol was designed [32] to objectively compare the performance of this system with respect to a GAITRite walking path. The experiment was recently conducted and a statistic analysis of results will be published soon.

10.6 Conclusions

This work presents the acquired knowledge from the study of gait activity in frail elders from a quantitative viewpoint, based on the development of several systems and applications that use sensors, smart devices, and computational resources to analyse gait and other factors related to frailty syndrome.

Gait is the most important activity to assess frailty. However, we have seen how other clinical factors should be studied to perform a complete analysis of this syndrome. The study of gait movement is divided into three important stages: data acquisition, segmentation and filtering, and analysis. This chapter shows each of these phases in detail. The source of gait data can be different depending on the system used and the purpose of the analysis. Accelerometer sensors, pressure sensors, or

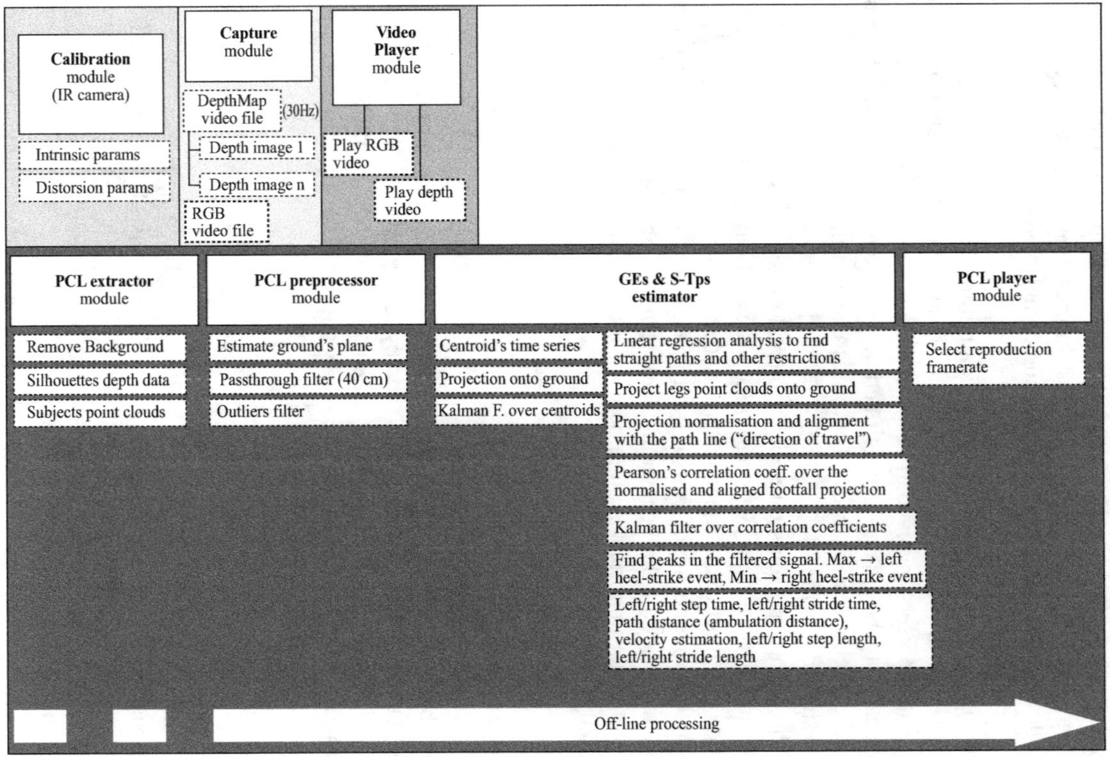

Figure 10.11 Overview of the computer vision system based on the utilisation of a structured light sensor

cameras are the main mechanisms presented in this work. Also, the placement of these elements is important in the collection of valuable data. They are usually attached to body parts and have communication features to transmit movement information. Developers must decide the most appropriate way of placing the device to make the system as unobtrusive as possible. Besides this, there are dedicated mechanisms and general purpose devices. In this case, devices have many features regarding storage, processing capabilities, and communication possibilities. However, we can combine both kinds of devices to create more complex and complete systems, facilitating the generation of more reliable results.

Gait analysis includes the application of data mining. Each process has several pros and cons to be considered, taking into account the aim of the study, environmental assumptions, and the users involved. The work presents and overview of the most common options to analyse data from gait activity. Obviously, gait is not the only factor to consider when assessing frailty. There are many factors from clinical domains that should be studied together.

Finally, the chapter presents three experimental systems developed to analyse gait and frailty in elderly people by using sensors, smartphones, and cameras. These systems provide a support to physicians in frailty detection and diagnosis.

The continuous monitoring of crucial frailty parameters such as those from gait activity, allow the study of the behaviour of movement, including gait variability over time, which is an important requirement for diagnosing several disabilities or diseases, not only frailty.

Acknowledgements

This work has been supported by the coordinated project *FRASE*, granted by the Spanish Ministerio de Ciencia e Innovación, and by the *UBIHEALTH* project under the International Research Staff Exchange Schema program (MC-IRSES 316337). This work has also been funded by Castilla-La Mancha University under the *Plan Propio de Investigación* programme.

References

[1] Woodhouse K, Wynne H, Baillie S, James O, Rawlins M. Who are the frail elderly? Q J Med. 1988;68:505–506.

[2] Gillick M. Long-term care options for the frail elderly. J Am Geriatr Soc. 1989;37:1198–12013.

[3] Rockwood K, Woodhouse K. Textbook of Geriatric Medicine and Gerontology: Expert Consult – Online and Print (Brocklehurst's Textbook of Geriatric Medicine & Gerontology). Edinburg; 1985.

[4] Rockwood K, Fox R, Stolee P, Robertson D, Beattie B. Frailty in elderly people: an evolving concept. Can Med Assoc J. 1994;150:489–495.

[5] Fried L, Tangen C, Walston J, Newman A, Hirsch C, Gottdiener J. Frailty in older adults: evidence for a phenotype. J Gerontol A Biol Sci Med. 2001;56:146–156.

[6] Hamerman D. Toward an understanding of frailty. Ann Intern Med;130: 945–950.

[7] Fontecha J, Hervás R, Navarro F, Bravo J. A Mobile and Ubiquitous Approach for Supporting Frailty Assessment in Elderly People. J Med Internet Res. 2013;15.

[8] Tinetti M. Performance-oriented assessment of mobility problems in elderly patients. J Am Geriatr Soc. 1986;34:119–126.

[9] Sutherland D. The evolution of clinical gait analysis: Part II Kinematics. Gait & Posture. 2002;16:159–179.

[10] Schepers HM, Asseldonk EHF, Baten CTM, Veltink PH. Ambulatory estimation of foot placement during walking using inertial sensors. J Biomech. 2010;43:3138–3143.

[11] Liu T, Inoue Y, Shibata K, Shiojima K, Han MM. Triaxial joint moment estimation using a wearable three-dimensional gait analysis system. Measurement. 2014;47:125–129.

[12] Senanayake C, Senanayake SM. A computational method for reliable gait event detection and abnormality detection for feedback in rehabilitation. Comput Methods Biomech Biomed Eng. 2011;14:863–874.

[13] Bamberg SJ, Benbasat AY, Scarborough DM, Krebs DE, Paradiso JA. Gait Analysis Using a Shoe-Integrated Wireless Sensor System. IEEE Trans Inf Technol Biomed. 2008;12:413–423.

[14] Joshi CD, Lahiri U, Thakor NV. Classification of gait phases from lower limb EMG: application to exoskeleton orthosis. In: *Proceedings of the IEEE Point-of-Care Healthcare Technologies (PHT)*; IEEE, January 2013. p. 228–231.

[15] Meng M, She Q, Gao Y, Luo Z. EMG signals based gait phases recognition using hidden Markov models. In: *Proceedings of the IEEE International Conference on Information and Automation (ICIA)*; IEEE, June 2010. p. 852–856.

[16] Fontecha J, Navarro FJ, Hervás R, Bravo J. Elderly frailty detection by using accelerometer-enabled smartphones and clinical information records. Personal and Ubiquitous Comput. 2013;17:1073–1083.

[17] Foerster F, Fahrenberg J. Motion pattern and posture: correctly assessed by calibrated accelerometers. Behav Res Methods Intrum Comput. 2000;32: 450–457.

[18] Rafi M, Hamid ME, Khan MS, Wahidabanu RS. A Parametric Approach To Gait Signature Extraction For Human Motion Identification. Int J Image Process. 2011;5:185–198.

[19] Yang S, Zhang JT, Novak AC, Brouwer B, Li Q. Estimation of spatio-temporal parameters for post-stroke hemiparetic gait using inertial sensors. Gait & Posture. 2013;37:354–358.

[20] Caillete F, Howard T. Real-Time Markerless Human Body Tracking with Multi-View3D Voxel Reconstruction. In: *Proceedings of the British Machine Vision Conference (BMVC)*, vol. 2; September 2004. p. 597–606.
[21] Stone E, Anderson D, Skubic M, Keller JM. Extracting Footfalls from Voxel Data. In: *Proceedings of the IEEE Eng Med Biol Soc.*; August 2010. p. 1119–1122.
[22] Scholmer T, Poppinga N, Henze N, Boll S. Gesture recognition with a wii controller. In: *Proceedings of the 2nd International Conference on Tangible and Embedded Interaction*; ACM, 2008. p. 11–14.
[23] Joselli M, Clua E. grmobile: A framework for touch and accelerometer gesture recognition for mobile games. In: *Proceedings of the VIII Brazilian Symposium on Games and Digital Entertainment*; IEEE, 2009. p. 141–150.
[24] Schloss F. Accelerometer noise. Sound Vib. 1993;27:22–23.
[25] Zijlstra W, Hof AL. Assessment of spatio-temporal gait parameters from trunk accelerations during human walking. Gait & Posture. 20013;18:1–10.
[26] McCamley J, Donati M, Grimpampi E, Mazzà C. An enhanced estimate of initial contact and final contect instants of time using lower trunk intertial sensor data. Gait & Posture. 2012;36:316–318.
[27] Mariani B, Hoskovec C, Rochat S, Büla C, Penders J, Aminian K. 3D gait assessment in young and elderly subjects using foot-worn inertial sensors. J Biomech. 2010;43.
[28] López-Nava IH, Muñoz Meléndez A. Towards Ubiquitous Acquisition and Processing of Gait Parameters. In: *Proceedings of the 9th Mexican International Conference on Artificial Intelligence (MICAI)* vol. 6437. Springer; 2010. p. 410–421.
[29] Fontecha J, Hervás R, Bravo J. Mobile services infrastructure for frailty diagnosis support based on Gower's similarity coefficient and treemaps. Mobile Inf Syst. 2014;10:127–146.
[30] González I, Fontecha J, Hervás R, Bravo J. An ambulatory system for gait monitoring based on wireless sensorized insoles. Sensors. 2015;15:16589–16613.
[31] Perry J. Gait Analysis: Normal and Pathological Function. Thorofare, NJ; Slack Inc.; 1992.
[32] López-Nava IH, González I, Muñoz Melendez A, Bravo J. Comparison of a Vision-based System and a Wearable Inertial-based System for a Quantitative Analysis and Calculation of Spatio-temporal Parameters. In: José B, Ramón H, Vladimir V, *Proceedings of the Ambient Intelligence for Health: First International Conference (AmIHEALTH)*, Springer; 2015. p. 116–122.

Chapter 11
Fall prevention and detection

Baldewijns G.[1,2,3], *Debard G.*[1,4,5], *Van Den Broeck B.*[1,2], *Mertens M.*[1,4,6], *Karsmakers P.*[1,2], *Croonenborghs T.*[1,6], *and Vanrumste B.*[1,2,3]

11.1 Introduction

In both research and clinical settings a fall incident is commonly defined as *unintentionally coming to rest on the ground, floor, or other lower level* [7]. Such a fall incident is experienced at least once a year by approximately one in three adults older than 65. This number increases with age and frailty level [20, 36].

Yearly, 37.3 million fall incidents have physiological consequences [1]. Additionally, psychological consequences, such as loss of mobility and independence often restrict the ability of older adults to perform daily activities [20, 36].

By detecting when a person has an elevated fall risk, and taking preventive measures when necessary, the number of fall incidents can be reduced. The first part of this chapter will therefore focus on automatic fall risk assessment techniques. Following this, the second part focuses on fall detection systems. As not all fall incidents can be prevented these systems can reduce the consequences of a fall incident by ensuring that timely aid is given.

11.2 Fall risk estimation

Several factors can influence the fall risk of a person. These can be categorized as biological, behavioural, environmental and socio-economic factors [1]. Falls mostly occur after an interaction between two or more of these factors [1]. Although not all factors are modifiable, tackling those that are can avoid such an interplay. Adapting the home environment of an older adult for instance will reduce the risk of future falls [4].

[1]KU Leuven Technology Campus Geel, AdvISe, Belgium
[2]KU Leuven, ESAT-STADIUS, Belgium
[3]iMinds Medical Information Technology Department, Gent, Belgium
[4]Thomas More Kempen, MOBILAB, Geel, Belgium
[5]KU Leuven, ESAT-PSI, Leuven, Belgium
[6]KU Leuven, Department of Computer Science, DTAI, Leuven, Belgium

Biological factors such as reduced muscle strength, overall frailty and a decline in cognitive functions can be measured. Often the effects of these factors can be reduced, for instance, by installing an exercise programme or by changing the medication regime. It is therefore important to detect when these biological factors come into play.

For this purpose, several clinical tests, such as the Timed-Up-and-Go test, Berg Balance Tests and the Dynamic Gait Index are already widely in use. A comprehensive overview of the different clinical fall risk assessment tests is given in [41]. These tools, however, have several disadvantages. They are mostly performed in an artificial setting [2], the participating older adult can be aware that he is under assessment (Hawthorne effect) [34] and specialized personnel is needed to take these tests. These disadvantages could lead to results which might not be representative of the actual fall risk in the home environment [2, 34].

Systems that can monitor the fall risk of older adults in real-world environments are not subject to these limitations. For this purpose several approaches to monitor fall risk-related parameters in the home environment are being researched, including wearable and contactless methods. Both are discussed in this chapter.

11.2.1 Using gait parameters to automatically assess the fall risk of a person

When researching fall risk assessment tools an important decision to make is which parameter(s) to monitor. As previous studies have shown that fear of falling and an increased fall risk are often associated with a decline in gait speed, this parameter is often used in both clinical [41] as well as research settings [4, 44, 47]. Gait, however, is a complex phenomenon, spatial (e.g. step and stride length) and temporal gait parameters (e.g. increased support time) can also be used as predictors of an increased fall risk as well as measures of gait variability [25, 50]. Lastly, there is the level of activity of a person. If a reduced activity level is measured this could be an indicator of a wide variety of health-related problems such as an elevated fall risk [35].

As a next step the type of sensor to measure these parameters needs to be chosen. For this purpose an overview of the different sensors used to monitor one or more of these parameters is given. For each measurable parameter one measurement system is discussed more in depth.

11.2.2 Sensors which measure fall risk-related parameters

A wide variety of sensors such as accelerometers [25, 19], radar [12, 44], 2D and 3D cameras [4, 47], audio sensors [51] and Passive InfraRed (PIR) sensors [24] are currently used for fall risk assessment. The use of video cameras to monitor gait speed, audio sensors to monitor other gait parameters and PIR sensors to monitor activity levels are discussed more in-depth. After this an overview of other sensors which are currently used for automatic fall risk assessment is given.

11.2.2.1 Monitoring gait velocity using video cameras

One approach currently under investigation by Baldewijns *et al.* [4] is the measurements of transfer times of elderly as an indicator of fall risk and other health-related problems. An in-depth study was performed by installing 2D cameras in the home environment of four participants. All participants were aged 74 or older and were living in a service flat or a retirement home. The transfer times were measured multiple times a day and trends in these transfer times were compared with health observations of a geriatric nurse who provided the medical follow-up of the participants.

The camera system acquired video data (12 fps) in the home of each participant for periods varying from 8 to 12 weeks. To isolate the participant in the resulting video frames, four preprocessing steps were taken. First, an estimation of the background (BG) is subtracted from the current video frame. After this, the shadows are removed using background correlation. This is followed by an erosion/dilation step combined with a connected components analysis. Lastly, a bounding box is drawn around the largest foreground object under the assumption that this is the person in the video [4].

After visual inspection of a one week sample of the video data, they defined a transfer zone per participant such as shown in Figure 11.1. Each time the participant crosses the transfer zone in the exact same way the time is measured. When the participant deviates from the predefined path or when the person stops during the transfer the walk is rejected. This automatic rejection of deviating walks results in a smaller variability in the measured times and thus a more accurate view in the gait

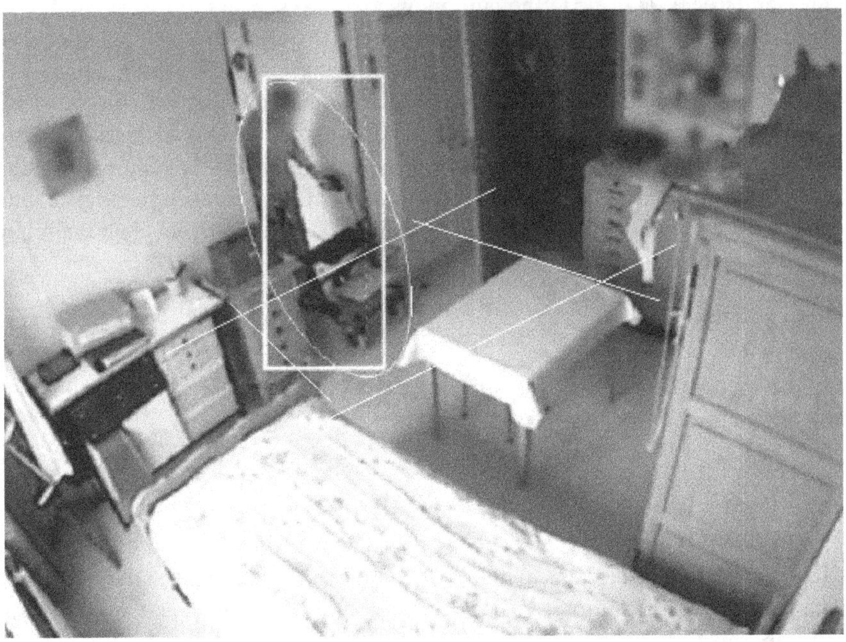

Figure 11.1 Camera view of predefined transfer zone with older adult performing the transfer

performance. It was opted to analyse transfer times instead of gait speed because of the differences in trajectories between participants. Some trajectories might include a 90° turn while others will be performed in a straight line. There can therefore be no comparison between the transfer times of different participants, hence an individual tool for monitoring gait performance is presented.

The transfer times were measured multiple times a day. Baldewijns *et al.* [4] however observed some noisy measurements. There are three main causes for this noise. Firstly, there is the noise which is caused by the video-preprocessing challenges faced in a real-life environment [14]. Next, there is the noise caused by the use of different walking aids which results in a large variation in transfer times [4, 43]. Lastly, there is the noise caused by the difference in walking speed depending on the time of day. Because of these noisy measurements a median of the measurements is calculated each day and used for further evaluation.

From the measured transfer times, Baldewijns *et al.* observed that a stable gait pattern and thus a low fall risk typically results in short transfer times with only small variations. When the fall risk of a person increases or the health of the person declines this is visible by an elongation of the transfer times and a larger variability between the transfer times. More information concerning this preliminary validation study can be found in [4].

11.2.2.2 Assessing temporal and spatial gait parameters through footstep detection algorithms using audio nodes

In Van Den Broeck *et al.* [51] the authors used audio sensors to assess the fall risk of a person. When having multiple recordings of the sound of a footstep captured at different sensor positions the footstep positions can be estimated. Given these positions both temporal and spatial gait parameters can be computed.

The key in estimating the location of an acoustic source lays in the time delay(s) measured between two (or more) microphone signals capturing the same source signal at different places in the room (see Figure 11.2), i.e. if the distance between a source on position s and two microphones is $D_1(s)$ and $D_2(s)$, respectively, then the relative arrival time at the microphone m will be

$$\tau_m(s) = D_m(s)/c, \tag{11.1}$$

with c the speed of sound through air (340.29 m/s at sea level). Thus, the time delay between both microphones will be

$$\tau_{1,2}(s) = (D_1(s) - D_2(s))/c. \tag{11.2}$$

Following this, when a time delay of τ is measured only a few source locations are possible since formula 11.2 has to hold. When increasing the number of microphones the possible source locations can even be limited to a single (the actual) position. To measure this time delay a cross-correlation between different microphone signals is performed. The delay corresponding to the highest correlation value indicates the most likely time delay [49].

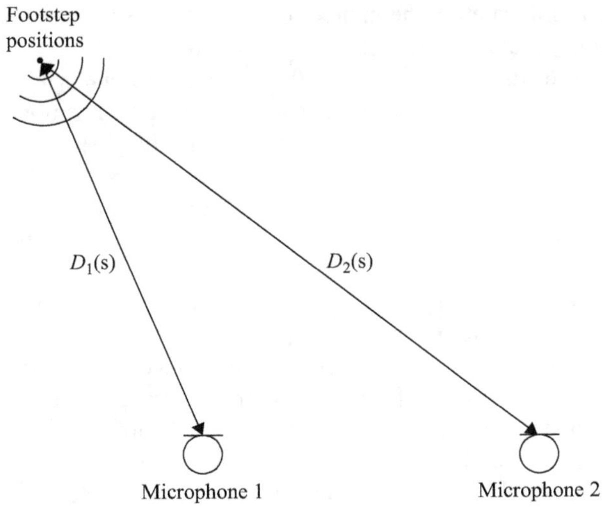

Figure 11.2 Example of acoustic footstep location estimation

Another method, called Steered Response Power (SRP) [49], is also used often. SRP balances all the time delays in the M microphone signals $x_m(t)$ as if the source was on a particular location s and adds them together

$$x_{BF}(s) = \sum_{m=1}^{M} x_m(t - \tau_m(s)). \tag{11.3}$$

This creates a delay-and-sum beam-former [49] cancelling out all acoustic sounds not coming from that particular location. Next SRP uses the energy of $x_{BF}(s)$ as a measure of sound energy on that particular location s

$$srp(s) = E(x_{BF}^2(s)). \tag{11.4}$$

with E the expectation operator. SRP can now scan a grid of possible footstep positions s and the position with the most energy will be used as the estimated footstep position.

In practice, the quality of the footstep position estimation will be dependent on the chosen algorithm and the spatial configuration of the microphones. The total number of microphones and their placement within an array will determine the accuracy of the SRP algorithm [9] and the placement of the array(s) in the room will determine the quality of the captured footstep sounds. Many different microphone set-ups are therefore being used.

For instance, Woo-han *et al.* [53] equipped a robot with two microphones to estimate footstep locations. Since only two microphones are used it was impossible to estimate the footstep location in a 3D space, instead it was able to estimate the azimuth (angle left-right) of the footstep. This is done by first estimating $\tau_{1,2}$, as in (11.2), by cross-correlation. Nonetheless, since a 3D position of the footstep is needed to extract the gait information the number of microphones needs to be increased to make this method suitable for the desired application.

Shoji *et al.* [46] used a square 0.4 × 0.4 m 4-microphone array attached to the ceiling (height 2.5 m). This set-up was able to estimate a 3D position of the footstep (considering footsteps only occurring at floor level). In [46] it is shown that the set-up can estimate footstep locations within a 4 m radius from the array with a maximum error of 20 cm. Despite this, it was also observed that the results degrade with increasing distance to the microphone array. This set-up is therefore only usable in small areas.

Van den Broeck *et al.* [51] proposed a set-up using multiple smaller arrays covering larger areas, trying to overcome this problem. Here, each array uses the SRP method for a number of potential footstep locations s that can be accurately estimated. Then the different SRP estimates from each microphone array are combined to obtain a 2D position estimate. This method is commonly called Global Coherence Field [10]. The strength of this method is that the number of arrays that are used is not fixed at a particular number which allows for a modular set-up that is able to cover all kind of areas. In [51] they moreover show that footstep locations closer than 1.5 m from the nearest array can be estimated with a median of errors smaller than 30 cm.

11.2.2.3 Measuring activity levels with PIR sensors

Motion sensors can provide essential information in assessing the health condition of the monitored person by evaluating the person's movement patterns and thereby they can support a fall risk estimate.

A PIR sensor senses heat radiated as infrared light from objects in its surroundings. Changes in temperature cause the sensor to trigger an event. Therefore, motion from e.g. persons and animals passing by the sensor are detected. Typically, the signal from the sensor is sent through a wireless network together with a time stamp indicating the time the movement was detected.

Often PIR sensors are implemented in a multi-modal system [6], combined with door contact switches [28], video surveillance systems [24, 42], radar [32], etc. Although these systems provide more information, multi-modality adds to the complexity of a monitoring system. Simple monitoring systems with only PIR sensors are less expensive, easy to install and flexible to extend. Such a system has typically one sensor per room or per zone. More sensors per zone give more information but this is likely to be more difficult to interpret [23].

Specifically related to fall risk estimation is the monitoring of walking speed. The PIR sensors allow for continuously assessing the time needed to traverse a path between two defined zones. Changes in the time needed for this trajectory can be indicative for an increased fall risk. Hagler *et al.* [24], for instance, place an array of such PIR sensors on the ceiling of a natural walking path in the home environment. The data resulting from this sensor array are used to measure the gait velocity of the older adult when walking past the sensor array.

Another approach when using PIR sensor is the monitoring of the activity level of a person. The overall number of motion events measured by the different PIR sensors in the home environment can, for instance, be used to assess the activity level of a person. When an important increase or decrease in the activity level of the monitored person is present this could indicate an underlying health problem. Hadidi *et al.* [23], for instance, observed that a decline in the number of events measured by PIR sensors was often accompanied by health-related problems. Moreover Hadidi *et al.* indicated that the number of motion events in a single room can also be used to detect health problems such as, for instance, bad sleep which in turn can cause an elevated fall risk.

11.2.2.4 Other sensors

Other possible sensors used to monitor gait-related variables are accelerometers, radar and 3D cameras.

Accelerometers are small and inexpensive sensors which measure acceleration (the rate of change of the velocity of an object) and can easily be worn on the body. They are often placed on the lower back of an older adult to monitor his movements [25]. Regarding accelerometers, walking speed is used by the majority of research groups for fall risk assessment [25]. Additionally, the monitoring of sit-to-stand transitions also yields good results for an automatic fall risk classification [25]. An overview of the different fall risk assessment algorithms using inertial sensors is given in [19, 25].

The majority in these papers reported studies, however, were validated in laboratory settings, therefore not proving the usability of the technology in real-world environments [19]. In contrast, Weis *et al.* [52] reported that monitoring older adults with accelerometers in their home environment can help identify past and future fallers. During this study older adults were monitored, with the accelerometer placed on the lower back, during three days in their home environment. Several measures such as step-to-step variability, number of steps taken and activity duration were proven useful to evaluate the fall risk of the monitored person.

Rantz *et al.* [44] and Cuddihy *et al.* [12] measure gait speed using a Doppler radar which transmits a 5.8-GHz pulse. The fall risk assessment algorithm relies on the fact that a difference in frequency in the return signal can be measured when a signal is reflected from a moving object. The return signal with the highest signal energy was reflected from the torso and is used to estimate the walking speed.

Stone *et al.* [47] on the other hand extract several gait parameters such as gait velocity and stride parameters from the raw-depth data of the Microsoft Kinect. In this study, foreground (FG) objects are identified from the depth image using a dynamic background subtraction technique. A tracker is subsequently used to track the moving objects across multiple frames. Walking paths are finally identified based on several criteria including path straightness, duration and distance. If the minimum requirements are met, walking speed is computed as well as different stride parameters.

Both Stone *et al.* [47] and Rantz *et al.* [44] validated their corresponding systems in a laboratory environment using the GAITrite walkway.[1] After this they installed 3D cameras, as well as radar sensors in the home environments of older adults and compared the gait velocity measurements from these sensors with the results of several

[1]The GAITrite system is discussed in depth in the gait analysis chapter

clinically proven fall risk assessment tools. This study showed a high correlation between the results of the sensor measurements and those from the fall risk assessment tools; both demonstrating the usefulness of measuring gait velocity in the home environment as well as the usability of such a system in the home environment.

11.2.3 Closing the loop

After the acquisition, the raw sensor data needs to be prepared for visualisation and automatic interpretation. Visualisation should only give the essential information to the caregiver with the least amount of overheads as possible. Automatic interpretation should result in the triggering of an alarm when there is an abnormality in the data. This automatic interpretation should simplify the reviewing process of the data by indicating the caregiver when the data should be inspected and thus when specific health-related problems are occurring.

11.2.3.1 Data visualisation

There are several ways of visually presenting data, each with its own scope, advantages and disadvantages.

Several commercially available products introduced an easy to understand user interface. Regarding the monitoring of Activities of Daily Living (ADL) both Tunstalls ADLife and the Quietcare system tries to cater for the needs of their users [38]. Regarding activity levels the ActivPal system gives an overview of the different activities the monitored person has performed during the day [38].

In research settings the design of user interfaces is a research field in itself. A lot of research effort has been invested in the design of systems which can visualise the daily life patterns of the monitored person such as the research off Mertens *et al.* [35], Robben *et al.* [45] and Munoz *et al.* [39]. Two visualisation methods proposed by Mertens *et al.* are discussed more in-depth.

The examples given in Figures 11.3 and 11.4 are taken from a monitoring research study called AMACS [35]. During this project, three persons were monitored using PIR sensors. In Figure 11.3 data from the sensors from each room is plotted as time series. This way, a reconstruction of the locations where the person spent his day is shown. In order to evaluate a person's habits over a longer period, sensor data is depicted in the form of a heatmap as shown in Figure 11.4. This representation is better suited for visualising medium term trends as will be explained below. In a heatmap, every row represents a day in a time slot resolution which can be chosen. Each time slot has a colour indicating the location where the subject has spent most of the time. Several days are stacked upon each other. This way, behavioural patterns become visible.

Incidents (e.g. fall incidents) will disrupt the regularity of the behaviour patterns and can indicate a possible health condition. This will be shown clearly on the heatmap. It is up to the caregiver to decide if further investigation is needed [21]. Obviously, every person has his own typical representation of behaviour. It is therefore impossible for a caregiver to detect changes in behaviour for every monitored person by the naked eye.

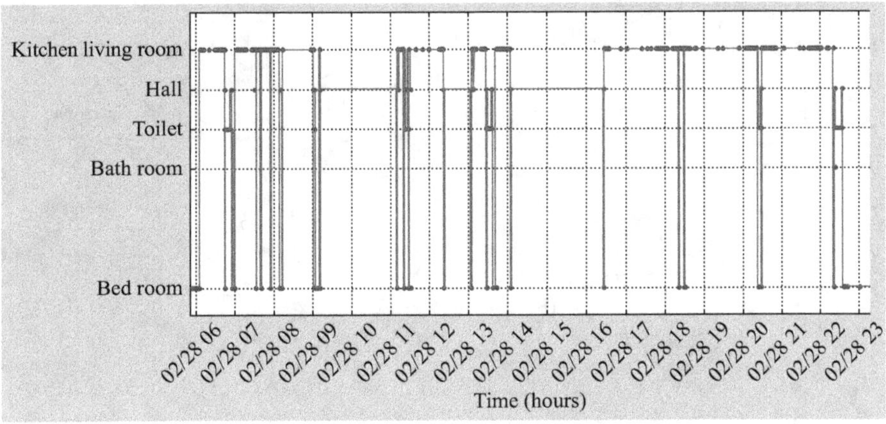

Figure 11.3 Reconstruction of one day based on motion events. Vertical axis represents the location, horizontal is time of day

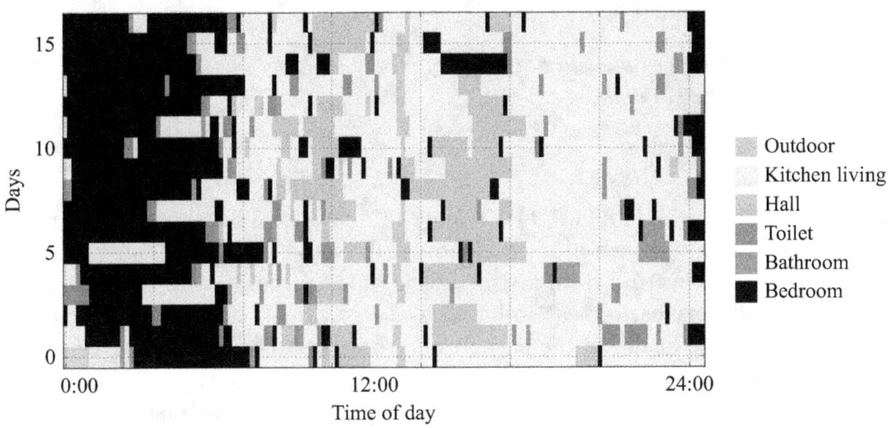

Figure 11.4 Heatmap with every row representing one day, shade of the different timeslots dependent on the location of the person on that particular time

Regarding gait-related parameters the research effort is mostly focussed on the visualisation of a single gait cycle for further clinical analysis such as, for instance, done by Deschamps *et al.* [16]. The visualisation of gait-related variables for the purpose of identifying overall health problems such as that suggested by Baldewijns *et al.* [4] is relatively new.

Baldewijns *et al.* [4] suggest the use of local regression models to visualise trends in the data. In this study the median of the measured transfer times of a participant, calculated per day, is plotted as an indicator of the gait speed of the monitored person. A local regression model is fitted on these medians visualising possible trends in the

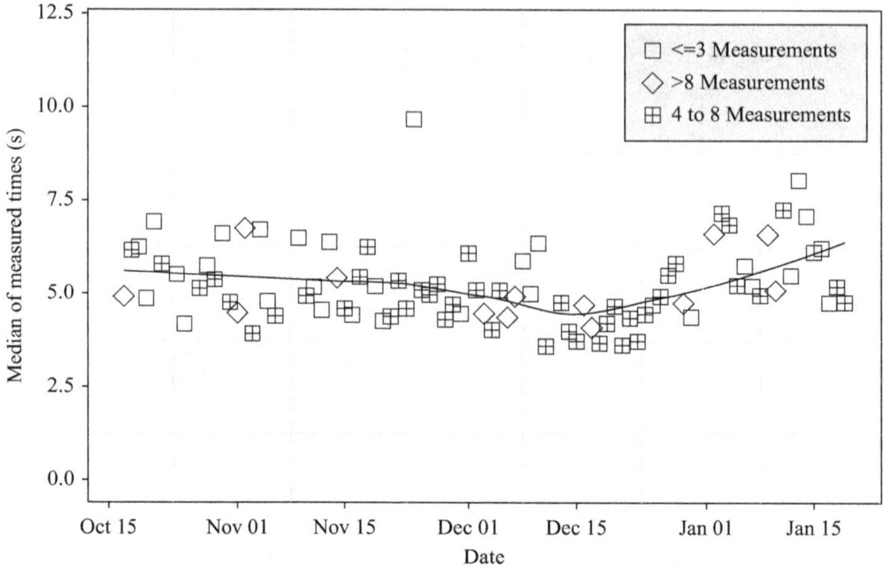

Figure 11.5 Median of measured transfer times calculated per day and fitted local regression model

data such as that shown in Figure 11.5. Figure 11.5 resulted after a monitoring period of three months of a 74-year-old participant. At the start of the monitored period the participant was recovering from a stroke which took place prior to the acquisition phase. This recovery period was followed by several small strokes which started a rapid decline in the health of the participant. These health-related observations are visible in the medians and the local regression model. The recovery period is visible as a decrease in the time needed to cross the transfer zone. In contrast the decline in health is visible by an increase in the needed time. Using a local regression model will facilitate the interpretation of the data and will therefore assist healthcare workers when visually inspecting the data. More details concerning this observational study can be found in [4].

11.2.3.2 Automatic data interpretation

A major drawback in current automatic fall risk evaluation tools is that the data resulting from these systems needs to be reviewed manually by healthcare workers for each patient individually. Even when the data visualisation is easy to interpret and the information is thus readily accessible this can be a very time-consuming process for which specialized knowledge is needed.

When reviewing gait-related variables Baldewijns *et al.* [5] propose the use of Statistical Process Control (SPC) techniques to detect when deviations in the transfer time occur. When such a deviation is detected an alarm can be sent to healthcare workers. This alarm eliminates the need to review the data manually and shortens the response time creating the opportunity to intervene before any major health problems arise.

SPC techniques are typically used to follow industrial processes while they are running using control charts. Baldewijns *et al.* [5] however proved that these control charts can also be used to detect deviating trends in the transfer times. Three different control charts were evaluated for the proposed application, namely: Tabular CUSUM, Standardized CUSUM and Exponentially Weighted Moving Average (EWMA) control charts. These three techniques are typically used to detect small shifts in the data by using information contained in the previous days as well as that from the current day.

Baldewijns *et al.* [5] generated an extensive simulation dataset for evaluation, optimization and validation of the different SPC techniques. The properties of this dataset were chosen based on the properties observed in the real-life datasets. Baldewijns *et al.* observed that the best performing method was the EWMA control chart. After optimization this control chart was able to detect 82% of all deviating trends in the transfer times within 10 days after the start of the trend.

EWMA calculates a weighted moving average of the medians of the transfer times using

$$z_i = \lambda x_i + (1 - \lambda)z_{(i-1)} \tag{11.5}$$

with λ the weighing factor chosen between 0 and 1. The starting value of z_0 is chosen the same as the target value μ_0. This weighted moving average in turn needs to remain between the upper and lower control limits defined by

$$\text{UCL} = \mu_0 + L\sigma\sqrt{\frac{\lambda}{2-\lambda}[1-(1-\lambda)^{2i}]} \tag{11.6}$$

and

$$\text{LCL} = \mu_0 - L\sigma\sqrt{\frac{\lambda}{2-\lambda}[1-(1-\lambda)^{2i}]} \tag{11.7}$$

with L determining the width of the control limits. An example EWMA control chart is given in Figure 11.6.

Baldewijns *et al.* [5] determined the target value and upper and lower control limits based on the first two weeks of measured transfer times data. The median of these first two weeks was chosen as the target value whereas a multiple of the standard deviation was chosen for the upper and lower control limits, therefore providing the opportunity to take into account the normal variations in the medians of the transfer times.

The advantage of this technique is that it is a generic technique. Although it was validated on transfer times it can be used equally well to detect changes in other gait-related parameters. However, a new optimization step will be needed. The system also works independent from the sensor used to measure the transfer times or the gait speed.

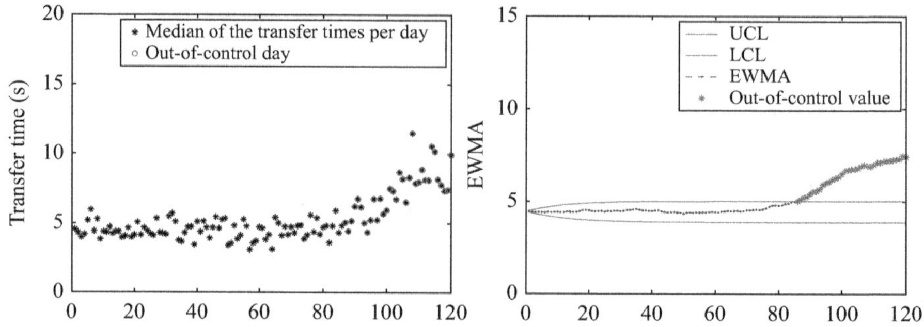

Simulated medians per day where an increase in transfer times is present after Day 80

EWMA control chart with a detected trend

Figure 11.6 Simulated medians and corresponding EWMA control chart where a negative trend (thus an elongation of the transfer times and an increase in the standard deviation) is present in the data after day 80

11.3 Fall detection

The previously mentioned fall risk estimation systems can reduce the overall number of fall incidents, but cannot eliminate them completely. To reduce the consequences of a fall incident, an accurate fall detection system is still needed. Similar to the fall risk estimation tools two main groups of fall detection sensors can be distinguished: those who need to be worn on the body and contactless sensors.

The main advantage of body-worn sensors it that they can detect a fall incident regardless of the location of the person and the number of persons in the room. However, they need to be worn at all times which can be a problem with people with dementia and it is sometimes experienced as stigmatizing. Contactless sensors such as 2D and 3D cameras and radar on the other hand are considered to be unobtrusive but the falls can only be detected in rooms in which the sensors are installed. Both types of sensors thus have their own advantages and disadvantages, leaving the user to decide their preference.

An extensive overview of the different fall detection algorithms published up to 2013 is given in [37]. The remainder of the chapter will therefore focus on more recent developments in this field.

11.3.1 Accelerometer-based fall detection systems

Many different approaches are being used to detect fall incidents either using only accelerometers or by combining accelerometer and gyroscope data. Typically the algorithms that distinguish between a fall or no fall use a set of parameters calculated on the accelerometer and gyroscope data. Such is, for instance, the case in [17] where a set of 6 different parameters are calculated and processed sequentially. Even if the number of parameters calculated by the algorithm is limited some type of impact detection is mostly present. In the majority of algorithms this impact detection is done by thresholding the acceleration sum vector (SV) calculated by

$$SV = \sqrt{a_x^2 + a_y^2 + a_z^2} \tag{11.8}$$

with x, y and z the acceleration in the three directions. This impact detection is often followed by an orientation detection either by comparing the position of the person before the impact with the position of the person after the fall incident or by following a change in orientation after the impact, indicating when a person is getting up after a fall incident and no alarm needs to be triggered as is done by Kung *et al.* [29].

The majority of accelerometer- and gyroscope-based algorithms report good results in both positive predictive value (PPV) and sensitivity on simulated data. Despite this, Kangas *et al.* [26] showed that large differences can be observed between the accelerometer data from real-life falls and falls that are simulated by healthy participants. In order to assess how the PPV and sensitivity of different accelerometer-based algorithms responds to a set of real-life falls, Bagala *et al.* [3] evaluated 13 different accelerometer-based fall detection algorithms on real-life accelerometer data. The results from this validation using real-life data are considerably worse than the reported results on the simulated datasets. The algorithms with the best results on the real-life dataset were the algorithms from Chen *et al.* [11] and from Bourke *et al.* [8].

11.3.2 Radar-based fall detection

Radar-based fall detection algorithms use the frequency information of the captured signal to detect fall events. Wu *et al.* [54], for instance, used the extreme frequency, extra frequency ratio and the length of the event period. These features are presented to a sparse Bayesian classifier which classifies an event as a fall or non-fall event.

Liu *et al.* [31] installed two Doppler radar sensors in the room. For each event a two second radar signal was used to calculate six Melfrequency cepstral coefficients. These coefficients are subsequently presented to three different classification algorithms being k-nearest neighbours, a support vector machine and naive Bayes. For each classifier a fall confidence level is calculated. These confidence levels are fused using fuzzy integrals to calculate a general confidence level which is thresholded to provide a classification.

In previously mentioned studies the radar signal is transformed into a single feature vector. Garripoli *et al.* [22] and Karsmakers *et al.* [27] in contrast represent an activity as a sequence of feature vectors. This sequence of feature vectors is subsequently classified as a fall or non-fall activity using the Least Squares Support Vector Machine framework with a global alignment kernel.

11.3.3 Video-based fall detection

Another sensor that has received a lot of research attention is the 2D camera. Typically, first a background subtraction technique is used to distinguish moving objects (the person) from their surroundings. After this different features such as the speed of the head of the person and a change in aspect ratio of the bounding box (BB) are used to distinguish a fall incident from other activities. This type of fall detector is for instance implemented by Lee and Chung [30] and by Nguyen *et al.* [40]. The accuracy of this type of fall detection algorithm however relies heavily on an accurate silhouette extraction. The accuracy of this silhouette extraction can in real-life situations be reduced by occlusions, overexposure, changes in illumination, etc. [14]. To overcome these difficulties Debard *et al.* [13] propose a combination of background subtraction (BS) and a particle filter (PF) to increase the robustness of the update of the background (BG) model.

The BS method used by Debard *et al.* [15] is based on an approximate median filter combined with cross-correlation to remove shadows from the image. Next an erosion/dilation step together with a connected components analysis removes noisy patches and finds the moving objects (the person) in the image. Lastly, a PF is used in combination with a person detector to increase the robustness of the update of the background model.

In general particle filters estimate the probability distribution $p(S_t|Z_1...Z_t)$ of a state vector S_t of a tracked object given Z_t. This probability can be approximated from a set of N weighted samples or particles. Debard *et al.* [13] use a Bootstrap filter. In their research each particle represents an ellipse, with centre coordinates, height, width, and angle and the speed of change of each of these values. The weight of each particle is determined by using a combination of foreground, weighted structural histogram, and person detection coefficients. The PF is initialized when a foreground object of over 4000 pixels is detected. Figure 11.7 shows an overview of the algorithm

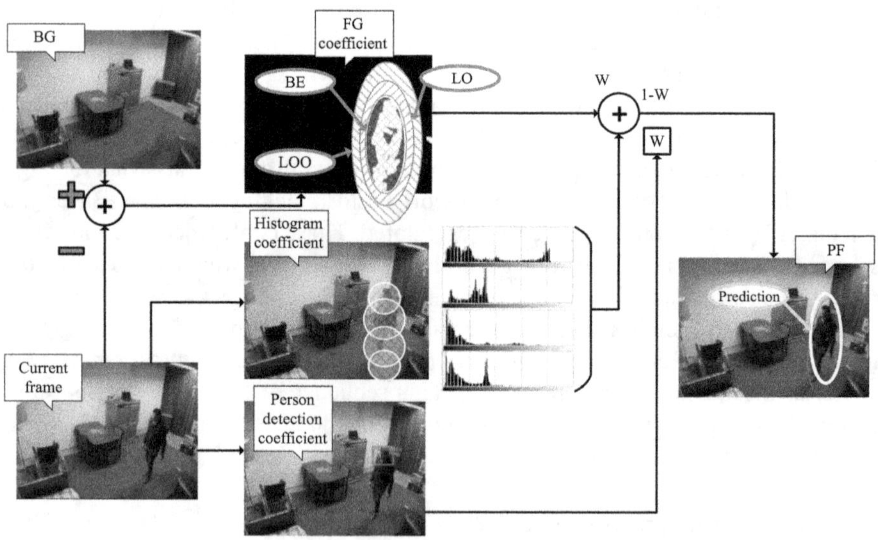

Figure 11.7 Overview of implementation of particle filter (PF) with the different measurement coefficients. (BG: background; FG: foreground; BE: bounding ellipse; LO: first outer layer; LOO: second outer layer; W: weight factor)

together with the different measurement coefficients used to determine the weight of the different particles.

Three coefficients are used to determine the weight of each particle and thus the probability that this particle represents the person in the field of view of the camera:

- The foreground (FG) coefficient C_F matches an ellipse to the FG that covered most of the FG pixels without including too many BG pixels.
- The histogram coefficient C_H is based on histogram matching of the image inside the bounding ellipse. A weighted structured histogram was used to make the histogram more distinctive. The ellipse was split into four parts representing head and shoulders, chest, abdomen and hips and lower legs. For each part an intensity histogram was calculated in which the centre had a higher weight than the edges to reduce the impact of possible BG pixels. Every frame in the histogram model was updated with 5% of the current prediction.
- A person detection coefficient is calculated every five frames using an upper-body detector [18]. This coefficient was used to replace low-weight particles and update the histogram model. Also the weight of the FG and the histogram coefficient were adapted accordingly to the score of the detector.

The final weight of each particle was calculated as a combination of C_F and C_H. The predicted ellipse for the tracked person in turn was calculated as the mean of the five best particles. Using this result the fall detection features were calculated.

The five features were: aspect ratio (AR), change in AR (CAR), fall angle (FA), centre speed (CS) and head speed [15]. Debard *et al.* create a feature vector that contains the mean and maximum values of the five fall features over different time slots from before, during, and after the fall incident and thus use multiple frames to assess if a fall incident has occurred. This feature vector in turn is used by an SVM to perform the needed classification.

11.3.4 Kinect-based fall detection

Some research groups are now considering the Microsoft Kinect for fall detection. The advantage of this system is that only one sensor is needed to provide a 3D image of the monitored room. Recent studies show good results when using the Kinect in a real-life environment.

Mastorakis and Makris calculate a 3D bounding box around the person using the depth image of the Kinect camera. From this bounding box the width, height and depth are calculated. During a fall the height of the bounding box will decrease while the depth and the width will expand. The velocity in which these changes occur is thresholded to distinguish between different fall phases such as 'fall initiated' and 'inactivity detected'. Only when the different phases are all detected in the right order an alarm is triggered [33].

Stone *et al.* implemented a two stage fall detection algorithm on the raw Kinect depth data. In the first stage the person's vertical state is detected in the individual depth frames by tracking the person over the different frames. After a person is found to be on the ground an ensemble of decision trees calculate a confidence level that this on the ground event was preceded by a fall incident. The system of Stone *et al.* is currently running real-life in the TigerPlace retirement community (Columbia, Missouri) since August 2013. The staff receives e-mail alerts when the system detects a fall with one of the residents [44, 48].

11.4 Conclusion

A wide variety of sensor types and algorithms are currently used for automatic fall risk estimation and fall detection.

Concerning automatic fall risk estimation in the home environment this chapter showed that gait-related parameters are often used as a measure of the fall risk of a person. Several sensor systems including 2D cameras, audio sensors, radar, 3D cameras, accelerometers, etc. can be used to measure these parameters. However, not only measuring these parameters but also visualisation and automatic interpretation of the data are important aspects to consider when developing an automatic fall risk assessment system. Some suggestions for both visualisation and interpretation of the data are therefore made in this chapter.

Similar sensors used for automatic fall risk estimation are also used in fall detection set-ups. This chapter gave an overview of the current state-of-the-art for both contactless and body-worn fall detection sensors. The accompanying algorithms are discussed. Visualisation, however, is not such as big an issue as with the automatic fall risk estimation tools. This was therefore not included in the chapter.

References

[1] World Health Organization. Ageing and Life Course Unit, *WHO global report on falls prevention in older age*, World Health Organization, Copenhagen, 2008.

[2] Daniel Austin, Tamara L Hayes, Jeffrey Kaye, Nora Mattek, and Misha Pavel, *On the disambiguation of passively measured in-home gait velocities from multi-person smart homes*, Journal of ambient intelligence and smart environments **3** (2011), no. 2, 165.

[3] Fabio Bagalà, Clemens Becker, Angelo Cappello, Lorenzo Chiari, Kamiar Aminian, Jeffrey M Hausdorff, Wiebren Zijlstra, and Jochen Klenk, *Evaluation of accelerometer-based fall detection algorithms on real-world falls*, PloS One **7** (2012), no. 5, e37062.

[4] Greet Baldewijns, Glen Debard, Marc Mertens, Els Devriendt, Koen Milisen, Jos Tournoy, Tom Croonenborghs, and Bart Vanrumste, *Automated in-home gait transfer time analysis using video cameras*, Journal of Ambient Intelligence and Smart Environments **8** (2015), no. 3, 273–286.

[5] Greet Baldewijns, Stijn Luca, William Nagels, Bart Vanrumste, and Tom Croonenborghs, *Automatic detection of health changes using statistical process control techniques on measured transfer times of elderly*, Engineering in

Medicine and Biology Society (EMBC), 2015 Annual International Conference of the IEEE, IEEE, Milano, (2015), 5046–5049.

[6] T.S. Barger, D.E. Brown, and M. Alwan, *Health-status monitoring through analysis of behavioral patterns*, IEEE Transactions on Systems, Man and **35** (2005), 22–27.

[7] H A Bischoff-Ferrari, B Dawson-Hughes, H B Staehelin, J E Orav, A E Stuck, R Theiler, J B Wong, A Egli, D P Kiel, and J Henschkowski, *Fall prevention with supplemental and active forms of vitamin d: a meta-analysis of randomised controlled trials*, BMJ **339** (2009), b3692.

[8] AK Bourke, P Van de Ven, M Gamble, R O'Connor, K Murphy, E Bogan, E McQuade, P Finucane, G ÓLaighin, and J Nelson, *Evaluation of waist-mounted tri-axial accelerometer based fall-detection algorithms during scripted and continuous unscripted activities*, Journal of Biomechanics **43** (2010), no. 15, 3051–3057.

[9] Darren W. Brandstein M., *Microphone arrays: signal processing techniques and applications*, Springer Science & Business Media, Berlin, 2001.

[10] Alessio Brutti, Maurizio Omologo, and Piergiorgio Svaizer, *Oriented global coherence field for the estimation of the head orientation in smart rooms equipped with distributed microphone arrays*, Ninth European Conference on Speech Communication and Technology, Lisbon, 2005.

[11] Jay Chen, Karric Kwong, Dennis Chang, Jerry Luk, and Ruzena Bajcsy, *Wearable sensors for reliable fall detection*, Engineering in Medicine and Biology Society, 2005. IEEE-EMBS 2005. 27th Annual International Conference of the, IEEE, Shanghai, 2006, pp. 3551–3554.

[12] Paul E Cuddihy, Tarik Yardibi, Zachary J Legenzoff, Liang Liu, Calvin E Phillips, Carmen Abbott, Colleen Galambos, James Keller, Mihail Popescu, Jessica Back, *et al.*, *Radar walking speed measurements of seniors in their apartments: Technology for fall prevention*, Engineering in Medicine and Biology Society (EMBC), 2012 Annual International Conference of the IEEE, IEEE, 2012, pp. 260–263.

[13] Glen Debard, Greet Baldewijns, Toon Goedemé, Tinne Tuytelaars, and Bart Vanrumste, *Camera-based fall detection using a particle filter*, Engineering in Medicine and Biology Society (EMBC), 2015 37th Annual International Conference of the IEEE, IEEE, 2015, pp. 6947–6950.

[14] Glen Debard, Peter Karsmakers, Mieke Deschodt, Ellen Vlaeyen, Eddy Dejaeger, Koen Milisen, Toon Goedemé, Bart Vanrumste, and Tinne Tuytelaars, *Camera-based fall detection on real world data*, Outdoor and large-scale real-world scene analysis, Springer, Berlin, 2012, pp. 356–375.

[15] Glen Debard, Marc Mertens, Mieke Deschodt, Ellen Vlaeyen, Els Devriendt, Eddy Dejaeger, Koen Milisen, Jos Tournoy, Tom Croonenborghs, Toon Goedemé, Tinne Tuytelaars, and Bart Vanrumste, *Camera-based fall detection using real-world versus simulated data: how far are we from the solution?*, Journal of Ambient Intelligence and Smart Environments **8** (2016), no. 2, 149–168.

[16] Kevin Deschamps, Filip Staes, Dirk Desmet, Philip Roosen, Giovanni Arnoldo Matricali, Noel Keijsers, Frank Nobels, Jos Tits, and Herman Bruyninckx, *A color-code based method for the interpretation of plantar pressure measurements in clinical gait analysis*, Gait & Posture **41** (2015), no. 3, 852–856.

[17] Mihail Dumitrache and Sever Pasca, *Fall detection algorithm based on triaxial accelerometer data*, E-Health and Bioengineering Conference (EHB), 2013, IEEE, 2013, pp. 1–4.

[18] Martin Eichner, Manuel Marin-Jimenez, Andrew Zisserman, and Vittorio Ferrari, *Articulated human pose estimation and search in (almost) unconstrained still images*, ETH Zurich, D-ITET, BIWI, Technical Report No **272** (2010).

[19] Andreas Ejupi, Stephen R Lord, and Kim Delbaere, *New methods for fall risk prediction*, Current Opinion in Clinical Nutrition & Metabolic Care **17** (2014), no. 5, 407–411.

[20] J. Fleming and C. Brayne, *Inability to get up after falling, subsequent time on floor, and summoning help: prospective cohort study in people over 90*, BMJ **337** (2008), a2227.

[21] Colleen Galambos, Marjorie Skubic, Shuang Wang, and Marilyn Rantz, *Management of dementia and depression utilizing in-home passive sensor data*, Gerontechnology **11** (2013), no. 3, 457–468.

[22] Carmine Garripoli, Marco Mercuri, Peter Karsmakers, Ping Jack Soh, Giovanni Crupi, Guy Vandenbosch, Calogero Pace, Paul Leroux, Dominique Schreurs, *et al.*, *Embedded dsp-based telehealth radar system for remote in-door fall detection*, IEEE Journal of Biomedical and Health Informatics **19** (2015), no. 1, 92–101.

[23] Tareq Hadidi and Norbert Noury, *A predictive analysis of the night-day activities level of older patient in a health smart home*, Ambient Assistive Health and Wellness Management in the Heart of the City, Springer, Berlin, 2009, pp. 290–293.

[24] Stuart Hagler, Daniel Austin, Tamara L Hayes, Jeffrey Kaye, and Misha Pavel, *Unobtrusive and ubiquitous in-home monitoring: a methodology for continuous assessment of gait velocity in elders*, IEEE Transactions on Biomedical Engineering **57** (2010), no. 4, 813–820.

[25] Jennifer Howcroft, Jonathan Kofman, and Edward D Lemaire, *Review of fall risk assessment in geriatric populations using inertial sensors*, Journal of Neuroengineering and Rehabilitation **10** (2013), no. 1, 10–91.

[26] Maarit Kangas, I Vikman, L Nyberg, R Korpelainen, J Lindblom, and T Jämsä, *Comparison of real-life accidental falls in older people with experimental falls in middle-aged test subjects*, Gait & posture **35** (2012), no. 3, 500–505.

[27] Peter Karsmakers, Tom Croonenborghs, Marco Mercuri, Dominique Schreurs, and Paul Leroux, *Automatic in-door fall detection based on microwave radar measurements*, Radar Conference (EuRAD), 2012 9th European, IEEE, Amsterdam, 2012, pp. 202–205.

[28] Jeffrey A Kaye, Shoshana A Maxwell, Nora Mattek, Tamara L Hayes, Hiroko Dodge, Misha Pavel, Holly B Jimison, Katherine Wild, Linda Boise, and

Tracy A Zitzelberger, *Intelligent Systems for Assessing Aging Changes: Home-Based, Unobtrusive, and Continuous Assessment of Aging*, Journal of Gerontology: Series B **66** (2011), 180–190.

[29] Hsu-Yang Kung, Chin-Yu Ou, Shin-Di Li, Chun-Hao Lin, Hong-Jie Chen, Yu-Lun Hsu, Miao-Han Chang, and Che-I Wu, *Efficient movement detection for human actions using triaxial accelerometer*, Consumer Electronics (ICCE), 2010 Digest of Technical Papers International Conference on, IEEE, 2010, pp. 113–114.

[30] Y Lee and Wan-Young Chung, *Automated abnormal behavior detection for ubiquitous healthcare application in daytime and nighttime*, 2012 IEEE-EMBS International Conference on Biomedical and Health Informatics (BHI), IEEE, San Diego, 2012, pp. 204–207.

[31] Liang Liu, Mihail Popescu, Marilyn Rantz, and Marjorie Skubic, *Fall detection using doppler radar and classifier fusion*, 2012 IEEE-EMBS International Conference on Biomedical and Health Informatics (BHI), IEEE, San Diego, 2012, pp. 180–183.

[32] Liang Liu, Mihail Popescu, Marjorie Skubic, and Marilyn Rantz, *An automatic fall detection framework using data fusion of Doppler radar and motion sensor network*, 2014 36th Annual International Conference of the IEEE Engineering in Medicine and Biology Society (EMBC), IEEE, 2014, pp. 5940–5943.

[33] Georgios Mastorakis and Dimitrios Makris, *Fall detection system using Kinect's infrared sensor*, Journal of Real-Time Image Processing **9** (2014), no. 4, 635–646.

[34] Jim McCambridge, John Witton, and Diana R Elbourne, *Systematic review of the hawthorne effect: new concepts are needed to study research participation effects*, Journal of Clinical Epidemiology **67** (2014), no. 3, 267–277.

[35] Marc Mertens, Glen Debard, Jonas Van Den Bergh, Toon Goedemé, Koen Milisen, Jos Tournoy, Jesse Davis, Tom Croonenborghs, and Bart Vanrumste, *Towards automatic monitoring of activities using contactless sensors*, Annual Belgian Dutch Conference on Machine learning, The Hague, Netherlands, vol. 20, 2011.

[36] K. Milisen, E. Detroch, K. Bellens, T. Braes, K. Dierickx, W. Smeulders, S. Teughels, E. Dejaeger, S. Boonen, and W. Pelemans, *Falls among community-dwelling elderly: a pilot study of prevalence, circumstances and consequences in flanders*, Tijdschr Gerontol Geriatr. **35(1)** (2004), 15–20.

[37] M. Mubashir, L. Shao, and L. Seed, *A survey on fall detection: Principles and approaches*, Neurocomputing **100** (2013), no. 0, 144 – 152, Special issue: Behaviours in video.

[38] M. Mulvenna, W. Carswell, P. McCullagh, J.C. Augusto, Huiru Zheng, P. Jeffers, Haiying Wang, and S. Martin, *Visualization of data for ambient assisted living services*, IEEE Communications Magazine **49** (2011), no. 1, 110–117.

[39] Andrés Muñoz, Emilio Serrano, Ana Villa, Mercedes Valdés, and Juan A Botía, *An approach for representing sensor data to validate alerts in ambient assisted living*, Sensors **12** (2012), no. 5, 6282–6306.

[40] Viet Dung Nguyen, Minh Thao Le, Anh Duc Do, Hoang Hai Duong, Toan Dat Thai, and Duc Hoa Tran, *An efficient camera-based surveillance for fall detection of elderly people*, 2014 IEEE 9th Conference on Industrial Electronics and Applications (ICIEA), IEEE, Hangzhou, 2014, pp. 994–997.

[41] Karen L Perell, Audrey Nelson, Ronald L Goldman, Stephen L Luther, Nicole Prieto-Lewis, and Laurence Z Rubenstein, *Fall risk assessment measures an analytic review*, The Journal of Gerontology Series A: Biological Sciences and Medical Sciences **56** (2001), no. 12, M761–M766.

[42] Andrea Prati, Roberto Vezzani, Luca Benini, Elisabetta Farella, and Piero Zappi, *An integrated multi-modal sensor network for video surveillance*, Proceedings of the Third ACM International Workshop on Video Surveillance & Sensor Networks (New York, NY, USA), VSSN '05, ACM, 2005, pp. 95–102.

[43] Steven Puttemans, Greet Baldewijns, Tom Croonenborghs, Bart Vanrumste, and Toon Goedemé, *Automated walking aid detector based on indoor video recordings*, Proceedings of the 37th Annual International Conference of the IEEE Engineering in Medicine and Biology Society, 2015, pp. 5040–5045.

[44] Marilyn Rantz, Marjorie Skubic, Carmen Abbott, Colleen Galambos, Mihail Popescu, James Keller, Erik Stone, Jessie Back, Steven J Miller, and Gregory F Petroski, *Automated in-home fall risk assessment and detection sensor system for elders*, The Gerontologist **55** (2015), no. Suppl 1, S78–S87.

[45] S. Robben, M. Boot, M. Kanis, and B. Krose, *Identifying and visualizing relevant deviations in longitudinal sensor patterns for care professionals*, 2013 7th International Conference on Pervasive Computing Technologies for Healthcare (PervasiveHealth), May 2013, pp. 416–419.

[46] Masanari Shoji, *Passive acoustic sensing of walking*, Intelligent Sensors, 2009 International Conference on Intelligent Sensors, Sensor Networks and Information Processing (ISSNIP), IEEE, Melbourne, 2009, pp. 219–224.

[47] Erik E Stone and Marjorie Skubic, *Unobtrusive, continuous, in-home gait measurement using the Microsoft Kinect*, IEEE Transactions on Biomedical Engineering, **60** (2013), no. 10, 2925–2932.

[48] ———, *Fall detection in homes of older adults using the Microsoft Kinect*, IEEE Journal of Biomedical and Health Informatics, **19** (2015), no. 1, 290–301.

[49] Ivan Jelev Tashev, *Sound capture and processing: practical approaches*, John Wiley & Sons, Chichester, 2009.

[50] Morag E Taylor, Marjolein M Ketels, Kim Delbaere, Stephen R Lord, A Stefanie Mikolaizak, and Jacqueline CT Close, *Gait impairment and falls in cognitively impaired older adults: an explanatory model of sensorimotor and neuropsychological mediators*, Age and Ageing **41** (2012), no. 5, 665–669.

[51] Bert Van Den Broeck, Lode Vuegen, Hugo Van hamme, Marc Moonen, Peter Karsmakers, and Bart Vanrumste, *Footstep localization based on in-home microphone-array signals*, Association for the Advancement of Assistive Technology in Europe (AAATE 2013), 2013.

[52] Aner Weiss, Marina Brozgol, Moran Dorfman, Talia Herman, Shirley Shema, Nir Giladi, and Jeffrey M Hausdorff, *Does the evaluation of gait quality during daily life provide insight into fall risk? a novel approach using 3-day*

accelerometer recordings, Neurorehabilitation and neural repair **27** (2013), no. 8, 742–752.

[53] Kyu-Dae B. Su-young J. Woo-han Y., Cheon-in O., *The impulse sound source tracking using kalman filter and the cross-correlation*, In proc. SICE-ICASE International Joint Conference 2006 (South Korea, Busan), 2006.

[54] Qisong Wu, Yimin D Zhang, Wenbing Tao, and Moeness G Amin, *Radar-based fall detection based on doppler time–frequency signatures for assisted living*, IET Radar, Sonar & Navigation **9** (2015), no. 2, 164–172.

Chapter 12
Supporting Activities of Daily Living

Colin Shewell[1], *Chris Nugent*[1], *Mark Donnelly*[1], *and Haiying Wang*[1]

12.1 Introduction

As a result of the increasing developments in medicine and health, the oldest (aged 65 plus) in society are now regarded the most rapidly expanding group [1]. This has resulted in a surge of age-related conditions, such as dementia. One potential approach towards reducing the associated burden placed on existing healthcare models of care delivery is to consider the adoption of Ambient Assisted Living (AAL). AAL promotes the potential to enable inhabitants to maintain the ability to remain within their own home for longer, allowing them to maintain an improved quality of life (QoL), thereby reducing the burden on formal care services and delaying the potential requirement to be re-situated within full-time care facilities [1]. AAL is typically realised through the use of sensor technology, to monitor the occupants activities and to afford support with task initiation or completion, if required. This chapter will focus on the use of AAL within smart environments towards supporting Activities of Daily Living (ADL) through the use of a smart environment. There are many methods in supporting ADL; however, they all share a common underlying methodology along with common technology, along with challenges to this area which require further research.

12.2 What is an activity of daily living

ADL is a term used to represent the set of common tasks that comprise of one's daily self-care requirements. The ADL concept was initially proposed by Dr Sidney Katz and his team at the Benjamin Rose Hospital in Cleveland and has since evolved into its present categorisation of activities [2]. ADLs can be separated into two main categories: Basic Activities of Daily Living (BADL) and Instrumental Activities

[1]Ulster University, Computer Science Research Institute and School of Computing and Mathematics, Ulster University, Newtownabbey, Co. Antrim, Northern Ireland, (UK)

of Daily Living (IADL). To maintain BADL, one requires a basic competency of self-care tasks such as bathing, dressing, eating, and toileting along with the care for personal devices such as a hearing aid. IADL typically requires more advanced skills as they require the use of higher functions such as social skills, use of electronic devices, and handling money. Activities categorised as a BADL include bathing, showering, toilet/personal hygiene, eating, and sleep. IADL activities include care of others, emergency responses, child rearing, financial/health management, meal preparation, and shopping. It is important to note that the ability to carry out individual ADL does not degrade in a linear fashion, for example, activities such as bathing and dressing become increasingly impaired as dementia progresses whereas activities such as toileting and feeding remain relatively intact even as their condition deteriorates [3]. A possible explanation for this could be due to different cognitive areas being associated with the performance on differing ADL, rather than all ADL [3].

The inability to carry out these activities can result in a loss of self-esteem and instil a deep sense of dependence to the person, along with a possible disturbance in family roles as partners are frequently required to assume the position of caregiver when the ability to carry out BADL/IADL is compromised. Supporting people in ADL allows them to experience a higher QoL as ADL performance results in a significant increase in QoL [3], promotes a sense of independence, and reduces the burden placed on caregivers.

Several scales are used to assess and individual's ability to perform ADLs. One such scale is the Katz Index [4], which provides a basis for measuring, predicting, and comparing decline and recovery of the person's condition and the level of support that will be required in order for them to successfully carry out ADL [5]. The Bristol ADL scale was developed in collaboration with caregivers to provide an assessment of people with mild dementia, living in the community [6]. A person's care requirements differ depending on the degree of cognitive decline with some people losing the ability to follow instructions, or forgetting the sequence or next step of a task part way through its completion. Their ability to maintain focus on a task may decline and they often stop recognising common objects or forget how to interact with them. The goal of supporting ADL through the use of AAL is for the affected person and their family to adapt to their life changes and to allow them to carry on with their daily lives as fully independent as possible with the beneficial side effect of reducing the stress placed on the healthcare system through an ageing society.

Technology offers increasing opportunities within the domain of AAL to provide increased support for those who require assistance with ADL. This is especially true in the earlier stages of cognitive decline when the person is still able to carry out tasks with a degree of independence and only require reminders or brief instructions on carrying out a task. Traditionally these reminders or instructions would have to be left by caregivers, either through one-to-one contact or through the use of reminders left throughout the environment such as post-it notes left on items or instructions left throughout the home. Table 12.1 shows a small comparative list of such support alongside technological means of offering comparative support.

Table 12.1 Comparison of differing provisions of care for ADL from a traditional and technological perspective

Traditional care	Technological care
Person to keep a diary for appointments	Automated calendar reminder to publish reminders/alerts to person's smart-phone or display
Important items to be kept in the same place	Alarmed receiver attached to important objects to aid in locating
Put labels on doors/cupboards	Wearable camera to recognise and remind persons of door/cupboard contents
Place important numbers by the phone	Phone with pre-stored numbers represented by familiar faces
Place note on back of door as reminder to take keys	Door sensor to remind occupant to take keys when door is opened
Label family photographs	Facial recognition to act as a reminder
Pin a weekly timetable to the wall	Automated calendar reminder to smart-phone or display
Write reminders to lock door at night, turn off gas, put rubbish out, etc.	Automated systems in house to take action at certain times, such as turning off cookers, locking doors, etc.

12.3 Leveraging smart environments to support ADLs

The solution of a smart environment has long been proposed as a means to ease this burden of an ageing population, which affords occupants who would normally require the assistance of carers, to be supported within their own home, through the use of technology based solutions and gain a larger degree of independence. This has been brought into further acceptance by the recent advent of consumer smart home appliances designed to be retro fitted into existing homes. A smart environment has been defined as being one that is 'able to acquire and apply knowledge about the environment and its inhabitants in order to improve their experience in that environment' [7]. In essence, a smart environment consists of distributed technology throughout an environment and encompasses room level equipments such as lighting sensors through to object-specific sensors such as automated switches. These technologies exist in order to gather information about an environment, which is then used to automate that environment, such as adjusting temperature via the heating system. This information can also be relayed back the person for specific interaction or to display reminders or alerts that is relevant to the person [8].

Regardless of the type of technology implemented the overall goal of a smart environment is to improve the QoL for those within the environment, often those with cognitive impairment, in order to offer greater levels of independence, and to reduce the need for/delay institutionalisation. This is achieved through wearable and environmental sensors that allows the facilitation of preventive care along with the monitoring of chronic conditions. It is in the purest sense an example of ubiquitous and pervasive computing which represents the idea of 'computing everywhere', in other words, making computing and communication effectively transparent [9].

12.4 Technology as an enabler

Selecting an appropriate technology to assist in ADL can be challenging with a wide range of competing technologies available. Generally, the function of these technologies can be broken down into four main applications [10]:

Ensure safety

These technologies ensure that the person is not put at risk due to declining memory from conditions such as dementia, or general ageing along with the general concern from persons and family members over persons being left alone [11]. Some possible applications include systems such as automated door locks that will activate at a certain time, it is also possible to have a camera installed that will only open the door when the person has confirmed that they know who the visitor is, facial recognition can also be used if the person has trouble remembering faces [12]. It is also possible to detect if the person may need medical assistance using technology to detect falls where the system will detect if the person has fallen and contact the relevant authorities, family, or carers [13].

Improve communication

Communication technology allows the person to keep in contact with friends and family to help avoid them feeling alone or isolated. Simple changes can include the replacement of conventional telephone buttons with picture buttons– where a friend/relatives face is printed on the button with their number stored to assist contact [14]. Facial recognition systems can also aid in this, using door or wearable cameras the person can be reminded of who the visitor is along with any relevant information, such as if this is a regular carer visit or if it's a scheduled appointment.

Multi-sensory stimulation

These technology aids aim to improve depression and loneliness, promotes physical well-being, along with improving relationships between people with dementia and their carers [15]. Examples of these include the creation of an individualised biographical reminiscence tool which provides videos and photographs from the person's past with the aim to try and trigger past memories from creating a familiar sense of belonging [16].

Memory enhancers

Due to the high occurrence of memory-related conditions amongst the elderly segment of society declining memory is a common issue [17]. In order to help alleviate these issues a range of reminder technology has been developed. These include electronic calendars where appointment reminders are prompted to the person through a smart-phone or other display or where important items can have alarmed receivers attached to them to aid in locating. These can range to complex systems where the persons activity is recognised and assistance is then offered if the person is determined to be struggling to complete an activity [18].

However it should be noted that persons may react differently to differing assistive technologies, while some persons may prefer a complex system, for example, one consisting of a system that monitors their medication intake and informs them accordingly, others may prefer a simple timed medicine dispenser that issues tablets at a set time each day [19]. There is also the problem that the condition of dementia can make persons apprehensive to try out new technologies with persons concerned over the complexity of such systems and their inability to use them, or reluctance to admit that they require assistance with ADL [11].

12.4.1 Emerging trends in sensor types

Sensors used to support ADL range from discrete state sensors to those that continuously record data, Table 12.2 presents a summary of commonly used sensor types, as reported in [20]. Contact/pressure sensors will record binary state data to inform whether an object has been interacted with, such as a door or bed respectively. Accelerometers and gyroscopes are normally combined to assess how active the occupant is, along with other measures such as gait analysis. There is also a range of sensing technology used for indoor location, as this can be key to supporting ADL due to certain activities having set locations that they are performed – such as cooking and bathing. Some examples of these are Bluetooth and Ultrasound beacons that allow a receiver and transmitters to determine location by measuring how long the signal takes to reach the receiver. Machine visions systems can also be used to determine location by using techniques, such as background subtraction, to establish the persons location, machine vision also allows you to determine what activity the occupant is carrying out [21].

An increasing technology consumer trend that is also witnessing adoption within the domain of smart environments and consumers is the use of wearable technology. Wearable technology offers new opportunities to AAL by enabling data to be gleaned not only from the environment but from the occupants of that environment.

Table 12.2 Brief overview of commonly used sensors within AAL

Sensor type	**Common use**
PIR	Person localisation/movement detection
Ultrasound	Person localisation
Bluetooth	Person localisation/object information
WiFi	Person localisation
Video	Person localisation/object detection/facial recognition/activity recognition
RFID	Contact/tag information
Pressure	Chair/bed/contact
Contact	Door/cupboard/opening/closing
Accelerometers/gyroscopes	Activity recognition/movement detection/object interaction.

Wearable sensors range from consumer activity monitoring devices produced by companies such as Fitbit and Withings which capture data relating to heart rate, sleep monitoring, and activity tracking, towards complex cutting edge technology that, for example, embed sensors within fabric. Beyond these, the emergence of head-mounted wearable technology, for example Google Glass, offers a new paradigm in wearable computing. These devices, such as Vuzix M100 [22], offer a first-person view of the environment, a eye-level display, along with on-board processing and communication capabilities. With real-time scene processing, such as object/facial/text recognition, the creation of supporting technology for the purposes of AAL [23] is allowed.

12.5 The role of ambient intelligence in making environments 'smart'

One of the areas within this domain that offers a lot of opportunities within the domain of AAL is Ambient Intelligence (AmI) [24], which is where the environment supports the occupants through the use of ambient sensors in place of the traditional input/output of a computer system [25]. A wide range of technologies are needed to enable AmI, generally a networked range of sensors along with computational facilities to interpret the sensor information and take action based on these readings. AmI can be thought of as an amalgamation of three areas of computer science, namely, ubiquitous/pervasive computing, sensor technology, and artificial intelligence [25].

AmI facilitates the continuous monitoring within the home environment which can aid in early detection of deteriorating health problems or detect a worsening in chronic health conditions that may not be easily detected within a clinical environment, such as unusual behaviour such as failing to take medication correctly when there is a visitor. There are multiple methods within AmI to achieve this goal, these include data-, knowledge-, and context-driven approaches.

12.5.1 Data-driven approaches

Data-driven approaches rely on collecting large amounts of data to 'learn' the occupants activities and habits [25]. Data mining is applied to the collected data in order to determine and collect patterns. Machine learning is a common technique used to reason on the data collected, this includes both supervised and unsupervised methods. Supervised methods require a set of labelled data on which to train on, for example, this consists of sensor data of the persons performing activities that can then be learnt in order to recognise these patterns in real time on unknown data. Supervised and unsupervised learning form the basis of data-driven approaches, where supervised approaches are given examples of activities made up from sensor data and unsupervised methods where no example activity data is provided.

A wide range of algorithms can be used to support supervised learning/activity recognition, these include Hidden Markov Models (HMM), decision trees, Support

Vector Machines, and ensemble approaches which can combine a number of algorithms [26]. Regardless of the algorithm chosen, there is a common set of steps involved in the training of a suitable model [26]. The first step is to acquire a training data set which will be used to train the system as to what sensor features make up an activity. This training set should be representative of the real-world application that the system will be deployed within and will include labelled annotations of what the person does and when. A test dataset should also be gathered in order to validate the generalisation ability from the training dataset. Once the algorithm has learnt the training dataset its performance will then be tested on the test dataset. It is common to have to repeat these steps with differing partitioning of the training and test dataset in order to refine the algorithm's ability to detect activities while avoiding the problem of over-fitting and improve generalisation [26].

Algorithms for unsupervised learning include K-Means, mixture models, and Bayes networks [26]. Unsupervised methods try to recognise and construct activities from unlabelled data, like supervised methods the first stage is to acquire a training dataset but in this case unlabelled. The next stage is to aggregate and transform the sensor data into features and then model them using density estimation or clustering methods. The goal of this method is to separate the data into differing clusters, so while it may not be aware what activity represents each cluster it can determine that cluster X is a different activity to cluster Y, with the goal of identifying groups of similar data within a larger dataset. One major challenge using a data driven approach, whether supervised or unsupervised, can be the requirement to obtain a vast amount of data for training purposes. This issue is further compounded if attempting to incorporate video and audio data into the reasoning process [24].

12.5.2 Knowledge-driven approaches for modelling ADLs

Knowledge-driven based systems rely on a series of rules that determines if an activity is being carried out. One advantage of knowledge-driven system is that it allows you to separate the activity detection from the supporting system thus allowing rules to be reused within the domain with only minimal customisation to unique needs that the person may require. There are differing methods of knowledge representation within smart environments, however, there is currently no accepted standard within the domain.

One form of knowledge-driven systems is the Event-Condition-Action (ECA) which is an architectural pattern for representing context awareness. This method consists of three modules: an Event, Condition, and Action modules. The Event module is responsible for gathering contextual information such as sensor data, the Condition module is responsible for the rules, and the Action module is responsible for executing the associated action with each rule [27].

These rules take the form of:

On (event expression)
If (condition)
Do (action)

The reading of these rules is as follows: *On* detecting a certain event check *If* a condition is true and if so then *Do* the specified action [27]. The *On* is defined as the rule trigger which is determined to be true if an event occurs that matches the event expression defined in *On*. The *If* section of the rule defines a conditional statement that has to be evaluated to true in order to trigger the final statement of the rule which specifies what action needs to be taken [25]. A simple example of this would be:

On (bedSensor == TRUE)
If (TV == ON)
Do (TV = OFF)

This reads as *On* detection that someone is in bed and *If* the TV is still on, then *Do* turn the TV off. An example of a traditional implementation of an ECA system is through the use of HomeRuleML [28] which is an XML-based schema to represent rules within a smart environment. HomeRuleML specifies a list of sensor IDs along with a conditional value which when true allows the system to determine if the conditions of an activity/rule have been specified along with an action statement for each rule.

Another form of knowledge-driven AmI is through the use of an ontology based system, which allows taxonomies and relationships between concepts to be defined [29]. Ontologies offer several advantages over traditional forms of knowledge representation, a well-defined ontology allows knowledge sharing a re-use, declarative semantics allow multiple policies to support context detection, and ontologies also provide complex inference mechanisms [30].

12.5.3 Context-driven

Context-aware approaches rely on information such as the date, time, persons location, roles of people present, as well as known objects [25]. Dey and Abowd define context as:

> '...any information that can be used to characterize the situation of an entity. An entity is a person, place, or object that is considered relevant to the interaction between a user and an application, including the user and application themselves...' [31]

Context can be used to represent two main categories within AAL—user-centric context and environmental context [25].

User-Centric Context

User-centric context revolves around the persons background, current behaviour, and emotional state [25]. Background factors such as a user's interests or medical conditions can have an effect on their current context or the actions that will need to be taken, for example, should the person be diabetic support will be required to ensure that blood glucose levels are checked at regular intervals and before meals. Current behaviour will factor in variables such as the persons current activity, if they are currently involved in an activity viewed as high importance, such as a discussion with their physician, then the person will not be notified/prompted with items that are considered to be of low importance. Lastly the person's emotional state is taken into account when determining user-centric context from multimodal sensors and analysis of user features, such as voice or tremors, as you may not want to further frustrate the person if they are in a poor emotional state.

Environmental Context

Environmental context revolves around the person's physical, social, and computational surroundings [32]. Physical factors include variables such as the current time, the persons physical location, and temperature. These can have a large impact on the type of support that is required, as certain activities will only take place within a certain time frame that is in a certain location–such as cooking dinner which would normally be within the hours of 1700–1900 within the kitchen [33]. Social factors will involve variables such as surrounding people, such as if the person is currently being visited by carers or family members, in which case they will most likely not wish to be disturbed [19]. The final environmental factor is that of the persons immediate computational surroundings, this takes into consideration equipment such as sensors or displays that are within close proximity to avoid problems such as sending notifications to displays which the person may not be able to see or that are in a different room to the person's current location [25].

In practice, there are certain contextual variables that are more important than others. Typically these include: location, identity, time, and the person's current activity. For example, location permits nearby objects, people, and activities to be determined and subsequently their identity then allows other background information to be inferred, such as contact details, birth date, list of friends, and relationships to other people within their environment. Dey and Abowd define three main categories for context aware applications [31]:

Presentation

Is the ability to display information that is relevant to the user, including contextual information, and not just a list of information that requires further user interaction. An example of this would be a mobile device that allows the display of friends or family members location and an awareness of their activity. Is an amalgamation of Schilit's proximate selection [34] and Pascoe's notion of presenting context [35].

Execution

Involves adapting the environment with additional information by associating particular data with an particular context. An example of this would be alerting a person when a visitor is within a certain distance of their home. Execution is based on Schilit's context-triggered actions, and Pascoe's contextual adoption.

Tagging

Tagging is the process in which the system records the actions and times that they were carried out. This can then be used later to help determine the persons behavioural habits and also to help determine if chronic conditions are deteriorating. This information can also be used in a data mining approach in order to recognise patterns.

A computer system that has knowledge of context is therefore able to sense, and react based on the person's requirements within an environment in order to improve their QoL.

12.6 Behaviour analysis

The goal of behaviour analysis within a smart environment is to detect gradual changes in behaviour as well as atypical behaviour. Atypical behaviour could be an early sign to a change in the condition of the patient's condition or a failure in sensor equipment. When analysing behaviour the following factors need to be considered [36]:

- Individuals within the environment will have differing routines and behaviour patterns, therefore personalised classifiers for behaviour recognition to better learn what features describe an activity for an individual.
- Behaviour will differ on different days of the week, with certain activities being carried out on set days–these correlations can be learned to better determine abnormal behaviour.
- Individuals' routines will change over time and will display differences in the activity they carry out, the time the activity is carried out, and the day of the week the activity is carried out.

When activities are carried out within a similar context, i.e. the same time, locations or carried out in a series we can then infer normal and abnormal behaviour. There are multiple methods for analysing behaviour within a smart environment:

Learning temporal relations

This method involves learning the temporal relations between events and form relations using temporal logic (before, after, meets, meet-by, overlaps, overlapped-by, starts, started-by, finishes, finished-by, during, contains, and equals). An example of this logic is an occupant taking pills after eating food which share the temporal relation 'after' between them, if this relationship is violated then abnormal behaviour is detected [37].

Mining sequential patterns

The basic principle behind sequential pattern mining is locating relevant sequential patterns within the data, these correlated patterns can give additional information to the connection between events in the data. In order for a pattern to be recognised it has to occur multiple times within part of the data stream to be determined a sequential pattern. An example of connection between events would be when the occupant awakes in the morning at 0900, they would then normally get dressed. If there was a change in this behaviour then it could be determined that the occupant's condition may be deteriorating [38].

Activity curves

Activity curves are a model that represents an occupant's generalised daily routines. They are used to model an occupant's regular behaviour using activity-labeled sensor data to learn a high-level representation of an occupant's daily routine. Due to the nature of some activities dominating particular times of the day, e.g. sleep activity at midnight, the day is segmented into equal-size windows and probability weighting is applied to each activity for that time window. To take account of the slight variations in time of activities, e.g. breakfast may be eaten at 0700 one day and 0730 the next, an aggregated activity curve is calculated of a window of x days [39].

12.7 Challenges for realising truly smart homes

There are many challenges in successfully supporting ADLs; these range from technical challenges, such as the heterogeneity of the data sources to challenges from a more human perspective, such as modelling the complexity of interleaved activities. Additionally, having multiple occupants within the same environment can present its own set of challenges along with the social, ethical, and security issues revolving around having such technologies within the home.

12.7.1 Interleaved Activities

The challenge with interleaved activities stems from the fact that in everyday life people do not carry out tasks in a sequential manner but rather switch between simultaneous tasks [40]. If the correct support is not given this could lead to confusing the occupants or them ignoring the system in the future.

Helaoui *et al.* distinguish between foreground and background activities, with foreground activities consisting of the activity that the occupant is carrying out at that point in time, any other interleaved activities that are being carried out concurrently are stepped into background activities once a new activity has been detected [41]. If we take the example scenario where the occupant was making a pot of tea while boiling eggs and setting the table at the same time. When the sensor activity shows the occupant operating the kettle then making tea becomes the foreground activity with cooking and setting the table moved to background activities, once the occupant then moves on to operate the cooker then making tea is stepped back into a background activity. At each time step only a single activity can be identified as a foreground activity, all other activities are stepped into background activities.

12.7.2 Multiple occupancy

Most of the current AAL solutions rely on the presence of only a single occupant within the environment [42]. If there are multiple occupants that require assistance with ADL then it can be difficult to give the correct support to the correct occupant as they require differing levels of support. A common problem with multiple occupancy is when two or more occupants interact with objects within the environment at the same point in time. If we consider an example where the oven door has been opened along with a cupboard door opened, there is insufficient evidence from the data to indicate which occupant has interacted with which object or even if this is indeed a case of multiple occupancy and both objects were coincidently within reach at that point in time.

This can lead onto further problems when attempting to learn behaviour as key components of this are when a specific user carries out a specific activity along with how they undertake that activity. This allows you to learn if an occupant is having particular trouble completing a specific ADL or if they are not undertaking the activity whatsoever. Due to the conflicts in sensor data due to having multiple occupants within the same environment it can be difficult to separate these events into user-specific events. Possible techniques to mitigate this issue can be user worn ID tags, such as RFID, that uniquely identify each user [43]. Video data can also be used to identify the occupants within the environment [44]; however, both these solutions introduce challenges of their own. This can be further compounded when the person has visitors or carers that may call in on a regular basis, as the visitors will not require support by the system but the persons may still require support.

12.8 Conclusion

This chapter has provided an overview of the technologies and applications involved within the domain of AAL. It has introduced the concept of ADL along with the concept of a smart environment and how technology can be applied to provide support in order to improve QoL for those that would normally require full time care or institutionalisation. A brief overview of the technology used has also been presented along with a comparison of the major techniques that can be applied to AmI, namely data, knowledge, and context driven approaches. Current methods of supporting ADL suffer from common problems such as the cost of retro fitting an environment along with the intrusiveness such an installation will incur. Other problems exist with current support of ADL such as the 'cold start' problem were a large amount of data needs to be collected for pattern recognition through data mining, when the system is initially installed there is no data to be processed therefore support will not be available. A number of challenges that AAL faces have also been presented and discussed, such as those of multiple occupancy, future challenges include the personalisation of support to each individual occupants needs, Each occupants condition will deteriorate at differing rates therefore support will need to be tailored to the individual. These challenges will need to be addressed in the future if the vision of AAL is to be achieved in a real-world setting.

References

[1] Kobayashi LC, Wardle J, von Wagner C. Internet use, social engagement and health literacy decline during ageing in a longitudinal cohort of older English adults. Journal of Epidemiology & Community Health. 2014;69(3):278–283. Available from: http://jech.bmj.com/cgi/doi/10.1136/jech-2014-204733.

[2] Noelker LS, Browdie R. Sidney Katz, MD. A new paradigm for chronic illness and long-term care. Gerontologist. 2014;54(1):13–20.

[3] Giebel CM, Sutcliffe C, Stolt M, *et al.* Deterioration of basic activities of daily living and their impact on quality of life across different cognitive stages of dementia: a European study. International Psychogeriatrics. 2014;26(08):1283–1293. Available from: http://www.journals.cambridge.org/abstract_S1041610214000775.

[4] Gurland BJ, Maurer MS. Life and Works of Sidney Katz, MD. A life marked by fundamental discovery. Journal of American Medical Directors Association. 2012;13(9):764–765.

[5] Wallace M, Shelkey M. Katz index of independence in activities of daily living (ADL). Urologic Nursing: Official Journal of the American Urological Association Allied. 2007;27(1):93–94.

[6] Bucks RS, Haworth J. Bristol activities of daily living scale: a critical evaluation. Expert Review of Neurotherapeutics. 2002;2(5):669–676.

[7] Cook DJ, Das SK. How smart are our environments? An updated look at the state of the art. Pervasive and Mobile Computing. 2007;3(2):53–73. Available from: http://linkinghub.elsevier.com/retrieve/pii/S1574119206000642.

[8] Wilson C, Hargreaves T, Hauxwell-Baldwin R. Smart homes and their users: a systematic analysis and key challenges. Personal and Ubiquitous Computing. 2014;19(2):463–476. Available from: http://link.springer.com/10.1007/s00779-014-0813-0.

[9] Weiser M. The computer for the 21st Century. Scientific American. 1991;265(3):94–104. Available from: https://www.ics.uci.edu/~corps/phaseii/Weiser-Computer21stCentury-SciAm.pdf.

[10] Cahill S, Macijauskiene J, Nygard AM, Faulkner JP, Hagen I. Technology in dementia care. Journal of Dementia Care. 2008;16(5):14.

[11] Mehrabian S, Extra J, Wu Y, Pino M, Traykov L, Rigaud A. The perceptions of cognitively impaired patients and their caregivers of a home telecare system. Medical Devices: Evidence and Research. 2015;8:21–29.

[12] Cleland I, Patterson T, Nugent C, Cruciani F, Black N, Paggetti C. Doorstep Security; using a technology based solution for the prevention of doorstep crime. In: International Workconference on Ambient Assisted Living. Springer International Publishing; 2014. p. 2–5.

[13] Yuan J, Tan KK, Lee TH, Choon G, Koh H. Power-efficient interrupt-driven algorithms for fall detection and classification of activities of daily living. Sensors. 2015;15(3):1377–1387.

[14] Mao HF, Chang LH, Yao G, Chen WY, Huang WNW. Indicators of perceived useful dementia care assistive technology: caregivers' perspectives. Geriatrics & Gerontology International. 2014;8(32):1–9.

[15] Chen Tj, Li Hj, Li J. The effects of reminiscence therapy on depressive symptoms of Chinese elderly: study protocol of a randomized controlled trial. BMC Psychiatry. 2012;12(1):189. Available from: http://www.pubmedcentral.nih.gov/articlerender.fcgi?artid=3507717&tool=pmcentrez&rendertype=abstract.

[16] Gibson G, Newton L, Pritchard G, Finch T, Brittain K, Robinson L. The provision of assistive technology products and services for people with

dementia in the United Kingdom. Dementia. 2014; p. 1–21. Available from: http://www.ncbi.nlm.nih.gov/pubmed/24803646.

[17] Christensen K, Doblhammer G, Rau R, Vaupel JW. Ageing populations: the challenges ahead. Lancet. 2009 oct;374(9696):1196–208. Available from: http://www.pubmedcentral.nih.gov/articlerender.fcgi?artid=2810516&tool=pmcentrez&rendertype=abstract.

[18] Czarnuch S, Cohen SR, Parameswaran V, Mihailidis A. A real-world deployment of the COACH prompting system. Journal of Ambient Intelligence and Smart Environments. 2013;5(5):463–478.

[19] Acampora BG, Ieee SM, Cook DJ, Ieee F, Rashidi P, Ieee M, *et al.* A Survey on ambient intelligence in healthcare. Proceedings of the IEEE. 2013; 101(12):1–25.

[20] Hightower J, Borriello G. Location systems for ubiquitous computing. Computer. 2001;34(8):57–66.

[21] Pirsiavash H, Ramanan D. Detecting activities of daily living in first-person camera views. In: 2012 IEEE Conference on Computer Vision and Pattern Recognition. Providence, RI: IEEE; 2012. p. 2847–2854. Available from: http://ieeexplore.ieee.org/lpdocs/epic03/wrapper.htm?arnumber=6248010.

[22] Vuzix. M100, Smart Glasses; 2014. Available from: http://www.vuzix.com/consumer/products_m100/.

[23] Ha K, Chen Z, Hu W, Richter W, Pillai P, Satyanarayanan M. Towards wearable cognitive assistance. In: Proceedings of the 12th annual international conference on Mobile systems, applications, and services. ACM; 2014. p. 68–81. Available from: http://dl.acm.org/citation.cfm?doid=2594368.2594383.

[24] Cook DJ, Augusto JC, Jakkula VR. Ambient intelligence: technologies, applications, and opportunities. Pervasive and Mobile Computing. 2009; 5(4):277–298. Available from: http://dx.doi.org/10.1016/j.pmcj.2009.04.001.

[25] Sadri F. Ambient intelligence. ACM Computing Surveys. 2011;43(4):1–66.

[26] Chen L, Khalil I. Activity recognition: approaches, practices and trends. In: Activity Recognition in Pervasive Intelligent Environments. vol. 4; 2011. p. 1–31. Available from: http://www.springerlink.com/index/P856L50744334014.pdf.

[27] Stavropoulos TG, Kontopoulos E, Bassiliades N, *et al.* Rule-based approaches for energy savings in an ambient intelligence environment. Pervasive and Mobile Computing. 2013;19:1–23. Available from: http://dx.doi.org/10.1016/j.pmcj.2014.05.001.

[28] Hallberg J, Nugent CD, Davies RJ, Synnes K, Donnelly MP, Member S, *et al.* HomeRuleML – A model for the exchange of decision support rules within smart environments. In: IEEE Automation Science and Engineering. Scottsdale; 2007. p. 513–520.

[29] Stavropoulos TG, Vrakas D, Vlahavas I. A survey of service composition in ambient intelligence environments. Artificial Intelligence Review. 2013;40(3):247–270.

[30] Babu ABKA, Sivakumar R. Development of ontology based middleware for context awareness in ambient intelligence. In: Hybrid Intelligent Systems. Kuwait: IEEE; 2014. p. 219–224.

[31] Dey AK, Abowd GD. Towards a better understanding of context and context-awareness. In: What, Who, Where, When, Why and How of Context-Awareness. Rotterdam; 2000. p. 304–307.

[32] Dey AK. Context-aware computing. In: Krumm J, editor. Ubiquitous Computing Fundamentals. 1st ed. Chapman and Hall/CRC; 2009. p. 321–352.

[33] Baldauf M. A survey on context-aware systems. International Journal of Ad Hoc and Ubiquitous Computing. 2007;2(4):263–277.

[34] Schilit BN, Adams N, Want R. Context-aware computing applications. In: Proceedings of the 1994 First Workshop on Mobile Computing Systems and Applications. IEEE Comput. Soc; 1994. p. 85–90. Available from: http://ieeexplore.ieee.org/xpls/abs_all.jsp?arnumber=4624429.

[35] Pascoe J. Adding generic contextual capabilities to wearable computers. In: International Symposium on Wearable Computers. vol. 44. IEEE; 1998. p. 92–99.

[36] Lago P, Jimenez-Guarin C, Roncancio C. A case study on the analysis of behavior patterns and pattern changes in smart environments. In: International Workconference on Ambient Assisted Living; 2014. p. 296–303.

[37] Jakkula V, Cook D. Learning temporal relations in smart home data. In: International Conference on Technology and Aging; 2007. Available from: http://www.eecs.wsu.edu/~cook/pubs/ta07.pdf.

[38] Hassani M, Beecks C, Daniel T, Seidl T. Mining Sequential Patterns of Event Streams in a Smart Home Application. In: LWA. October; 2015. p. 159–170.

[39] Dawadi PN, Cook DJ, Schmitter-edgecombe M. Modeling patterns of activities using activity curves. Pervasive and Mobile Computing. 2015;25.

[40] Gu T, Wu Z, Tao X, Pung HK, Lu J. epSICAR: An emerging patterns based approach to sequential, interleaved and concurrent activity recognition. In: 7th Annual IEEE International Conference on Pervasive Computing and Communications, PerCom 2009; 2009. p. 1–9.

[41] Helaoui R. Recognizing interleaved and concurrent activities: a statistical-relational approach. In: Pervasive Computing and Communications (PerCom), 2011 IEEE International Conference on. IEEE. Seattle: IEEE; 2011. p. 1–9. Available from: http://ieeexplore.ieee.org/xpls/abs_all.jsp?arnumber=5767586.

[42] Giroux S, Castebrunet M, Boissier O, Rialle V. A multiagent approach to personalization and assistance to multiple persons in a smart home. In: Artificial Intelligence Applied to Assistive Technologies and Smart Environments. Artifical Intelligence Applied to Assistive Technologies and Smart Environments; 2014. p. 11–15.

[43] Lotfi A, Langensiepen C, Mahmoud SM, Akhlaghinia MJ. Smart homes for the elderly dementia sufferers: identification and prediction of abnormal behaviour. Journal of Ambient Intelligence and Humanized Computing. 2012;3(3):205–218.

[44] Brumitt B, Meyers B, Krumm J, Kern A, Shafer S. Handheld and Ubiquitous Computing. Thomas P, Gellersen HW, editors. Springer Berlin Heidelberg; 2000.

Chapter 13

Outdoor mobility assistance–technologies helping on the way

René Hempel[1], Christoph Stahl[3], Birgit Stockinger[2], Ferdinand Kemeth[2], and Thorsten Vaupel[2]

13.1 Introduction

In our society, the individual quality of life in all age groups is significantly influenced by a person's own mobility [1]. Mobility is a core requirement for many activities through which people are able to, for example, earn their livelihood or participate in society on a social level. An individual's desire for mobility barely changes as they grow older but their requirements and possibilities do [2]. A decreasing amount of individual flexibility notably hinders a person's use of the means of transport available to them. Whether and to what extent people can meet their needs depends not only on the available transportation means. In particular, factors such as health, exercise and cognitive capacity, personal sense of safety, driving competence and especially the access to information allow or restrict the mobility [3].

Assistive systems, either analogue or digital, can help the user finding the right way and transportation mode. But a seamless mobility can only be established, if the user has continuous and ubiquitous access to all required information. However, providing this information is a complex process involving different networks, systems and technologies as shown in the following scenario.

13.1.1 Scenario

Dorothy uses her tablet computer to find a route to the medical eye specialist that her family doctor suggested recently. One of the many positive aspects of her new flat in the citizen's residence is the bus station in front that makes a ride into town so convenient. If only the timetable at the station would be readable! But since her oldest son explained her step by step how to use the app of the local public transportation

[1]Institut f. Automation u. Kommunikation e.V., Werner-Heisenberg-Str. 1, 39106 Magdeburg, Germany
[2]Fraunhofer Institute for Integrated Circuits IIS, Nordostpark 84, 90411 Nuremberg, Germany
[3]German Research Center for Artificial Intelligence (DFKI), Enrique-Schmidt-Str. 5, 28359 Bremen, Germany

services, she is able to get the information even in her room. The app now suggests to take the bus to the central station in 17 minutes, then take the subway and finally walk 8 minutes. Dorothy is confident to manage the walking distance with her walker and some navigational aid provided by her tablet computer. She decides to take the next bus towards the city centre and mounts her tablet on the walker. Arriving at the central train station, it's always difficult for her to find the right escalator to the subway between the shops in the underground level. Unfortunately her mobile device cannot help her either, since the map only covers the street level. At least a friendly shop owner explains her the way. Arriving at the destination station, Dorothy has no idea which exit to take, so she just takes the next elevator up. What a nice day for a walk in the city! But in the sunshine, she can hardly read the map on her tablet computer. And once the digital map indicates her position, she realises that she went to the opposite direction and has to cross two main roads. Returning to the underground is not promising either as she always loses orientation in the maze of tunnels. Oh, it would be wonderful if the system could suggest crossings with lowered kerbs. Even though her beautiful red walker is made of lightweight materials, it strains her body to move it up and down the pavement. On her way to the doctor, she has to stop and check the map frequently for the next turning point. She can hardly hear the voice of the tiny device at home. There, above the noise of the trucks and buses, it is impossible. Could not someone invent something more practical for her walker? 'You have reached your destination' the voice confirms. Standing in front of the medical centre she is both relieved and deeply impressed by the number of specialists that work here. Only four minutes left until her appointment. The detour at the metro station took much more time than expected and her tablet computer seems to be useless again. But wait, she remembers that she can also use it to make a phone call. The doctor will be able to explain her the route through the building!

13.1.2 Current challenges

The scenario above is meant to illustrate the current situation for urban areas, where multimodal navigational aid has already become ubiquitous. Apps for navigational aid are pre-installed on most mobile devices, such as smartphones and tablets and provide free maps, e.g. from Apple or Google. Both companies have also started to integrate information about public transportation services into their maps, but the service providers are usually reluctant to provide their data.

Instead, most public transportation services provide their own free apps for routing services, some even allow for mobile ticketing and payment. The downside is that the users have to install and learn new apps for every city they visit. Concerning the usability of these apps, elderly users can learn to benefit from mobile devices with some training e.g. by relatives. Smartphones and tablets can be mounted on mobility platforms, such as wheelchairs and walkers, but only if the weather conditions permit it. Glossy displays are difficult to read outdoors in bright sunlight and only a few devices are water-resistant. Mounting devices on a walker can also cause problems in crowded trains or unsafe areas. The use of maps is limited for elderly users with diminished vision (font sizes are typically too small) and too many Points Of Interest

(POIs) can be confusing. Spoken navigational instructions can be difficult to understand due to traffic noise and age-related hearing loss (insufficient volume/sound quality of the mobile device). Current maps are primarily designed for car drivers, so they usually don't include smaller footpaths for pedestrians. They mostly provide no information about lowered kerbs for wheelchairs and walkers and they do not show the precise location of all entrances to underground stations. Indoor maps are currently only available for selected places, such as airports, shopping malls or train stations.

In the following sections, we will therefore present an overview of the state of the art regarding assistive systems for outdoor mobility. Beside current challenges in information networking, available products and important European research projects are presented here. Eventually, the core aspects of seamless mobility as well as recommended actions are outlined.

13.2 Towards seamless mobility

The term *Seamless Mobility* is commonly used in the area of wireless communication systems. Usually, *seamless* denotes the coupling of different radio access technologies in order to maintain continuous Internet connectivity and information flow [4]. Transferred to a person's individual mobility, *Seamless (Multimodal) Mobility* characterises the combination of different transportation modes to implement door-to-door mobility. Furthermore it involves smartphones and mobile apps to provide information and access to all transportation modes [3]. The lack of information and data as well as the use of parallel and unconnected systems and services today are the main obstacles constraining the realisation of seamless mobility. Therefore, the following chapter will give an insight into the main technologies and methods forming the main pillars of Seamless Mobility.

13.2.1 Localisation and positioning

Global Navigation Satellite Systems (GNSS)

Global Positioning System (GPS) has become a widely used synonym for satellite navigation systems, although today a range of other systems too are in operation, such as the Russian GLONASS (fully operational since 2011), the European Galileo (testing since 2013, fully operational in 2020) and the Chinese Beidou (testing since 2011, fully operational in 2020). Some modern chipsets for mobile devices are compatible with all three satellite systems. Since the US government has deactivated the selective availability in 2000, NAVSTAR-GPS can be freely received worldwide and it often provides a position estimation that is within 10 metres of the real position. This is accurate enough for car navigation systems, which match the position with the road network. For pedestrians however, the accuracy is not sufficient to estimate on which side of the road the user is. It is also not possible to guide a blind user along the side walk.

The positioning precision can be improved by *Differential GPS* (D-GPS) where ground stations measure the current time difference caused by atmospheric effects in real-time. Satellite-based augmentation systems (EGNOS in Europe, WAAS in USA) distribute correctional data by satellite and achieve a resolution of 3 metres. Regional D-GPS systems, such as the *Satellite Positioning Service of the German Landesvermessung* (SAPOS), use a network of reference stations and provide the data via the Internet.

In conjunction with professional GPS receivers that are capable of measuring the signal carrier phase, *Real Time Kinematic* (RTK) systems achieve an accuracy of 1 centimetre. This technology is not yet available in mobile devices, but the low-cost RTK receiving module Piksi is available from SWIFT Navigation. Such RTK systems could be used as a reference to evaluate the accuracy of other positioning systems. One major disadvantage of GPS receivers is the time needed (up to 12 minutes) to download all satellite's ephemeris data after a cold start. Most modern smartphones use *Assisted GPS* (A-GPS) instead, which receives the required data by the Internet within seconds.

Infrastructure-based positioning

GNSS requires a line of sight connection; any obstacles like walls or roofs attenuate or even obliterate the signals sent by satellites. Therefore complementary technologies are necessary to determine one's position inside of buildings or in the so-called urban canyons. Infrastructure-based *radio frequency* (RF) positioning systems are using a locally installed infrastructure in order to estimate the location of a person or device. We distinguish between pre-installed infrastructure nodes and installing a separate network for locating purposes only. The most common RF technologies used for positioning are the *Global System for Mobile Communications* (GSM) and its successors (GPRS, EDGE, UMTS and LTE), which ensure a very widespread signal coverage worldwide. In urban areas, usually *wireless local area network* (WLAN) hotspots are available in most buildings and at public places; furthermore many private access points are installed. In addition, *Bluetooth Low Energy* (BLE) is the upcoming wireless personal area network technology. For indoor positioning, Apple recently introduced the iBeacon and Google the similar Eddystone standard.

The positioning approaches can be categorised in estimating distances or angles between sending and receiving units. A third category uses pattern recognition.

In order to estimate the distance between two RF nodes, systems measure the *Time of Flight* (ToF) of a transmitted signal and calculate the distance by multiplying the time with the propagation speed. But most set-ups do not include high synchronised clock. Measuring the *Round-Trip Time* (RTT) supersedes absolute clock synchronisation, thus only time differences are recorded: The node sends an RF signal to the transponder, which reflects the signal back to the sender. The sender measures the time used for this round-trip and calculates the distance with the help of the propagation speed. A more complex but more accurate approach is based on *Time Difference of Arrival* (TDoA), where all infrastructure nodes are time-synchronised. These nodes receive a broadcast signal sent by a transponder and forward the exact reception time to a central unit, where the difference between the reception times is being evaluated and thus the distance is calculated [5].

Beside the distance measurement technologies, the most common way of evaluating the direction of a mobile transmitter is the *Angle of Arrival* (AoA) measurement: A mobile transponder sends out a radio signal, which is received at an antenna array consisting of multiple antenna elements. Each array element detects the signal at a different time and thus with a different phase. These phase differences are transformed to an angle using algorithms like *Multiple Signal Classification* (MUSIC) [6]. The *Fingerprinting* technique belongs to the third category that uses pattern recognition. The system requires a training phase where signal strengths of the surrounding transponders are measured at specific positions. The *receivable signal strength indicator* (RSSI) measurements with the corresponding transponder identifiers and the position of the measurement make up the fingerprint database. In the locating phase live measurements are compared with the database using pattern recognition to identify the most similar fingerprints as the receiver's position.

Dead reckoning

Whereas the aforementioned technologies calculate the coordinates of the user by signal measurements, *dead reckoning* methods use motion sensors to estimate change in position over time from a known starting position. *Inertial sensors* became cheap and tiny using the *microelectromechanical systems* (MEMS) technology and measure acceleration and rotation angles in three axes. Integrating frequent measurements over time allows estimating the current position and orientation in three dimensions (3D), which is how planes navigate in the air for example. An Earth magnetic field sensor can be added to improve the orientation information and a barometer can provide information about the altitude [7]. Furthermore, wheeled devices, such as cars or robots, can use rotary encoders to measure wheel rotations. *Odometry* uses this distance data to incrementally update the estimated position and orientation over time. *Visual odometry* uses sequential camera images to estimate the distance travelled, which is for example applied for navigating blind users in the research project IMAGO [8]. Route models are created as 3D point clouds through image acquisition, feature extraction and matching between images. The proposed system yields a high positioning accuracy of 80 per cent of samples being located within 1.6 metres. Thus, the system is usable for pedestrian navigation, especially for visually impaired persons.

Combination of different technologies

We have presented a variety of methods and sensors that are available today to estimate one's position and all have their individual strengths and weaknesses: most technologies do not cover all kinds of location to provide seamless positioning solutions. The accuracy of most methods decreases or even collapses suddenly while approaching to the coverage boundaries, so its early detection is necessary. Figure 13.1 shows an example for a seamless scenario realised in [9], where elderly people were observed while wearing a special watch: For a rural area, GNSS was sufficient to find a person when lost. In order to cover the urban area and indoor scenarios, a WLAN fingerprinting technology helps paramedics to search in the right room. A more accurate system using a Bluetooth-based AoA technology was installed in selected rooms, in

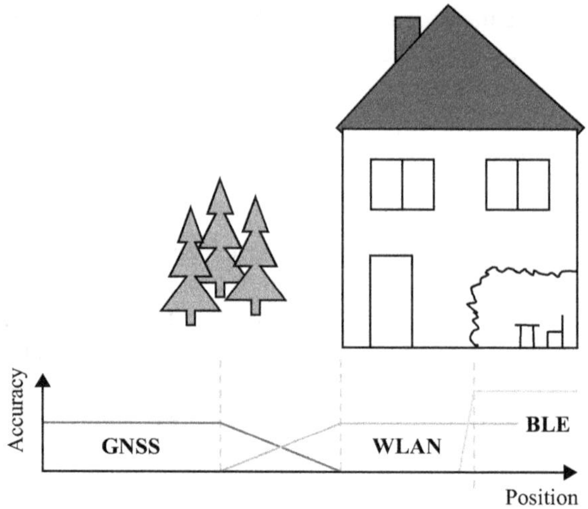

Figure 13.1 Seamless positioning

order to detect an emergency situation like: 'Person is lying next to the bed and does not move!'

The challenge of seamless positioning can be met by combining multiple approaches: *sensor fusion*. Well-known scientific methods for sensor fusion are particle or Kalman filters. Nevertheless characterising each technology mathematically is difficult but necessary for application of these filters.

13.2.2 Environment model and routing

The link between basic sensor technologies and applications, which fulfil needs of the user, is the model of the environment and routing graphs. They allow for various visualisation approaches containing vectorial maps, bitmap renderings and satellite photos. This provides context for location information based on positioning technologies.

Models abstract environments into digital representation and formats (maps) for positioning technologies. There are common solutions especially for car navigation systems. Nowadays *Personal Navigation Assistants* (PNA) like the products by Garmin, TomTom or Navigon are widespread and well-established. They are insufficient for pedestrian navigation systems as they lack detailed information like lowered kerbs in addition to the road network as well as accuracy even if this information was available. Assistive systems are often isolated applications based on specific models. Those models are not standardised and mostly mutually incompatible. Use cases like described in our scenario require more complex models which are built on isolated models to create additional benefits. These models contain specific information depending on the technologies used and the intended purpose. In case of positioning technologies one important aspect of the environment model is their coordinate system.

Coordinate systems

Historically it is of major interest to measure the dimensions of our surroundings. Several mapping approaches lead to many different representations in form of geodetic systems. Typical worldwide coordinate systems like *World Geodetic System 1984* (WGS84) or *European Terrestrial Reference System 1989* (ETRS89) cover Earth's surface completely. They use parametrised ellipsoids to mathematically approximate sea-level; together with several reference points on Earth, it is possible to define a datum for each and every location. Beyond those globally used systems, countries or regional activities often use ellipsoids fitting better locally. A transformation between both is typically easy, but it is erroneous if the details of a datum or its source are unknown.

Many applications use 2D Cartesian coordinate systems, whose possibilities of linear movements on maps match the capabilities of display-based map apps intuitively. Therefore they use a projection to transform locations on an ellipsoid onto a flat plane. Common systems are the globally used *Universal Transverse Mercator* (UTM) and *Gauss-Krüger* (GK) with locally more specific ellipsoids–an example for the application specific choice with better fit.

2D coordinates are not sufficient for indoor scenarios if the floors of a building need to be distinguished. Therefore mostly application-specific coordinate systems are used indoors. Many technical systems use proprietary representations for maps, calculations and visualisation. Surprisingly, even professional guides use pixel-based coordinates for the presentation and for all internal algorithms, as well. Those approaches are limited to a building or an area; often each level is modelled separately. Compatibility with outdoor data is not guaranteed. Sometimes even if mathematically possible, the transformation between an app-centric indoor model and common outdoor systems contains too many errors to form a suitable, combined system.

Databases and data acquisition

There are various databases for geographic information. We distinguish commercial databases and non-profit databases. Prominent companies which provide commercial databases, where data is usually gathered by employees, are Google, Apple and HERE. They offer for example map functions on smartphones based on these huge databases (compare Section 13.3). Restrictions on the use and availability of commercial databases have led to non-profit organisations like OpenStreetMap (OSM) who resort to the idea of crowdsourcing [10] for data acquisition: Volunteers gather data. In the beginnings interested individuals used their GPS tracking devices to collect data on the road network. Due to its free character the database became popular and its coverage increased. For an individual high-precision navigation the use of this data – based on the road network – is limited [11]. An increasing amount of governmental data is available; combining several data sources is an opportunity to gain the necessary level of detail.

Nowadays it is even possible to define many additional data types like footpaths and buildings or even indoor maps [12]. Crowdsourcing seems to be a popular approach for fast data acquisition of huge areas [13]. It is a method to complement existing data and keep it up to date especially for mapping pedestrian related information (ibid.). As keeping the data up-to-date is a generic problem, even commercial providers apply this basic idea. Although huge amounts of data are gathered it is difficult to ensure the quality of gathered data.

Special interest groups like wheelchair users adapted this approach. Their participants are experts and user at the same time. This constellation allows to collect the relevant data and qualifies the information in a single step. Upon utilising these strengths it is possible to build and provide highly tailored assistance applications fulfilling the user's demands. A famous example is the Wheelmap project; even though many projects rely on these databases, no standardised definitions for specific environmental data types exist. In addition to the basic map data representing streets, buildings and even indoor environments, it is necessary but difficult to gather barrier's data. One needs to record and keep track of a broad variety of data. The coordinate is basic, but is it sufficient to measure the real world items to derive the accessibility for a certain user group? Obviously not in every case; a special work-flow and a common classification is crucial. A suitable opportunity is the employment of trained personnel equipped with a supporting tool. For instance apps like the ones developed in the research projects ACCESS [14] or WikiNavi [15] to gather environmental information resulting in barriers depending on a person's mobility.

However, it is still challenging to provide wide-range navigation systems for pedestrians [16]. There is a lack of methods and models for creating data base containing a much more detailed data set compared to car navigation systems [17]. On the other hand, the local geodata derived from scientific projects are only available for research and not for commercial use.

Routing and routing strategies

Where routing for vehicles can easily be done relying on the road network as a model, routing for pedestrians is much more challenging. Map matching and routing on such a model with allowed paths (*positive map*) are not sufficient any more. A switch to models with impassable areas (*negative map*) is essential; new routing graphs are necessary–a generation should be possible but is poorly supported by the database's formats and information. The general limitations become crucial in assistance applications for people with any desired nuances of disabilities.

Furthermore pedestrians benefit from systems providing information and routing indoors and outdoors. Considering large public buildings and interchanges a technical support system would help the majority of people. The dimensions mentioned above get more complicated, because the combination of global coordinates with small scale drawings for buildings can cause errors. The limitation of 2D calculations, e.g. during the route preparation (possibly with height), have to be extended to fully 3D approaches. This is more complex in algorithms and the problem's space itself. On the one hand, 3D models also allow for three-dimensional visualisation; upcoming

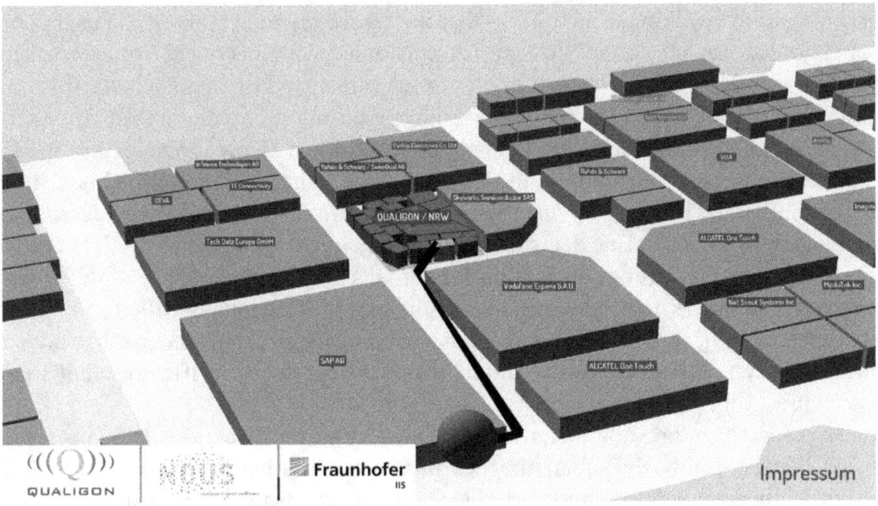

Figure 13.2 3D visualisation and route. Reproduced by permission of QUALIGON GmbH and Nous Wissensmanagement GmbH

products utilise this to replace the classical map approach (see figure 13.2). Nonetheless breaks in the media (device or application) should be avoided and are a challenge for the sensor technologies, too.

Classic routing approaches use a weight of edges in the routing graph to calculate a route matching certain presets (e.g. fastest, shortest). Advanced assistants additionally use barriers within the calculations to fulfil the needs of special user groups. An additional challenge remains: it is necessary to find a match between the (current) abilities of a person and the given situations (including temporary characteristics). Therefore it is common to project the barriers of an environment on the background of disabilities. This is not sufficient, but beyond ongoing research a path from coarse categorisations to a situational characterisation including an individual matching with the person's current profile needs to be found. The goal for a future foundation for route calculations should be a new mapping of environmental characteristics onto the abilities of individuals.

13.2.3 Human–computer interaction

A navigation task usually consists of two steps: first the user must select a destination, then the system guides the user with turn-by-turn directions. A destination can either be specified by typing its address or picked from a map. Entering an address on a mobile device can be difficult for elderly users because they must know how to enter text using a virtual keyboard. Car navigation systems typically also include databases of POIs and smartphones integrate navigation apps with web search, contacts and calendars. The map usually focuses on the current location of the user and rotates according to the walking direction, so that the environment in front of the user is up. Some systems can also show POI icons on the map, which is not recommended for elderly users, since the icons tend to clutter the screen and distract the user from the navigational task. If a route is set, the map highlights the streets along the path. As the user approaches an intersection in the street network, an arrow is shown and speech output announces the next action to take, e.g. 'turn right in 50 metres'. This style of interaction is known as *turn-by-turn* instructions. It works well for driving tasks, since the road intersections define decision points where information is required. In typical pedestrian areas however, such as squares, gardens or reception halls, people directly walk from landmark to landmark. This pathfinding behaviour is called *piloting* in research (see [18], for example) and is not adequately supported by current products. Further, speech output is not well suited outside the car due to traffic noise and maps lose contrast on LCD screens in bright sunlight.

In research, interesting alternatives to turn-by-turn instructions have been presented that continuously signal the bearing of the destination to the user. Stahl [19] developed an auditory landmark display that presents information about POIs in a zoo using headphones and *spatial audio*. The system virtually plays animal sounds from the direction where they are located, so that spatial information is provided without distracting the visitor's attention from watching the animals. Robinson *et al.* [20] implemented a system that provides directional assistance in the form of *vibrotactile feedback* through a handheld pointing device. Pedestrians can casually scan the environment to discover the direction of their destination. The interaction is designed to 'free users from the need to look at a screen or be micro-managed by listening to turn-by-turn instructions'. It also has benefits in situations with no defined waypoints and where the user prefers the exploration of a city over a shortest path. Interestingly, all 24 participants of their user study successfully completed the navigation task and found the end point with only the vibrotactile feedback to guide them. A similar approach has been taken by [21] with Tacticycle, a navigational aid for bicycle trips. Vibration patterns in the handlebars indicate the direction of the destination relative to the current driving direction. This approach has been adapted by Krieg-Brückner [22] for rollator handlebars that give directional cues by light and vibration. Jakab and Varga [23] use a tactile belt worn by blind runners and guides them running on an all-weather running track and Ross and Blasch [24] experimented with pathfinding aid to help blind people cross a street. Besides spatial audio signals, they used speech about where the destination is relative to the participant (e.g. destination is at 'one o' clock'). They also used tappers (placed near the right and left shoulders and in between) to indicate the relative position of the destination to the user.

13.3 Products

As the previous chapters have shown, navigation and routing can be quite complex tasks to fulfil. Therefore it is easy to understand that not all users are equally able to orient themselves, to follow a given path or even to plan their way to the mall. Hence, several assistive systems are available on the market supporting users in different

Figure 13.3 Garmin eTrex® Touch 35. Reproduced by permission of Garmin Deutschland GmbH

stages of navigation. The following chapter presents a selection of these devices and systems. Anyhow, there are a lot more and many research projects (see Section 13.4) are working on innovative solutions for the future.

13.3.1 Pedestrian navigation

Pedestrian Navigation is a very complex application field. Therefore only devices and systems exist which are able to navigate in specific environments. For example, the two market leaders Garmin and TomTom offer several devices dedicated to e.g. cycling, hiking or swimming (see figure 13.3). In addition to these specialised products, a lot of smartphone applications are already offering pedestrian navigation. In 2010, Google announced the activation of their pedestrian navigation service in Google Maps. Apple Maps (Apple) and HERE (Audi AG, BMW Group, Daimler AG) have joined ever since. Pushed by the release of the Apple Watch in 2014, the wide spread of smartwatches allows for presenting navigational information on the pedestrian's wrist. Companies like Garmin already have modified their apps to visualise e.g. the user's current position as well as giving turn-by-turn instructions.

13.3.2 Wearable GPS tracker

Users with mild dementia suffer from memory loss and can feel disoriented even on familiar routes. At the same time, dementia patients have increasing problems to handle complex devices, such as navigation systems. As a consequence, simplified tracking devices are gaining popularity among this user group. They continuously estimate the position of the user by satellite navigation and transmit it to relatives

Figure 13.4 Teltonika GH4000 Tracker. Reproduced by permission of HQ JSC Teltonika

or care givers by the GSM network. *Geofencing* applications automatically raise an alarm if the user leaves a predefined spatial region. Most products come in the shape of badges or wrist watches, so that they can be worn comfortably every day. In order to raise acceptance by the users, some devices integrate additional functions, such as an *emergency call* button and *fall detection* sensor. Schneider and Henneberger [25] evaluated several products of this class (e.g. Xexun TK203, Laipac S911 and Teltonika GH3000; see Figure 13.4) in terms of battery time, time to first position fix, location accuracy and fall detection accuracy.

13.3.3 Mobility aids

The original *rollator* has been invented by Aina Wifalk for herself in 1978 as a four-wheeled walking aid with hand brakes and a built-in seat. Modern rollators have reduced weight below 8 kg, such as the TOPRO Troja that is on the market since 2002 and has won several prizes. It has large wheels and a kerb lifter help to cross the street and are easy to fold for travelling. The new generation of wheeled walkers also features modern and colourful designs that aim to make them less stigmatising for their users, such as the Red Dot Design-awarded Gemino series from Sunrise Medical. Further, modern rollators offer optional accessories for a variety of uses, e.g. shopping bags, trays, umbrellas, lights and locks to protect the rather pricey models from theft.

A *pedal electric cycle* (pedelec) is a bicycle that assists the rider's pedalling by a electric motor, where the maximum speed is usually restricted to 25 km/h (15 mph) for safety reasons. In many countries, such as Germany, the rider of a pedelec does not need a driving license or insurance, in contrast to more powerful *electric bikes* (e-bikes). Electric support is nowadays also available for *tricycles* that provide more stability and safety than bicycles and are gaining popularity among seniors. Some also provide a lower seating position that makes it more easy and comfortable to enter the bike.

Electric wheelchairs are usually designed for the needs of disabled persons and controlled by a joystick on the armrest, such as Xeno by Otto Bock. *Mobility scooters* are quite similar in their functionality, but their different design with a steering column and storage space in front makes them more attractive for elderly users. However, they require some upper-body strength for an upright posture. In some touristic destinations, such as Tenerife, they can even be rented.

13.3.4 Solutions for blind and visually impaired people

Visually handicapped people usually can rely on many aids to orientation when moving outdoors and indoors respectively. Mostly guidance systems (tactile, optical, acoustical), walking indicators (e.g. groove plates), barrier-free signs, tactile maps as well as reading aids or magnifiers are used. Additionally, more and more intelligent devices are available. Actually, recent white canes and glasses provide object and obstacle detection using ultrasound (AmbuTech iGlasses, I-Cane Mobilo, SmartCane). Furthermore, some even provide GPS navigation with tactile feedback instead of audio response. Specific navigation systems like Kapten Mobility (see Figure 13.5) or Trekker Breeze+ (see Figure 13.6) guide the user via voice output and are able to present relevant information e.g. for public transport or POIs. Using a smartphone, one can also rely on applications like Blindsquare to store and find favourite places.

Figure 13.5 Kapten Mobility. Reproduced by permission of KAPSYS

Figure 13.6 Trekker Breeze+ handheld talking GPS. Reproduced by permission of the HumanWare Group

13.4 Research activities

A lot of research has been conducted over the last years on a European level. Especially in Germany, where the term of *Ambient Assisted Living* (AAL, also referenced as *Active and Assisted Living*) was invented [26], several national funded projects searched for improvements of pedestrian navigation methods and systems. Most notably, during the *AAL Joint Programme* started in 2008 by the European Commission (EC) [27], many projects have developed concepts and prototypic systems to help people staying active and mobile. Table 13.1 lists some of the most important projects co-funded by the EC along with a short description.

Naturally, that list is far from complete. A thorough list would clearly go beyond the scope of this chapter. Therefore the following paragraphs give additional references to further readings regarding AAL and mobility aspects of elderly people. First of all, the *Active and Assisted Living Programme*[1] gives an overview of all funded projects since 2008. The European Joint Programming Initiative *More Years, Better Lives*[2] additionally is dedicated to the coordination and collaboration between research programmes related to demographic change. Focussing on mobility, the *Trans-European Transport Network* (TEN-T)[3] helps with an overview of funded projects to improve mobility in European cities. Furthermore, the Federal Ministry of Education and Research Germany (BMBF)[4] as well as the Federal Ministry for

[1] http://www.aal-europe.eu/our-projects/
[2] http://www.jp-demographic.eu/
[3] http://ec.europa.eu/transport/newsletters/2015/05-04/articles/inea_research_grant_en.htm
[4] http://www.mtidw.de/ueberblick-bekanntmachungen/mobil-bis-ins-hohe-alter

Table 13.1 Overview of selected European research projects on mobility

Project	Short description
Alice – Assistance for Better Mobility and Improved Cognition of Elderly Blind and Visually Impaired [28]	The navigational assistant is able to offer visually impaired users a cognitive description based on a fusion of perceptions gathered from a range of sensors
ASSAM – Assistants for Safe Mobility [22]	The goal of the project is to develop modular assistance systems for all platforms (walker, wheelchair, tricycle)
Assistant – Aiding Sustainable Independent Senior Travellers to Navigate in Town [29]	Aid to use public transport anywhere, feedback when to exit or change direction
BLINDTRACK – Blind or partially sighted runners will be tracked using wireless tracking technology [23]	The information will be transmitted to the athlete via WLAN, with a vibrating belt producing warning signals as required and indicating when and how to change lane
Guiding Light – Light-Assistance-System to support spatio-temporal orientation of elderly people [30]	Intelligent lighting assistance for maintaining and improving indoor and outdoor mobility of older people at different stages of ageing process
iWalkActive – An Active Rollator for Active People [31]	The aim is to offer a open walker platform that goes beyond that of a conventional walking aid
ICITYFORALL – Intelligible City For All [32]	The objective of the project is the enhancement of the sense of security and self-confidence of presbycusis people
MobileSage – Situated Adaptive Guidance for the Mobile Elderly [33]	Provide context-sensitive, personalised and location-sensitive tools to elderly people which allow them to carry out and solve everyday tasks
T&Tnet – Travel & Transport solutions through emotional-social networking [34]	Provide personalised context-based multi modal and multinational social journey planning with affective capabilities and an easy to follow adaptive real-time guidance
VIRGILIUS [35]	Help seniors orient themselves and move around autonomously in indoor and outdoor areas by using a navigation system for mobile devices

Economic Affairs and Energy (BMWi)[5] have funded several projects on seamless mobility. In addition to European AAL activities, the BMBF launched its own AAL funding programme named *Age-based assistive systems for a healthy and autonomous living*[6] in 2008 (discontinued).

[5]http://www.tuvpt.de/index.php?id=foerderung000101
[6]http://www.mtidw.de/ueberblick-bekanntmachungen/selbstbestimmt-leben

Moreover, there are many insightful documents available to broaden the view on AAL and the ongoing activities in other countries. Among them, the Assistant Secretary for Planning and Evaluation, USA, published a study on *Aging Services Technology*[7] in 2012. The survey from Parisa Rashidi and Alex Mihailidis in the *IEEE Journal of Biomedical and Health Informatics*[8] highlights a collection of AAL tools for older adults. Key transport challenges in the UK are addressed in a study on *The Future of Transport in an Ageing Society*[9] released in June 2015. Also in that year, the Austrian Ministry for Transport, Innovation and Technology published a policy document on AAL, the *AAL Vision Austria*.[10]

Connecting different devices and technologies as a key aspect of AAL directly leads to standardisation and interoperability issues. On an international level, the *System Committee Active Assisted Living*[11] within the International Electrotechnical Commission (IEC) fosters cross-vendor interoperability in all AAL areas. As one input to that, the German Commission for Electrical, Electronic and Information (DKE) published the *German AAL Standardization Roadmap*[12] in 2012. The AAL Open Association (AALOA)[13] as well as the open platform and reference specification for AAL, *unversiAAL*,[14] further contribute to the idea of a connected AAL world.

13.5 Summary

This chapter on outdoor mobility showed that the realisation of seamless and intermodal mobility requires not only a lot of technologies but also the systematic integration and combination of different navigational methods. First of all the positioning of pedestrians is constrained by a high inaccuracy of GNSS in urban areas. By entering public or private buildings, there is a need for a seamless change to indoor positioning technologies, which in turn are only available in selected areas. Additionally, both—the indoor as well as the outdoor environment–need to be modelled precisely and with a high level of detail. At last, the lack of interoperability still remains. Although the use of crowdsourcing concepts could lead to a higher data quality in the meantime, there are only just a few standardisation approaches. Even efficient and intuitive human machine interfaces for pedestrian navigation are still

[7] http://www.norc.org/PDFs/Projects/Aging%20Services/ASTSRptCong.pdf

[8] http://www2.csd.uoc.gr/hy564/files/material/03/survey%20on%20aal%20tools%20for%20older%20adults.pdf

[9] http://www.ilcuk.org.uk/index.php/publications/publication_details/the_future_of_transport_in_an_aging_society

[10] http://www.aal.at/sites/default/files/AAL_Vision_%C3%96_Positionspapier_final_online_27042015.pdf

[11] http://www.iec.ch/dyn/www/f?p=103:186:0::::FSP_ORG_ID,FSP_LANG_ID:11827,25

[12] https://www.vde.com/en/dke/std/Documents/German%20AAL%20Standardization%20Roadmap%20%5B1%5D.pdf

[13] http://aaloa.org/

[14] http://www.universaal.org/index.php/en/

subject to research. Hence, these are the most important aspects to consider when implementing navigational assistive systems:

- *Positioning:* A single positioning technology does not sufficiently fulfil the needs of navigation in all environments. A seamless crossover between all or at least the best one is one of the important topics of today's sensor fusion and system development activities.
- *Modelling and routing:* The transition between indoor and outdoor results in a transition of different coordinate systems and different visualisations. Therefore the underlying models and databases are crucial in their compatibility, coverage and access.
- *Data:* Crowdsourcing and Open Data approaches promise a fast and flexible initiation of an assistive solution. The lack of standardisation and accessibility may restrict initiatives anyway. Even more important, effective solutions to acquire great amounts of highly detailed data are still missing. Another major transition to the characterisation of one's abilities and environmental characteristics is required. Subsequent research activities can lead to the development of an individual matching of both.
- *Human–Machine Interface:* The desired mobility assistant needs to adapt to the user, the user's behaviour and especially to the user's abilities. It should not require to switch the running application or even to change the device. The way of navigation and visualisation has to adapt fully automatically and seamlessly in any case.

Finally, there is a wide range of ongoing (research) activities but not even one ubiquitous product on the market covering all of the user's needs.

References

[1] Mollenkopf H. Significane of Out-of-Home Mobility in Modern Society. In: Mollenkopf H, Marcellini F, Ruoppila I, Széman Z, Tacken M, editors. Enhancing Mobility in Later Life. IOS Press, Amsterdam. 2005. p. 1–10.

[2] Rantanen T. Promoting Mobility in Older People. Journal of Preventive Medicine and Public Health. 2013;46(1):50–54.

[3] Klug S. Multimodal personal mobility; 2013. https://eu-smartcities.eu/sites/all/files/Multimodal%20personal%20mobility%20january.pdf, Accessed: 2016-01-22.

[4] Emmelmann M, Wiethoelter S, Koepsel A, Kappler C, Wolisz A. Moving towards Seamless Mobility – State of the Art and Emerging Aspects in Standardization Bodies. Springer Journal on Wireless Personal Communication. 2007;43(3):803–816.

[5] Heuberger A, Elst G, Hanke R, Heppner J, Kirsch K. Microelectronic Systems: Circuits, Systems and Applications. SpringerLink: Bücher. Springer Berlin Heidelberg; 2011.

[6] Tuncer TE, Friedlander B, Yasar TK. Classical and Modern Direction-of-Arrival Estimation. Elsevier, Academic Press, Burlington, MA, USA; 2009.

[7] Brugger M, Kemeth F. Locating rate adaptation by evaluating movement specific parameters. In: Arslan T, Keymeulen D, Merodio D, Benkrid K, Erdogan AT, Patel UD, editors. AHS. IEEE; 2010. p. 127–133.

[8] Jonas SM, Sirazitdinova E, Lensen J, Kochanov D, Mayzek H, de Heus T, et al. IMAGO: Image-guided navigation for visually impaired people. Journal of Ambient Intelligence and Smart Environments. 2015 September;7(5): 679–692.

[9] Faßbinder M. IS-02: SmartSensors-D – Nahtlose Lokalisierung;. http://www.medical-valley-emn.de/projekt/02-smartsensors-d-nahtlose-lokalisierung, Accessed: 2016-01-22.

[10] Brabham DC. Crowdsourcing. MIT Press, Cambridge; 2013.

[11] Neis P, Zielstra D. Recent Developments and Future Trends in Volunteered Geographic Information Research: The Case of OpenStreetMap. Future Internet. 2014;6(1):76–106.

[12] Puyol MG, Frassl M, Robertson P. Collaborative Mapping for Pedestrian Navigation in Security Applications. Communications in Computer and Information Science. 2012;318:49–60.

[13] Karimia HA, Kasemsuppakorn P. Pedestrian network map generation approaches and recommendation. International Journal of Geographical Information Science. 2012;27(5):947–962.

[14] Rupp R. Access–Navigation auf barrierefreien Wegen;. http://www.fp-access.de/, Accessed: 2016-01-22.

[15] Hempel R, Stellmacher T, Scheffler R, Hug K, Nickenig HP. Mobility for All–Results from the joint research project WikiNavi. In: Proceedings of the 8th German AAL Conference. VDE Verlag GmbH; 2015.

[16] Karimi HA, Jiang M, Zhu R. Pedestrian Navigation Services: Challenges and Current Trends. GEOMATICA. 2014;67.

[17] Karimi HA, Zhang L, Benner JG. Personalized Accessibility Maps (PAMs) for Communities with Special Needs. vol. 7820 of Web and Wireless Geographical Information Systems, Lecture Notes in Computer Science. Springer-Verlag, Berlin Heidelberg; 2013.

[18] Golledge RG. Wayfinding behavior: cognitive mapping and other spatial processes. Johns Hopkins University Press, Baltimore, MD, USA; 1999. XVIII, 428 S; 24 cm : ill.

[19] Stahl C. The Roaring Navigator: A Group Guide for the Zoo with Shared Auditory Landmark Display. In: Proceedings of the 9th International Conference on Human Computer Interaction with Mobile Devices and Services. MobileHCI '07. New York, NY, USA: ACM; 2007. p. 383–386.

[20] Robinson S, Jones M, Eslambolchilar P, Murray-Smith R, Lindborg M. "I Did It My Way": Moving Away from the Tyranny of Turn-by-turn Pedestrian Navigation. In: Proceedings of the 12th International Conference on Human Computer Interaction with Mobile Devices and Services. MobileHCI '10. New York, NY, USA: ACM; 2010. p. 341–344.

[21] Poppinga B, Pielot M, Boll S. Tacticycle: A Tactile Display for Supporting Tourists on a Bicycle Trip. In: Proceedings of the 11th International Conference on Human-Computer Interaction with Mobile Devices and Services. MobileHCI '09. New York, NY, USA: ACM; 2009. p. 41:1–41:4.

[22] Krieg-Brückner B. ASSAM–Assistants for Safe Mobility;. http://assam-project.eu/, Accessed: 2016-01-22.

[23] Ágnes Jakab, Varga M. BLINDTRACK–Guiding system for visually impaired for individual running on the track;. http://blindtrack.eu/, Accessed: 2016-01-22.

[24] Ross DA, Blasch BB. Wearable Interfaces for Orientation and Wayfinding. In: Proceedings of the Fourth International ACM Conference on Assistive Technologies. Assets '00. New York, NY, USA: ACM; 2000. p. 193–200.

[25] Schneider C, Henneberger S. Electronic Spatial Assistance for People with Dementia: Choosing the Right Device. Technologies. 2014;2(2):96. Accessed: 2016-01-22. Available from: http://www.mdpi.com/2227-7080/2/2/96.

[26] Huch M: Preparation and Start-Up Phase of the European AAL Joint Programme. In: Augusto JC et al. (Eds.): Handbook of Ambient Assisted Living – Technology for Healthcare, Rehabilitation and Well-being. Vol 11 of Ambient Intelligence and Smart Environments, p. 602. IOS Press, Amsterdam; 2012.

[27] Active and Assisted Living Programme–About the programme;. http://www.aal-europe.eu/about/, Accessed: 2016-01-22.

[28] Vizintin M. Alice Project;. http://www.alice-project.eu/, Accessed: 2016-01-22.

[29] Transport & Travel Research Ltd. Assistant;. http://aal-assistant.eu/, Accessed: 2016-01-22.

[30] Kempter G. Guiding Light: Research Projects – User Centered Technologies Research;. http://guidinglight.labs.fhv.at/, Accessed: 2016-01-22.

[31] iHomeLab. iWalkActive;. http://www.iwalkactive.eu/, Accessed: 2016-01-22.

[32] Laboratoire d'Intégration des Systèmes et des Technologies du Centre d'Energie Atomique CEA. Intelligible City For All;. http://www.icityforall.eu/, Accessed: 2016-01-22.

[33] Solheim I. MobileSage – Situated Adaptive Guidance for the Mobile Elderly;. http://mobilesage.eu/en, Accessed: 2016-01-22.

[34] Ingeniería y Soluciones Informáticas S L . T&Tnet – Travel & Transport solutions through emotional-social NETworking;. http://ttnet-aal.eu/, Accessed: 2016-01-22.

[35] Pilloni P. VIRGILIUS project;. http://www.virgilius.eu/, Accessed: 2016-01-22.

Chapter 14

Location and orientation technologies based on Wi-Fi systems for people with disabilities in indoor environment

Javier Coret Gorgonio[1], *Javier Pérez Bou*[2], *and Francisco Alcantud Marín*[1]

14.1 Introduction

Most location applications are focused on outdoor environments. These applications are based on Global Navigation Satellite Systems (hereinafter GNSS). This is a set of technologies that provides geo-spatial positioning with global coverage independently and thus offers an estimation of the user location. Through a network of satellites, a GNSS receiver can determine its position in four dimensions (longitude, latitude, altitude, and time). The technology has been used for many civilian and military applications.

However, there are still some location systems limitations based on GNSS. One of the most important problems is the inaccuracy in certain areas, such as urban environments or indoor buildings. A GNSS receiver can be located where a receiver acquires information from at least four satellites. In urban areas tall buildings may block the line of sight between the satellites and the receiver. The same is true in rural areas with trees. Both obstacles can produce wave diffraction and, consequently, a poor positioning of people or objects.

In wave propagation, there is signal degradation due to multipath effect. This effect must be considered and its influence is higher in indoor environments where the diffraction waves are compounded by different building materials. The result is less accuracy in location and orientation in indoor systems.

Disabled people have benefited from the location and orientation systems in open spaces. These applications facilitate people location and tracking in urban routes with higher autonomy. However, research and development of location technologies in indoor environments is far from finding optimal solutions. Actually, we can say that

[1]Universitat de València, Unidad de Investigación ACCESO, Facultad de Psicología (planta baja), Avd. Blasco Ibáñez, 21 46010 Valencia, Spain
[2]Universidad Católica de Valencia San Vicente Martir

Figure 14.1 GPS geolocation system for people with disabilities MIMOV[1]

this is an area that is constantly growing and evolving. An example is shown in Figure 14.1.

It is estimated that the use of these technologies (localisation inside buildings) will evolve quickly. This is reflected in the study published by the company Markets and Markets in 2013, "Indoor Location Market: Global Advancements, Market Forecasts and Analysis (2013–2018)." This study suggest an expected a growth of 42.1 per cent from 2013 to 2018 in the market for location, positioning, and navigation technology in indoor environments.

On the other hand, Location Based Services, hereinafter LBS, are expected to be the market's largest profit sector with a current preference for road location – as shown in the study GNSS_Market Report_2013_web. Issue 3.[2]

These issues justify the need for systems that provide position, location, and orientation with the least possible error, which should be available to all users in public buildings in the same way that it has been achieved in outdoor environments.

In this chapter we review the developments made in the technology of location and positioning inside buildings and pilot tests that have been carried. First, as a general framework we will begin by reviewing the key elements that influence in technological development (the activity, the operator, the environment, and the technology). In the next section, we describe the most important advances and development carried out

[1]http://www.mimov.com, Last accessed: March 2016.

[2]http://www.gsa.europa.eu/sites/default/files/GNSS_Market%20Report_2013_web.pdf, Last accessed: March 2016.

on people with different disabilities. Finally, in the concluding section we try to draw the lines that we believe should continue future development.

14.2 Assistive technology

Cook and Hussey [1] define "Assistive Technology" as any item, total or partial equipment, or any system acquired commercially or adapted to a person who is used to increase or improve functional capabilities of individuals with disabilities or to establish or modify behaviours. Scientific literature has used different terms to define the field of action of technological care for people with disabilities. Rehabilitation Technology, Assistive Technology, Adaptive Technology, and Access Technology are some of these. The development of these concepts, and the effort to mitigate the costs of removing barriers and adaptation of standard resources, has also led to more generic approaches such as "design for all."

Based on the HAAT ("Human Activity Assistive Technology") model, Cook and Hussey [1] define the role of assistive technologies in the everyday life of people affected by disabilities (Figure 14.2). From our perspective, this model can serve as a context for explaining the role of assistive technologies in developing prototypes of solutions focused on people with disabilities.

This model is based on the interaction of four basic components: the activity, the human factor, the context, and the assistive technology. Each one of the HAAT model components takes a role in the system. The system begins with the need or desire of

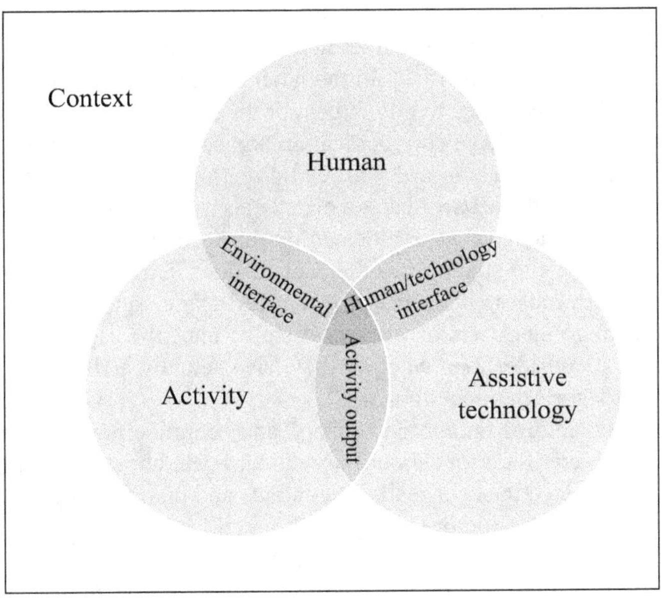

Figure 14.2 HAAT Model

a person to perform an activity. The activity checks the target system and it is carried out in context. This activity could be cooking, writing, playing tennis, or, as in our case, going to a specific location within a building. The combination of activity and human context will determine the skills required to achieve the goal.

The person uses assistive technologies in cases where they lack the necessary skills to carry out the activity. Obviously, the use of the technology also requires certain skills. Either way, the end-user needs to adjust the individual capabilities and the technologies to help them to achieve the goal of the activity.

14.2.1 The activity

Activity is the main element of HAAT and it defines the goal. The activity is the process of doing something, and it represents the functional outcome of human performance. The activity is performed as part of our daily life and it's necessary for human existence and the human learning. The activity is influenced and even determined by the lived society and culture. Any activity can be broken down into separate tasks. In turn, these tasks can be divided into subtasks. These subtasks are made up by tasks of minor order up to coming to sequences of automated actions of global execution. Those involved in the activity tasks need to possess a range of skills that could be either observable or mental.

In most cases, a disabled person differs from the average when carrying out the sequence and procedure of tasks and subtasks of the activity. The same applies to the necessary support or accessibility measures. Analysing the activity from the perspective of the necessary skills enables the provision of the necessary alternatives in the case of a disabled user.

In the example of movements inside a building, it is possible to analyse the steps we already know and whether the user knows the building or the person is accessing it for the first time (see Figure 14.3). In the latter case, the user will usually look to the building staff for assistance to guide them or the person will consult information panels. The user needs to have communication and social skills or the capability of accessing information panels in order to read directories. In many cases this is not possible. Once informed, the user makes a plan that is divided into steps. For example, if a person wishes to achieve a goal located on the second floor, the first step is to locate the elevator or the stairs up to the second floor. Once its location has been confirmed, the user must make a plan for the next stage, again using information panels or by asking other users or workers. Finally, once the user has reached their goal, he/she must consider then whether they have actually arrived at their desired destination or at some other location.

Task analysis requires the motor, sensory, and cognitive functions necessary to carry it off. Therefore, the users should not discriminate about where the directory is placed. Once in place the user reads its contents and plans the route to the target, etc. If any of the sensory, motor, or cognitive functions are ineffective, the system will look for compensation mechanisms for all potential users to accomplish their objectives. Braille or flat-relief directories of the surface of the buildings are used as

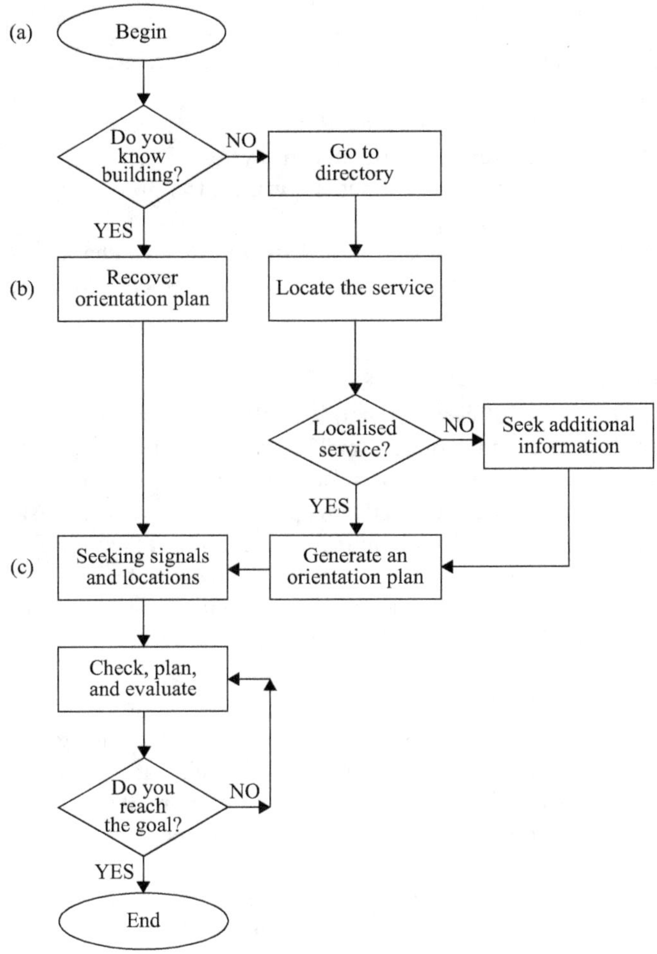

Figure 14.3 Task analysis activity for a basic movement in a building

resources for the blind. There are three main steps in a task analysis activity to achieve a basic movement in a building, as can be seen in Figure 14.3:

(a) Begin: the individual is at the front door of the building.
(b) If the user already knows the building, only an action plan is needed. In cases where the user does not know the building, support in the form of static directories and information boards or dynamic information, such as asking other users of the building, is required. The ultimate goal is to restore or create a plan of action that involves various tasks that need to be undertaken to achieve the desired goal.
(c) A monitoring plan requires continuous evaluation of the subtasks, rectification of any errors, and further planning to reach the final goal.

14.2.2 The human factor

In the HAAT model a second component appears directly related to the first, it is the human operator.

In every human activity the operator is always the action developer. The underlying skills of the operator directly influence the system. These skills can be grouped into three main categories: sensory input (Input), central processing (Agency), and responses or outcomes (Output).

Using the World Health Organization classification [2], we characterise the potential weaknesses that may contribute within the model described as the "operator" when interacting with information technology.

14.2.2.1 Visual impairment

We distinguish three types of visual impairment: blindness, partial vision, and reduced vision. Blindness implies a total or near total loss of forms of visual perception. Partial vision refers to a great dependence on information from other senses. This suggests the ability to use some features of visual perception. Finally, reduced vision can create obstacles for a person in situations requiring a high level of the use of vision.

Visual impairment poses a major impact on access to technologies in that all information that uses graphics, text, and images will be inaccessible for visually impaired people, thus limiting their ability to act.

14.2.2.2 Hearing impairment

Hearing impairment means a total or partial loss of the ability to perceive aural information. Deafness is severe or profound hearing loss. It is usually defined as a situation in which the loss is above 90 dB. When it is from 70 to 90 dB it is referred to as a severe loss. In cases between 40 and 70 dB, the person is considered to suffer moderate loss.

The limitations of people with hearing impairment in the use of technology are characterised by limitations for information using sounds

14.2.2.3 Disabled in speech production

This refers to any reduction in a person's ability to use the voice in a functional and intelligible way. The origin of this disability can be congenital, as in the case of a severe congenital such as dysphasia, or by a speech distortion due to lack of muscle control (dysarthria). It may also be an acquired disability such as loss of expressive language skills (expressive aphasia) caused by stroke or brain tumour.

This disabled impacts on the person's ability to communicate and, by extension, on the use of technologies that are based on communication, for example, the telephone.

14.2.2.4 Disabled in language comprehension

This refers to a loss or reduction in the ability to understand language. In congenital disorders, poor understanding also has implications for the ability of expression. It is a deficiency linked to one of the elements that make up communication and influences the use of technological media that use audio and written messages.

14.2.2.5 Intellectual disability

These include all limitations that result from an inadequate or insufficient cognitive process to be able to address the different requirements of the person. In most situations, people with the inability to process information or content, face great difficulty in analysing, assimilating, and responding appropriately. The lengthy and complex procedures, the large number of elements to be considered, the need for flexible response strategies, etc., that are involved in the use of most technological resources present an insurmountable barrier.

The human functions of planning, evaluation, working memory, selective attention, and behavioural control are part of the "executive function." Planning routes and itineraries that involve spatial representations of environments are stored in our memory. When a person arrives in a new location or it has no information in their memory, referrals and support are required in this environment, such as building directories or physical signals in which we trust to determine whether we are on the right track. These representations are constructed by previous experiences located in memory [3].

14.2.2.6 Physical impairment: locomotion disability

A reduced function of legs and feet implies dependence on a wheelchair or other walking aid apparatus (crutches, cane, etc.). Legs or feet are not required in the use of most technological tools, but the difficulty or inability to move does imply limitation to their proper use. For example, the need to use hands and arms to control the wheelchair may require adaptations to the technological application that rely on these resources.

14.2.2.7 Physical impairment: upper limbs disability

These include the loss of arms or hands, or a reduced ability to use these limbs due to limiting force or coordination. Disability in arms and hands has a significant impact on the use of any type of objects, especially those that are technological in nature. Most technologies require some kind of manual operation, thus a reduction or loss of capacity in the upper limbs involves usage limitations.

14.2.2.8 Multiple disabilities

Some people have multiple disabilities, such as combined visual and hearing impairment, hearing impairment, and problems with control movement of arms and hands. People with intellectual disabilities are often also affected by other deficiencies. In these cases, the access limitations are compounded and have a multiplier effect on the difficulties in managing technological tools.

A particular example is older people. With age, some physical, sensory, and cognitive abilities can become modified, regardless of the level of technological literacy that older people may have.

14.2.3 The context

The third element in HAAT model is the context or environment where the activity takes place. The context includes four basic considerations: (1) setting/environment (private home, factory, and community), (2) social context (even with strangers), (3) cultural context, and (4) physical context (temperature, humidity, light, etc.). It is often the case that the context where the activity takes place can be problematic and may be the determining factor in the success or failure of assistive technologies system.

The first meaning of context refers to the setting or scene in which assistive technologies will be used. However, the scene is more than just a place, it is a combination of an environment where tasks are carried out (for example, study, eat, shopping), a set of rules governing the tasks (to be in the location by 8:30am, to turn on the lights at 8pm, silent ...), and a level of adaptation (formal, informal, playful, serious).

Physical context refers to the environmental conditions that exist where the system is used. Three parameters are generally used: temperature, noise, and light. Many materials are temperature sensitive and these are affected by excessive heat or cold. In the case of use that is based on radio communication technologies, we include building characteristics and construction materials as in the existence of magnetic fields produced by machinery or other diffracting waves.

14.2.4 The technology

The fourth component in the HAAT model is assistive technology. Assistive technologies are included in a context interface where there is a relationship with the human element, or the Human Machine Interface, that results in a method of response or activity. In our case, the context interface is mainly technological and it assumes communication between the mobile device carried by a disabled person and the network via Wi-Fi antennas. The human–computer interface is limited to options that are currently on the market: based terminals Android, iOS, Windows Phone, RIM, and others system.

An important research line in indoor systems is to develop a locating and tracing technology that is comparable to the success achieved in outdoor environments. The main problem for indoor environments is the great variability. A shopping centre is an unstructured building with considerable variability but some administrative offices are fully structured buildings. These buildings can be of various heights, the spaces can be static or dynamic, and these can change according to functionality. Surfaces can have different textures and there may be different entrances to the building.

Moreover, the contents of the buildings may also be influential, for example, in terms of furniture changing its positions. Finally, there is an additional problem based on the technological elements already installed in the building (wiring, lighting, power supplies, magnetic fields, etc.).

There is a wide range of public buildings and this factor increases the difficulty in finding a technical solution that allows a person's position to be calculated with reliability and an acceptable level of error. In addition, the solution should be accessible to different groups of people with diverse functionalities.

Thus, people with motor impairment or wheelchair users, have an additional limitation because their hands are often employed managing the chair, thus making it difficult to use any normal mobile device.

Users with visual disabilities often use mobile taking it by the ear so that no need to use headphones. The use of headphones inhibits users from others acoustic signals.

People with disabilities in the upper extremities can usually handle electromagnetic devices. These devices can influence location and orientation systems.

Furthermore, the monitoring system cannot be based on the characteristics of an average person. The positioning of sensors or antennas must take the speed into account. The route selection has to be intelligible and flexible and it must report on possible problems of accessibility, providing the user with alternative routes to reach their goal.

Certainly the greatest challenge is for people with intellectual disabilities and older people with cognitive impairment. Planning functions, immediate memory function, recognition and evaluation functions, or guidance functions may be significantly affected to varying levels and with numerous combinations. Thus, technological elements must facilitate orientation tasks without introducing cognitively complex tasks.

Finally, indoor navigation and location-based services can be viewed as a jigsaw puzzle. In this case there are a number of fragments such as tracking systems, building maps and contents, navigation algorithms, and user and interface devices. The needs and preferences of the users are the connecting links.

14.3 Indoor location system

Indoor location systems are applications that collect information from wireless networks signals and check it against previously stored information on a device or server. Through this process we can locate the mobile carrier on a map or plan and provide information to the user. Different elements that may influence the effectiveness of this solution are hidden within this simple idea. Once the objective of informing the user the place location, we can propose routes for reaching a particular destination or inform them as to which services exist in the building in which it is located.

14.3.1 Content (maps of buildings, signs, and identification of spaces)

Public buildings are, by definition, complex and multifunctional, from airports, hospitals, schools, colleges, museums, through to management centres, municipalities, ministries, and so on. Moreover, these may be publicly or privately owned, such as shopping centres or supermarkets. From an architectural point of view, we can find buildings with one or several floors. In terms of content, we can find buildings or spaces with defined functions and other spaces whose use changes at different times.

In general, most of the buildings have a main directory at the entrance. That directory shows the available services and their location. A static signs system tells us where we are – ground floor, first floor, library, restaurant, etc. – and the direction we should go in – elevator, services, emergency doors, evacuation route, etc. If these directories are computerised, these can provide people with information about location and about the services that are connected to these areas.

Table 14.1 Worldwide smartphone operating system market

Period/OS	Android (%)	iOS (%)	Windows Phone (%)	BlackBerry OS (%)	Others (%)
2015Q2	82.8	13.9	2.6	0.3	0.4
2014Q2	84.8	11.6	2.5	0.5	0.7
2013Q2	79.8	12.9	3.4	2.8	1.2
2012Q2	69.3	16.6	3.1	4.9	6.1

Source: IDC, August 2015[3]

14.3.2 Mobile devices

The use of mobile, Internet access, and developed applications, has increased in recent years. This has facilitated an improvement in oral and written communication for those people with cognitive, sensory – vision or hearing – or motor impairments. Furthermore, the development of these applications is improving world interaction and providing some very important advantages, one of which is an increase in autonomy. Different disabilities, however, affect the user's mobile experience and mobile applications.

According to a study by the International Data Corporation (IDC), in the second quarter of 2015 (2Q15) mobiles companies worldwide have sold about 341.5 million smartphones. This indicates an increase of 13.0 per cent from 2Q14. Referring to worldwide smartphone operating system market share; this study indicates that the dominant operating system is Android. Table 14.1 shows different OS and its growth.

These smartphones are also supplemented by other devices that interact with applications such as GoogleGlass, SmartWatch, shoes, or gloves that incorporate sensors, Bluetooth. These devices provide new services to end-users and increase the number of services that can be accessed.

14.3.3 Location and navigation algorithms based on wireless technologies

Most current location applications are focused on locations in open space or outdoors, using Systems Global Navigation Satellite System, GNSS. These are satellite navigation systems that provide geo-spatial positioning with global coverage. That is, these systems accurately estimate user's location with global coverage. This is achieved through a network of satellites and a GNSS receiver. In this way it is possible to determine the position of a user in four dimensions – longitude, latitude, altitude, and time. Most of applications have been designed for military purposes and subsequently made available to civilians. To achieve this, a GNSS receiver can be located by acquiring information from at least four satellites. In urban areas the line of sight between satellites and receiver is obstructed by tall buildings and nearby trees. This causes wave diffraction resulting in the poor positioning of objects or people.

[3]http://www.idc.com/prodserv/smartphone-os-market-share.jsp, Last accessed: March 2016.

Recent studies have concluded that for detection in urban environments or indoors power levels are required between −160 dBW[4] and −200 dBW. GPS receivers usually have a noise level of around −131 dBW [4]. Therefore, in indoor environments, the GNSS signal is weak and cannot penetrate the walls. This produces precision failures in certain areas, such as inside buildings or urban environments. Consequently, an incorrect positioning and subsequent guidance of people takes place in these environments. This is the main limitation of tracking systems based on GNSS indoors.

In addition, the multipath effect causes signal degradation including the problem of whether the receiver is able to capture sufficient satellite signals. This effect is greater in indoor than in outdoor locations because the reflected signal is often greater than the direct view.

It should also be noted that satellite-based technology requires a high-energy consumption, and that in resource-limited mobile devices, the battery is quickly consumed [5].

As for people who need assistance, researchers have proposed and developed different systems for satellite-based aids such as the Satellite Based Augmentation System, hereinafter SBAS. Satellite system approaches are increased, improving both the horizontal and vertical positioning receiver. In this way, information about the quality of the signal is provided, performing clock, ionospheric, and ephemeris corrections. It evaluates and transmits SBAS parameters that limit the corrections ambiguity [6]. Table 14.2 shows the most important SBAS systems across the world.

Table 14.2 SBAS systems across the world

Location	System
Europe	European Geostationary Navigation Overlay Service, EGNOS[5]
United States of America	Wide Area Augmentation System, WAAS; Global Differential GPS, GDGPS; International GNSS Service, IGS[6]
Japan	Multi-Functional Satellite Augmentation System, MSAS[7]; Quasi-Zenith Satellite System, QZSS[8]
India	GPS and GEO Augmented Navigation, GAGAN[9]
China	BeiDou System, BeiDou[10]
Australia	Ground Based Augmentation System, GBAS[11]

[4]dbW is the power unit referred to in decibels (dB) that produces 1 Watt (W). Last accessed: March 2016.
[5]http://www.egnos-portal.eu/, Last accessed: March 2016.
[6]http://www.gps.gov, Last accessed: March 2016.
[7]http://www.mlit.go.jp/koku/15_hf_000105.html, Last accessed: March 2016.
[8]http://www.qzs.jp/en, Last accessed: March 2016.
[9]http://www.gpsindia.com, Last accessed: March 2016.
[10]http://en.beidou.gov.cn, Last accessed: March 2016.
[11]http://www.faa.gov, Last accessed: March 2016.

We can say that GNSS is a useful and accurate outdoor system, but it is not for indoor or urban environments. In the latter cases, other technologies or mixed systems are required. This has led to further research over recent years [7–8].

14.3.4 Wireless networks inside buildings

From the perspective of radio signal propagation, indoor environment is extremely complex. It is full of obstacles as well as interference from other devices. According to some studies, the free space loss can be quantified in a simple mode. In contrast, in a building or in urban environments, attenuation becomes highly complex making a loss determination difficult. The main causes of these losses are attenuation, the difference in materials, wall thickness, and the differences in frequencies used for communication between the transmitter and the receiver. Therefore, in indoor environments we can find loss due to free space loss, attenuation, and scattering [9]. Several studies have found similar conclusions as to the difficulty of calculating these losses in buildings [9–12].

To reduce location errors due to these effects, several location algorithms as trilateration and triangulation are used [13–15]. In these studies, we talk about algorithms such as Time of Arrival (TOA), Time Difference of Arrival (TDOA), Angle of Arrival (AOA), Phase of Arrival (POA), Reflectron Time of Flight (RTOF), Received Signal Strength (RSS), Fingerprinting, or others. It is concluded that the losses calculation in indoor environments is not trivial, thus experimentation with end-users is necessary.

One of the most used communication technologies by transmission properties is wireless technology communication. Through this, the connection of several devices is performed using radio waves or infrared light. Wireless networks provide access to dynamic devices within a coverage area. The disadvantage in mobile devices is the speed of transmission. In wireless networks it is not higher as in traditional wired or optic fibre networks. It also has the disadvantage of being insecure as all information transmitted by radio is likely to be heard by unauthorised people.

As an intermediate solution, mixed networks are adopted. In these types, wired networks or optic fibre and wireless networks technologies are mixed, thus providing a high transmission rate to the access node, making the backbone network. A wireless network is created using access points, which provides mobility for the device. This mobility allows scrolling within a coverage area that is delimited by the range of the access points in the wireless network.

The main technologies for wireless communication are based on Wi-Fi, GSM networks, Zigbee, RFID technology, infrared communication, Bluetooth technology or based on GNSS satellites, among others. [16–18]

Many of the main studies are based on Wireless-Fidelity technology or Wi-Fi that can be used as a viable, alternative positioning infrastructure [19–20].

In the basic operation in a Wireless Local Area Network (WLAN) each Wi-Fi access point generates a radio signal with a unique identifier, the Media Access Control or MAC of the wireless device. This allows the identification of specific access points to mobile devices. Consequently, any device with access to a wireless network can be located at any time, allowing use by anyone with a mobile terminal capable of accessing the Wi-Fi network. The main reason for the use of this technology is the rapid increase in the number of access points in public areas and users depth penetration. Typical Wi-Fi location accuracy is 10 m in metropolitan areas, showing greater accuracy than other mobile locating technologies. This is because Wi-Fi access points are closer to the base stations than in other mobile networks.

Definitively, which we want is to estimate the situation of a device inside a building. To achieve it, it is necessary to realise a process that allows defining the exact location and the characteristics of each one of the Access Points that the device will be able to detect in every zone.

The device needs an algorithm that allows estimating his position comparing the signs Wi-Fi that receives and the previous knowledge that has both of the positions and of the characteristics of the points of access.

Keep in mind, however, that the Wi-Fi system was not designed for a location signal based on power measurements, the measure is, therefore, imprecise. In fact it has not been thoroughly investigated as researchers have focused more on the design of location algorithms. The main methods used in positioning Wi-Fi are triangulation [21–23], statistical methods [24–26], and based on Fingerprinting methods [24, 27–29]. Each method has its own advantages and estimation errors.

In recent years mixed systems have been able to locate smartphones inside buildings where there is a GNSS signal, geolocation, and guiding people who are carrying or using mobile devices [30, 31].

In such studies the location is determined using Location Based Services, hereinafter LBS. Virrantus and colleagues defined it as information services accessible with mobile devices through its network to acquire the mobile device's location [32]. On the other hand, according to the definition provided by the Open Geospatial Consortium, it is a wireless internet service that uses geographic information to serve a mobile user [33].

According to another line of research, it is a method for obtaining geographical data and information services to be provided to users through mobile telecommunication networks [34]. Thus, it can be defined as a convergence of new information and communications technologies with spatial and GIS databases that allow mobile internet access.

Thus LBS can be defined as a service that provides location information (latitude and longitude, civil addresses, etc.) using positioning technologies that respond to requests for location services for mobile terminal users through different communications networks.

14.3.5 Indoor location for people with disabilities

A range of studies has been conducted for people with disabilities, such as that carried out by Intel Research investigators. These have focused on the use of wireless networks to help people with physical disabilities in navigation, and information was gathered on the application of open code referred to as Lab Pleases. In this application, there are observed Wi-Fi access points that surround it and it uses the power signal that it receives and the proximity of the signal to estimate the Wi-Fi device position [35].

This system maintains the privacy of the users; likewise it maintains the systems based on GNSS.

An important challenge for people with cognitive disabilities, in locations both inside and outside of buildings, is how they continue to be oriented through path recovery routines, as well as their movements in unknown areas. To improve the quality of life of people with mental disability, a range of studies have been conducted that are based on Bluetooth sensors by means of photos that show the user direction to be taken. These sensors have settled on the points of interest, intersections, elevators, exits, or other significant points. Once installed, tests have been conducted with real users [36]. Later studies continuing along this line, have produced an application whose instructions are visualised by means of photos or videos. It was observed a progress of the videos visualisation over 25–28 per cent with regard to the indications by means of photos [37].

For users with low or no visibility the main areas of concern are navigation and orientation without assistance in buildings that are unknown or unfamiliar. The greater challenge for these users problem is that they do not have easy access to the maps of buildings, signs, and other devices of orientation. In recent years studies have been conducted that guide people, using, for example, gloves that exchange data with the Android mobile via Bluetooth which communicates via Wi-Fi with the developed server, although it must be said that these require hardware installation. The system accepts different types of events that, among others, include help or replay [38].

The choice of the user interface for blind people is very important. For this reason there have been studies in which interviews have been conducted with users with low or no visibility to identify specific problems that they encounter [39]. This research has indicated that people with visual impairment prefer: navigate using verbal commands; walking near walls rather than across rooms; the quantity of way changes; the route should be minimised with short straight segments and changes with angles of 90° between the segments. Blind users can easily identify the doors, walls, and stairs with their canes. In the study mentioned earlier, a low cost system of navigation has been designed for interior use that is compatible with Near Field Communication Technology (NFC) and is based on the use of mobile terminals. The proposed navigation system allows users to imagine the layout of the rooms (such as the dimensions, the relative position of the points of interest). This information is stored in RFID tags where messages audible to the users are left either when approaching speakers or through sensors that are added to the user's own jacket. In similar researches, other studies that develop indoor localisation systems for blind people are based on RFID technology. These labels carry the identity of a nearby object to which the RFID receiver is sending radio waves. To transmit the identity, information required by RFID transmitter so that any RFID receiver knows the transmitted information and therefore the exact location of the point where it is [8, 39, 40].

We can also find studies, based on the same technology, in which users enter the building and direct the camera on their mobile device to a Quick Response (QR) code available at the points of interest. In this way they receive navigation information and instructions to be followed through an augmented reality browser (ARB) that shows the navigation direction [41].

Another type of works of interest is those that go in the line of the use of games like the consisting one in trying to avoid the obstacles that they can find the blind subjects. These consist of a series of sensors that express a vibratory pulse in sensor direction (there are eight sensors) that signify the location of obstacles. Short vibrations if the object is properly or large vibrations if it is not, indicating "Take care" [42]. The system was tested on subjects without visual problems in two series: one using sight and obeying the instructions of the sensors, and other with bandaged eyes. Both iterations obtained a positive result and managed to move around without clashing with any obstacles.

14.4 Conclusions and future lines

Indoor location systems that use smartphones:

- Produce cumulative errors in mobile devices since they have low cost integrated sensors (accelerometer, compass, etc.) which mean that they are of low quality and that each of the devices carried out mistake.
- The application requires a high level of accuracy for detecting the effects of diffraction, reflection, scattering, and others presented in indoor environments, imperceptible to humans.
- Continued use of internal sensors and the processing needs of the location system consume extra energy requiring that devices with energy constraints (such as mobile phones of last generation) are continuously connected to networks in order to process so many applications.
- The use of a mobile device takes place in multiple positions and possible orientations and keeping the phone in your hand, pocket, or in a protective case whilst walking, affects signal reception and, therefore, the power required for communication.
- Some areas that are particularly sensitive for visually impaired users, such as stairs and pillars, have many reflections and diffractions, thus the system may not be localised and oriented properly to the user.

For these reasons several researchers have focused on a mixed system consisting of the integration of technology with the 802.15 and 802.11. To achieve this, beacons have been included that increase the precision obtained with the Wi-Fi system on which it is based. These beacons are fully compatible with the system and these are based on IEEE 802.15 technology, also known as Bluetooth.

In terms of future lines:

It is necessary to conduct an in-depth study of the line of the inertial systems of mobile devices in a way that will enhance their use in different positions.

In general, the Wi-Fi system is used and the above-mentioned system recognises the users that it has connected in the field of interest. Thus, the achievement of a database appears in the systems of orientation, which includes the users that the habitual routes gain access, the access frequency, as well as the needs of the user based on the shortcomings he/she possesses. In this way, whenever he/she gains access to the building a route is provided that he/she usually uses or that is more appropriate to his/her needs. To do this, when accessing this area, the MAC will detect the mobile device that was previously registered, and will provide the latest route for the user from their current location.

Current public Wi-Fi networks pose some problems in their design, such as the lack of conformity in terms of the number of channels used, as well as the random frequency-hopping in the various networks, and, therefore, independent systems must be developed to address these problems.

In addition, you must standardise regulations for installation in public buildings, for example, the distances between access points, the same height, distances between an access point and beacon. This set of rules, should be created and should be included among the standards of accessibility of public spaces. It should extend the use of these spaces, also generating rules on the use and location of the furniture, advertising panels or education that could act as beacons screens, and Wi-Fi signal (to minimise the occurrence of obstacles that can act as screens). This rule would have a double meaning, could be used to improve signal propagation and in turn, help users ambulation marking obstacles.

Another idea is the possibility that the user could select the route that he/she wants to follow by choosing between the available alternative routes in the database, being able to change the route or to repeating, the route used on a previous occasion. If the route seems complicated, or if he/she knows of another route that could provide access with precision or accuracy, it can be provided and remain registered in the database to be offered to other users with the same needs.

On the other hand, as has been found in other studies, if a person in wheelchair, for example, perceives that there is another user with similar need following the same route, he/she should be able to choose between continue along the route or to look for an alternative route. The purpose is twofold: first, the route may show a path that passes through a room or hallway that does not support the simultaneous passage of two wheelchairs. Second, the user may not be in the mood to make conversation with another user, and can, thus, decide to avoid an encounter. We can extrapolate this idea to a mixed system using GSM technology in the exterior at those points with low power and Wi-Fi in adverse or complicated environments. In this form it provides access in both indoor and outdoor environments as it can cover the whole campus.

Another possibility for improving the designed system is the use of augmented reality that allows one to define a physical real-world environment through a technological device combined with virtual elements in a manner that creates a mixed reality in real time, adding the virtual part to the real. Thus, we add vision to the computer, as well as the recognition of objects that are available to devices, providing information about the actual physical environment and objects that is stored and can be retrieved for later access, facilitating access to the various departments and travel routes as reliably as possible.

A possibility would be to use the information captured by the cameras on mobile devices, as well as from other cameras and devices, such as sensors, that are in the field of interest and connected to the network, to realise the indications of the ideal ways in real-time. If the information mentioned above was in the database it might generate the route according to the needs of the user, as well as of the real environment of the area of interest.

Companies are currently promoting this type of augmented reality technology, which has gone from being practically inaccessible to users – because of the enormous cost of purchasing devices and the development of new applications – to dissemination to fully accessible devices. Thus, with a minimum investment of a few hundred Euros you can have augmented reality technology using hardware such as Oculus Rift[12] or I AM CARDBOARD®.

As can be seen, the technology that has been created so far is the main issue for users in that most applications are designed for users with "normal" vision.

Overcoming this problem requires the development of an auditory centred system so that 3D sounds are captured by mobile devices using Bluetooth technology. Although a more suitable system would be RFID, such as the ones designed in most of systems implemented at present for this type of user, due to its high precision this technology suffers from the need to acquire devices adapted to receive signs, as well as the high cost of the RFID sensors as compared to Bluetooth.

A further idea that may be of interest would be the creation of a system of proximity to hands, body, and head, which measures the distances between objects in a manner similar to a traditional sound but that issues sounds in the style of Morse code, increasing the frequency of emission in the direction of the obstacle and being soft or null in the direction that should be taken. Obstacles could be indicated using a repetitive sound that warns of collision or even a sound accompanied by a vibration at high frequency.

References

[1] Cook A, Hussey S. "Assistive technologies: Principles and practice." St. Louis: Mosby Year Book, Inc., 1995.

[2] Cunningham C, Coombs N. *Information access and adaptive technology*. Greenwood Publishing Group Inc. Westport, CT, USA, 1996.

[3] Wang RF, Brockmole JR. "Human navigation in nested environments." *Journal of Experimental Psychology: Learning, Memory, and Cognition* 2003; Vol. 29, No. 3, P. 398–404.

[4] Marron J. *Multi sensor system for pedestrian tracking and activity recognition in indoor environments*. Graduate School Thesis and Dissertation. University of South Florida. Scholar Commons; 2014.

[5] Carroll A, Heiser G. "An analysis of power consumption in a smartphone." *Proceedings of the 2010 USENIX Annual Technical Conference*, Boston, MA, USA, 2010; P. 271–285.

[6] Dai D, Walter T, Enge P, Powell JD. "Interoperation of distributed SBASS: theory, experience and future perspectives." *Proceedings of the 11th International*

[12] https://www.oculus.com, Last accessed: March 2016.

Technical Meeting of the Satellite Division of the Institute of Navigation (ION GPS 1998), Nashville, TN, USA, 1998. P. 1355–1364.

[7] Choo JH, Cheong SN, Lee YL. “Design and development of NFC smartphone indoor interactive navigation system.” *Journal of World Applied Sciences* 2014; Vol. 29, No. 6, P.738–742.

[8] Chumkamon S, Tuvaphanthaphiphat P, Keeratiwintakorn P. “A blind navigation system using RFID for indoor environments.” *Proceedings of the 2008 5th International Conference on Electrical Engineering/Electronics, Computer, Telecommunications and Information Technology (ECTI-CON 2008)*, Krabi, Thailand, 2008, Vol. 2, P. 765–768.

[9] Li B, Dempster A, Rizos C, Barnes J. “Hybrid method for localization using WLAN.” *Proceedings of SSC 2005 Spatial Intelligence, Innovation and Praxis*; Melbourne, Australia, 2005. P. 341–350.

[10] Ghassemzadeh S, Greenstein L, Kavcic A, Sveinsson T, Tarokh V. “UWB indoor path loss model for residential and commercial buildings.” *Proceedings of the Vehicular Technology Conference, 2003 IEEE 58th*, 2003, Vol. 5, P. 3115–3119.

[11] Motley A, Keenan J. “Personal communication radio coverage in buildings at 900 MHz and 1700 MHz.” *Journal of Electronics Letters* 1988; Vol. 24, No. 12, P. 763–764.

[12] Seidel SY, Rappaport T, Feuerstein MJ, Blackard K, Grindstaff L. “The impact of surrounding buildings on propagation for wireless in-building personal communications system design.” *Proceedings of the Vehicular Technology Conference, 1992 IEEE 42nd*, 1992, P. 814–818.

[13] Legarretaetxebarria A. *Sistema de localización y seguimiento de personas en interiores mediante cámara PTZ basado en las tecnologías kinect y ubisense.* Master Thesis and Dissertation. Universidad Del País Vasco; 2011.

[14] Roxin A, Gaber J, Wack M, Moh ANS. “Survey of wireless geolocation techniques.” *IEEE Globecom Workshops*, Washington, DC, USA, P. 1–9, 2007.

[15] Sayed AH, Tarighat A, Khajehnouri N. “Network-based wireless location: challenges faced in developing techniques for accurate wireless location information.” *Signal Processing Magazine, IEEE 2005*; Vol. 22, No. 4, P. 24–40.

[16] Crow BP, Widjaja I, Kim JG, Sakai PT. “IEEE 802.11 wireless local area networks.” *Communications Magazine, IEEE* 1997; Vol. 35, No. 9, P. 116–126.

[17] LAN/MAN Standards Committee. *Part 15.4: Wireless medium access control (MAC) and physical layer (PHY) specifications for low-rate wireless personal area networks (LR-WPANs).* IEEE Computer Society, 2003.

[18] Infrared Data Association. *Point and shoot profile. Version, 1* 2000; P. 1–34.

[19] Alizadeh-Shabdiz F, Pahlavan K. *Estimation of Position using WLAN Access Point Radio Propagation Characteristics in a WLAN Positioning System* November 2007, US7916661.

[20] del Prado Pavon J, Choi S. “Link adaptation strategy for IEEE 802.11 WLAN via received signal strength measurement.” *Proceedings of the IEEE International Communications Conference, 2003. ICC’03*; May 2003, Vol. 2, Anchorage, AK. P. 1108–1113.

[21] Mazuelas S, Lago FA, González D, *et al.* 'Dynamic estimation of optimum path loss model in a RSS positioning system.' *Proceeding of the Position, Location and Navigation Symposium, 2008 IEEE/ION*; May 2008, Monterey, CA, P. 679–684.

[22] Vegni AM, Di Nepi A, Neri A, Vegni C. "Local positioning services on IEEE 802.11 networks." *Proceeding of the Applied Electromagnetics and Communications, 2007. ICECom 2007. 19th International Conference*, September 2007; Dubrovnik, Croatia. P. 1–4.

[23] Yang J, Chen Y. "Indoor localization using improved rss-based lateration methods." *Proceeding of The Global Telecommunications Conference, 2009. GLOBECOM 2009. IEEE*, November–December2009, Honolulu, HI, P. 1–6.

[24] Kjærgaard MB, Munk CV. "Hyperbolic location fingerprinting: a calibration-free solution for handling differences in signal strength (concise contribution)." *Proceeding of the Pervasive Computing and Communications, 2008. PerCom 2008. Sixth Annual IEEE International Conference,* Mar 2008, Hong Kong, P. 110–116.

[25] Roos T, Myllymäki P, Tirri H, Misikangas P, Sievänen J. "A probabilistic approach to WLAN user location estimation." *International Journal of Wireless Information Networks 2002*; Vol. 9, No. 3: P. 155–164.

[26] Seco F, Koutsou K, Ramos F, Jiménez AR. "Localización personal en entornos interiores con tecnología RFID." *Journal of Revista Iberoamericana de Automática e Informática Industrial RIAI 2013*; Vol. 10, No. 3,P. 313–324.

[27] Brunato M, Battiti R. "Statistical learning theory for location fingerprinting in wireless LANs." *Journal of Computer Networks 2005*; Vol. 47, No. 6, P. 825–845.

[28] Kim J, Choi H, Lee D, Kang S. "A 2-stage hybrid position estimation framework in RF fingerprint WPS." *Journal of Wireless Networks 2014*; Vol. 20, No. 6, P. 1–16.

[29] Mirowski P, Whiting P, Steck H, *et al.* "Probability kernel regression for WiFi localisation." *Journal of Location Based Services 2012*; Vol. 6, No. 2, P. 81–100.

[30] Canedo-Rodríguez A, Álvarez-Santos V, Regueiro C, Iglesias R, Barro S, Presedo J. "Particle filter robot localisation through robust fusion of laser, WiFi, compass, and a network of external cameras." *Journal of Information Fusion* 2015; Vol.27, P. 170–188.

[31] Coret J, Perez J, Alcantud F. "Sistemas de orientación en el interior edificios de concurrencia pública. prototipo ISMO." *Proceeding of VI Congreso Internacional De Diseño, Redes De Investigación Y Tecnología Para Todos, DRT4ALL 2015*, Madrid, Spain, September 2015, P. 313–341.

[32] Virrantaus K, Markkula J, Garmash A, *et al.* "Developing GIS-supported location-based services." *Proceedings of* the *Web Information Systems Engineering, 2001. The Second International Conference on*, Kyoto, Japan, December 2001, Vol. 2, P. 66–75.

[33] Open G. Consortium Inc.; Date: 3-May-2005; *Reference Number of this OpenGIS® Project Document: OGC 04-094*, 2005.

[34] Shiode N, Li C, Batty M, Longley P, Maguire D. *The impact and penetration of location-based services.* Centre for Advanced Spatial Analysis (UCL), 2002, London, UK.

[35] Kölsch D. *The place lab project.* January 2006. Available from http://pi4.informatik.uni-mannheim.de/pi4.data/content/courses/2005-ws/seminar/Ausarbeitung_Das_PlaceLab_Projekt.pdf

[36] Chang Y, Chu Y, Chen C, Wang T. "Mobile computing for indoor wayfinding based on bluetooth sensors for individuals with cognitive impairments." *Proceeding of: Wireless Pervasive Computing, 2008. ISWPC 2008. 3rd International Symposium,* Santorini, Greece, May 2008, P. 623–627.

[37] Chang Y, Wang T. "Comparing picture and video prompting in autonomous indoor wayfinding for individuals with cognitive impairments." *Journal of Personal and Ubiquitous Computing 2010*; Vol. 14, No. 8, P. 737–747.

[38] Ganz A, Gandhi SR, Schafer J, *et al.* "PERCEPT: Indoor navigation for the blind and visually impaired." *Proceeding of the Engineering in Medicine and Biology Society, EMBC, 2011 Annual International Conference of the IEEE,* Boston, MA, USA, 2011, P. 856–859.

[39] Ivanov R. "Indoor navigation system for visually impaired." *Proceedings of the 11th International Conference on Computer Systems and Technologies and Workshop for PhD Students in Computing on International Conference on Computer Systems and Technologies,* New York, NY, USA, 2010, P. 143–149.

[40] Alghamdi S, Van Schyndel R, Alahmadi A. "Indoor navigational aid using active RFID and QR-code for sighted and blind people." *Proceeding of the Intelligent Sensors, Sensor Networks and Information Processing, 2013 IEEE Eighth International Conference*, Melbourne, VIC, Australia, 2013, P. 18–22.

[41] Magableh B, Barrett S. "Self-adaptive application for indoor wayfinding for individuals with cognitive impairments." *Proceeding of the Computer-Based Medical Systems (CBMS), 2011 24th International Symposium*, Bristol, UK, 2011, P. 1–6.

[42] Moller K, Toth F, Wang L, *et al.* "Enhanced perception for visually impaired people." *Proceeding of the Bioinformatics and Biomedical Engineering, 2009. ICBBE 2009. 3rd International Conference*, Beijing, China, 2009, P. 1–4.

Chapter 15
Enabling health, well-being and engagement in life through ambient-assisted living technologies – an occupational therapist's lens

Debra Young[1]

Moving beyond obvious markets of entertainment and security systems, home technology can enable greater independence and safety across the lifespan. New developments in technology can enable older adults to live independently and safely at home. These innovations also empower and provide peace of mind for caregivers. Increasing access to technologies that are embedded within the environment allows for remote control of our homes, while the expansion of home health technologies permits all generations to be actively involved in their own, as well as their loved one's health. This chapter will discuss and explore ambient-assisted living (AAL) technologies that support productive ageing vis-à-vis universal design (UD), sensory needs and changes which occur with age and disability, as well as in the context of the societal and demographic drivers creating demand. Product usability analysis through an occupational therapist lens, considering UD principles to feature match solutions to enable health, well-being, social connectedness and engagement, is to be addressed.

15.1 Common societal drivers

15.1.1 Ageing population

The world is witnessing a global ageing epidemic. According to the United Nations *World Population Prospects* report, the number of people over the age of 60 is expected to more than double by 2050 and more than triple by 2100. Further, all major areas of the world, with the exception of Africa, will have nearly a quarter or more of their population aged 60 or over [1]. The old-age dependence ratio, the number of people age 65 and older per one hundred working age people, is rising around the world, with significant projected proportional increases for Japan, South Korea, Spain and Italy [2, 3].

[1]Owner, EmpowerAbility® LLC, 107 Syracuse Drive, Newark, Delaware 19713 (USA)

Although we are gaining years of life, the quality of that longevity, according to The *World Health Organization (WHO) Report on Ageing and Health*, is not necessarily healthy [4]. The report analysed the prevalence of older adults aged 65–74 years and 75 years or older, across countries needing assistance for the completion of activities of daily living (ADLs) (e.g. eating, bathing, dressing getting in and out of bed and using the toilet) and instrumental activities of daily living (IADLs) (e.g. telephone use, medication management, money management, grocery shopping, meal prep and using a map). The report indicates that although there is a marked difference among the countries related to the percentage of older adults requiring assistance with at least one ADL, it is the impact of age that is significant. The analysis demonstrated that older adults aged 65 years to 74 years have far less a need for assistance than those aged 75 years or older with both ADLs and IADLs [4]. According to the Centers for Disease Control and Prevention, nearly 50% of adults living in the United States with two or more chronic health conditions, such as heart disease, and 34.9% of all adults are obese [5, 6].

The steady increase in our older adult population, coupled with older adults living longer and requiring more assistance, is driving caregiver demand. More than 65 million people, 29% of the adult population in the United States provide care for a chronically ill, disabled or ageing family member or friend during any given year and spend an average of 20 hours per week providing care for their loved one [7]. The International Alliance of Carer Organizations has also tallied caregiving statistics by nation, with total number of caregivers in Canada at 8.1 million, in the United Kingdom at 6.5 million, Australia with 2.86 million and Sweden with 1.3 million [8]. Although more men are assuming the role of caregiver, women still largely dominate this arena globally [7].

The market is seeing a boom in innovations that are not only geared towards assisting those millions of people to safely remain in their homes for as long as possible, but to also empower and provide peace of mind for caregivers. From gero-technology systems to mobile service robots (MSRs), the integration of AAL technologies is certainly changing the future of caregiving.

15.1.2 Desire to age in place

Many equate home with identity. The concept of home creates a sense of familiarity and comfort, a link to one's circle of support. Therefore it comes to no surprise that many older adults want to remain in their familiar environments and live there as independently and safety for as long as possible. This is the desire to age in place. Within the European Union (EU), research is showing that residential mobility between private homes is usually found to decline with age, but it is also time dependent as the more years an older adult spends in their home, the less likely they are to leave it [9, 10]. In the United States, an American Association of Retired Persons (AARP) survey on ageing in place found that nearly 90% of people over age 65 indicated a desire to remain in their homes for as long as possible, and 80% believe their current home is where they will always live [11]. This global synergy demands attention.

15.1.3 Housing and community

The ageing population is living longer, living with chronic health conditions, and living in their ageing homes. The *American Housing Survey* found that the median age of the more than 124 million homes in the United States housing stock is 36 years old [12]. Given the characteristics of the existing housing stock, many older adults will have to make modifications to their homes to accommodate the physical limitations that arise with age [13]. Commonality can be found in the EU. According to the 2008 Survey of Health, Ageing and Retirement in Europe, in most countries in the EU, there is a clear deficit of special provisions in housing that assist persons with physical limitations and other health concerns, thus, creating increased risk of those residents having to relocate [14].

Community mobility is integral to health and wellness and is associated with independence and autonomy. For older adults, limited access to the community because driving is no longer a safe option can lead to depression, isolation, decreased participation in meaningful activities, decreased ability to care for oneself, and diminished wellness [15, 16]. This transportation challenge combined with urban sprawl creates an even wider gap, further limiting social connectedness and participation in meaningful activities. AAL technologies that support ageing in place are an option to bridge this gap.

15.2 Role of occupational therapy

Occupational therapists have long-standing expertise in providing services that incorporate assistive technology and environmental modifications (EM). A science-driven education affords occupational therapists a holistic knowledge of human development across the lifespan, the impact of age-related changes, illness, injury and/or disability and the role of the environment on an individual's participation in meaningful activities. Occupational therapists look at the environments in which a person must act, the occupations (activities) they need to engage in, and the personal factors that impact their performance in these occupations in various environments. This is the Person–Environment–Occupation Model (PEOM).

15.2.1 Person–Environment–Occupation Model

The PEOM was developed by Law *et al.* [17] and is a framework to analyse the dynamic relationship between people, the environment and their occupations and roles. Developed from environmental-behavioural theories, PEOM is a transactive approach and assumes an interdependence of person and environment [17]. In this approach, it is acknowledged that behaviour is influenced and cannot be separated by contextual influences, temporal factors and physical and psychological characteristics [17]. The model looks at an individual holistically, taking into consideration that all areas of a person's life may impact occupational performance. Further, the model guides an occupational therapist's clinical reasoning, providing support and justification for what we choose and why.

The PEOM consists of four major constructs: person, environment, occupation, and occupational performance. The first construct, the 'Person', is inclusive of an array of intrinsic factors (e.g. psychological, cognitive, physiological, spiritual and neurobehavioural) that compose one's set of skills and abilities. The person brings to the context their own set of attributes, skills, knowledge and experience. The basic assumption of the model is that the person is continually developing and is intrinsically motivated [17]. The client is a unique being who assumes multiple roles and cannot be separated from contextual influences [17].

The second construct, the 'Environment' considers that participation is always impacted by extrinsic characteristics of the environment in which it occurs. These extrinsic factors of the environment are the external physical and social conditions that surround the client along with the cultural, personal, temporal and/or virtual contexts within which occupational performance takes place [18].

The third construct, 'Occupation' refers to self-directed meaningful tasks and activities engaged in throughout a lifespan [17]. Occupations are engaged in to fulfil intrinsic need for self-maintenance, expression, and life satisfaction and they are carried out within multiple environments and contexts to satisfy individual roles [17]. The temporal aspects that encompass the occupational routines of an individual over time (e.g. day, week, lifespan) are an important consideration as it the interaction of all three components that varies across time [17].

The fourth construct is 'Occupational Performance and Participation'. This is the act of doing and accomplishing a selected action (performance skill), activity or occupation [19–21] that results from the dynamic transaction among the client, the context and the activity [18]. Occupational performance (Figure 15.1), in the PEOM is the

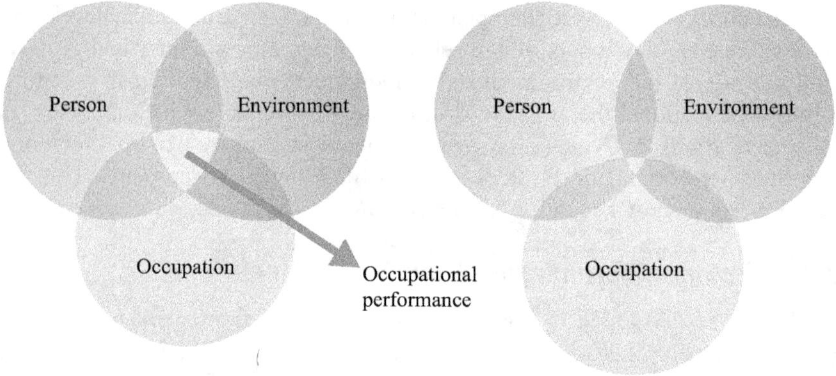

Figure 15.1 Person-Environment-Occupation Model (PEOM) with concentric circles showing "fit." (From Foundations of Theory and Practice for the Occupational Therapy Assistant (pg. 272), by Amy Wagenfeld, 2015, Philadelphia: Wolters Kluwer. (Adapted from Law M, et al. 'The person-environment-occupation model: A transactive approach to occupational performance'. Canadian Journal of Occupational Therapy 1996;63(1):9–23.)

outcome of the interaction of the person, the environment, and one's chosen occupation [17]. Visual representation of this model is depicted via three concentric circles that dynamically interact with one another over time. The constructs demonstrate better 'fit' when there is more overlap. Conversely, the less the circles (constructs) overlap, 'fit' is minimised, ultimately compromising occupational performance.

15.2.2 Distinct value of occupational therapy

Occupational therapists offer distinct value to the design of spaces and products with the human lifespan, development and varying levels of ability in mind. Experts in activity analysis, occupational therapists analyse the demands of an activity (occupation) to understand the specific body structures, body functions, performance skills, and performance patterns that are required and to determine the demands the activity (occupation) makes on the client [18]. Bringing knowledge of products and systems, both assistive and mainstream, occupational therapists analyse usability and feature-match products and systems to meet a client's current and potential long-term needs. This feature matching comes from a comprehensive client-centred evaluation including physical, sensory and cognitive abilities and the subsequent interaction with a particular piece of equipment, within a space.

15.3 Universal design

UD is the design of products and environments to be usable by all people, to the greatest extent possible, without the need for adaptation or specialised design [22]. UD accommodates a wide range of user abilities, regardless of age, ability, or preference. It encompasses a variety of human factors, addressing usability holistically and integrating features into the environment that provide ease of use, convenience, and safety for all. UD features, when done well, seamlessly integrate into the overall design of the space or product, rendering them virtually invisible [23].

There are seven principles of UD that assist with evaluation of a space; guidance for a design process; and education regarding features that provide increased usability of products, services, and environments [24]. These seven principles are discussed in the following sections.

15.3.1 Equitable use

It means that products or environments should be usable by everyone. Ideally, the means by which people use an environment, service, or product should be the same. If it cannot be, the design must provide options to ensure a user's privacy, security, safety, and convenience. Furthermore, the environment, service, or product should not isolate or stigmatise any group of users [25]. It is the difference between having a separate entrance to a building or separate sink to use in a bathroom, versus entering through the same doors or having access to any sink as all other patrons.

15.3.2 Flexibility of use

The design should allow people to use the features in multiple ways. It should accommodate age, for handedness as well as the user's pace [25]. A technology system that allows for access via voice or touch as well as within the actual object/device versus on a cell phone, tablet or computer is an example of flexibility of use, as it is usable by people of varying size, age, and ability.

15.3.3 Simple and intuitive use

This is the third principle. The purpose of each design feature, service, or product and how to use it needs to be easily understood and intuitively obvious so that it operates as anticipated and can be used spontaneously [25]. Consider AAL technologies that are integrated directly into the common objects we use on a daily basis, versus the creation of a new stand-alone product that provides similar data and assistance. Users continue to use the product/device no differently than previously, without the addition of new learning. The product therefore is familiar and easy to use, drawing upon prior knowledge, and does not require much thought.

15.3.4 Perceptible information

The design provides all essential information in a variety of representations (e.g. written, pictorial, tactile, auditory) to ensure effective communication for all users. For instance, written information must be presented with enough background colour contrast so that it is distinguishable from its surroundings [25]. An example of *perceptible information* is a thermostat having a larger display both with numbers/text and pictorial representations as well as colour coding to allow the user to discriminate if heat or air conditioning is being delivered.

15.3.5 Tolerance for error

It is the fifth principle. The design should eliminate or shield features that could prove hazardous or inconvenience a user. When potentially dangerous conditions are unavoidable, users should receive warnings as they approach the design feature. The design should also anticipate accidental or unintended actions [25]. One example is a medication management system that safeguards a missed dosage of medication by locking it away in a separate area of the device and/or making it inaccessible to the user in order to eliminate the concern that it would be taken at the wrong time and/or as an accidental double dose with the next timed medication. Further safeguarding would allow for communication with a caregiver regarding the missed medication dose.

15.3.6 Low physical effort

This is the sixth principle. The design should incorporate features that require little or no physical effort. If a low level of force is required, users should be able to apply the necessary force without assuming an awkward or dangerous position [25]. An AAL technology example is the ability to lock and unlock doors and/or turn lights on and off via a mobile device versus having to manipulate the items with hands. Domestic robots eliminate virtually all effort from having to complete certain housekeeping tasks such as vacuuming the floor, other than having to activate the button to turn the robot on and off.

15.3.7 Size and space for approach and use

The design should allow for adequate space to enable anyone to use the design. One example is the use of a television versus a tablet or cell phone to participate in a video chat. The increased size of the screen is easier for viewing for young and ageing eyes, while it also eliminates the need to have to hold, prop, or mount a mobile device during use. Further, it allows for multiple people to access and view simultaneously, with less need for crowing to view the screen.

The objective of UD, also known as *lifespan design, inclusive design, and/or design for all*, is designing for all people, not specifically for persons with illness, injury, or disability. UD is an approach that accommodates a wide range of user abilities, regardless of age, ability, or preference. By incorporating ergonomics, UD features consider the interaction of the user within their environment subsequently reducing daily stresses and strains that can lead to repetitive motion injuries. UD features provide convenience, integrating work simplification principles to promote energy conservation. Analysing product usability with a UD lens provides insight into the flexibility of the product to accommodate a diversity of users. Having knowledge of the spectrum of available consumer products, assistive and mainstream, is integral to an EM plan. By integrating UD into the EM plan, there is potential for improved health outcomes for all persons within the home environment. UD, as an intervention, is a method to achieve targeted outcomes of prevention, health and well-being by addressing the needs of clients without illness, injury or disability, and providing enriched contextual and activity experiences that will enhance performance for all people in the natural contexts of life [18]. Occupational therapists' expertise in activity analysis is cohesive to this process.

15.4 AAL: enabling health, well-being and engagement in life

Our technology world is ever-changing and becoming more integrated into every-day life.

While we may be used to certain devices within our environment being automated (e.g. automated teller machine), many ordinary objects are being reinvented with digital sensing, computing and communications capabilities [26]. Objects within our homes are now being powered with a 'digital voice' giving feedback about their status along with data relevant to the surrounding environment. According to Gartner, Inc., 6.4 billion connected 'things' will be in use by the end of 2016. In 2020, 13.5 billion connected 'things' will be in use [27]. These AAL technologies can enable greater independence and safety across the lifespan.

In occupational therapy practice, identification and use of assistive technologies, application of UD principles and the recommendation of changes to an environment or activity are preparatory methods utilised concurrently with occupations to support daily occupational performance [18]. This includes assessment, selection, provision and education and training in use of devices [18].

Some of the devices are stand alone, performing one function and not easily integrated into a system that controls multiple services. While others are all-inclusive, providing a variety of services and flexible enough to start with one service with the ability to expand services based on a user's changing wants or needs. These devices can be further classified into systems that are wearable, systems that are embedded in the user's environment as well as systems that allow for both set-ups. Increasing access to technologies permits all generations to be more actively involved in their own, as well as their loved one's health and well-being.

15.4.1 Well-being

Technology for well-being has been around for quite some time. However, many of these technologies relied on the user. Whether it was a pendant, watch or belt clip for fall detection and prevention or a pair of sneakers or radio frequency id tag on a shirt, the success of the device was based on it being worn. AAL technologies are embedded within the user's environment, removing the user as the variable to the success of carryover of use. Compliance is taken out of the equation.

15.4.1.1 Fall detection/prevention

According to the World Health Organization (WHO), each year 424,000 individuals die from falls globally. In fact, falls are the second leading cause of accidental or unintentional injury deaths worldwide. Furthermore, of those falls, adults aged 65 years and older suffer the greatest number of fatal falls [28]. The integration of EMs, including technology, can help decrease the risk of falls, provide caregiver peace of mind, reduce costs to society and create a safer home environment.

Systems that are embedded into the user's environment utilise wireless motion sensors that are strategically placed throughout the home. These sensors send information to a laptop, mobile phone or other designated device about the user's movements within the home. Individualised parameters are set for the sensors to send alerts based on specific conditions, one such being unusual activity in a room and/or lack of activity in a room, which may be the result of a fall. Wireless sensor tags are available for more than just detecting motion. Monitoring temperature, detecting water and humidity along with locating lost items are also possibilities.

Beyond the use of sensors, current research is also exploring the idea of integrating the sensors into the design of the product from the beginning, versus retrofitting a space after the fact. One pilot research study tested the idea of a smart floor that is able to detect human falls by monitoring the vibration patterns in the floor [29]. Taking this concept one step further, the University of Manchester scientists more recently created a magic carpet. The scientists have integrated plastic optical fibres into the underlay of the carpet, with tiny electronics located at the carpet edges that act like sensors. These sensors can then analyse a person's gait, detect incidence of a fall and relay this information to a computer [30]. Another option that allows for solid

Figure 15.2 SensFloor® by © Future-Shape GmbH © Witex Flooring Products GmbH (http://future-shape.com/de/)

surface application is the SensFloor® by Future Shape (Figure 15.2). SensFloor® is a large area sensor system, based on a 3-mm thick textile underlay [31]. Currently this technology is being utilised in long-term care facilities, but has merit for use in the residential space. The technology has the capability to not only detect a fall and send an alert for help, but it can also be used as an environmental control by turning lights on at night when someone gets out of bed and as an occupancy sensor, turning lights off once exiting the bathroom. Adding another level of safety, the flooring sensors can also detect if someone has entered your home in the event that you are away.

15.4.1.2 Smart home automation for safety and remote monitoring

Home automation is certainly not new, but with the popularity of mobile technologies, we are witnessing an explosion of interest in this market. More and more technology is going mobile and has multigenerational benefits. The continuously growing market brings along innovations that are becoming more appealing, convenient and meeting common needs and interests across generations. New technologies such as standalone smart thermostats are touted as being able to save energy and money, but can be controlled from both inside as well as both from inside as well as outside our homes with a cell phone, tablet or other mobile device. Smart thermostats also have larger displays with both text and pictorial representations as well as an easier interface to manipulate than a standard thermostat (Figure 15.3). Other devices, both as standalone and integrated systems, allow for the monitoring of stove use, smoke, as well as the temperature in the home via wireless sensors. Many of these systems can

Figure 15.3 Nest learning thermostat (https://nest.com/thermostat/meet-nest-thermostat/)

also expand to include door opening, lighting control and even control of window treatments. Where security systems were once separate from home automation, we are now seeing the marriage of these two markets for a fully integrated system.

Moving beyond the familiar market of home automation, remote monitoring involves using technology to keep caregivers, or family members, apprised of older adults' status. The goal of these technologies is to enable independent living and create safer living environments. Taking into consideration a user's typical routine and familiar trajectory around their home, AAL technologies not only provide convenience but also support ageing in place. Consider this scenario: when you wake up in the morning, you swing your legs out of bed and the pressure of your feet on the floor sends a signal to your coffee pot to start brewing that first cup, simultaneously turning on the lights that lead to the bathroom. It also sends a text message to your adult children alerting them that you have gotten out of bed. As you pour your first cup of coffee, your coffee mug sends you a text message as a reminder to take your medication. Not a mobile phone user? Your medication bottle automatically blinks at the time you are to take your medication. As you move to your favourite chair to watch television, a motion sensor sends a message to your family members alerting them of your location in the home. If you sit in your favourite chair too long, another notification is sent. This is one scenario that captures a small amount of remote monitoring that can be accomplished with AAL.

From wander management to detecting a possible flood in the basement, the possibilities for use with sensors embedded within the home environment for remote monitoring are endless. Having knowledge of a person's performance patterns is integral to this process. Performance patterns are the habits, routines, roles and rituals used in the process of engaging in occupations that can support or hinder occupational performance [18]. Targeting specific areas of the home that correspond with a person's typical behaviours allows for the determination of sensor placement and subsequent collection of data that can be sent to remote caregivers. A sensor placed inside the refrigerator door is a notification that someone is up and has started their day. A sensor placed on the garage door may be an alert to wandering. A pressure sensor located within a couch or favourite chair may signal a person has fallen asleep in that room versus their bed. Further, consider adding sensors in areas of the home that may otherwise be difficult to access, a water/moisture sensor in the basement to detect the potential presence of flooding.

Although sensors have the capability to provide an alert or notification, a mechanism for managing the potential issue is also necessary. It is imperative to have a thorough knowledge of what devices are currently on the market and to take the time to analyse these products and feature-match the technology to a person's current and potential future needs. For instance, a temperature sensor over the stove will alert a caregiver of a high-temperature reading, possibly indicating that the stove has been left on while cooking. However, placement of a sensor may not be enough to meet a person's needs. There are systems that not only provide for the ability to detect excessive heat that may be indicative of a stove fire but also respond by shutting the stove off or extinguishing the fire.

Taking fire safety one step further, our traditional home smoke alarm is now working harder and smarter. Harnessing the ability to not just detect, but to assess the ambient conditions within the environment, one company has empowered a smoke and carbon monoxide detector with the ability to give a warning before a full alarm (Figure 15.4). The warning provides detailed information about where smoke or carbon monoxide has been detected [32]. Where with traditional alarm systems we are alerted to smoke and potential fire without the knowledge of its origin, this system may signal residents of the beginning of toast burning in the kitchen or a house fire starting in the garage. Having knowledge of smoke being detected in a specific location in the house may make the difference in choosing one's fire escape route. The system can also send notifications to a smartphone, providing you information concerning the ambient conditions of one's home in real time whether you are at home or away, while also providing the ability to silence alarms, as appropriate.

The use of sensors embedded within an environment for remote monitoring frequently goes undetected by the residents as they are small and non-intrusive. They allow residents to move freely about their home without an obvious visual reminder that they are being monitored. Upping the ante is the integration of video. When receiving a notification is not enough, the addition of video cameras, either within or around the house, is the next step in remote caregiving. Current technology allows for both a visual and the ability to communicate through the camera via a smart phone, tablet or computer (Figure 15.5). Motion detection can be set up to send alerts to a mobile device, assisting with both remote monitoring and a basic level of security. Moving from the addition of separate video cameras to a more ambient product is the advent of the video doorbell. Typical human behaviour when the doorbell rings

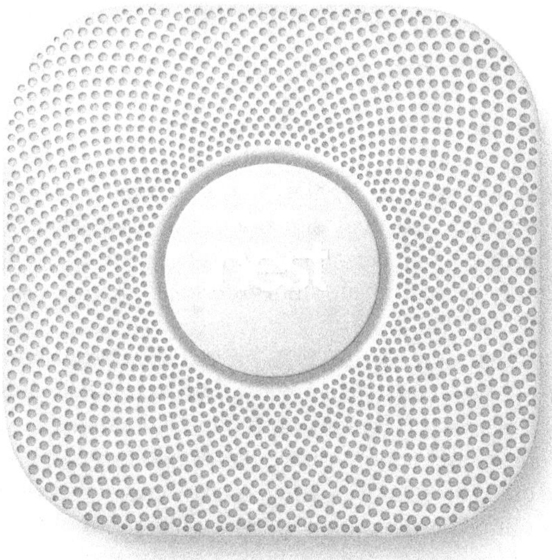

Figure 15.4 Nest Protect smoke and carbon monoxide detector (https://nest.com/smoke-co-alarm/meet-nest-protect/)

Figure 15.5 Nest cam (https://nest.com/camera/meet-nest-cam/)

Figure 15.6 Ring video doorbell (https://ring.com)

is to either look out of a peep hole or through a side light, prior to opening the door. A video doorbell provides the option to see who is at the door without even leaving your couch (Figure 15.6). For persons who may not be able to answer the door in a timely fashion and/or may not be able to answer the door at all, this provides the ability to not only know who is at the door but to also communicate with the visitor via an integrated speaker. Factor in the ability to network a smart doorbell with a smart lock and you have an AAL technology system that allows control of answering the door, whether it is to greet the person who is delivering a package or providing access to a caregiver without having to provide a key.

15.4.1.3 Case study

The individual is a 49-year-old male with medical history of multiple sclerosis diagnosis 20 years ago. He currently lives with his wife, who works full time. The client's wife expressed concern about leaving her husband home alone during the day. The couple had been relying on a system in which the husband was supposed to call his wife whenever he moved about the house, specifically, when he needed to get up from the couch to go to the bathroom. This system was not consistently working. The husband would too frequently forget to call his wife. In an attempt to resolve this situation, the wife purchased a personal emergency response system (PERS), a wearable wrist band/watch device. It was a wearable device in the form of a wrist band/watch. This solution also proved to be ineffective as the husband would not consistently wear the device. In addition to discussing ways to create a safer physical environment, the occupational therapist discussed how the wife could monitor her husband's safety when she was out of the house. AAL technology was recommended. The idea of technology embedded into the environment to provide remote monitoring decreased reliance on the client either sending a message or wearing a device. In this scenario, the AAL provided that if the individual walked past the sensors embedded within his route to and from the bathroom, it sent a signal to the wife via text message. Further, it would also provide the wife a text message if the sensor had not been activated after a certain period of time. This was a notification to the caregiver that a change occurred in the user's routine, such as not leaving the bathroom after a typical period of time. Having this system in place automated the alerts, removing responsibility from the user, and provided the wife peace of mind.

15.4.2 Health

Over the last few years we have seen a surge in popularity with wearable health and fitness devices. Garnering multigenerational appeal, smart wristbands, fitness monitors and their subsequent apps started as novel and quickly became a typical part of everyday life. According to GfK, 51 million wearables are forecast to be purchased globally in 2015 and are projected to continue to grow. This is three times the 2014 sales [33]. The advent of health tracking may have been initially promoted to the fitness enthusiast, however, the continued expansion of these technologies has found a market for allowing older adults to manage and monitor their own health from within their own home.

AAL technologies, whether wearable or embedded within the user's environment, are now empowering a person to take a more active role in their health and provide a significant opportunity to manage chronic diseases like diabetes and hypertension. They also provide the convenience of completing medical tasks within one's own home, versus having to find transportation to a medical appointment [23]. Tablets and smart phones are now connecting to a variety of medical peripherals such as blood pressure cuffs, scales, glucometers, medication dispensers, spirometers, temperature readers and even pulse oximeters to record the readings, analyse and graph the data, and send the information to a doctor or caregiver (Figure 15.7).

Similar to personal emergency response systems, health and wellness technologies are moving in the same trajectory and becoming integrated into everyday objects within the home environment. Sleep and health monitoring, traditionally managed by a wristband, is now integrated into a smart bed. The natural act of going to sleep allows for the collection of data such as heart and breathing rates, pressure, motion and presence. The data is wirelessly transmitted to a computer or mobile device for remote monitoring by medical staff and/or a caregiver.

Moving beyond obtaining data for traditional vitals, the availability of sensors to analyse indoor air quality adds another level of health and wellness monitoring. Indoor air pollution poses a health risk to all people, but even more so to vulnerable populations such as children, older adults and persons with chronic illnesses. Globally, 4.3 million deaths were attributed to household air pollution, in 2012 [34]. Integrating air quality devices and systems into the home environment creates an awareness of one's own personal exposure to household air pollution and allows for making changes

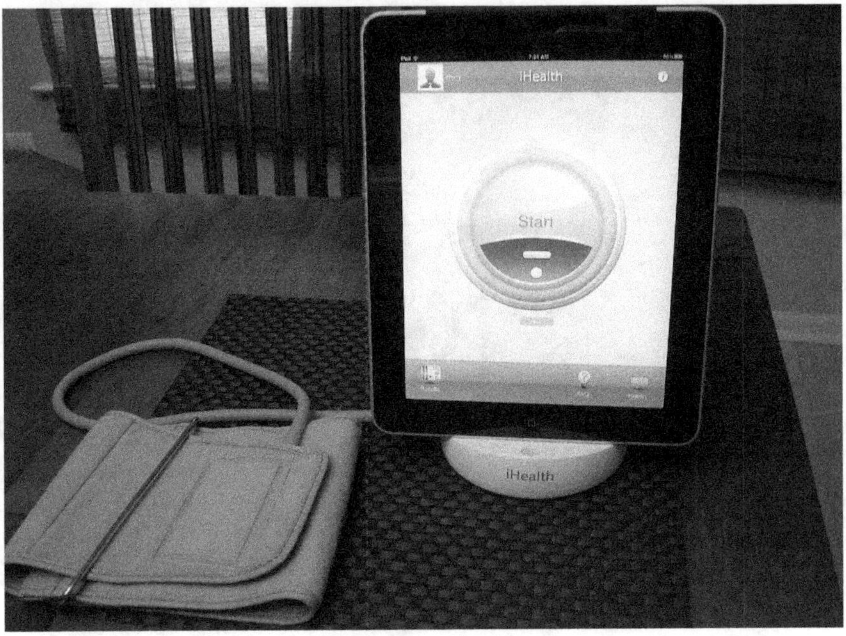

Figure 15.7 Tablet with blood pressure cuff and iHealth app

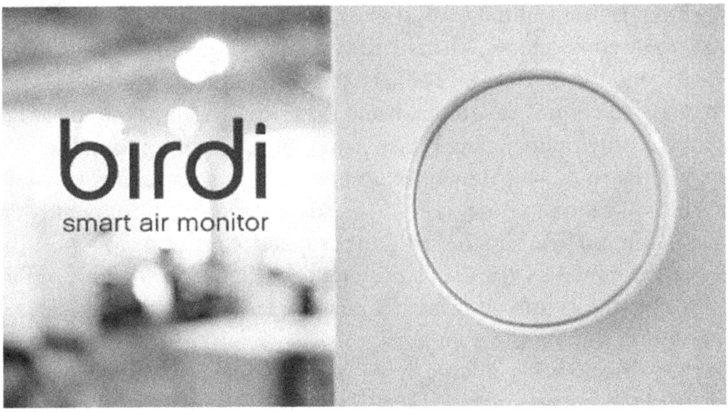

Figure 15.8 Birdi smart air monitor (http://getbirdi.com)

as appropriate. Development of stand-alone devices to be added to spaces within the home was the first step. The ambient version of this stand alone is currently coming to us in the form of a smoke detector (Figure 15.8). Taking a device that is already analysing air quality for smoke and carbon monoxide, and making it smarter to analyse for dust pollution, pollen, and humidity makes it a hard-working ambient system that enables health and wellness.

Figure 15.9 Ecovent smart vent (https://www.ecoventsystems.com)

The next natural step in the evolution of this technology would be for the entire system to talk to its components. Visualise a system where multifaceted smoke alarms, air ducts, vents, and the thermostat connect and communicate with the HVAC system about the status of the home environment and the HVAC system responds accordingly. We are already seeing many smart HVAC components coming to the market. Smart thermostats will learn a user's temperature preferences as it relates to daily routine. Smart vents monitor temperature, humidity and air pressure subsequently communicating this information to a hub that will direct specific air vents to adjust the airflow in a space as determined by the user's preferences (Figure 15.9). This ambient technology provides comfort and convenience to all users, but further provides an added a layer of safety for vulnerable populations.

The market already has available many Wi-Fi scales that provide the convenience of tracking weight, body fat and body mass index while sending this data wirelessly to an app that helps individuals analyse and keep track of this information. In addition to these measurements, it is also possible to measure one's pulse at the same time (Figure 15.10). One step on the scale and the information is retrieved. This is convenient for persons who are able to step onto a scale, but what about persons who have difficulty with balance; difficulty with stepping up and down; or even difficulty with standing altogether? The future of AAL technologies is ripe for the creation of a smart floor within the bathroom that allows for the tiles to take weight, body fat and body mass index measures, simply by standing on top of them – no scale needed. This would provide many people a means to obtain this data who might not be able to otherwise;

Figure 15.10 Withings smart body analyzer (http://www.withings.com/us/en/products/smart-body-analyzer)

such as those persons who utilise wheelchairs for mobility and do not have access to an accessible scale. Having access to this information is key to alerting doctors and/or caregivers of potential changes in a person's health.

Consider the health and wellness power of adding remote monitoring to a familiar household item that all people utilise multiple times a day: the toilet. One company has already moved in this direction, creating a toilet that has the ability to analyse blood sugar, blood pressure, body fat and weight as well as urine temperature and measurement. A urine-catcher within the toilet aids in the measurement of most of the above-mentioned body functions and then sends this information to a computer for analysis. Expanding the capabilities of this toilet from detecting to reporting, the company pushed the envelope to also consider health promotion and prevention via the creation of a cookpad. The toilet, once it completes its analysis, will communicate with the cookpad in the kitchen, and wirelessly send it recipes [35]. The toilet's analysis provides insight into the user's eating habits and subsequently recommends recipes to the cooktop that allow for meeting that user's specific nutritional needs.

15.4.2.1 Mental health

Occupational therapists observe and analyse performance skills to understand the transactions among client factors (e.g. specific capacities, characteristics, or beliefs that reside within the person and that influence performance in occupations) – which is inclusive of the physiological and psychological function of body systems [18], context and environment, and activity or occupational demands, which support or hinder performance skills and occupational performance [18, 36, 37]. Occupational therapists address the social, emotional, psychological and physical aspects of interaction between an individual and their environment [38]. A person's environment can have a direct impact on their mental health, subsequently impacting participation in occupations. This concept extends beyond the design of an interior environment and includes the ability to go outside. Research demonstrates a positive correlation of the health benefits of interacting with nature and psychological health and well-being. Reported benefits include positive/improved mood, increased prosocial behaviour, increased happiness, increased vitality and improved quality of life [39]. Further benefits reported include positive effect on cognitive ability or function, positive effect on physical functional and/or physical health including perceived health and well-being [39]. Translating research to application, one company has created LED ceiling light panels that feature realistic outdoor nature scenes. The panels bring the outside world inside to people who may have decreased access to their windows or to the outside world overall. They provide the connection to nature that otherwise may not be easily accessible or even possible. Lack of daylight or windows results in the absence of cues about the natural world, which is a form of sensory deprivation [40]. By providing ambient visual access to nature, the theoretical outcome is that it will also provide the psychological benefits that come along with it. Further, the addition of simulated natural daylight conditions in a space can help alleviate the potential for sensory deprivation and resulting feelings of anxiety and depression.

15.4.2.2 Case study: health monitoring at home

The client, a 74-year-old retired homemaker, lives in a suburban area and does not have an active driver's licence. Her medical history is significant for colon cancer and she has recently undergone a cholecystectomy and colon resection. After surgery, the client was diagnosed with deconditioning which required a short-term stay in a sub-acute rehabilitation. Further, the client also developed post-surgical hypertension, which required her to be placed on a blood pressure medication. The client's discharge plan was to transition to her son's home, where she desired to age in place. To assist with this transition, the client's family reached out to a private practice occupational therapist that specialised in EMs for a home accessibility assessment of the son's residence.

Upon discharge, a post-surgical wound in the client's abdomen justified a need for her to receive home health nursing care three days a week. The client, who previously did not have blood pressure issues was not fond of taking medication and inquired with her doctor if she would be able to return to her baseline of being medication free. In an effort to safely wean the medication, the doctor requested daily blood pressures be taken at home to monitor for any changes and determine feasibility for the elimination of the medication. The home health nurse was only available to complete three of the seven requested blood pressure readings per week.

Management of the client's hypertension was integrated into the occupational therapist's EM plan. Determining supports to maximise the client's independence with health status monitoring, for return to baseline blood pressure status as well as for future prevention, was a prioritised goal for the client [23]. The occupational therapist noted that the client was familiar with and comfortable with using a tablet device for accessing the Internet and other meaningful applications.

The occupational therapist trained the client and her family on the use of a system that utilised a blood pressure cuff that attaches to a tablet. The system also included a health app that monitors, tracks and stores the blood pressure readings. The client and her family demonstrated independent and safe carryover of the device, and understanding of the health application, meeting her goal and effectively integrating the management of her hypertension into her daily routine [23]. The final outcome was the successful elimination of the blood pressure medication due to the client's daily home monitoring demonstrating consistently safe blood pressure readings. The integration of this AAL technology into the client's EM plan created a health-promoting habit and routine while also promoting the client's health, wellness, prevention, and productive ageing [23].

15.4.3 Social interaction

Research shows that participation in social activities results in improved physical health and functioning, in terms of ADLs [41, 42] better performance, even if living with a chronic disease [43], and lower mortality rates [44, 45]. Furthermore, engaging in social activities has also been shown to improve quality of life [46–48].

According to Mary Madden, senior research specialist at the Pew Research Center, 'Social media has the potential to bridge generational gaps. There are few other spaces—online or offline—where tweens, teens, sandwich generation members, grandparents, friends and neighbors regularly intersect and communicate across the same network' [49]. As of February 2012, the Pew Internet and American Life Project Research Center reports that one-third (34%) of the Internet users aged 65 and older use social networking sites and 18% do so on a typical day. Among adults who are currently aged 76 and older, 20% of the Internet users use social networking sites and 8% do so on a typical day. Pew further reports that email continues to be the backbone of online communication for older adults, with 86% of the Internet users ages 65 and older using email, 48% doing so on a typical day [50].

Technologies that integrate features that allow for social interaction are becoming increasingly available. We live in a connected world where having a mobile device on your person at any given moment, including next to you when you sleep, has become commonplace. Multi-modal communication options (talk, text, video) provide flexible access for older adults to keep connected with friends and family, as well as provide peace of mind for caregivers. The addition of social media networking and computer games provides a new approach to recreation and leisure and create new experiences for keeping the older adult mind engaged [15].

Video chatting is a social opportunity available to anyone with a computer, tablet or cell phone. Although this is a readily available technology, this method of communication relies on the user having some knowledge and comfort with using technology. This leaves an entire population who do not own, know how to use or feel comfortable with these devices out of the communication loop. The power of an AAL technology device for communication lies in a common household item: the television. From the add-on Skype-ready cameras to the integrated Skype-ready televisions, consumers have the ability to video chat from the comfort of home and without the added stress of learning a new technology device. Although there is still a learning curve to using Skype, the familiarity of using one's own television is surely less daunting. Further, the larger television screen and louder speakers, in comparison to tablets and cell phones, may prove to be of benefit to ageing eyes and ears. Video chat has multigenerational benefits, helping all people to better connect and communicate with peers, family and friends.

15.4.4 Engagement in life

While many technologies are created to provide a level of convenience to the majority of users, these systems and devices also provide support for persons who may have difficulty with or cannot complete daily tasks. There is a large spectrum of technologies to support daily occupations, such as ADLs and IADLs. ADLs, as defined by the occupational therapy practice framework, are activities oriented towards taking care of one's own body [18, 51]. IADLs are those activities to support daily life within the home and community that often require more complex interactions than those used in ADLs [18]. From monitoring the completion of, issuing reminders for, and/or performing the task for the user, this section will discuss AAL technologies that support engagement in life.

15.4.4.1 Activities of daily living

One of the most basic ADLs is feeding. Drinking enough fluids to ensure personal hydration is important to all people. Water represents a critical nutrient, the absence of which will be lethal within days [52]. Each person has individualised daily water intake requirements; add in factors such as age, illness, activity level and medications and these requirements may change. In an effort to help users track daily water intake, a smart water bottle is born (Figure 15.11). The technology is within a module that is fitted to the bottom of a water bottle. The module is able to track the amount of water consumed and will send the information wirelessly to an app [53]. The app allows for determining how much water on average the user should drink in one day and creates this as a daily goal. The daily goal is also adjusted based on the local weather. A familiar and frequently used daily item is empowered by AAL technology to support ADLs, health and well-being.

Along these same lines we are also seeing technologies that support older adult's better eating habits to decrease risk of malnutrition. According to the WHO, older adults are particularly vulnerable to malnutrition. This is compounded by common diseases affecting older adults – cardiovascular and cerebrovascular disease, diabetes, osteoporosis and cancer – which are all diet-affected [54]. Diet4Elders is a system that aims to use AAL technology to monitor older adults eating habits and provide guidance on proper nutrition for their individualised needs. A diet is created for each user and delivered to their home via the use of a tablet with smart food ordering [55]. The end goal of this AAL technology system is to help older adults better feed themselves and decrease risk of malnutrition. Another programme, framework for optimising the process of feeding, takes this concept one step further by embedding the

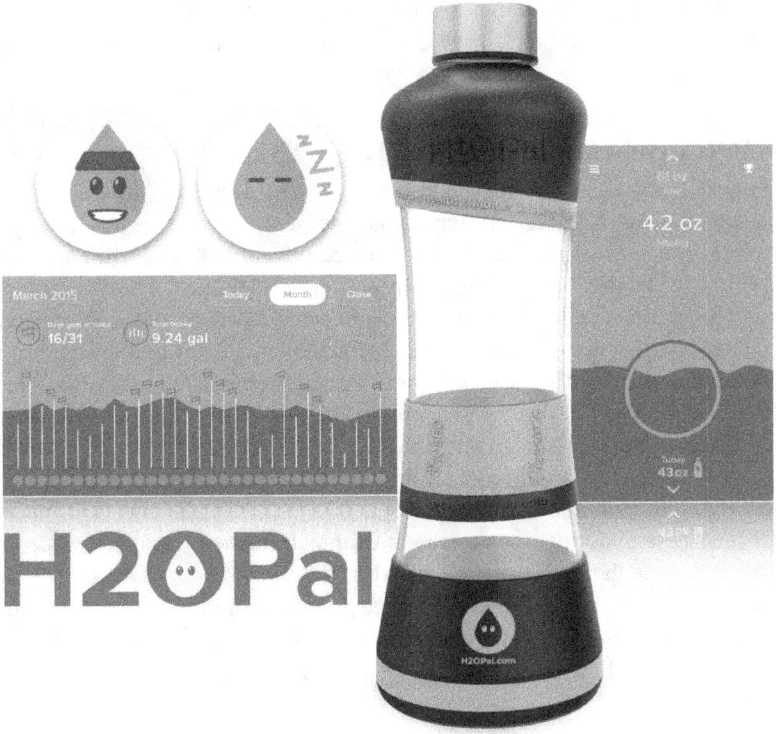

Figure 15.11 H_2O-Pal hydration tracker (https://www.h2opal.com)

technology into the kitchen environment: networking a kitchen via smart appliances and connections to Internet grocery shopping and food access [56]. Although these systems are addressing the basic ADL of feeding, they are also addressing the bigger picture, which is the IADL of food access and grocery shopping.

Functional mobility, which is the movement from one position or place to another during the performance of everyday activities, such as in bed mobility, functional ambulation, wheelchair mobility and transfers (e.g. wheelchair, bed, car, shower, tub, toilet, chair), is another ADL where AAL technologies are proving to be valuable [18].

The market is booming with sensors to monitor bed mobility; pressure sensors (discussed previously in this chapter) in mattresses along with occupancy sensors and floor mats provide alerts about a person's movement in and out of bed. However, the feedback from these sensors is primarily geared towards meeting a caregiver's needs for remote monitoring, versus providing feedback and support for the person. Smart technology is now addressing the user's functional mobility needs by interfacing with the mobility device being utilised. Rolling walkers are being outfitted with an AAL system that provides the user with indoor and outdoor navigation assistance.

One company has expanded upon its mobile robot solutions to incorporate this technology into power wheelchairs, providing the user collision free power wheelchair navigation [57].

15.4.4.2 Instrumental activities of daily living

Among those millions of 'things' that are currently connected to the Internet are technologies that support IADLs. Health management and maintenance, an IADL, is the development, management and maintenance of routines for health and wellness promotion [18]. Medication management and the creation of a medication routine is one such example. Currently, there are many stand-alone medication management systems on the market that provide reminders for taking medication. Similar in concept to the smart water bottle discussed earlier in the chapter, one company has integrated cell phone technology into the actual pill bottle, versus creating a separate system (Figure 15.12). When it is time to take the medication, the pill bottle will flash to alert the user. If the user takes the medication, the bottle senses when the cap has been removed and will send a signal to the cloud. The bottle is also aware of its weight, and will send data about the measurement of the amount of medication currently in the bottle each time the cap is removed [58]. If the user does not take the medication when prompted, the bottle will then flash red, provide an auditory beep cue and either place a phone call, send an email, or text message as a reminder. These alerts can be sent to the user as well as family members and other caregivers.

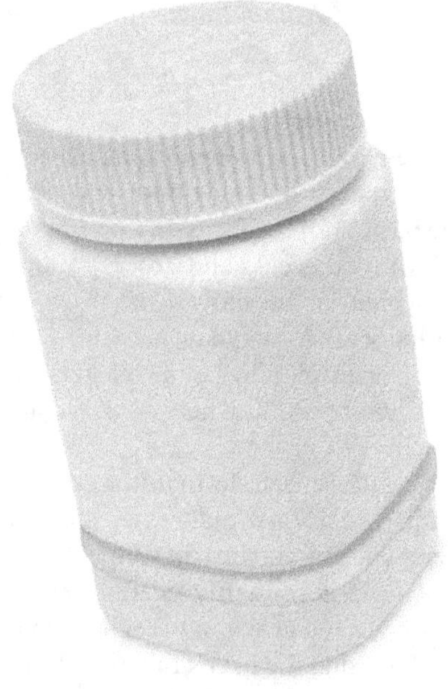

Figure 15.12 AdhereTech wireless pill bottle (https://adheretech.com)

Many home management technologies have been available for the past decade. A welcome IADL assist, people have become accustomed to the idea of domestic robots assisting with vacuuming and washing the floors. Continuing this trend, the market now has robots that will assist with exterior home management activities such as mowing and watering the grass and landscape.

Although these technologies provide convenience and fill a need for persons who may not be able to complete home management tasks (IADLs) independently, they are one-dimensional, providing assistance only for the task for which they were designed. The simplicity of a device with one purpose has merit. However, a person's changing or growing needs must be considered when making the determination of if a single use device is appropriate, or if an expandable system that can start small but allow for the addition of devices to meet both current and future needs, is most appropriate. The need for easy interaction with a central hub to conveniently manage multiple Internet connected standalone devices within a home is quickly realised. From turning on the lights to playing your favourite music, Amazon Echo is a voice-activated device that becomes a home personal assistant (Figure 15.13). Echo is a modern-day reminder of the environmental control units (ECUs) of yesteryear; big, clumsy devices that utilised a combination of x-10, infrared, and radio frequency transmission to control devices within the environment. Unlike Echo, ECUs, more commonly referred to as electronic aids to daily living (EADLs), were created and marketed to persons with

Figure 15.13 Amazon Echo (http://www.amazon.com/Amazon-SK705DI-Echo/dp/B00X4WHP5E)

Figure 15.14 LG HomeChat smart appliance network (http://www.lghomechat.com/us/)

disabilities. Specifically, for those persons with limited physical abilities secondary to tetraplegia, or a progressive neurological disease, an ECU's ability to start, stop or adjust electric or electronic devices within the environment was very important. Echo, however, provides universal appeal, basking in the sunshine of a world that enjoys using technology on a daily basis. It translates voice commands into convenience for all users, but is the difference of having control of one's environment to persons who otherwise would not.

By harnessing the power of current cell phone and tablet technology, the next generation of MSRs has arrived. Beyond providing physical support for IADLs, MSRs are evolving, acting as companions more than assistants, providing psychosocial support and caregiver peace of mind. These robots can control home automation functions (e.g. turning on your television, lights), provide video monitoring to view its surroundings or video chat with a person either in the home or remotely, as well as keep a schedule and provide daily reminders. Other features include facial recognition, the ability to control other Internet-controlled devices within the home via voice or through the display and provide access to interactive games.

MSRs and voice command hubs, although provide convenience and value across generations, are still the middleman between a user and their home. Removing the middleman and communicating directly with a network of common household items is evolving. Smart appliances can transform how IADLs are completed. A refrigerator that informs you of its contents, including if food items, is about to expire. A smart oven can search recipes and auto-set the appropriate cooking mode. A washer and dryer can remotely start a load of laundry while also giving you on demand updates of its status. A robot vacuum that can be remotely programmed and provide status of task completion [59]. This is LG HomeChat: ambient technology that allows for communication with the home through an app (Figure 15.14). Next stop, a larger network of items embedded within the environment that can respond to communication via voice directly, versus through a hub, while also being able to provide ambient feedback, communicate with each other, and respond as well.

15.5 Concluding statement

AAL technologies can enable older adults to live safely and independently at home. Embedding technologies within the environment facilitates active involvement of all generations in their own, as well as their loved one's health while empowering and providing peace of mind for caregivers. Occupational therapists are uniquely qualified and well suited to consult with computer scientists along with mechanical, industrial and biomedical engineers with regard to AAL technologies to support productive ageing.

References

[1] United Nations Department of Economic and Social Affairs/Population Division. *World population prospects: The 2015 revision, key findings and advance tables* [online]. New York: The United Nations; 2015. Available from http://esa.un.org/unpd/wpp/Publications/Files/Key_Findings_WPP_2015.pdf [Accessed 20 March 2016]

[2] Pew Research Center. *Attitudes about aging: A global perspective* [online]. Washington D.C.: Pew Research Center; 2014. Available from http://www.pewglobal.org/2014/01/30/attitudes-about-aging-a-global-perspective/ [Accessed 20 March 2016]

[3] Stokes, B. *The countries that will be most impacted by aging population* [online]. Washington D.C.: Pew Research Center; 2014. Available from http://www.pewresearch.org/fact-tank/2014/02/04/the-countries-that-will-be-most-impacted-by-aging-population/ [Accessed 20 March 2016]

[4] World Health Organization. *Report on ageing and health* [online]. Luxembourg: World Health Organization; 2015. Available from http://apps.who.int/iris/bitstream/10665/186463/1/9789240694811_eng.pdf?ua=1 [Accessed 20 March 2016]

[5] Centers for Disease Control and Prevention (CDC). *Chronic disease overview* [online]. Atlanta: CDC; 2014. Available from http://www.cdc.gov/chronicdisease/overview/ [Accessed 30 March 2016]

[6] Centers for Disease Control and Prevention (CDC). *Adult obesity facts* [online]. Atlanta: CDC; Available from http://www.cdc.gov/obesity/data/adult.html [Accessed 20 March 2016]

[7] The National Alliance for Caregiving and AARP. *Caregiving in the U.S* [online]. Washington, D.C.: National Alliance for Caregiving; 2015. Available from http://www.caregiving.org/wp-content/uploads/2015/05/2015_CaregivingintheUS_Executive-Summary-June-4_WEB.pdf [Accessed 20 March 2016]

[8] International Alliance of Carer Organizations (IACO). *Caregiver statistics by nation* [online]. IACO; 2015. Available from http://www.internationalcarers.org/carer-facts/global-carer-stats/ [Accessed 20 March 2016]

[9] Angelini, V. and Laferrère, A 'Home, houses and residential mobility' in Börsch-Supan, A., Brugiavini, A., Jurges, H. *et al.* (eds.). *Health, Ageing and Retirement in Europe (2004–2007) – Starting the Longitudinal Dimension.* Mannheim: MEA; 2008. pp. 99–107.

[10] Boehm, T.P. and Schlottmann, A. 'Comparison of household mobility for owned manufactured, traditional owned and rental units using the American Housing Survey'. *Journal of Housing Economics*. 2006;15:126–142.

[11] Farber, N., Shinkle, D., Lynott, J., Fox-Grage, W., and Harrell, R. *A state survey of livability policies and practices* [online]. Washington D.C: AARP; 2011. Available from http://www.ncsl.org/documents/transportation/Aging-in-Place-2011.pdf [Accessed 20 March 2016]

[12] Sullivan, B. *American housing survey* [online]. Washington D.C.: U.S. Department of Housing and Urban Development; 2009. Available from http://portal.hud.gov/hudportal/HUD?src=/press/press_releases_media_advisories/2010/HUDNo.10-138 [Accessed 20 March 2016]

[13] Joint Center for Housing Studies of Harvard University. *Housing America's older adults: Meeting the needs of an aging population* [online]. Cambridge: Joint Center for Housing Studies of Harvard University; 2014. Available from http://www.aarp.org/content/dam/aarp/livable-communities/documents-2014/Harvard-Housing-Americas-Older-Adults-2014.pdf [Accessed 20 March 2016]

[14] Kohli, M., Künemund H., and Vogel C. (2008). 'Staying or moving? Housing and residential mobility' in Börsch-Supan A., Brugiavini, A., Jurges, H. *et al.* (eds.). *Health, Ageing and Retirement in Europe (2004–2007) – Starting the longitudinal dimension,* Mannheim: MEA. pp. 108–113.

[15] Young, D. *Automating the quest to age in place: Supporting productive aging and independence through gerontechnology and home modifications for aging in place* [online]. Leawood, Kansas: *Rehab Management Magazine*; 2012. Available from http://www.rehabpub.com/2012/11/automating-the-quest-to-age-in-place/ [Accessed 20 March 2016]

[16] Dickerson, A.E., Molnar, L., Eby, D., *et al.* 'Transportation and aging: A research agenda for advancing safe mobility'. *The Gerontologist.* 2007;47:578–590.

[17] Law, M., Cooper, B., Strong, S., Stewart, D., Rigby, P and Letts. L. 'The person-environment-occupation model: A transactive approach to occupational performance'. *Canadian Journal of Occupational Therapy*. 1996;63(1): 9–23.

[18] American Occupational Therapy Association. 'Occupational therapy practice framework: Domain and process' (3rd ed.). *American Journal of Occupational Therapy*. 2014; 68(Suppl.1), S1–S48.

[19] Fisher, A.G. *Occupational Therapy Intervention Process Model: A model for planning and implementing top-down, client-centered, and occupation-based interventions*. Fort Collins, CO: Three Star Press; 2009.

[20] Fisher, A.G. and Griswold, L.A. 'Performance skills: Implementing performance analyses to evaluate quality of occupational performance' in Boyt Schell, B.A., Gillen, G. and Scaffa, M. (eds.). *Willard and Spackman's Occupational Therapy* (12th ed.). Philadelphia: Lippincott Williams & Wilkins; 2014. pp. 249–264.

[21] Kielhofner, G. *The Model of Human Occupation: Theory and Application* (4th ed.). Philadelphia: Lippincott Williams & Wilkins; 2008.

[22] Mace, R. *Universal Design, Barrier free Environments for Everyone.* Los Angeles: Designers West; 33(1):147–152; 1985.

[23] Young, D. 'Technology for managing high blood pressure at home'. *AOTA OT Practice*. 2013;18(10):5–6.

[24] Young, D., Van Oss, T., and Wagenfeld, A. 'Universal design for a lifetime: Interprofessional collaboration and the role of occupational therapy in environmental modifications'. *AOTA OT Practice.* 2014;19(13): CE1–CE8.

[25] Connell, B.R., Jones, M.L., Mace, R., *et al. The principles of universal design* [online]. 1997. Available from https://www.ncsu.edu/ncsu/design/cud/about_ud/udprinciplestext.htm [Accessed 20 March 2016]

[26] Gartner, Inc. *Gartner says 4.9 billion connected 'things' will be in use in 2015* [online]. Barcelona: Gartner, Inc.; 2014. Available from http://www.gartner.com/newsroom/id/2905717 [Accessed 20 March 2016]

[27] Business Wire. *Gartner says 6.4 billion connected "things" will be in use in 2016, up 30 percent from 2015* [online]. New York: Business Wire; 2015. Available from http://www.businesswire.com/news/home/20151110006335/en/Gartner-6.4-Billion-Connected-2016-30-Percent [Accessed 20 March 2016]

[28] World Health Organization. *Falls fact sheet* [online]. World Health Organization; Geneva: World Health Organization; 2012. Available from http://www.who.int/mediacentre/factsheets/fs344/en/ [Accessed 20 March 2016]

[29] Alwan M., Rajendran P.J., Kell S., *et al*. 'A smart and passive floor-vibration based fell detector for elderly'. *Proceedings of the International Conference on Information and Communication Technologies*; Berkeley, California, April 2006, pp. 1003–1007.

[30] The University of Manchester. *Magic carpet can help prevent falls* [online]. Manchester: The University of Manchester; 2012. Available from http://www.manchester.ac.uk/discover/news/article/?id=8648 [Accessed 20 March 2016]

[31] Future Shape. *SensFloor—large area sensor system* [online]. Hohenkirchen: Future Shape; 2005. Available from http://future-shape.com [Accessed 20 March 2016]

[32] Nest. *Nest Protect* [online]. Palo Alto, California: Nest; 2016. Available from https://nest.com/smoke-co-alarm/meet-nest-protect/ [Accessed 20 March 2016]

[33] PR Newswire. *GfK forecasts 51 million wearables will be bought globally in 2015* [online]. Nuremburg: PR Newswire; 2015. Available from http://www.gfk.com/news-and-events/press-room/press-releases/pages/gfk-forecasts-51-million-wearables-sold-globally-2015.aspx [Accessed 20 March 2016]

[34] World Health Organization. *Burden of disease from household air pollution* [online]. Geneva: World Health Organization; 2012. Available from http://www.who.int/phe/health_topics/outdoorair/databases/HAP_BoD_results_March2014.pdf?ua=1 [Accessed 20 March 2016]

[35] Masaki Matunaga. (2012, July 9). *BBP2012 Final presentation: TOTO intelligence toilet with cookpad.* [Video File]. Available from https://www.youtube.com/watch?v=3GOKGyoJloA [Access 20 March 2016]

[36] Chisholm, D. and Boyt Schell, B.A. 'Overview of the occupational therapy process and outcomes' in Boyt Schell, B.A., Gillen G., and Scaffa M. (eds.). *Willard and Spackman's Occupational Therapy* (12th edn.). Philadelphia: Lippincott Williams & Wilkins; 2014. pp. 266–280.

[37] Hagedorn, R. *Tools for Practice in Occupational Therapy: A Structured Approach to Core Skills and Process*. Edinburgh: Churchill Livingstone; 2000.

[38] American Occupational Therapy Association. *Occupational therapy in mental health act* [online]. Bethesda: American Occupational Therapy Association; 2015. Available from: https://www.aota.org/-/media/Corporate/Files/Advocacy/Federal/Tips-and-Tools/AOTA%20NHSC%201%20Pager-%202015.pdf [Accessed 20 March 2016]

[39] Sandifer P.A., Sutton-Grier A.E., and Ward B.P. 'Exploring connections among nature, biodiversity, ecosystem services and human health and well-being: Opportunities to enhance health and biodiversity conservation'. *Ecosystem Services*. 2015;12:1–15.

[40] Boubekri, M. *Daylighting design: Planning strategies and best practice solutions*. Berlin: Birkhauser; 2014. p. 36.

[41] Mendes de Leon, C.F., Glass, T.A., Beckett, L.A., Seeman, T.E., Evans, D. A., and Berkman, L.F. 'Social networks and disability transitions across eight intervals of yearly data in the New Haven EPESE'. *Journals of Gerontology, Series B: Psychological Sciences and Social Sciences.* 1999;54:162–172.

[42] Mendes de Leon, C.F., Glass, T.A., and Berkman, L.F. 'Social engagement and disability in a community population of older adults—The New Haven EPESE'. *American Journal of Epidemiology.* 2003;157:633–642.

[43] Seeman, T.E. and Chen, X. 'Risk and protective factors for physical functioning in older adults with and without chronic conditions: MacArthur studies of successful aging'. *Journals of Gerontology, Series B: Psychological Sciences and Social Sciences.* 2002;57:135–144.

[44] Avlund, K., Damsgarrd, M.T., and Holstein, B.E. (1998). 'Social relations and mortality: An eleven year follow up study of 70-year old men and women in Denmark'. *Social Science and Medicine.* 1998;47:635–643.

[45] Giles, L.C., Glonek, G.F., Luszcz, M.A., and Andrews, G.R. 'Effect of social networks on 10-year survival in very old Australians: The Australian longitudinal study of aging'. *Journal of Epidemiology and Community Health.* 2005;59:574–579.

[46] Dahan-Oliel, N., Gelinas, I., and Mazer, B. 'Social participation in the elderly: What does the literature tell us?' *Critical Reviews in Physical and Rehabilitation Medicine.* 2008;20:159–176.

[47] Silverstein, M. and Parker, M.G. 'Leisure activities and quality of life among the oldest old in Sweden'. *Research on Aging.* 2002;24:528–548.

[48] Stav, W.B., Hallenen, T., Lane, J., and Arbesman, M. 'Systematic review of occupational engagement health outcomes among community-dwelling older adults'. *American Journal of Occupational Therapy*. 2012;66(3):301–310.

[49] Pew Research Center. *Older adults and social media* [online]. Washington D.C.: Pew Research Center; 2010. Available from http://www.pewinternet.org/2010/08/27/older-adults-and-social-media-3/ [Accessed 20 March 2016]

[50] Pew Research Center. *Older adults and Internet use* [online]. Washington D.C.: Pew Research Center; 2012. Available from http://www.pewinternet.org/2012/06/06/older-adults-and-internet-use/ [Accessed 20 March 2016]

[51] Rogers, J.C. and Holm, M.B. 'Assessment of self-care' in Bonder, B.R. and Wagner, M.B. (eds.). *Functional Performance in Older Adults*. Philadelphia: F.A. Davis; 1994. pp. 181–202.

[52] Popkin, B.M., D'Anci, K.E., and Rosenberg, I.H. 'Water, hydration and health'. *Nutrition Reviews*. 2010;68(8):439–458.

[53] Marjanovic, M. *H_2O-Pal* [online]. Out of Galaxy, Inc.; 2015. Available from https://www.h2opal.com [Accessed 20 March 2016]

[54] World Health Organization. *Nutrition for older persons*. Geneva: World Health Organization; 2016. Available from http://www.who.int/nutrition/topics/ageing/en/index1.html [Accessed 20 March 2016]

[55] Active and Assisted Living Programme. *Diet4Elders* [online]. Brussels: Available from http://www.diet4elders.eu/en [Accessed 20 March 2016]

[56] Active and Assisted Living Programme. *Framework for optimizing the process of feeding (FOOD)* [online]. Brussels: Available from http://www.food-aal.eu [Accessed 20 March 2016]

[57] GeckoSystems Intl. Corp. *GeckoSystems presents the SafePath™ wheelchair* [online]. Georgia: GeckoSystems. Available from http://www. geckosystems.com/markets/wheelchair.php [Accessed 20 March 2016]

[58] Stein, J. *Smart wireless pill bottles* [online]. New York: AdhereTech; 2015. Available from https://adheretech.com [Accessed 20 March 2016]

[59] LG Electronics. *LG HomeChat* [online]. Seoul: LG Electronics; 2014. Available from http://www.lghomechat.com/us/ [Accessed 20 March 2016]

Chapter 16
Tablet-based clinical decision support system for hospitalised older adults

Mirza Mansoor Baig[1], *Hamid GholamHosseini*[1], *and Martin J. Connolly*[2]

16.1 Introduction

Today, clinicians, nurses and family members can receive instant alert/messages about health information of their patients on a smartphone, tablet, laptop, personal digital assistant or personal computer (PC). The most commonly performed monitoring is the collection of patients' vital data using state-of-the-art medical devices, sensors and wearable textiles which collect and transmit the data to a remote server or processing unit for analysis and storage and for generating alerts to other devices [1].

Recent research is highly focused on the area of remote/mobile and wireless patient monitoring using body sensors, wireless devices and/or wearable systems. Such systems have been established as expanding areas of research using their advanced features and capabilities to turn 'data' into 'useful information'. It is reported that an ideal vital signs monitoring system should be able to: (i) collect high-quality data via medical devices/sensors; (ii) interpret and present collected data in a meaningful and valuable manner; (iii) facilitate decision support via expert knowledge into real health situations; and (iv) perform appropriate actions with clinicians' feedback to the patient on the basis of collected data [2, 3]. Such systems are being designed and developed for every possible healthcare scenario, i.e. emergency departments and remote monitoring of indoor and outdoor locations [4]. Remote monitoring systems, in particular, play an important role when patient and doctor are at a distance and such systems are capable of reducing healthcare-related costs and enhancing the quality of patient healthcare delivery [4].

Moreover, the world is witnessing an increase in the use of ubiquitous devices (smartphones, tablets, laptops, etc.). A smartphone generally includes advanced functionality beyond making phone calls and sending text messages. Users will have plenty

[1]School of Engineering, Computer and Mathematical Sciences, Auckland University of Technology, Private Bag 92006, Auckland 1142, New Zealand
[2]North Shore Hospital, University of Auckland, Takapuna, Auckland, New Zealand

of PC features in their handheld smartphone, accessible with the touch of a button or swipe of a finger [5]. To clarify the use of the terminology, we distinguish 'vital signs' such as the measurement of heart rate (HR), blood pressure (BP), pulse (P), and temperature (*T*) from 'physical signs', which we interpret as the state of abnormal vital signs, such as bradycardia, tachycardia, hypertension, hypotension and hypxaxemia.

16.2 Need for a decision support and monitoring system

There is an increasing interest in potentially preventable causes of in-hospital morbidity and mortality [6]. Evidence suggests that the management of many critically ill patients can be improved with the result that some cardiac arrests, deaths and intensive care unit (ICU) admissions may be avoided [7]. It is reported that prior to cardiac or respiratory arrest, up to 84% of patients have significant physiological deterioration (vital signs) [8]. Often insufficient action is taken, despite up to 60% of arrests on general hospital wards having potentially correctable antecedent events, such as hypoxia and hypotension [8].

To assist in the early detection of physical signs, many hospitals now use an 'early warning score' (EWS) that allocates points to routine vital signs measurements on the basis of their changes from a 'Normal' range [9]. These points provide the EWS, and the outcome from the weighted values requires the appropriate set action to be taken by the ward staff. The process by which EWS is obtained involves accurate vital signs collection, the correct recording (manually) of a weighted value according to the degree of change and the arithmetical addition of weighted values to form the EWS. Each of these stages can introduce error, which may influence the EWS. Errors may also occur in the transcription of raw or derived data on to paper charts. There is always the chance of overscoring that may lead to the unnecessary calling of medical staff. Underscoring is also possible which may lead to a delay in the detection of patient health deterioration [9]. Some of the studies reported that as many as 80% of ward patients have physiological parameters outside normal ranges as much as the 24 h preceding ICU admission [10].

There is thus an argument for a reliable automated (computerised) early detection and vital signs monitoring system in hospitals' general wards. The current chapter aims to present the development of remote/wireless vital signs monitoring and early detection system in order to detect multiple physical signs in older adults in hospital ward settings.

Clinical decision support systems (CDSS) that analyse data to help healthcare providers make clinical decisions are facing some challenges in acquiring up-to-date and accurate information continuously for decision-making. This becomes even more complex when there is no single best option and/or when decisions are based on patient-centred care [11–13].

The success of tablet-based decision support systems in hospital care settings is still debatable [5]. There are very limited studies on successful implementation and/or clinically trailed CDSS in wider hospital care settings. On the other hand, there are reports on the design, development (prototype) and accuracy (by simulation) of the CDSS in hospital care as well as other clinical and non-clinical settings, which lack real system-related issues such as implementation, consultation and user accessibility [14, 15]. Providing tablet-based clinical decision support is a challenging task, because limitations specific to mobile application developed and implemented in hospital care settings [16].

16.3 Background and literature review

16.3.1 Clinical decision support for inpatient care

Manual paper/hard copy-based 'tools' and 'scales' are being automated with easy to use graphical user interface (GUI) via tablet-based applications to integrate the advanced technology into medical settings. A tablet-based wound care application [17] was developed to replicate the paper-based Pressure Ulcer Scale for Healing (PUSH tool) [18], Braden Scale [19] and the Bates–Jensen tool [20]. This system automated the existing tools and let clinicians to create, view, access, delete and re-assess patient records related to wound care. Critical design consideration has been given to the tablet application such as simplicity of GUI, minimising visual elements on any given screen to reduce clutter, using colour cues to focus information and converging on critical information. It is also reported that the number of steps to complete a common task related to wound care are significantly reduced using tablet-based tools. The application was developed for a Nexus 4 smartphone and Nexus 7 tablet and tested with eight nurses, age range of 31–60 years. There was a strong perceived correlation between the paper-based forms and the wound care application (app) in terms of content and data entry expectations, with scores of 4.60/5.00 for the Braden Scale and 4.57/5.00 for the PUSH tool [17]. However, the effective use of clinical data stored by the entities/departments/organisations is a great concern for the nationalised data-sharing schemes regarding security and privacy [21].

16.3.2 Mobile-based clinical decision support for assistive care for older adults

A randomised controlled trial was conducted to compare the efficacy of a computerised physical activity program [22], where older adult patients are given a tablet computer for the initial two months to enter their daily activity data using a simple GUI. However, 263 older adults were recruited from three outpatient clinics at Boston Medical Center with average age of 65 years. It was reported that there was a higher rate of pedometer data capture in the control group after 12 months of trial (52.5% vs. 39.6%). It was also notable that the tablet decline rate of the users was at an average of 35.8 ± 19.7 times during the 60-day in-home intervention phase, and after the first week (from an average of 4.7 to 4.0 sessions/week) and then gradually declining (to 3.3 sessions/week). Most of mobile health (mHealth) applications are designed without clinician/end user consultations. Therefore, they lead to issues such as application usage, especially for older adults who may have difficulty understanding the application and/or handling the tablet application [23] – which potentially reducing the global adoption of tablet-based applications [24, 25].

Decision support systems assist clinicians by providing evidence so that they can make better point of care decision. Unfortunately, reviewing tailored information with each patient is costly in terms of time and resources [26]. In an attempt to engage clinicians and patients during their hospital stay, a tablet application was developed to provide interactive information to patients [27]. This application was integrated into Microsoft's Health Vault-online patient information storage application. The information queried from the electronic health record (EHR) was updated in real time. In an evaluation of this device, the interviewed patients (all were male cardiothoracic surgery patients; mean age 55.4 years and the average hospital length of stay at the time the interview was 5.4 days) believed that the tablet computer application was very useful, especially in terms of having access to their hospital medication history. A total of five patients rated this as 'Useful' or 'Very Useful' (4 or 5, respectively, on a 5-point scale). The integration of such applications to other EHR models (online or local) poses security and privacy issues [27].

Decision support systems need structured data in order to interpret the outcome. Structured data could be originating from the EHR integration, online repositories or other medical databases [14, 24]. A tablet application was developed towards clinical work flow optimisation in hospital emergency department [14]. The application consists of activity and role categories that reflect current practice in emergency departments. The activities included in the tool are registration, assessment, treatment, status and updates. These activities are defined at a broad level and include observable behaviours and cognitive actions. It is reported that this approach made a positive impact on the patient-record/data-related hospital emergency department work-flow. The application was able to capture the sequence of activity 'role' for 108 patient care episodes and achieved high inter-ratter reliability assessment (Kappa value 0.79) [14]. In a similar study, Holzinger *et al.* [15] developed tablet (iPad)-based application 'MoCoMED Graz' to engage patients with additional information related to their visit via an automated questionnaire [28]. It was reported that, by using tablet-based automated data entry/monitoring application, 20 medical doctors (age median 43 years) needed an average of 10 min (range 8–14 min) to enter the data via the paper-based questionnaire, and by using the proposed application MoCoMED Graz, the data-entry time was reduced by approximately 90% (median: 1 min, min: 1 min, max: 4 min) [28]. This strongly suggests that data feeding and retrieval can be made easy by using such automated applications. However, continuous incoming data from each patient requires huge space, either in local storage (device itself or local server) or in the cloud. This brings an additional problem of 'big data' analyses where large amount of data is transferred to clinicians without any analysis [29], Table 16.1 summarises the above reviewed systems.

Table 16.1 Summary of the reviewed tablet-based monitoring systems

Name	Application	Hardware	Trial	Outcome
Friesen *et al.* [17]	Chronic wound care	Nexus 7 tablet/Nexus 4 smartphone	8 nurses	Comfort with smartphone/tablet interfaces and with touchscreens was self-assessed at 4.57/5.00 (range = 4.0–5.0; SD = 0.53) and 4.71/5.00 (range = 4.0–5.0; SD = 0.49)
Bickmore *et al.* [22]	Automated exercise coach for older adults	Tablet computer (name not mentioned)	263 older adults	Participants walked more steps at 2 months (adjusted mean 4,041 vs. 3,499 steps/day, $P = .01$)
Vawdrey *et al.* [27]	Patient participation in hospital care	Apple's iPad device	Five male patients (mean age was 55.4 years)	All five patients rated this as 'Useful' or 'Very Useful' (4 or 5, respectively, on a 5-point scale)
Ozkaynak and Brennan [14]	Patient-oriented workflow in hospital emergency departments	Tablet computer (name not mentioned)	108 patients	The application achieved high inter-rater reliability assessment (kappa 0.79).
Holzinger *et al.* [15]	Reengineer workflows in the hospital care	Apple's iPad device	20 medical professionals (median age 43 years)	Using the proposed application the data-entry time was reduced by approximately 90% (median: 1 min, min: 1 min, max: 4 min) when compared to the paper-based model.

16.4 Current issues and challenges in hospital care

In the past 30 years, there have been significant advancements towards decision support tools, initially with paper-based manual tools, scales, scores and charts, gradually transferred to computerised and semi-automated systems [30]. Currently, there are emerging tablet applications fully integrated to other EHR systems to support clinicians in their decision-making. Despite the significant advancement in decision support technologies, there are still gaps which are preventing wider adoption in healthcare settings [30]. Health innovations using information and communications technology (ICT) (such as tablet applications) offer great advantages, but at the same time they pose a security and privacy threat to the patients' personal information, clinician and end-user acceptability, data storage, data access, data organisation and

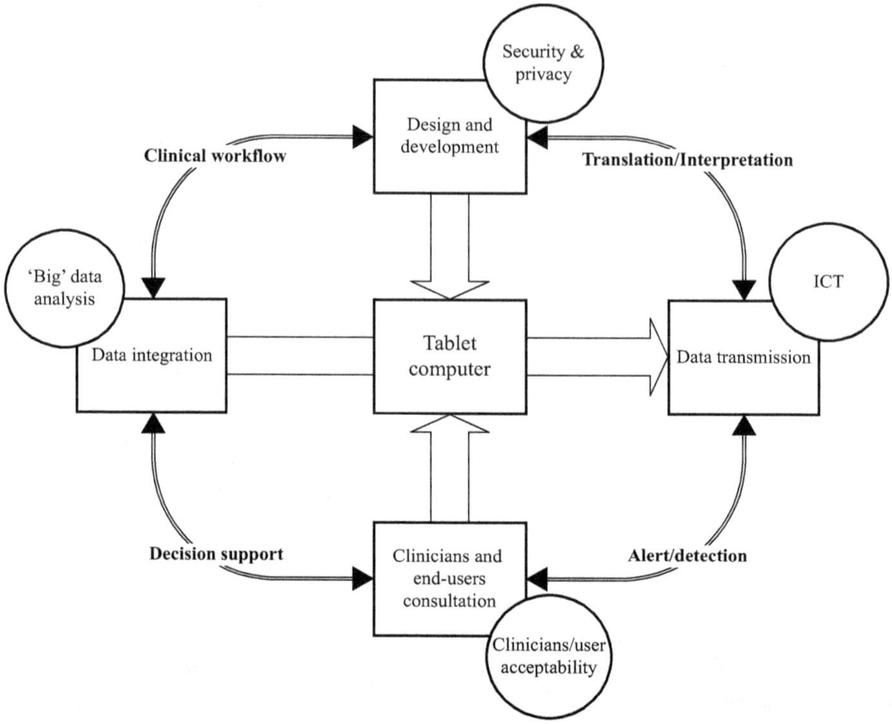

Figure 16.1 Diagram related to tablet-based applications and their issues and challenges

its processing [12, 29]. Data processing on local device impacts on the device's battery life, which is currently one of the biggest issues [31, 32]. In the context of this review, it is necessary to identify critical issues related to decision support systems in particular. Therefore, we excluded common tablet-related issues which are extensively covered by many researchers in literature such as power consumption [11], battery life [31, 33, 34], user acceptance [35] and clinical data security and privacy [13, 36–41]. Instead, we focus on the issues related to the decision support ability of tablets and their acceptance in medical settings such as data processing and analysis [42, 43] and ICT and infrastructure, required for optimal integration of tablet applications. Figure 16.1 shows the diagram related to tablet-based applications and their issues and challenges. The major issues and challenges are categorised into four main divisions: (1) clinician/user consultation; (2) transmission and processing; (3) data integration; (4) design and development [44], as discussed below.

16.4.1 Data processing and analysis

Medical standards compel the use of high-quality data in tablet applications for high accuracy and reliable communications. Various techniques have been applied to collect high-quality data [45–51]. It is necessary for the vital sign monitoring and/or clinical decision support applications to capture and transfer the highest possible data quality and at early stage of application development, a theoretical framework be set to manage physiological parameters. At each stage of the development, the feedback of medical professionals should also be considered and discussed in every possible aspect in relation to the end user (patient) [52, 53].

Personalised and predictive medicine can greatly benefit from data integration and its machine learning model (big data analytics) [42]. Today, most health systems use a combination of rules-based approaches as well as manual methods to best interpret the medical data. Big data analysis seems to be a solution for the medical data. Generation of data is increasing rapidly due to the extensive use of electronic medical devices and storing of digital data. The increasing rate of data generation is much faster than expanding rate of the world's data storage capacity [42]. However, big data analysis is at its early stages and needs more time to achieve clinical acceptance [42].

User-centred design is essential in order to incorporate these perceptions into the product, especially at the earlier stages of the project development. When considering user's needs, the solution designer should consider a number of factors such as task analysis, surveys, interviews and focus groups to address user acceptability issues [54]. Highly sophisticated data analysis techniques become irrelevant if users do not understand the operation and its user interface [55].

Data processing and analysis algorithms implemented on tablets in real time may have intense impact on the devices' battery life. Such algorithms cannot be implemented for continuous monitoring applications due to their large power consumptions. Moreover, due to the limited hardware and processing power of the device, sophisticated data processing and interpretation are limited when compared with high-end PCs [52]. In case of using the cloud-based service, the collected data can be pushed to the cloud for processing, and only alerts are transmitted back (generated by the cloud-based applications) and pushed to various devices including tablet computers. In such applications, the alert will be delayed depending upon the Internet connectivity and cloud-processing capabilities [56, 57].

16.4.2 ICT and infrastructure

ICT plays a very important role in the success of tablet-based technology. Mostly, tablet applications are developed either as a web-based or local machine-based. In both cases, they will be connected to another software for data transfer. However, the local tablet applications usually do not have data integration capabilities with other clinical databases to fetch patient data for processing; they mostly relay on local patient data and limited to specific area or a particular issue. Local machine-based applications have the advantage of patients' information security and privacy, but have two critical drawbacks: limited memory space of tablet computers and patients' data access [58]. On the contrary, fully connected applications have the advantage of collecting and analysing more patients' data from other data repositories in order to predict/interpret the trend of data for the purpose of clinical decision support.

Such applications could also be connected to cloud-based environment for information transmission but require an ICT infrastructure – with cost implications at the initial stage [23, 59]. Another issue is the privacy and security of patients' personal information [36, 40, 60, 61]. It is suggested that there should be balance between clinical use and ICT dependence for tablet applications [62, 63]. For example, if the medical device or hospital internet connectivity fails then the whole application will be effected, because normally all data is stored in the cloud due to the tablet's limited space [2, 64–66].

It is recommended that, in a limited resource setting, it is optimal to use the local tablet-based applications for specific healthcare area so that huge ICT-related costs could be saved [44]. Tablet applications with data integration and cloud-based connectivity and storage introduce challenges regarding security and privacy of patient's personal and clinical data. The Washington Heights/Inwood Informatics Infrastructure for Comparative Effectiveness Research (WICER) project [59] attempts to minimise such barrier by using advanced data encryption process where data is encrypted on the tablet computer before transmission to the cloud server, and data were only decrypted after being retrieved by the target organisation. To address the issue of integration/connectivity, the study uses a local application on the device rather than using the tablet computer as a browser for web-based forms. This application not only reduced workflow issues, because of connectivity, but also carried the risk of storing information on the device [59], which again points towards the ICT infrastructure to store and handle huge amount of incoming patient data. However, storing information on the hand-held device introduces the risk of losing the device (easier to lose the mobile device than desktop work station) and one cannot physically steal or lose the cloud. Table 16.2 summarises the above-mentioned four challenges according to the current literature.

Table 16.2 Issue and challenges identified in tablet-based monitoring systems

Name	**ICT**	**Data analysis**	**Clinician/User acceptability**	**Security and privacy**
WICER [59]	No	Yes	No	Yes
Bickmore *et al.* [22]	No	Yes	Yes	No
Friesen *et al.* [17]	No	Yes	Yes	Yes
Holzinger *et al.* [15]	Yes	No	Yes	No
Ozkaynak and Brennan [14]	No	Yes	Yes	No
Fromme *et al.* [23]	No	Yes	Yes	No
Chatterjee [58]	Yes	No	No	Yes
Gao *et al.* [67]	Yes	Yes	No	No
Bajorek and Nowak [68]	Yes	No	No	Yes
Ng *et al.* [69]	Yes	No	No	No

16.5 Vital signs interpretation

16.5.1 Individualised monitoring

The majority of systems used today have adopted a generalised monitoring model based on either set threshold ranges or standard deviation (SD) changes which are implemented specifically for certain age groups (older adults, adults and children) [70, 71] and/or particular illness/health issue(s) [72, 73]. The proposed model has adopted individualised monitoring because of the fact that physiological parameters are different in each individual, hence threshold or SD-based monitoring models often generate high false alarms [74] and eventually reduce the reliability of the overall system.

The model uses the individual data for the interpretation called 'individualised monitoring' and the whole data set only serves as the outline boundary of the framework for the whole age group. The unique feature of the individualised monitoring module is that its adaptive boundary limits will be changing throughout the monitoring phase. Every 10th recording, or every 10 min, the engine updates the limits and compares this with the previous ones so that any considerable changes can be detected. The adaptive limits have an accuracy advantage over the set limits; in cases of transient hypertension where BP will be high (higher than normal for that particular patient) and upon treatment (medication) BP may be normal, this does not mean that the patient will always have normal BP from now on. While set individual limits will detect the transient hypertension, it resets the status to normal upon/after medication because no attempt has been made to update/change the set limits. Whereas the proposed adaptive limits will detect the transient hypertension and upon continuous update of the limits, will have a higher accuracy for transient or persistent hypertension detection if that particular health issue persists in the future for that patient (iterative optimisation).

For every 10 recordings there is a window to capture the adaptive limits. For 11–20 recordings, the 'MAX1' is the maximum limit update, at 'MAX1' the current maximum (C_{MAX}) value is checked against the maximum value for the last 10 recordings and any considerable changes will be detected. Similar processing has been carried out for mean (ME_1 to ME_n) and minimum (MIN_1 to MIN_n) values. In the 21–30 recordings window, there is a considerable change detected when calculating the difference between 'MAX2' and current 'c_{MAX}'. In this case, the BP (systolic) time stamped data is sent to the next module (evidence-based/features extraction) for further processing. The same process is followed for all the vital sign(s).

The manual data entry is performed by entering the scores (numbers) only to avoid the complexity of text and image analysis. The data of medical devices is the primary source for the rules engine and analysis; however, the manual data entry make it possible to limit the entry icon to the numbers only (based on the scope of this research).

16.5.2 Weighted parameters

Each time the information is collected from the credibly sourced evidence, a score is assigned to the corresponding health event as true positive. In this framework, there are primarily seven health events which are defined from E1 to E7, where,

E1: Bradycardia
E2: Tachycardia
E3: Hypotension
E4: Hypertension
E5: Hypoxaemia
E6: Fever
E7: Hypothermia.

For instance, consider hypertension, which is directly related to BP. Similarly all the other direct links are established based on the medical literature and standard vital signs and physical signs relationship. Let us consider the vital sign BP, for E3 (hypotension) and E4 (hypertension); BP is the direct association (directly linked vital sign). From the gathered evidence of the above modules, the weighted scoring for indirect linking suggested that BP (at least in the in-patient setting) is also often associated with E2 (Tachycardia) and E5 (Hypoxaemia). In this case, E2 and E5 have multiple vital signs, both direct (E2-HR and E5-SpO_2) and indirect (E2-BP and E5-BP).

16.5.3 Multilayer diagnosis

Early detection of physical sign(s) can reduce adverse events [75–77]. The multilayer diagnostic module can improve the interpretation engine performance in the early detection of multiple physical signs by dividing the outcome into two priorities, instead of a simple 'yes' or 'no' classification. The whole data set is completely divided into four states, i.e. normal (pre-event), P2 (priority 2), P1 (priority 1) and normal (post-event) states, respectively. The sign of deterioration (P2) can be detected before the actual event (P1) occurs. In this case, the hypertension is detected in the area of 'P2' which is before the actual alarm point of 'P1'.

The multi-priority approach is adopted using two different techniques: P1 alarm is a fuzzy logic-based model, which employs the fuzzy inference model to detect the P1 level alarms and P2 warning is a classifier which uses expert rules with weighted parameters to detect the P2 level warnings.

Priority-1: For interpretation of seven different physical signs using four basic physiological parameters (BP, P, SpO_2 and T). Fuzzy logic modelling is used to map several degrees of membership functions from each vital sign to several physical signs. Due to its non-crispiness and flexibility, it can achieve low false alarms.

Priority-2: An optimised classifier using SD (calculated from each patient) either side of the mean value of physiological parameters. This mode can reduce false alarms by continuous changing of limits and boundaries.

Both priorities use their own classification rule base to detect the respective priority (P1 or P2) using a different set of limits and ranges with no overlap. Priorities

Table 16.3 Vital signs data handling and classification using two priorities

Priority	Type of monitoring	Type of messages	Outcome message*
Priority 1	Fuzzy logic-based interpretation of physical signs	Alert	'Possible physical sign(s)'
Priority 2	Change of SD with weighted parameter for each parameter	Warning message No alert	'Possible physical sign(s)' with change in data

*Data represents the appropriate abnormal vital sign(s) and its related possible physical sign(s). Possible physical signs are bradycardia, tachycardia, hypertension, hypotension, hypoxaemia, fever or hypothermia.

are assigned with different alert mechanisms such as messages/warnings and alarms that can be transmitted to the clinician's or nurse's devices. Table 16.3 shows vital signs handling according to the proposed two priorities.

16.5.4 Physical signs detection using multiple vital signs

Now, let us consider E1–E7 physical signs, using the combination of four vital signs: HR, BP, SpO_2 and T with two levels of priority (P1 and P2). In this case, let us consider all possible scenarios where the physical signs extracted from any physiological measurement include additional and finer fuzzy sets, with the inclusion of *very low (VL), low (L), high (H)* and *very high (VH)*. For example, priority-1 Hypotension (E3P1) is possible when the combination of vital signs indicates very low BP (VL_{bp}) and very high HR (VH_{hr}). In some other cases such as hypoxaemia, it is necessary to consider low SpO_2, but also, HR and BP which can be high or normal or low. In this case, the proposed model gives more weight to SpO_2 for 'Low' and assigns normal weight to HR and BP. Prioritisation of outcomes gives clinicians a higher degree of control over the system's outcome because of two levels of priorities. This structure potentially provides a higher reliability and stability to compare with only single criteria for generating alarms.

The terms 'very high', 'high', 'very low' and 'low' are automatically/continuously optimised using the individualised monitoring module, where each vital sign is divided into four abnormal categories with respect to that particular patient. In the case of E6 (fever), the system uses two inputs HR and T. In direct association with fever is the body temperature; hence in this case the T has more weight when predicting 'fever'.

The proposed system has been tested for both real-time as well as offline data. A multilayer concept has been introduced to enhance the overall reliability and accuracy of the proposed system. The multilayer approach has the potential of reducing the false alarms because early detection of physical sign(s) as the priority-2 warning does not generate alarm but only a warning. Then at the time of deterioration health event, the priority-1 alarm will be activated and trigger the alarm. This mechanism is best utilised in this context by feeding a multiple input–output combinational relationships in real time.

16.6 Results and validation

Kappa analysis [78] was used to measure the level of agreement/disagreement between the proposed system and a medical expert (professor of geriatric medicine), i.e. as the measure of how accurately the system can mimic human performance. Accuracy is generally used to assess the performance of classifiers. However, on its own it is not a realistic metric that should be used to assess classifiers' performance for the used data set, as the influence of negative samples on overall accuracy is much higher than that of positive samples. Precision, as it pertains to agreement between observers (inter-observer agreement), is often reported as a Kappa analysis [78].

16.6.1 Real-time testing of physical signs detection model

On an average each patient was visited three times during the one year trial. During each visit, four sets of vital signs were collected. Each set contained seven measurements: two BP measurements – one standing and one sitting with the time difference of one minute on average, HR, oxygen saturation, ear temperature and blood glucose readings. Overall, a total of $1_{\text{patient}} \times 3_{\text{visits}} \times 4_{\text{data sets}} \times 6_{\text{recordings}} = 72$ recordings for one patient were collected. Finally, for 30 patients, a total of $(72_{\text{recordings}} \times 30_{\text{patients}})$, which is equal to 2160 recordings were collected.

The proposed system raised a total of 356 alarms and out of these, 127 alarms were promoted to P1 (priority-1) alarms and the rest were P2 warnings (229). It is important to mention that only P1 alarms are considered for Kappa analysis and compared with a medical expert's diagnosis. From another viewpoint, P2 (229) represents 64% of total alarms (356) when compared to 36% P1 (127), which suggests that for every P1 there was 1.8 P2, indicating that the proposed system successfully detects P1 early. The detection of P2 before and after P1 demonstrates that early detection of multiple physical signs as well as high predictability of the proposed system, which could be useful for clinician for an early detection of patient's physical signs.

Table 16.4 below details the P1 alarms categorised according to the physical signs and compares it with the medical expert's diagnosis for the same data.

Table 16.4 P1 alarms generated by the proposed system compared with the medical expert's diagnosis

Physical signs	**Proposed system**	**Medical expert**
Hypotension	8	8
Hypertension	11	9
Tachycardia	12	12
Hypoxaemia	15	15
Hypothermia	6	3
Total	52	47

The results show that in 47 out of 52 instances, the system and the expert were positive (true positive) and for the remaining five instances the system was positive but the expert was negative (false positive). There were no instances recorded where the system was negative and the expert was positive (0 – false negative) and the rest of the alarms were characterised as true negative (75) where the system and expert were both negative; this classification is defined in Table 16.5.

Based on values from Table 16.5, Kappa analysis was carried out using the standard agreements between the two diagnoses may be affected by chance. Kappa (κ) is a measurement of agreement between the expert and the system which has been corrected for error by chance. Kappa (κ) was calculated by subtracting the proportion of readings that are expected to agree by chance (P_e) from the overall agreement (P_o) and dividing the remainder by the number of cases on which agreement is not expected to occur by chance. The Kappa analysis results for the proposed system's performance are described in Table 16.6; P_o, P_{pos} and P_{neg} are overall, positive and negative agreements, respectively. SE represents the standard error and $CI_{95\%}$ is 95% confidence intervals for Kappa.

The overall Kappa value is calculated as $\kappa = 0.91$ and the strength of agreement beyond chance lies in the range of 0.81 to 1 described as 'almost perfect'. The Kappa-based statistical analysis showed a substantial level of agreement ($\kappa = 0.91$ or 91%) between the expert's and the system's diagnoses. Table 16.6 shows that the proposed system achieved an overall agreement of 96% with Kappa κ with a value of 91%. Now, the quantitative categories were calculated such as accuracy, sensitivity, specificity and predictability using the standard equations (refer Table 16.5 for TP, TN, FP and FN values). The proposed system achieved an accuracy of 96%, sensitivity of 100%, specificity of 93.75% and predictability of 90.38% compared with the diagnosis of an expert for the same physical signs.

Table 16.5 TP, TN, FP and FN values extracted from 20 patients' data for Kappa analysis

System/Expert	Expert (+ve)	Expert (−ve)	Total
System (+ve)	47 (TP)	5 (FP)	52
System (−ve)	0 (FN)	75 (TN)	75
Total	47	80	127

Table 16.6 Results from Kappa analysis and agreement evaluation

Overall agreement	Positive agreement	Negative agreement	Agreement by chance	Standard error	95% Confidence Intervals for κ
P_o	P_{pos}	P_{neg}	P_e	SE	$CI_{95\%}$
0.96	0.94	0.96	0.52	0.03	0.98 and 0.83

16.7 Discussion

Tablet computer applications such as electronic medical records, nursing communication tools and personal health records are designed to connect healthcare delivery systems. The design, implementation and evaluation of these technologies require rigorous and systematic workflow methodologies. Zheng *et al.* [79] identified that there is a discrepancy across the findings of health IT studies in terms of their impact on workflow. As they speculated the discordance may be due in part to oversimplified workflow modelling practices. The patient-orientated workflow methodology and the available automated tool can assist researchers with collecting data that can greatly help clinicians in decision-making [79].

Current tablet applications lack flexibility in terms of personalisation of data, interactive approach and wide data accessibility [69]. It is predicted that ICT can overcome these limitations by capturing individual clinical data, translating them into meaningful data and providing handy advantages for end-users as well as medical professionals [69].

Ng *et al.* [69] recommended five main characteristics of tablet-based applications in order to be clinically acceptable:

1. ICT experts and infrastructure to support such applications
2. Evaluation of applications including testing and user training
3. Careful planning to develop robust and safe application
4. Stakeholder involvement towards decision support tools
5. Traditional face-to-face interaction should not be reduced or compromised and final decision-making should be based on the clinicians' experience and knowledge and not on the tablet application's recommendations

Tablet computer technology is a possible solution to many healthcare problems as it is now widely used in every healthcare department [80]. Tablet computers can be used by both patients and clinicians to discuss the options and decisions within or outside the consultations. There are concerns that this may marginalise older people who may not be IT-savvy [69]. The use of Internet among older people has increased in recent years [22] and older patients who are making decisions about their chronic illnesses may benefit from tablet computer technology [22]. In addition, the use of ICT can make decision-making more interactive. The use of animation, videos, audio-visual aids and personalised options allows patients, especially those with a lower education level, to be more engaged in the decision-making process [16, 81, 82]. Information can be presented to the patient in stages or controlled by users to avoid information overload. The use of navigation function allows users to select information that is relevant to them (target specific) [58].

Very few studies have reported that older people and senior clinicians are happy about using tablet computers in healthcare [83], whereas most studies have reported that senior/experienced medical professional are less likely to believe that tablet computers create better decision support or significantly improve patient–clinicians relationship [80]. Similar results can also be found in a previous study in the psychology literature investigating generational differences in the adoption rate of new technology [80]. It is now widely believed that younger generations have been exposed to technology from an early age and a large percentage of their education has utilised technological media. Therefore, they exhibit favourable attitudes toward innovative technology, thus resulting in its rapid adoption [84]. Another study reveals that new technology is initially utilised in hospital care, because it is popular in the workplace among other colleagues. It is not until the device is used for some time in practice that experienced medical professional recognise its full potential (when compared to younger medical professionals) and it is possible that due to the greater knowledge and experience they need these devices less than their younger colleagues [84]. Due to this technological exposure, it may be assumed that less experienced (younger) physicians are more rapidly adopting tablet computer technology into hospital care/clinical practice than other more experienced medical professional.

16.8 Conclusion

A common concluding statement, 'using tablet computer in hospital care reduces the amount of time clinicians spent at work stations and hence more time can be given at the bedside' [85] is an unjustified conclusion without any concrete evidence available in the literature. Outcome of such statements can lead to confusions and many patients might be upset, offended or even worried to see their physicians 'checking their mobile device' in the middle of a consultation (i.e. misinterpret that the physician is in fact involved in a social text conversation or may be worried to see their physician 'looking things up' as he/she has not got a clue what to do). The true justification for such conclusion needs (1) evidence of tablet application tested/evaluated in hospital care settings involving senior clinicians, (2) continuous involvement of medical professional and end-user in the app design and (3) accurate data analysis and interpretation provided to the medical professional via tablet application in reasonable time frame. Existing studies have emphasised the tablet's efficiency as a communication tool, educational tool, reference source and decision resource for both patients and providers [16, 80]. Furthermore, recent survey results have demonstrated a positive patient perception of physicians who use tablet computers in the clinical setting regardless of patient age, race, gender or income [86].

Emerging tablet-based healthcare apps are changing the face of the 21st century medical practice, especially hospital care. The new technologies will allow remote monitoring of patients and their access to healthcare, health data collection, patient identification and medicine administration and compliance. Data information exchange with the providers and organisations access to medical record, amongst several other benefits that would continue to accrue with the fast pace development of medical and allied healthcare technologies.

This review provides an overview of the literature on a broad range of tablet-based healthcare applications in hospital care settings. Most of the systems described are either in their development stage or simulation stage but not actually implemented in the hospital care (exceptions to systems conducted clinical trials). Further focused research is required in such applications with respect to their clinical implementation, end-user acceptability, evaluation by medical professionals and security and privacy.

Two core 'areas of concerns' related to tablet computer technology have appeared as the product of this work, and they may serve as a useful guide for researchers, engineers and medical professionals towards improving hospital care in using tablet computer-based healthcare applications. First, clinicians and patients are the under-used resources in tablet applications design and development; early collaboration of developers and clinicians along with the target users are critical towards the successful healthcare applications being implemented in the hospital care settings. Second, appropriate use/representation of data including data analysis, interpretation, translation, integration and processing should be transparent and presented to the medical professional in a simple and easy format. Due to the 'big data' coming from multiple eHealth sources, it is now a critical task to represent the raw data as meaningful data to the clinician, so that the decision support can be enabled efficiently. Currently, most systems lack clinical implementation; very few applications are available today which incorporate and address all the critical aspects of the tablet applications including design, development, implementation and user/clinicians collaboration.

16.9 Conflict of interest statement

Authors declare no conflict of interest.

References

[1] Klingeberg, T. and Schilling, M. 'Mobile wearable device for long term monitoring of vital signs'. *Computer Methods and Programs in Biomedicine.* 2012;**106**(2):pp. 89–96.

[2] Lau, F., Kuziemsky, C., Price, M. and Gardner, J. 'A review on systematic reviews of health information system studies'. *JAMIA*. 2010;**17**(6): pp. 637–645.

[3] Chan, M., Estève, D., Escriba, C. and Campo, E. 'A review of smart homes—Present state and future challenges'. *Computer Methods and Programs in Biomedicine.* 2008;**91**(1): pp. 55–81.

[4] Cook, R. 'Exploring the benefits and challenges of telehealth'. *Nursing times.* 2012;**108**(24):pp. 16–17.

[5] Reiswig, J. 'Apps: finding the best'. *Journal of the Medical Library Association: JMLA*. 2011;**99**(4):pp. 326–327.

[6] Vincent, C., Neale, G., and Woloshynowych, M. 'Adverse events in British hospitals: preliminary retrospective record review'. *British Medical Journal.* 2001;**322**(7285):pp. 517–519.

[7] Kause, J., Smith, G., Prytherch, D., Parr, M., Flabouris, A. and Hillman, K. 'A comparison of antecedents to cardiac arrests, deaths and emergency intensive care admissions in Australia and New Zealand, and

the United Kingdom—the ACADEMIA study'. *Resuscitation*. 2004;**62**(3): pp. 275–282.

[8] Franklin, C. and Mathew, J. 'Developing strategies to prevent inhospital cardiac arrest: analyzing responses of physicians and nurses in the hours before the event'. *Critical Care Medicine*. 1994;**22**(2):pp. 244–247.

[9] Prytherch, D.R., Smith, G.B., Schmidt, P., Featherstone, P.I. 'Calculating early warning scores—a classroom comparison of pen and paper and handheld computer methods'. *Resuscitation*. 2006;**70**(2):pp. 173–178.

[10] Goldhill, D., Worthington, L., Mulcahy, A., Tarling, M. and Sumner, A. 'The patient-at-risk team: identifying and managing seriously ill ward patients'. *Anaesthesia* 1999;**54**(9):pp. 853–860.

[11] Baig, M.M., GholamHosseini, H., and Connolly, M.J. 'Mobile healthcare applications: system design review, critical issues and challenges'. *Australasian Physical & Engineering Sciences in Medicine*. 2014; **38**: pp. 1–16.

[12] Baig, M.M., Gholamhosseini, H., and Connolly, M.J. 'A comprehensive survey of wearable and wireless ECG monitoring systems for older adults'. *Medical & Biological Engineering & Computing*. 2013;**51**(5):pp. 485–495.

[13] Baig, M.M. and Gholamhosseini, H. 'Smart health monitoring systems: an overview of design and modeling'. *Journal of Medical Systems*. 2013;**37**(2):pp. 1–14.

[14] Ozkaynak, M. and Brennan, P. 'An Observation Tool for Studying Patient-oriented Workflow in Hospital Emergency Departments'*. *Methods of Information in Medicine*. 2013;**52**(20130906).

[15] Holzinger, A., Kosec, P., Schwantzer, G., Debevc, M., Hofmann-Wellenhof, R. and Frühauf, J. 'Design and development of a mobile computer application to reengineer workflows in the hospital and the methodology to evaluate its effectiveness'. *Journal of Biomedical Informatics*. 2011;**44**(6): pp. 968–977.

[16] McColl, L.D., Rideout, P.E., Parmar, T.N. and Abba-Aji, A. 'Peer support intervention through mobile application: An integrative literature review and future directions'. *Canadian Psychology/Psychologie Canadienne*. 2014;**55**(4):pp. 250.

[17] Friesen, M.R., Hamel, C., and McLeod, R.D. 'A mHealth application for chronic wound care: Findings of a user trial'. *International Journal of Environmental Research and Public Health*. 2013;**10**(11):pp. 6199–6214.

[18] Black, J., Baharestani, M.M., Cuddigan, J., *et al.* 'National Pressure Ulcer Advisory Panel's updated pressure ulcer staging system'. *Advances in Skin & Wound Care*. 2007;**20**(5): pp. 269–274.

[19] Bergstrom, N., Braden, B.J., Laguzza, A. and Holman, V. 'The Braden scale for predicting pressure sore risk'. *Nursing Research*. 1987;**36**(4):pp. 205–210.

[20] Harris, C., Bates–Jensen, B., Parslow, N., Raizman, R., Singh, M. and Ketchen, R. 'Bates-Jensen wound assessment tool: Pictorial guide validation project'. *Journal of Wound Ostomy & Continence Nursing*. 2010;**37**(3): pp. 253–259.

[21] Luxton, D.D., Kayl, R.A., and Mishkind, M.C. 'mHealth data security: The need for HIPAA-compliant standardization'. *Telemedicine and e-Health*. 2012;**18**(4):pp. 284–288.

[22] Bickmore, T.W., Silliman, R.A., Nelson, K., *et al*. 'A randomized controlled trial of an automated exercise coach for older adults'. *Journal of the American Geriatrics Society*. 2013;**61**(10):pp. 1676–1683.

[23] Fromme, E.K., Kenworthy-Heinige, T., and Hribar, M. 'Developing an easy-to-use tablet computer application for assessing patient-reported outcomes in patients with cancer'. *Supportive Care in Cancer*. 2011;**19**(6):pp. 815–822.

[24] MobileSmith. 'Mobile apps as tools of cost reduction in healthcare' in *MobileSmith*. Mobile Smith Inc: USA; 2014. pp. 1–9.

[25] Market Research. 'mhealth apps and solutions market by connected devices – Global trends and forecast to 2018'. *Market Research*: Rockville, MD, USA; 2013. pp. 277.

[26] McQueen, A., Cress, C., and Tothy, A. 'Using a tablet computer during pediatric procedures: a case series and review of the "apps"'. *Pediatric Emergency Care*. 2012;**28**(7):pp. 712–714.

[27] Vawdrey, D.K., Wilcox, L.G. Collins, S.A., *et al*. 'A tablet computer application for patients to participate in their hospital care'. *AMIA Annual Symposium Proceedings*; American Medical Informatics Association, 2011.

[28] Gell, G., Leodolter, W., Köle, W., and Leitner, H. 'HIS purchase projects in public hospitals of Styria, Austria'. *International Journal of Medical Informatics*. 2000;**58**:pp. 147–155.

[29] Lohr, S. *The Age of Big Data*. New York Times, 2012. pp. **11**.

[30] Garg, A.X., Adhikari, K., McDonald, H., *et al*. 'Effects of computerized clinical decision support systems on practitioner performance and patient outcomes: a systematic review'. *Journal of the American Medical Association*. 2005;**293**(10): pp. 1223–1238.

[31] Long, Y., Joonsung, B., Seulki, L., Taehwan, R., Kiseok, S., and Hoi-Jun, Y. 'A 3.9 mW 25-electrode reconfigured sensor for wearable cardiac monitoring system'. *IEEE Journal of Solid-State Circuits*. 2011;**46**(1): pp. 353–364.

[32] Yoo, J., Long, Y., Seulki, L., Yongsang, K., and Hoi-Jun, Y. 'A 5.2 mW self-configured wearable body sensor network controller and a 12 μW wirelessly powered sensor for a continuous health monitoring system'. *IEEE Journal of Solid-State Circuits*. 2010;**45**(1):pp. 178–188.

[33] Ullah, S., Higgins, H., Braem, B., *et al*. 'A comprehensive survey of wireless body area networks'. *Journal of Medical Systems*. 2012;**36**(3):pp. 1065–1094.

[34] Pandey, S., Voorsluys, W., Niu, S., Khandoker, A., and Buyya, R. 'An autonomic cloud environment for hosting ECG data analysis services'. *Future Generation Computer Systems*. 2012;**28**(1): pp. 147–154.

[35] Baig, M.M., GholamHosseini, H., Linden, M., and Connolly, M.J. 'Review of vital signs monitoring systems – Patient's acceptability, issues and challenges'. *Neuroscience and Biomedical Engineering*. 2014;**2**(1):pp. 2–13.

[36] Al Ameen, M., Liu, J., and Kwak, K. 'Security and privacy issues in wireless sensor networks for healthcare applications'. *Journal of Medical Systems*. 2012;**36**(1):pp. 93–101.

[37] Carrión, I., Fernández-Alemán, J., and Toval, A. 'Usable Privacy and Security in Personal Health Records'. In *Human-Computer Interaction – INTERACT 2011*. vol. 6949, in Campos P., *et al*., (eds.). Berlin/Heidelberg: Springer; 2011. pp. 36–43.

[38] Ming, L., Wenjing, L., and Kui, R. 'Data security and privacy in wireless body area networks'. *IEEE Journal on Wireless Communications*. 2010; **17**(1):pp. 51–58.

[39] Xiaodong, L., Rongxing, L., Xuemin, S., Nemoto, Y., and Kato, N. 'Sage: a strong privacy-preserving scheme against global eavesdropping for ehealth systems'. *IEEE Journal on Selected Areas in Communications*. 2009;**27**(4):pp. 365–378.

[40] Kargl, F., Lawrence, E., Fischer, M., and Lim, Y.Y. 'Security, privacy and legal issues in pervasive ehealth monitoring systems'. *7th International Conference on Mobile Business, 2008. ICMB '08*. 2008; Barcelona: IEEE.

[41] Ng, H., Sim, M., and Tan, C. 'Security issues of wireless sensor networks in healthcare applications'. *BT Technology Journal*. 2006;**24**(2):pp. 138–144.

[42] Murdoch, T.B. and Detsky, A.S. 'The inevitable application of big data to health care'. *Journal of the American Medical Association*. 2013;**309**(13): pp. 1351–1352.

[43] Groves, P., Kayyali, B., Knott, D., and Van Kuiken, S. *The Big Data Revolution in Healthcare: Accelerating Value and Innovation*. New York (NY): McKinsey Global Institute; 2013.

[44] Yu, P., Wu, M.X., Yu, H., and Xiao, G.Q. 'The challenges for the adoption of M-health. in Service Operations and Logistics, and Informatics, 2006'. *IEEE International Conference on SOLI'06*. IEEE, 2006.

[45] Khandoker, A.H., Karmakar, C.K., and Palaniswami, M. 'Comparison of pulse rate variability with heart rate variability during obstructive sleep apnea'. *Medical Engineering & Physics*. 2011;**33**(2):pp. 204–209.

[46] Dong-Her, S., Hsiu-Sen, C., Binshan, L., and Shih-Bin, L. 'An embedded mobile ECG reasoning system for elderly patients'. *IEEE Transactions on Information Technology in Biomedicine*. 2010;**14**(3):pp. 854–865.

[47] Rashid, R.A., Rahim, M.R.A., Sarijari, M.A., and Mahalin, N. 'Design and implementation of wireless biomedical sensor networks for ECG home health monitoring' *ICED 2008, International Conference on Electronic Design*, 2008.

[48] Yong Gyu, L., Ko Keun, K., and Kwang Suk, P. 'ECG recording on a bed during sleep without direct skin-contact'. *IEEE Transactions on Biomedical Engineering*. 2007;**54**(4):pp. 718–725.

[49] Tatara, E. and Cinar A. 'Interpreting ECG data by integrating statistical and artificial intelligence tools'. *Engineering in Medicine and Biology Magazine. IEEE*, 2002;**21**(1):pp. 36–41.

[50] Garcia, J., Martinez, I., Sornmo, L., Olmos, S., Mur, A., and Laguna, P. 'Remote processing server for ECG-based clinical diagnosis support'. *IEEE Transactions on Information Technology in Biomedicine*. 2002;**6**(4): pp. 277–284.

[51] Hernandez, A.I., Mora, F., Villegas, M., Passariello, G., and Carrault, G. 'Real-time ECG transmission via Internet for nonclinical applications'. *IEEE Transactions on Information Technology in Biomedicine*. 2001;**5**(3): pp. 253–257.

[52] Dumbill, E. (2013). A Revolution That Will Transform How We Live, Work, and Think: An Interview with the Authors of Big Data. Big Data, **1**(2), 73–77.

[53] Chawla, N.V. and Davis, D.A. 'Bringing big data to personalized healthcare: a patient-centered framework'. *Journal of General Internal Medicine*. 2013;**28**(3):pp. 660–665.

[54] Schleyer, T., Mattsson, U., Ni Riordain, R., *et al*. 'Advancing oral medicine through informatics and information technology: a proposed framework and strategy'. *Oral Diseases*. 2011;**17**(Supplement S1):pp. 85–94.

[55] Dilmaghani, R.S., Bobarshad, H., Ghavami, M., Choobkar, S., and Wolfe, C. 'Wireless sensor networks for monitoring physiological signals of multiple patients'. *IEEE Transactions on Biomedical Circuits and Systems*. 2011;**5**(4):pp. 347–356.

[56] Raghupathi, W. and Raghupathi, V. 'Big data analytics in healthcare: promise and potential'. *Health Information Science and Systems*. 2014;**2**(1):pp. 3.

[57] Groves, P., Kayyali, B., Knott, D., and Van Kuiken, S. *The 'Big Data' Revolution in Healthcare.* Seattle, Washington: McKinsey Quarterly; 2013.

[58] Chatterjee, S., Chakraborty, S., Sarker, S., Sarker, S., and Lau, F.Y. 'Examining the success factors for mobile work in healthcare: A deductive study'. *Decision Support Systems*. 2009;**46**(3): pp. 620–633.

[59] Wilcox, A.B., Gallagher, K.D., Boden-Albala, B., and Bakken, S.R. 'Research data collection methods: from paper to tablet computers'. *Medical Care*. 2012;**50**:pp. S68–S73.

[60] Kotz, D., Avancha, S., and Baxi, A. 'A privacy framework for mobile health and home-care systems', *Proceedings of the First ACM Workshop on Security and Privacy in Medical and Home-care Systems*; Chicago, Illinois, USA: ACM, 2009:pp. 1–12.

[61] Latré, B., Braem, B., Moerman, I., Blondia, C., and Demeester, P. 'A survey on wireless body area networks'. *Wireless Networks*. 2011;**17**(1):pp. 1–18.

[62] Kay, M. *mHealth: New Horizons for Health through Mobile Technologies.* World Health Organization, USA; 2011, pp. 66–71.

[63] Jovanov, E. and Milenkovic, A. 'Body area networks for ubiquitous healthcare applications: Opportunities and challenges'. *Journal of Medical Systems* 2011;**35**(5):pp. 1245–1254.

[64] Mosa, A.S.M., Yoo, I., and Sheets, L. 'A systematic review of healthcare applications for smartphones'. *BMC Medical Informatics and Decision Making*. 2012;**12**(67):pp. 1–31.

[65] Ehrenreich, B., Righter, B., Rocke, D.A., Dixon, L., and Himelhoch, S. 'Are mobile phones and handheld computers being used to enhance delivery of psychiatric treatment?: A systematic review'. *The Journal of Nervous and Mental Disease*. 2011;**199**(11):pp. 886–891.

[66] Brownsell, S., Bradley, D., Blackburn, S., Cardinaux, F., and Hawley, M.S. 'A systematic review of lifestyle monitoring technologies'. *Journal of Telemedicine and Telecare*. 2011;**17**(4):pp. 185–189.

[67] Gao, T., Greenspan, D., Welsh, M., Juang, R., and Alm, A. 'Vital signs monitoring and patient tracking over a wireless network'. *27th Annual International Conference of the Engineering in Medicine and Biology Society, 2005. IEEE-EMBS 2005*; 2006, IEEE.

[68] Bajorek, M. and Nowak, J. 'The role of a mobile device in a home monitoring healthcare system'. *2011 Federated Conference on Computer Science and Information Systems (FedCSIS)*. 2011, IEEE.

[69] Ng, C.J., Lee, Y.K., Lee, P.Y., and Abdullah, K.L. 'Health innovations in patient decision support: Bridging the gaps and challenges'. *The Australasian Medical Journal*. 2013;**6**(2):pp. 95.

[70] MOTIVA. *Motiva – Improving people's lives*. [cited 2013; Available from: http://www.healthcare.philips.com/us_en/products/telehealth/Products/motiva.wpd. last access: March 2016.

[71] Mitra, U., Emken, B.A., Lee, S., *et al*. 'KNOWME: A case study in wireless body area sensor network design'. *IEEE Communications Magazine*. 2012;**50**(5):pp. 116–125.

[72] Longo, D.L., Kasper, D.L., Jameson, J.L., Fauci, A.S., Hauser, S.L., and Loscalzo, J. *Clinical Problems of Aging: Introduction*. McGraw-Hill, USA; 2012.

[73] Taub, D.M., Leeb, S.B., Lupton, E.C., Hinman, R.T., Zeisel, J., and Blackler, S. 'The Escort System: A Safety Monitor for People Living with Alzheimer's Disease'. *IEEE Pervasive Computing*. 2011;**10**(2):pp. 68–77.

[74] Imhoff, M. and Kuhls S. 'Alarm algorithms in critical care monitoring'. *International Anesthesia Research Society*. 2006;**102**(5):pp. 1525–1537.

[75] Baig, M.M. and GholamHosseini, H. 'A remote monitoring system with early diagnosis of hypertension and hypotension'. *Point-of-Care Healthcare Technologies (PHT), 2013 IEEE*, 2013.

[76] Griffiths, J.R. and Kidney E.M. 'Current use of early warning scores in UK emergency departments'. *Emergency Medicine Journal*. 2012;**29**(1): pp. 65–66.

[77] Kyriacos, U., Jelsma J., and Jordan S. 'Monitoring vital signs using early warning scoring systems: A review of the literature'. *Journal of Nursing Management*. 2011;**19**(3):pp. 311–330.

[78] Kundel, H.L. and Polansky M. 'Measurement of observer agreement'. *Radiology*. 2003;**228**(2):pp. 303–308.

[79] Zheng, K., Haftel, H.M., Hirschl, R.B., O'Reilly, M., and Hanauer, D.A. 'Quantifying the impact of health IT implementations on clinical

workflow: A new methodological perspective'. *Journal of the American Medical Informatics Association*. 2010;**17**(4):pp. 454–461.

[80] Sclafani, J., Tirrell, T.F., and Franko, O.I. 'Mobile tablet use among academic physicians and trainees'. *Journal of Medical Systems*. 2013;**37**(1):pp. 1–6.

[81] Mosa, A.S.M., Yoo, I., and Sheets, L. A systematic review of healthcare applications for smartphones. *BMC Medical Informatics and Decision Making*. 2012;**12**(1):pp. 67.

[82] Brown, I. and Adams, A.A. 'The ethical challenges of ubiquitous healthcare'. *International Review of Information Ethics*. 2007;**8**(12):pp. 53–60.

[83] Want, R. 'People first, computers second'. *IEEE Pervasive Computing*. 2007;**6**(1):pp. 4–6.

[84] Morris, M.G. and Venkatesh, V. 'Age differences in technology adoption decisions: Implications for a changing work force'. *Personnel Psychology*. 2000;**53**(2):pp. 375–403.

[85] Horng, S., Goss, F.R., Chen, R.S., and Nathanson, L.A. 'Prospective pilot study of a tablet computer in an emergency department'. *International Journal of Medical Informatics*. 2012;**81**(5):pp. 314–319.

[86] Strayer, S.M., Semler, M.W., Kington, M.L., and Tanabe, K.O. 'Patient attitudes toward physician use of tablet computers in the exam room'. *Family Medicine*. 2010;**42**(9):p. 643.

Part 3

Associated Issues and Case Studies

Chapter 17

Smart, age-friendly cities and communities: the emergence of socio-technological solutions in the Central and Eastern Europe

Andrzej Klimczuk[1] *and Łukasz Tomczyk*[2]

17.1 Introduction

The primary goal of the chapter is to introduce a background of the development of ambient-assisted living (AAL) technologies and applications. The chapter is an attempt to propose an integrated approach to concepts of (1) smart cities and (2) age-friendly cities and communities. Although these ideas are widely promoted by the European Union (EU) and the World Health Organisation (WHO), they are perceived as separate. Meanwhile, these concepts are closely intermingled in theory and practice with regard to the promotion of healthy and active ageing, a universal design, usability and accessibility of age-friendly environments, reducing the digital divide and robotic divide and reducing social isolation of older adults.

The first two parts of this chapter will outline the concepts, premises and examples of practical solutions for smart and age-friendly cities and communities (SAFCC). The discussion will include the presentation of the factors shaping the local space, which foster efficiency of older people in the performance of activities of daily living (ADLs). The chapter will provide a brief description of three perspectives on the development of SAFCC, namely, (1) the needs of older adults, (2) supporting factors and barriers to the implementation of AAL solutions in local ageing policies and (3) examples of technological solutions, in particular, associated with the modernisation of public spaces and solutions related to the concept of 'ageing in place 2.0.'

The further parts of this chapter will describe the needs of the older adults that influence the design and implementation of SAFCC in the Central and Eastern Europe (CEE). We focus on countries of this region not only due to the rapid ageing of populations but also due to infrastructure deficiencies and institutional barriers that clearly show the need to integrate concepts of SAFCC. To this end, the brief description of indicators of SAFCC and differences between selected cities will be presented.

[1]Warsaw School of Economics, Warszawa, Poland
[2]Pedagogical University of Cracow, Krakow, Poland

The proposed analysis will include both 'soft factors' (aimed at activation of older citizens and the development of cultural and educational activities) and 'hard factors' (physical accessibility, infrastructure and technological solutions). This review will be possible by the description of theoretical assumptions affecting the need to create local ageing policies in relation to the cities and communities of the fastest ageing populations in the EU.

With reference to the model of age-friendly cities and communities [1] and an eclectic and multidimensional definition of smart cities [2], we will briefly analyse examples of technological solutions related to the AAL such as gerontechnologies and social innovations in eight fields: outdoor spaces and buildings, transportation, housing, social participation, respect and social inclusion, civic participation and employment, communication and information, and community support and health services.

The conclusion will underline the need for participatory formulation of AAL technologies and applications with older adults and the need for advocacy to promote the AAL in association with the concept of the 'silver economy.'

17.2 The smart cities and communities in the context of the population ageing: An overview of the concept

At the beginning of the twenty-first century, the discourse on the revitalisation of urban spaces is particularly important. Local administration, especially in urban areas, must meet the growing expectations of different groups of people (stakeholders) both as citizens and, among others, consumers, workers, entrepreneurs, investors and members of families and communities. Furthermore, in the context of demographic change, urban renewal must include, among others, issues such as the population ageing, increasing the life expectancy, increasing pressure on the needs of people with disabilities regarding the elimination of architectural barriers and accessibility of public infrastructure, depopulation of some districts as well as unfavourable changes in birth rates, fertility and mortality.

According to the United Nations (UN), particularly noticeable is the process of 'double ageing' of the population that is the rapid increase of old–old (75–84 years) and oldest–old (over 85 years) age groups in the population. The population of people over 80 years old is expected to grow worldwide from 120 million in 2013 to 392 million in 2050 [3]. In urban areas, these changes will be reflected primarily in the increased demand for care services, health services and social services [4–6]. However, it should be noted that the population ageing not only significantly reduces the potential for the informal care provided mainly by families. This process also leads to the emergence of the 'sandwich generation,' that is defined mostly as people in middle age (40–59 years), who must reconcile their caring responsibilities towards their children and parents with maintaining a professional activity [7, 8].

The population ageing entails many consequences (social, economic, political and cultural), but it is also a challenge to create new solutions so that societies can function better in the future. Attention, in particular, should be paid to older generations because the rate of population growth in the retirement age in CEE according to demographic projections will vigorously increase [9]. Poland, the Czech Republic, Hungary and Slovakia (the countries of the Visegrad Group) in particular in recent years started to feel the effects of demographic changes through phenomena such as depopulation of large cities. This process is occurring mainly in centres, which concentrated industry during the period of real socialism, e.g., in Poland in the years 1995–2013 only six cities were characterised by an increase of population: Kraków, Olsztyn, Rzeszów, Warsaw, Zielona Góra and Białystok [10]. However, the progressive urbanisation in CEE enforces implementation of new socio-technical solutions in the local urban systems together with the full transfer of the idea of active ageing in various areas of life. The active ageing allows people to realise their potential and participate in social life [11]. This approach points to the need for a multidimensional approach to support older adults. An age-friendly environment consisting of a relevant resource for educational and learning institutions is one of the factors that allow the activation of older people [cf. 12].

The discussed demographic changes force the search for innovative solutions to employment, health, life–work balance, participation in culture, recreation and sport as well as the delivery of public services to facilitate the implementation of various tasks and goals of stakeholders (e.g. transport, telecommunications, social housing and environmental protection). Thus, the population ageing is an important stimulus for the creation of technological innovation and social innovation. Such innovations include product solutions and services defined at the beginning of the twenty-first century as 'calm technologies' that do not require user's attention; 'ambient technologies' that are embedded in user's environment; and 'smart technologies' that facilitate the provision of complex solutions to the user in the simplest way possible [13, 14]. The term 'smart' is used in particular in relation to devices and processes that are using the latest technology in such a way that, despite its inherent complexity may be intuitively used by users, thus may respond to human needs in an efficient and activating way.

The main goals of the Strategy 'Europe 2020' [15] that were adopted by the EU in 2010 are striving for the development of smart, sustainable and inclusive growth. The European Commission's report, 'The Growing Silver Economy in Europe' [16] points out that 110 European regions consider healthy and active ageing as their priority of smart specialisation in the coming years. This means that these regions intend to focus on support supplied to industries of the 'silver economy,' that is a system of production and distribution of goods and services aimed at ageing populations [17]. These sectors include, among others, cosmetics and fashion, silver tourism, service robotics (including AAL solutions), the health and wellness sector (including medical equipment, pharmaceuticals and telemedicine), functional food, security, culture and lifelong learning, entertainment, smart and autonomous transport (e.g. self-steering cars), assistive technologies, and banking and financial services dedicated to older adults.

In the context of the silver economy and smart growth, the concept of smart cities may be understood in particular as urban areas that widely utilise information and communication technologies (ICTs) to organise and provide all urban functions, e.g., to reduce costs of infrastructure maintenance (e.g. roads, bridges, subways, airports, seaports, public transport and sewerage), consumption of resources (e.g. gas, electricity and water supply), better use the free spaces as well as to engage citizens in local governance [18]. Terms similar to the smart city are a cyber city, cyberville, digital city, electronic communities, electronic spaces, flexicity, information city, knowledge-based city, teletopia, telicity, intelligent city, virtual city and wired city [19, 20]. At this point, we need only to mention that in 2012, the European Innovation Partnership for Smart Cities and Communities [21, 22] was launched to connect cities, industry, SMEs, finance sector, research and other stakeholders in activities related to, among others, sharing knowledge, create international partnerships as well as develop and implement smart city solutions.

A special position in the concept of smart cities can be given to (1) smart homes and technologies compatible with the paradigms of (2) AAL and (3) ageing in place 2.0, which aim to facilitate activity during ageing mainly with electronic aids for ADLs. The concept of smart homes includes solutions such as mobile applications and web-based control, monitoring, and automation functions of housing, such as lighting, heating, ventilation, alarm system, video surveillance, fire alarm system, emergency power supply and management system for production and storage of renewable energy [6, 23]. While the concept of the AAL refers to the wider application of technology in living environments to extend the sole use of them by older adults as well as increasing their autonomy and certainty, reducing risky and monotonous actions, monitoring health and providing health care, increasing the sense of security and saving resources [24]. The last-mentioned concept – ageing in place 2.0 – focuses on the empowering older citizens by systems of integrated services that need two categories of infrastructure investment: (1) connections to the world outside the home (e.g. transportation and community infrastructure and (2) home design, devices and assistive technologies that will reduce risk and facilitate-wise and effective use of resources [25]. Examples of AAL and ageing in place 2.0 include intelligent living spaces, detecting and mitigating problems of older adults, such as memory loss, wandering, and health issues; smart products such as floors equipped with sensors and shoes with actuators to detect or prevent them falling, pendants for starting the alarms of social assistance or equipment to remind about the time to take medicines; and digital information services, including, among others, the sphere of security, monitoring of health (e-health) and remote care (telecare).

The idea of smart cities as a response to the new needs of residents and civilisational challenges (e.g. population ageing, mobility and environmental protection) includes all concepts mentioned above. The idea of the city 'smarter' than in the past assumes the most optimal use of the potential of the various capitals of the city and citizens as well as the integration of resources through the creation of new solutions and implementation of best practices for improving the quality of life [26]. Local administration or non-governmental organisations (NGOs) and community leaders activate these processes. The smart city can be defined by services such as:

- Smart power grids to enable end users to reduce the amount of used energy power.
- Traffic control to reduce traffic jams and, therefore, the increased liquidity of traffic and environmental protection.
- Supervision of pedestrian traffic to raise the level of security.

- Remote monitoring of patients' health, among others, by telecare that will optimise the involvement of medical personnel and prompt assistance.
- Vehicle telematics systems associated with the technical and information infrastructure for the use of galvanic cell-powered cars.
- Registry to monitor the activity of urban municipal services including public transport system [26].

One of the visions of smart cities assumes that such type of the agglomeration allows greater operational efficiency due to the use of similar instruments and algorithms, which are designed to increase the competitiveness of enterprises [2]. However, despite the increasing wealth of technical solutions aimed at improving the quality of life, an idea of the smart city cannot be implemented without reference to the needs of residents. Efficient organisation of urban space requires an approach that highlights the human agency of residents, because the use of 'technocentric approach' creates a sense of dehumanisation and loss of the sense of autonomy, and thus inhibits even the most innovative solutions [cf. 26]. In the concept of smart cities, the need becomes a starting point for creating solutions that create new, intelligent solutions for creating sustainable spaces.

Another, the supplementary approach assumes that smart cities should be based on the human capital and social capital of inhabitants [27]. In a situation where the state, local governments or private companies are not able to innovatively respond to the needs of individual residents, their potential is activated independently through bottom-up initiatives. Examples include time banks, local currencies, barter initiatives, neighbourhood care circles, cohousing, car sharing and all kinds of social enterprises [17]. This approach to the smart cities is also related to the notion of 'creative class' that is category of knowledge workers, intellectuals, artists and other free professions that play major role in restructuring cities and regions in relation to the needs of cultural and creative industries (e.g. science, engineering, research, arts, design and media) [28]. Thus, the SAFCC are already described as spaces that foster the 'creative ageing,' that is, engaging older adults in creative activity, which aims to shift from highlighting their problems to underline their potential [29, 17].

The smart cities also focus on the integration of data about the activities of communities and big data analysis that allows the creation of 'tailored solutions' on the local level. Municipal institutions collect huge amounts of data in digital form, which, by their respective linking and processing are an enormous source of information for those responsible for creating urban spaces. Increasing and improving the quality of databases is possible thanks to the wider use of the concept of the 'Internet of things,' which is expanding the network through the devices and equipment (e.g. home electronics, machines and sensors) that can communicate with people and with each other, e.g., during the provision of energy services, reducing traffic congestion, reducing the consumption of street lighting, monitoring pollution and noise, supporting the urban horticulture, and waste management [30, 31]. The smart home system, e.g., can remember an about watering lawns and turning off the iron. It can also memorise the behaviour of the household and during holiday simulate their presence by turning on and off lights and shading windows. The system can also at the right time open the service door for the cleaners. Smart electricity meter can work with household appliances for efficient energy management. Selected devices such as washing machines or dishwashers can be switched off at the energy summit and turned beyond. Therefore, the smart cities appear not only in the prism of new solutions based on increasing and improving the service offer by the public institutions but also as the better use of existing data.

Given the difficulties of unifying the concept of smart cities, scholars from Vienna University of Technology for measuring the scale and characteristics of this phenomenon offered an eclectic and multidimensional definition [2]. Thus, the smart city can be defined through the development of the six fields:

- The first field is a 'smart economy,' which is understood as competitiveness. This field is measured by indicators that describe innovation climate, entrepreneurship, the image of the local economy, productivity, labour market flexibility, embeddedness of the city in networks of international cooperation and the ability to transform the city continuously.
- The second field is called as 'smart people' that are the human and social capital of the city residents. To measure these field indicators such as the level of education, the tendency for lifelong learning, social and ethnic diversity, flexibility, creativity, cosmopolitanism, openness and participation in public life are used.
- The third dimension is the 'smart governance,' which is widely understood as social participation. Indicators of this field are the participation of citizens in decision-making bodies, the scope of public and social services, transparency of governance, public policies, and political perspectives.
- The fourth field is 'smart mobility' related both to the transport and to the use of ICTs. The indicators are the availability of local, national and international accessibility of the city, the presence of ICT infrastructure, sustainable, innovative and secure transport systems.
- The fifth field is a 'smart environment' that is an approach to natural resources. Indicators of this dimension are the attractiveness of natural conditions, the level of the pollution, environmental protection and sustainable management of resources.
- The last field is 'smart living' that is defined as a quality of life. This area includes factors that affect the lives of the inhabitants of the city, such as the availability of cultural and educational institutions, health conditions, personal safety, quality of housing, tourist attractiveness and social cohesion.

In conclusion, the discussed definition of smart city emphasises not only the close links between cities and new technologies but also the pursuit of the sustainable development by emphasising the importance of simultaneous support for the economic growth, social development and the environment protection.

17.3 The smart and age-friendly cities and communities from the perspective of the oldest residents

The concept of age-friendly cities and communities is about the creation of conditions for activities to promote healthy and active ageing [cf. 32]. This idea assumes allowing people of all ages and, in particular, older people to continue working, social and political engagement, and the implementation of their plans. Such cities and communities should provide both a sense of security, and adapt their infrastructure, resources, development goals, and operational programmes to the needs and opportunities of an ageing population.

The age-friendly city and community [1] through the extensive inclusion and support of activities of various groups of stakeholders in public policies picks also aim to change the perception of older adults and ageing, in particular, by younger generations [33]. The concept of age-friendly cities and communities aims to improve the quality of life, not only among the representatives of the older generation but also among other social groups that have a large resource of free time and specific physical conditions. In other words, it is a space that by the adoption of the principles of universal design [34] is to be friendly to everyone regardless of age, in particular for groups such as children and young people, their parents, mothers with prams, people with disabilities.

The age-friendly cities and communities take into account the diversity among the seniors (inhomogeneity due to, e.g., health, education, income and model of the family), creates optimal conditions for active people and allows the smooth functioning of those less mobile, supports equal rights through the development of various capitals of older adults, ease of restrictions arising from the biological course of ageing, integrates all age groups and strengthens intergenerational dialogue and solidarity and fosters a sense of community [cf. 33].

In the last years in the literature, the concept of age-friendly cities and communities was developed by various authors [35–37]. However, at this point we want to underline the WHO's framework that focus on global, standardised and a bottom-up participatory approach. The WHO [1] has developed a set of indicators and the checklist of essential features favouring the practical implementation of the idea of the age-friendly cities and communities. This was possible due to the programme for the development of standards to adapt cities to the needs of older adults that was launched in 2005. The indicators of age-friendly cities and communities were divided into solutions that promote the active ageing in eight dimensions:

1. Outdoor spaces and buildings (e.g. cycle paths have to be separated from pavements; buildings are well signed outside and inside with, among others, accessible elevators, and non-slip floors)
2. Transportation (e.g. driver education and refresher courses are promoted; priority parking and drop-off spots for people with special needs are available and respected)
3. Housing (e.g. sufficient and affordable housing for frail and physically challenged older people)
4. Social participation (e.g. activities and events can be attended alone or with a companion; good information about activities and events is provided, including details about the accessibility of facilities and transportation options for older people)
5. Respect and social inclusion (e.g. older people are visible in the media, and are depicted positively and without stereotyping)

6. Civic participation and employment (e.g. workplaces are adapted to meet the needs of physically challenged people; training in post-retirement options is provided for older workers)
7. Communication and information (e.g. public and commercial services provide friendly, person-to-person service on request; electronic equipment has large buttons and big lettering)
8. Community and health services (e.g. emergency planning takes into account the vulnerabilities and capacities of older people; health and social services are conveniently located and accessible by all means of transport).

The list of indicators may be divided into three categories. The first three areas are related to the physical environment of older adults and their mobility. The next three fields create the social environment of older adults. The last two areas are related to socioeconomic factors of quality of life of older adults and focused on possibilities of receiving adequate support.

The list of indicators is not closed and is not intended to compare which city and community are more age-friendly, but it makes possible to make a detailed self-assessment and identify deficiencies that require intervention and corrective action. The list also allows policy makers to determine what factors interfere with residents in a given, specific city, and the environment [33]. The guidelines and the list do not show, however, how to tackle identified problems and do not show any effects of introduced programmes [cf. 38, 39].

The WHO in 2010 established the initiative of the Global Network of Age-friendly Cities and Communities. Centres that are candidates for membership in the network are committed to the development and implementation of reform programmes in the dimensions mentioned above. Programmes are evaluated and supported by the WHO, and the main criterion is the involvement of older people in all stages of action – not only, e.g., as members of the boards to generate some opinions but also as animators of projects and those who are monitoring their progress, participate in the evaluation and improve future operations. In 2011, Dublin in Ireland and in 2013 Quebec City in Canada hosted international conferences on age-friendly cities and communities to collect as much as possible a diverse range of stakeholders of this social movement and to identify common problems in the implementation of the concept and strategies to break down the barriers of their development [40].

In 2015, the WHO also proposed guidelines for measuring the age-friendliness of cities [41]. The framework further divides indicators related to previously described eight dimensions into five categories. The first category includes indicators of inputs that are resources and structures used during public interventions. The second category refers to outputs which are the effects of interventions regarding age-friendly environments such as planning and land use, the design of public spaces, housing and transportation design. The third category includes indicators of outcomes that are short-term changes generated by interventions such as accessibility of public spaces, affordability of housing and participation in decision-making. The fourth category refers to indicators of impact that are long-term changes such as increased health and well-being. The last set of indicators refers to equity and is crosscutting the framework

Table 17.1 Dimensions of a smart and age-friendly city

Age-friendly city	Smart city
– Outdoor spaces and buildings	– Smart environment
– Transportation	– Smart economy – Smart mobility
– Housing	– Smart economy – Smart living
– Social participation	– Smart people – Smart governance
– Respect and social inclusion	– Smart people – Smart governance – Smart living
– Civic participation and employment	– Smart people – Smart economy
– Communication and information	– Smart economy – Smart mobility
– Community and health services	– Smart governance – Smart living

Source: Own elaboration based on References 2 and 1.

as a principle of interventions and may be measured by comparing differences between two reference groups in terms of gender, age, wealth and neighbourhood. This framework may be used to monitor and evaluate interventions, compare achievements of various cities and communities and to collect best practices.

At this point, we may compare the features of age-friendly cities and communities to the elements of a smart city (Table 17.1). However, the introduction of the more integrated theory of SAFCC is still the challenge that needs more research and discussion.

Researchers from Intel Corporation [42] tried to combine the concepts of the smart city and the age-friendly city in the context of demographic challenges and urbanisation in China. They combined the use of ICTs with changes of the model of care for older adults towards support them in the local community, neighbourhoods and ageing in place. Thus, their idea aims to allow reducing the cost of social and health care by decreasing the growing demand for nursing homes and long-term care facilities.

Intel's researchers as the current challenges in achieving the full benefits of the implementation of the smart and age-friendly cities recognise the demand for services, which is so high that it brings a negative impact on their quality; insufficient level of an integrated evaluation system of care services; systems that are not integrated enough to respond to individual needs; and infrastructure that is outdated in many places, which restricts the use of the potential of mobile technology [42]. Moreover, the researchers note that the implementation of the idea of smart and age-friendly cities will allow societies to achieve benefits such as the improvement of communication and active ageing in local communities, the increase in personal safety, the autonomy and independence of older adults and the facilitation of their further activity and productivity also in the labour market. Intel's analysts also point out that in smart cities it is crucial to highlight the recognition of their multisectoral character for the quality of service provision for older people. A key role in the organisation of services may be given to local authorities that can also provide services, which limits their quality. Alternatively, the government may delegate tasks of providing services to private entities or NGOs. In another perspective, the services of the smart city for older adults may be developed by market logic, in which access to services may be restricted to the poorer groups, and the quality of services will be very diverse if there will be no industry self-regulation mechanisms.

Also, the creators of smart care services should take into account some stakeholders who, in varying degrees, can participate in the information flow and anticipation and meeting the needs of seniors [42]. The information system should connect not only the devices but also diverse stakeholders relevant to older adults as recipients of services. The stakeholders considered here are platform manager, suppliers of the individual services (e.g. volunteers and nurses), services partners, payers (e.g. local government), a family of older adult, a provider of ICTs, health care system and rescue organisations (e.g. ambulance, gas, power, and heating maintainers and providers).

Intel's researchers recommend the use of the three stages of building a platform for smart and age-friendly city and community: (1) to meet the basic requirements, e.g., creating services, to ensure their compliance with the standards, and integration; (2) open development consisting of further social sharing of access to the platform, increasing the impact and generating added value; and (3) making decisions through the use of knowledge of the local government, service providers and seniors.

However, it should be noted that the disadvantage of a concept proposed by Intel Corporation is that it focuses on care and health services and to some degree on housing [cf. 23]. This concept, however, is certainly a good starting point for constructing systems that will also include other fields of SAFCC.

17.4 Shaping of smart and age-friendly cities and communities on the example of the CEE

In this section on the basis of the typology of age-friendly characteristics proposed by the WHO and review of definitions of a smart city, we will discuss examples of solutions from CEE countries that focus on the development of SAFCC. So far usually in the literature are discussed examples from Western Europe, North America and Asia-Pacific region such as Amsterdam, Barcelona, Copenhagen, Helsinki, Singapore, Songdo, Vancouver and Vienna. Moreover, these cities have been usually described only in separated terms of smart cities or age-friendly cities [38, 43, 44, 19, 45–50, 5, 51, 23, 52–56].

The approach presented in this chapter tries to take a closer look at the cities and communities in countries that are characterised not only by rapidly ageing populations and mainly medium-sized cities but also by shortcomings of the infrastructure that makes the development and implementation of AAL and ageing in place 2.0 solutions even more challenging and significant [57–59].

17.4.1 The smart and age-friendly cities and communities in connection with the active ageing index and European welfare systems

Scholars from Vienna University of Technology [60, 2] conducted three editions of ranking of medium-sized smart cities in the EU in 2007, 2013 and 2014. The third edition of the ranking compares 77 medium-sized cities with a population between 100,000 and 500,000 inhabitants. Top 10 positions in the total ranking that summarises all six dimensions of the smart city are (1) Luxembourg, Luxembourg, (2) Aarhus, Denmark, (3) Umeå, Sweden, (4) Eskilstuna, Sweden, (5) Aalborg, Denmark, (6) Jönköping, Sweden, (7) Odense, Denmark, (8) Jyväskylä, Finland, (9) Tampere, Finland and (10) Salzburg, Austria. Thus, all best practices of medium-sized smart cities are outside CEE countries.

Interesting findings may be generated by the comparison of the ranking of medium-sized smart cities (SCR) with the active ageing index (AAI). The AAI includes comparisons between the 28 EU Member States, including indicators divided into two dimensions and four domains and further subdivided by gender: actual experiences of active ageing (employment; participation in society; independent, healthy and secure living); and capacity and enabling environment for active ageing [61]. Moreover, a positive correlation of the AAI with GDP per capita was noted, which suggests that the countries with relatively higher standards-of-living are generating better capacity and enabling environment for active ageing [61].

To understand the differences in the development of SAFCC, we may divide the results of the SCR and AAI according to the typologies of welfare states. A well-known concept of different welfare regimes was proposed by Gøsta Esping-Andersen [62] and includes the decommodification index and impact of benefits and social services on social stratification. This theory explains that in countries with a social-democratic regime there is a high degree of decommodification, universal care and nursing services provided by public entities, high wage replacement by social benefits, full employment policy and the strong position of labour unions, e.g., the Nordic countries. In a conservative corporatist regime, the state supports the family, social security depends on the position of individuals in the labour market, social care insurance and nursing care services primarily delivered by informal carers, NGOs, and local communities, e.g., Austria, France, Germany, and Italy. The last group of countries represents a liberal regime, which is characterised by individualism in the field of social security, low decommodification and social benefits, benefits based on income, widespread commercialisation of care and nursing services, competition between different suppliers and a small role of labour unions, e.g., the United Kingdom and the United States. Yuri Kazepov [63] expanded Esping-Andersen typology by two types of welfare states. The first, the familistic system, focuses on fewer resources that are targeted to family policies and overload of the family in social caring responsibilities, the gender segmentation in the labour market, local differentiation of particularistic interests and different forms of governance, and weak prevention of poverty, e.g., Spain and Italy. The second, welfare system in transition countries, refers to an emerging model in the CEE countries. The reforms are taken here in the context of financial constraints and with ambivalent consequences. Some countries focus on privatisation while other invests in the coordinated market and social policies. In addition, there were significant reforms of the territorial organisation oriented to move away from central regulation towards more decentralised levels.

The comparison of selected countries that represent various welfare systems and the AAI (Table 17.2) shows that social-democratic and liberal regimes had relatively better cope with the population ageing than corporative regimes, familistic regimes and regimes in transition. The red colour used in the table suggests the highest ranks in comparison to other countries and regions as well as awareness of the challenges of the population ageing. In contrast, the green colour shows the lowest positions in the ranks. The further comparison of the SCR and AAI confirms that medium-sized smart cities in the social-democratic and liberal regimes may offer a higher quality of life for older adults (Table 17.3). These differences might be important to study the variety of models of the silver economy, AAL solutions and specific fields of the SAFCC.

17.4.2 Outdoor spaces, buildings and housing

The first example of creating a smart and age-friendly urban environment in CEE countries and at the same time welcoming public space, not only for seniors but also for other age groups is the development of spaces that allow popularisation of physical activity, e.g., tennis, health paths and bike trails. In an increasing number of Polish cities can be seen specialised devices for developing physical activity among older people. Most of the Polish Universities of the Third Age (UTA) have sports sections [12] that benefit from newly created urban resources available in the public parks, along rivers or near the market, senior clubs, UTA and various cultural facilities. Part of the sports facilities developed in Poland is created mainly through social participation programmes, among others, under local civic budgets.

Many interesting examples of building dedicated to seniors and of the smart economy and smart living can be found in the property development market in CEE countries. The property development and organisations specialised in care for older adults to perceive the market potential inherent in older age groups of society [65]. Adaptation of buildings and housing to meet the needs of older adults are primarily the removal of existing obstacles to overcome, such as architectural barriers, lack of elevators, narrow corridors and entrances that make difficult moving for people with disabilities.

A valuable solution that is fully compatible with the idea of a smart and age-friendly city is in a building designed by the Jewish community in Prague, Czech Republic (district of Vinohrady) [66]. Near the old town hall from 1911, in 2008 was built, the modern complex with more than 100 places for older people of Jewish origin. The building is inhabited not only by persons who need professional rehabilitation but also by the active seniors. The building received international awards for its

Table 17.2 The active ageing index (AAI) and European welfare systems in comparative perspective

Welfare systems	Employment				Participation in society				Independent, healthy and secure living				Capacity and enabling environment for active ageing				Overall			
	Index		Rank		Index		Rank		Index		Rank		Index		Rank		Index		Rank	
	2012	**2014**	**2012**	**2014**	**2012**	**2014**	**2012**	**2014**	**2012**	**2014**	**2012**	**2014**	**2012**	**2014**	**2012**	**2014**	**2012**	**2014**	**2012**	**2014**
Liberal (The United Kingdom)	35.5	35.8	3	4	20.0	20.0	11	11	75.7	73.7	6	11	61.4	61.3	5	5	39.3	39.2	4	5
Social-Democratic (Sweden)	41.0	43.4	**1**	**1**	22.6	22.6	4	4	78.7	78.6	**2**	**4**	69.5	69.2	**1**	**1**	44.0	44.8	**1**	**1**
Corporative (France)	21.0	24.1	21	16	22.4	22.4	5	5	74.6	75.9	9	6	57.8	59.1	9	9	34.2	35.7	12	9
Familistic (Italy)	20.9	23.0	**22**	19	24.1	24.1	**2**	**2**	69.9	69.0	19	17	52.8	53.4	15	15	33.3	34.0	15	14
In transition (Poland)	17.8	22.4	**24**	**20**	12.2	12.2	**27**	**28**	67.5	64.9	**21**	**24**	46.7	47.9	**22**	**22**	27.3	28.2	**27**	**27**
EU27/EU28	27.1	27.9	–	–	18.1	18.1	–	–	71.7	70.6	–	–	54.2	54.4	–	–	33.8	34.0	–	–

Source: Own elaboration based on References 63, 64 and 61.

Table 17.3 The active ageing index (AAI), European welfare systems and domains of medium-sized smart cities ranking (SCR) in 2014

Welfare systems	City	Domains of medium-sized smart cities ranking (SCR)							Average country rank in the SCR	Country rank in the AAI
		Smart economy	Smart people	Smart governance	Smart mobility	Smart environment	Smart living	Total		
Social-Democratic (Sweden)	Umeå	24	5	**2**	34	**1**	13	**3**	**4**	**1**
Social-Democratic (Sweden)	Eskilstuna	21	**1**	7	24	3	41	4	**4**	**1**
Social-Democratic (Sweden)	Jönköping	32	13	3	11	2	26	6	**4**	**1**
Corporative (France)	Montpellier	29	20	16	46	4	30	19	25	9
Corporative (France)	Nancy	44	37	13	23	11	20	23	25	9
Corporative (France)	Poitiers	47	38	10	44	8	15	25	25	9
Corporative (France)	Clermont-Ferrand	38	40	18	36	10	17	26	25	9
Liberal (the United Kingdom)	Cardiff	12	32	44	19	58	16	31	37	5
Liberal (the United Kingdom)	Aberdeen	8	25	52	42	43	32	34	37	5
Corporative (France)	Dijon	52	41	21	41	7	22	35	25	9
Liberal (the United Kingdom)	Portsmouth	9	30	42	13	65	53	36	37	5
Liberal (the United Kingdom)	Leicester	**7**	35	46	**5**	67	63	39	37	5
Liberal (the United Kingdom)	Stoke-On-Trent	31	42	47	26	55	60	45	37	5
Familistic (Italy)	Venezia	33	63	68	32	59	**8**	46	54	14
Familistic (Italy)	Verona	25	57	61	45	66	28	49	54	14
Familistic (Italy)	Trento	43	48	74	62	30	39	52	54	14
Familistic (Italy)	Trieste	37	58	66	66	31	47	53	54	14
In Transition (Poland)	Rzeszów	63	64	49	56	56	50	55	**62**	**27**
In Transition (Poland)	Szczecin	64	62	50	55	47	65	56	**62**	**27**
Familistic (Italy)	Perugia	57	54	65	**68**	53	42	57	54	14
Familistic (Italy)	Padova	34	44	**75**	60	**73**	45	61	54	14
In Transition (Poland)	Bydgoszcz	60	68	54	50	64	64	62	**62**	**27**
Familistic (Italy)	Ancona	54	66	72	67	29	62	65	54	14
In Transition (Poland)	Białystok	68	67	45	61	72	61	66	**62**	**27**
In Transition (Poland)	Kielce	**71**	65	51	65	68	58	68	**62**	**27**
In Transition (Poland)	Suwałki	67	**70**	55	57	71	**68**	**70**	**62**	**27**

Source: Own elaboration based on References 60, 63 and 61.

functionality, original architecture and innovative solutions. The new building is integrated with the old town hall and creates a bridge between history and modernity. In the old building, there is a 'hospoda' (place for feasts) and kindergarten (also for non-members of the Jewish community). This type of construction proves that the facility for older people does not necessarily need to resemble the traditional caring houses, but may also arouse admiration regarding architecture and promote intergenerational solidarity.

In comparison, a different solution that may be disseminated in CEE countries was developed in Burlington, Canada [67]. The HomeShare initiative is a model of cohousing that assumes cooperation between two people representing various households who reside together. Apartments and homes are divided between older adults into private space and common areas including the living room and kitchen. This model allows people to exchange services, share responsibilities, offer companionship, reduce isolation and limit accommodation expenses.

At this point, it is worth noting some other smart solutions implemented, in particular in Poland after 2014. In cities of Bydgoszcz, Rzeszów and Poznań are ongoing projects for the creation of smart transport systems. Poznań, Ostrołęka and Warsaw are creating smart heating networks. Bydgoszcz, Szczecin and Kraków are creating remote lighting management systems [68]. However, Gdańsk is considered the most advanced example of 'smart city' in Poland due to early implementation of many innovative solutions during the project 'ACCUS' [68]. This project helped to create, among others, smart traffic control, urban lighting and automatic identification of vehicles aimed at raising road safety.

The aforementioned examples may be further developed as in Brabant, Netherlands [4]. This city is crucial in the implementation of a regional innovation strategy that aims to foster the creation of new types of infrastructure and technologies for older adults in relation to the concept of active and healthy ageing. In other words, cities in CEE countries may use smart infrastructure investments for the development of local clusters and ecosystems of providing services, and regional models of the silver economy. For comparison in Hong Kong, NGOs created partnership with gerontology research institutes and local government to establish research project that will establish intergenerational programmes and institutions in various fields described in the WHO's framework of age-friendly cities [69].

17.4.3 Social participation, civic participation and employment

According to the WHO, one of the elements of age-friendly cities is the social participation of their citizens. This feature also provides the potential of smart people and smart living in a given space. There are two examples of these ideas in Polish cities. The first is related to selecting projects under separate municipal budgets dedicated to NGOs. Among them, there are also institutions performing tasks for older adults. Votes for civil society projects usually take place once a year in the traditional way (e.g. in the municipal office) or via the Internet.

Another model of cooperation between local governments with seniors involves the establishment of 'senior councils,' which are designed to act as consultants and advisory bodies. Councils are an attempt to activate older people through the involvement in co-decision and governance regarding the implementation of tasks of local government units. So far in Poland, youth councils only have been operated, but without any legitimacy. At the end of 2013 under the amendment to the Local Government Law, the concept of senior councils was added. This situation not only allows the co-decision of older adults in the governance of the cities, municipalities or villages but also supports the promotion of values such as intergenerational solidarity and strengthening of the civic participation. Until 2014, at least 76 senior councils were launched in Poland [70]. Most of these entities have a common range of activities. The senior council in Oświęcim, Poland, e.g., has been operating since 2012, and its tasks include [71]:

- Monitoring needs of seniors and initiating projects aimed at social inclusion of older citizens and fully meet the needs of this social group.
- Take measures to use the potential of older people and their time for public initiatives, in particular in the fields of culture, sport and education.
- The consultation of legal solutions concerning the situation of older adults or likely to affect it and matters submitted to the council by the city's mayor.
- Advice on the issues of older people, particularly in the field of social security, health care, social support and care services.
- Informing the inhabitants of the city about specific directions of activities undertaken by the municipal services and NGOs regarding older citizens.
- The cooperation with organisations and institutions that deal with the problems of older adults.
- The participation of representatives of the senior council in sessions of the city council.
- Submitting annual reports on the activities of the council to the mayor.

Thus, senior councils with legitimacy can create space for civic-administrative communication that allows efficient creation of innovative solutions and services for the ageing population. Undoubtedly, this type of new civic formation is not able to represent the interests of all seniors but provides an opportunity to perceive better the needs of older adults by officials responsible for managing the city.

Civil engagement of older adults may be stimulated in a less direct way [72]. Cité Seniors in Geneva, Switzerland, is an example of the information and meeting centre. The facility provides space, events, workshops and training courses to exchange ideas and learn including computer skills and creative arts. The facility is related to a platform of NGOs run by older adults that include more than 35,000 members.

Social participation in SAFCC can also manifest itself in various ways. An example might be a solution in which the Hungarian NGOs engage in cooperation with the city regarding the care services for older adults. The NGO called as the Maltese Assistance Centre with the financial support of Gyôr City specialises in helping older people who are deprived of care and support. Volunteers and a dedicated team under the agreement with the city provide assist in the homes of older citizens and special facilities. These activities are carried out through long-term cooperation with the city authorities [73].

The described examples may be also compared to solutions developed in Manchester and Newcastle in the United Kingdom [4]. In these cities strategies have been created that consider citizen involvement in decision-making. These documents also foster the development of centres of excellence in research on ageing that have been established at local universities. Another example is Livorno, Italy, where the emphasis was put on promoting the activity of older workers through training and apprenticeships, mid-life career advice and job guarantees [4]. More focus on the productive ageing was also put in Toyama, Japan [4]. The city has developed a network of Silver Human Resources Centres that are offering job placement services and temporary jobs to maintain the employment of older workers in the agricultural industry and maintain population density.

17.4.4 Transportation

Support of activities for older people is also related to public transport. The SAFCC in this field are characterised by not only the introduction of free communication for seniors but also a specific model of providing information about the available transport connections. Friendly signboards with clear fonts, information on delays and the transparency of the graphic on the carrier's website facilitate the movement of older adults in urban space.

During the implementation of activities in the field of mobility should be defined stages or events in life that cause or may cause a change. In the case of older adults, this can be retirement; will to be a grandfather or grandmother; moving to a smaller and/or better-adapted housing; illness or accident; death of a spouse; loss of licence; or the relationship of older adults with their grandchildren [74]. For older adults facilitation of movement in public space are not high curbs, even surface roads and sidewalks, safe and equipped with regulation of light and sound to pass through the road, densely spaced benches enable short rest, good and understandable signage, public transport equipped with benches and roofed shelters, as well as the presence of public toilets [75].

The 'Active Senior's Card' [76] launched in Białystok (Poland) entitles older adults with discounts for public cultural, sport and recreation institutions, and in private companies that have joined the programme. This card is also integrated with public transport card so that the surcharge can act as authorising a special ticket for multiple trips with all bus lines in the city.

Another solution that supports the development of smart mobility of older people is urban taxis dedicated to seniors. Pioneers in this field are Czech cities such as Cesky Krumlov, Plzeň and Bohumin [77]. The condition of using the service is a check in the city and pays a small fee (Plzeň – 30kc is about 1.2 Euro for the whole taxi ride within the city). Seniors may move dedicated taxis in the city from 6:00 to 22:00. Ordering a taxi is possible at the earliest three weeks before the planned trip.

17.4.5 Communication and information

Another field that characterises SAFCC is the communication or the support for smart mobility and the smart economy. The efficient transfer of information between institutions in the city and its citizens can be realised in many ways. The seniors from CEE countries are characterised by increasing rates of use of new media. However, they still show low levels of involvement in the use of network services, although a growing number of older adults use cell phones and e-mails [78].

The use of new media in communication between citizens and the authorities appears in more and more cities [78]. Tychy, Poland, may serve as an example of a city where every citizen, regardless of age, has the possibility of free registration at the town hall or through a special form (e-government system). With these solutions, senior receives notifications on the mobile phone associated with important events, including the status of the settlement of administrative cases. Also, every senior has access to SMS service to current and free information about cultural and sports events and obstacles on the road. A smart city is a city that provides up-to-date relevant information useful in everyday life.

The 'Open Gdańsk' project [68] may be pointed as an example of enabling the inhabitants of the city by the free access to all non-confidential information generated by the city and its organisational units. The system offers access among others, to the register of city's expenses, petitions, maps of urban plots, real estate waiting for the management, selection of sites for planting trees, information about the waiting time in offices, timetables of public transport, base of waste collection centres, base of tourist accommodation facilities, the location of hotspots, the API base of cultural events, contact details for social animators, data on traffic on the bike paths and fault reporting system.

Social innovations in providing services related to communication for older adults in cities may be even simpler. The Telephone Rings at 5 in Setúbal, Portugal, is a project that combines companionship and mental stimulation for those who are not able to leave their homes alone [72]. Every day group of four older adults is invited by a volunteer moderator for discussion by a telephone. Moreover, 'virtual' guided tours and debates on the current events are offered. Participants receive information and photos of places and the moderator leads a virtual visit. Similar programmes were established in Manitoba, Canada; and Oakland, California.

17.4.6 Respect, social inclusion, community services, and health services

An interesting example of social integration and activation of seniors in Poland [79, 12], Czech Republic [80] and Slovakia [12] are UTA, senior clubs, and academies of the third age. These entities at the same time demonstrate the potential of smart people, smart governance and smart living. As part of their activities, such institutions work with local community centres, universities, NGOs and religious organisations. Organisations that educate seniors in addition to the traditional activities that are similar to courses in universities work closely with local entities. These organisations, therefore, fit into the model of the smart city, as most of them allows the activation of seniors who are living near to the institutions and responds to the educational needs of senior citizens. Moreover, older adults in the UTA discuss problems of the local community, because lecturers are often people known in a particular place or region.

According to the WHO definition, the age-friendly cities and communities promote social inclusion. One of the notable examples in this field is time banks set up by NGOs and municipal social welfare centres. Such social innovations are especially disseminated in Japan's cities as 'fureai kippu,' which refers to 'care tickets' that are earned and exchanged by individuals when they are helping older adults in their community [59]. This system established by the Sawayaka Welfare Foundation in recent years in various forms is also spreading in cities in the Asia-Pacific region.

Senior's Time Bank in Częstochowa, Poland, may serve as an example from CEE countries of a group of people who want to provide services to each other [81]. Each of the members has different skills and may exchange them. Every hour of work is saved in a personal account, increasing the balance. Every hour of receiving help, diminish the status of accounts. The rule is that every working hour is equal, regardless of the type of work. Many time banks have been created as a support networks in the neighbourhoods and the mutual provision of various services. For the members of the time bank ceases to be a problem, e.g., walking a dog during an illness or care for older parents and children in unforeseen situations. It becomes easier to get a lift to the city and coming back. The members of the time bank are less likely to spend money at the hairdresser, instrument lessons, language lessons, massage, plumbing and repair services. As emphasised by the founders of the so-called time bank, this social innovation is based on equity and social engagement of older people.

In recent years, a variety of telecare or e-health solutions has been developed. One of the interesting solutions developed at the local level was established in Cape Town, South Africa [82]. AgeWell programme aims at promoting the health of older adults by decreasing isolation. Pilot sites are hiring seniors to be companions for others. After training, they are matched to help specific individuals based on characteristics such as gender, language and interests. The whole process is accompanied by mobile health technologies to increase the wellness of users as well as to support them by relevant medical and social service providers.

Given the assumption of active ageing, and the fact that seniors are living longer and postulate that the role of that group changes in society from passive recipients of services to an increasingly socially engaged citizens [83], the time banks and similar initiatives become a symptom of smart ageing. The SAFCC are seen here as organisms with a powerful resource of human, social and cultural capital inherent also in older citizens. These capitals due to a large amount of free time of older adults and their desire for social involvement may be used to solve a variety of problems in local communities. Thus, it is important to draw attention to social innovation, entities of social economy and NGOs sector in relation to an ageing population.

17.5 Conclusion

The examples presented in this chapter show the multidimensionality of the implementation of SAFCC. This concept is based on the synergy of the human factor of an ageing population (particularly various capitals of older citizens), technical solutions including AAL addressed to not only older adults, and the factors uniting technology and society such as local government and leaders, NGOs and commercial entities offering technological innovation.

The SAFCC are a conglomerate of assumptions involving the right combination of 'soft factors' with 'hard solutions.' Such cities and communities are based on the needs of the ageing population and proper diagnosis of these needs conducted by institutions established for doing so. The ageing policy, which includes the creation of smart cities and communities, can be considered as an element of social policy towards the entire life cycle. Thus, the adopted solutions and the achieved outcomes from this public policy need to serve to 'society for all ages' and thus, be consistent with the UN's policies. In this way, it is possible to omit both the 'gerontocentric' and the 'technocentric' model of ageing policy. Activities undertaken in regard to SAFCC are aimed at strengthening intergenerational solidarity, enabling them a mutual investment in each other, and sharing achievements accordance with the principles of reciprocity and equality [33].

The analysis of the selected concepts and case studies of SAFCC allows us to formulate three other areas of research. First, it is reasonable to attempt to create more integrated theoretical approaches to social and technological innovation associated with ageing in the cities and communities. Second, it is important to study organisations of older people that focus on the dissemination and implementation of innovations and to identify barriers to their development and impact on the ageing policy. Third, comparative studies of different implementations of the concept of SAFCC may be valuable if they include relations with the adopted models of the welfare state and models of the development of the silver economy.

References

[1] WHO. *Global Age-Friendly Cities: A Guide*. Geneva: WHO; 2007.

[2] Giffinger R, Fertner C., Kramar H., Kalasek R., Pichler-Milanović N., Meijers E. *Smart Cities: Ranking of European Medium-sized Cities*. Vienna: Centre of Regional Science, Vienna UT; 2007.

[3] UNPD. *World Population Prospects: The 2012 Revision Volume I: Comprehensive Tables*. New York: United Nations Population Division; 2013.

[4] Cox E., Henderson G., Baker R. *Silver Cities: Realising the Potential of Our Growing Older Population*. Manchester: IPPR North; 2014.

[5] OECD. *Ageing in Cities*. Paris: OECD; 2015.

[6] OECD. *Ageing, Housing and Urban Development*. Paris: OECD; 2003.

[7] Abaya C. *Welcome to the sandwich generation*; 1995. Available from www.sandwichgeneration.com [Accessed 7 Aug 2015].

[8] Miller D.A. 'The "sandwich" generation: Adult children of the aging'. *Social Work* 1981;26(5):419–423.

[9] Zrałek M. 'Tworzenie warunków sprzyjających aktywności ludzi starszych [Creation of favorable conditions for active aging]' in *Osoby starsze w przestrzeni życia społecznego* [*Older People in the Space of Social Life*]. Katowice: Regionalny Ośrodek Polityki Społecznej Województwa Śląskiego; 2014. pp. 27–39.

[10] Szukalski P. 'Depopulacja dużych miast w Polsce [The Depopulation of Large Cities in Poland]'. *Demografia i Gerontologia Społeczna – Biuletyn Informacyjny* 2014;(7):1–5.

[11] WHO. *Active Ageing: A Policy Framework*. Geneva: WHO; 2002.

[12] Maćkowicz J., Wnęk-Gozdek J. '"It's never too late to learn"–How Does the Polish U3A Change the Quality of Life for Seniors?' *Educational Gerontology* 2015. Available from doi:10.1080/03601277.2015.1085789.

[13] Brenner S.W. *Law in an Era of "Smart" Technology*. New York: Oxford University Press; 2007.

[14] Weiser M., Brown J.S. 'Designing calm technology'. *PowerGrid Journal* 1996; 1(1):75–85.

[15] European Commission. *Europe 2020: A Strategy for Smart, Sustainable and Inclusive Growth*: COM(2010) 2020. Brussels; 2010.

[16] European Commission. *Growing the Silver Economy in Europe: Background Paper*. Brussels: European Commission; 2015.

[17] Klimczuk A. *Economic Foundations for Creative Ageing Policy, Volume I: Context and Considerations*. New York, Basingstoke: Palgrave Macmillan; 2015.

[18] Batty M., Axhausen K.W., Giannotti F., *et al*. 'Smart cities of the future'. *European Physical Journal-Special Topics* 2012; 214(1):481–518.

[19] Cocchia A. 'Smart and digital city: A systematic literature review' in Dameri R.P., Rosenthal-Sabroux C. (eds.). *Smart City*. Cham: Springer; 2014. pp. 13–43.

[20] Komninos N. 'Intelligent Cities' in Anttiroiko A., Mälkiä M. (eds.). *Encyclopedia of Digital Government*. Hershey PA: Idea Group Reference; 2007. pp. 1100–1104.

[21] European Commission. *Smart Cities and Communities*. Available from http://ec.europa.eu/eip/smartcities/about-partnership/what-is-it/index_en.htm [Accessed 19 Apr 2015].

[22] *The European Innovation Partnership on Smart Cities and Communities*. Available from https://eu-smartcities.eu [Accessed 20 Sep 2015].

[23] Skouby K.E., Kivimäki A., Haukipuro L., Lynggaard P., Windekilde I. 'Smart cities and the ageing population'. *Outlook: Visions and Research Directions for the Wireless World* 2014;(12):1–13.

[24] Steg H., Strese H., Loroff C., Hull J., Schmidt S. *Europe is Facing a Demographic Challenge: Ambient Assisted Living Offers Solutions*. VDI/VDE/IT, Berlin; 2006.

[25] MetLife Mature Market Institute. *The MetLife Report on Aging in Place 2.0: Rethinking Solutions to the Home Care Challenge*. New York: MetLife Mature Market Institute; 2010.

[26] Bendyk E., Kosieliński S., Krupis R., Rutkowski P. *Potencjał Gdańska w zakresie inteligentnego miasta 2012* [*Gdańsk Potential in the Field of Smart City in 2012*]. Gdańsk: Instytut Mikromakro; 2012.

[27] Caragliu A., Del Bo C., Nijkamp P. 'Smart cities in Europe'. *Journal of Urban Technology* 2011;18(2):65–82.

[28] Florida R. *The Rise of the Creative Class: and How It's Transforming Work, Leisure, Community and Everyday Life*. New York: Perseus Book Group; 2002.

[29] Bronzini F., Barbini N., Marinelli G., Pantaloni M., Giordano E. 'Active ageing and public space: The creative city 3.0' in Longhi S., Siciliano P., Germani M., Monteriù A. (eds.). *Ambient Assisted Living: Italian Forum 2013*. Cham: Springer; 2014. pp. 361–371.

[30] Perera C., Zaslavsky A., Christen P., Georgakopoulos D. 'Sensing as a service model for smart cities supported by internet of things'. *Transactions on Emerging Telecommunications Technologies* 2014;25(1):81–93.

[31] Zanella A., Bui N., Castellani A., Vangelista L., Zorzi M. 'Internet of things for smart cities'. *Internet of Things Journal, IEEE* 2014;1(1):22–32.

[32] Szarota Z. *Starzenie się i starość w wymiarze instytucjonalnego wsparcia na przykładzie Krakowa* [*Aging and Old Age in the Dimension of Institutional Support on the Example of Krakow*]. Kraków: Wyd. Naukowe Uniwersytetu Pedagogicznego; Uniwersytet Pedagogiczny im. Komisji Edukacji Narodowej w Krakowie; 2010.

[33] Warsaw City Council. *Programme Warszawa przyjazna seniorom na lata 2013–2020* [*Warsaw Senior Friendly Programme for 2013–2020*]. Warszawa: Rada m.St. Warszawy; 2013.

[34] Gassmann O., Reepmeyer G. 'Universal design – Innovations for all ages' in Kohlbacher F., Herstatt C. (eds.). *The Silver Market Phenomenon: Business Opportunities in an Era of Demographic Change*. Berlin, Heidelberg: Springer; 2008. pp. 125–40 .

[35] Handler S. *An Alternative Age-friendly Handbook*. Manchester: University of Manchester Library; 2014.

[36] O'Hehir J. *Age-Friendly Cities and Communities: A Literature Review*. Adelaide, SA: University of South Australia; 2014.

[37] Phillipson C. 'Developing age-friendly communities: New approaches to growing old in urban environments' in Settersten R.A, Angel J.L. (eds.). *Handbook of Sociology of Aging*. New York, Dordrecht, Heidelberg, London: Springer; 2011. pp. 279–293.

[38] AGE Platform Europe. *Towards Smart, Sustainable and Inclusive Places for All Ages*. Brussels: AGE Platform Europe; 2010.

[39] Lui C., Everingham J., Warburton J., Cuthill M., Bartlett H. 'What makes a community age-friendly: A review of international literature'. *Australasian Journal on Ageing* 2009;28(3):116–121.

[40] WHO. *Age-friendly World*; 2015. Available from http://agefriendlyworld.org [Accessed 1 Aug 2015].

[41] WHO. *Measuring the Age-friendliness of Cities: A Guide to Using Core Indicators*. Kobe: WHO; 2015.

[42] Liang H., Liang Y., Jian S. *Building a Smart, Age-Friendly Community, Home Based Elder Care Information Platform*: Intel Corporation; 2013. Available from www.intel.com/content/dam/www/public/us/en/documents/case-studies/building-smart-age-friendly-community-exec-summary.pdf [Accessed 6 Apr 2015].

[43] Buffel T., Phillipson C. 'Ageing in urban environments: Developing 'age-friendly' cities'. *Critical Social Policy* 2012;32(4):597–617.

[44] China Academy of Information and Communications Technology. 'EU-China Policy Dialogues Support Facility II' (eds.). *Comparative Study of Smart Cities in Europe and China 2014*. Berlin, Heidelberg: Springer; 2016.

[45] Cohen B. *The Smartest Cities in The World*; 2014. Available from www.fastcoexist.com/3038765/fast-cities/the-smartest-cities-in-the-world [Accessed 20 Apr 2015].

[46] Deakin M. *Smart Cities: Governing, Modelling and Analysing the Transition*. London: Routledge; 2014.

[47] European Commission. *A Compilation of Good Practises: Innovation for Age-friendly Buildings, Cities and Environments*. Brussels: European Commission; 2013.

[48] Fitzgerald K.G., Caro F.G. 'An overview of age-friendly cities and communities around the world'. *Journal of Aging & Social Policy* 2014;26(1–2):1–18.

[49] Geissbühler A., Demongeot J., Mokhtari M., Abdulrazak B., Aloulou H. (eds.). *Inclusive Smart Cities and e-Health: 13th International Conference on Smart Homes and Health Telematics*. Cham: Springer; 2015.

[50] Green G. 'Age-friendly Cities of Europe'. *Journal of Urban Health* 2013;90(1):116–128.

[51] Riva Sanseverino E., Riva Sanseverino R., Vaccaro V., Zizzo G. (eds.). *Smart Rules for Smart Cities: Managing Efficient Cities in Euro-Mediterranean Countries*. Cham: Springer; 2014.

[52] Tamai H. 'Fujitsu's approach to smart cities'. *Fujitsu Scientific & Technical Journal* 2014;50(2):3–10.

[53] Thite M. 'Smart cities: Implications of urban planning for human resource development'. *Human Resource Development International* 2011;14(5): 623–631.

[54] *Top7 Intelligent Communities of the Year*. Available from www.intelligentcommunity.org/index.php?src=gendocs&ref=Top7&category=Events [Accessed 10 Apr 2015].

[55] Townsend A.M. *Smart Cities: Big Data, Civic Hackers, and the Quest for a New Utopia*. New York, London: W.W. Norton & Company; 2013.

[56] Vinod Kumar, T.M. (ed.). *E-Governance for Smart Cities*. Singapore: Springer; 2015.

[57] Błędowski P., Szatur-Jaworska B., Szweda-Lewandowska Z., Kubicki P. *Raport na temat sytuacji osób starszych w Polsce* [*Report on the Situation of Older People in Poland*]. Warszawa: Instytut Pracy i Spraw Socjalnych; 2012.

[58] Corsi M., Samek Lodovici M., Botti F, D'Ippoliti C. *Active Ageing and Gender Equality Policies: The Employment and Social Inclusion of Women and Men of Late Working and Early Retirement Age*. Luxembourg: Office for Official Publications of the European Communities; 2012.

[59] Hayashi M. Japan's Fureai Kippu Time-banking in Elderly Care: Origins, Development, Challenges and Impact'. *International Journal of Community Currency Research* 2012;16:30–44.

[60] *European Smart Cities*. Available from www.smart-cities.eu [Accessed 20 Apr 2015].

[61] Zaidi A., Gasior K., Hofmarcher M.M., *et al. Active Ageing Index 2012: Concept, Methodology and Final Results: Methodology Report Submitted to European Commission's DG Employment, Social Affairs and Inclusion, and to Population Unit*, UNECE, for the project: 'Active Ageing Index (AAI). Geneva: UNECE; 2013.

[62] Esping-Andersen G. *The Three Worlds of Welfare Capitalism*. Princeton, NJ: Princeton University Press; 1990.

[63] Kazepov Y. 'Rescaling social policies towards multilevel governance in Europe: Some reflections on processes at stake and actors involved' in Kazepov Y. (ed.). *Rescaling Social Policies: Towards Multilevel Governance in Europe*. Farnham, Burlington: Ashgate; 2010. pp. 35–72.

[64] Zaidi A. *Policy Brief: AAI 2014: Active Ageing Index for 28 European Union Countries*. Available from www1.unece.org/stat/platform/download/attachments/76287849/Policy_Brief_AAI_2014_November2014_FINAL.PDF [Accessed 14 Jan 2015].

[65] Kirejczyk K, Brzeski W, Kozłowski E. *Budownictwo senioralne w Polsce. Perspektywy rozwoju* [*Construction Sector for Older Adults in Poland. Prospects for Development*]. Warszawa: REAS; 2015.

[66] *DSP Hagibor*; 2015. Available from www.dsphagibor.cz/cs/ [Accessed 1 Sep 2015].

[67] *Burlington Age Friendly Seniors Council*. Available from: http://burlington agefriendly.ca [Accessed 15 Jan 2016].

[68] *Portal komunalny: Inteligentne miasta* [*Communal Portal: Smart Cities*]; 2015. Available from http://portalkomunalny.pl/category/aktualnosci/inteligentne-miasta/ [Accessed 1 Sep 2015].

[69] *Help Build Hong Kong into an Age-Friendly City Project*. Available from: www.ioa.cuhk.edu.hk/en-GB/community-outreach/age-friendly-city-project [Accessed 15 Jan 2016].

[70] Starzyk K., Wiśnicka M. *ZOOM na rady seniorów. Diagnoza funkcjonowania* [*ZOOM on the Advice of Seniors: Diagnosis of Functioning*]. Warszawa: Towarzystwo Inicjatyw Twórczych "ę"; 2014.

[71] Kwiecień K. *Seniorzy mają swoją reprezentację w samorządzie* [*Seniors Are Represented in Local Government*]: Oswiecimskie24.pl; 2012. Available from https://oswiecimskie24.pl/newsy,7568-powiat-oswiecim--seniorzy-maja-swoja-reprezentacje-w-samorzadzie [Accessed 17 Oct 2015].

[72] WHO. *World Report on Ageing and Health*. Geneva: WHO; 2015.

[73] Széman Z. 'Elderly-friendly housing model: Results of an action research' in Széman Z. (ed.). *Challenges of Ageing Societies in the Visegrad Countries*. Budapest: Hungarian Charity Service of the Order of Malta; 2013. pp. 130–137.

[74] Fiedler M., Fenton B., Decker B., Rupprecht S. *Staying Mobile: a Guide to Mobility Management in Ageing Societies*. Szentendre: The Regional Environmental Center for Central and Eastern Europe; 2011.

[75] Aspinall P., Bell S., Thompson C.W. (eds.). 'Open Space: People Space 3'. *An International Conference on Research into Inclusive Outdoor Environments for All*. Edinburgh: OPENspace Research Centre; 2011.

[76] *Karta Aktywnego Seniora* [*Active Senior's Card*]; 2015. Available from www.seniorzy.bialystok.pl [Accessed 1 Sep 2015].

[77] Pecuch M. *O službu Senior expres je zájem* [*About the Senior Express Service*]; 2015. Available from www.plzen.eu/obcan/aktuality/z-mesta/o-sluzbu-senior-expres-je-zajem.aspx [Accessed 1 Sep 2015].

[78] Tomczyk Ł. *Vzdělávání seniorů v oblasti nových médií* [*Senior Education in New Media*]. Praha: Asociace Institucí Vzdělávání Dospělých ČR; 2015.

[79] Klimczuk A. 'Kierunki rozwoju uniwersytetów trzeciego wieku w Polsce [Development Directions of the Third Age Universities in Poland]'. *E-mentor* 2013;(51):72–77.

[80] Kocianová R. *Analysis and Comparison of Forms and Methods for the Education of Older Adults in the V4 Countries*. Praha: Filozofická fakulta Univerzity Karlovy; 2013.

[81] MOPS. *Bank Czasu Seniora* [*Seniors' Time Bank*]. Available from www.mops.czestochowa.pl/mops/projekty-i-programmey/bank-czasu-seniora/ [Accessed 14 Sep 2015].

[82] *AgeWell Global*. Available from: www.agewellglobal.com [Accessed 15 Jan 2016].

[83] Tomczyk Ł., Klimczuk A. *Aging in the Social Space*. Kraków, Białystok: The Association of Social Gerontologists; 2015.

Chapter 18

Towards accessible ambient-assisted living environments

Constantine Stephanidis[1,2]

18.1 Introduction

There is overwhelming evidence that the population of the developed world is aging [1]. The number of older persons was 841 million in 2013, which is four times higher than the 202 million that lived in 1950. Life expectancy at older age is also increasing. The older population will almost triple by 2050, when it is expected to surpass the two billion mark. In the next 40 years, life expectancy at age 60 is projected to increase by two years on average, from 20 years to 22 years for the world as a whole from 19 years to 21 years in the less developed regions, and from 17 years to 19 years in the least developed countries. During the same period, life expectancy at age 60 in the more developed regions is expected to rise from 23 years to 26 years.

As a consequence of the above, older people are increasingly becoming the dominant group of customers of a variety of products and services (both in terms of number and buying power). Recent advances in information and communication technologies (ICT) have great potential for meeting the needs of older people and help them stay healthier, live independently for longer, counteract reduced capabilities due to age, and remain active for longer.

Ambient-assisted living (AAL) refers to intelligent systems of assistance for a better, healthier, and safer life in the preferred living environment and covers concepts, products, and services that interlink and improve new technologies and the social environment, with a focus on older people [2].

The roots of AAL are in traditional assistive technologies for people with disabilities, universal design approaches to accessibility, usability, and ultimately acceptability of interactive technologies, as well as in the emerging computing paradigm of ambient intelligence (AmI), which offers the new possibility of providing intelligent, unobtrusive, and ubiquitous forms of assistance to older people and to citizens in general [3].

[1]Foundation for Research and Technology – Hellas (FORTH) – Institute of Computer Science, N. Plastira 100, Vassilika Vouton, GR-700 13 Heraklion, Crete, Greece
[2]University of Crete, Department of Computer Science, Heraklion, Crete, Greece

Enabling technologies for AAL include sensing, reasoning, actuation and control, communication, and interaction. The latter is particularly relevant in the perspective of making ICT for aging accessible and usable for the largest possible population and taking into account the requirements of older users.

AAL relies heavily on AmI to provide seamless and unobtrusive (i.e., natural) interaction in the human environment, thus moving away from more traditional assistive technologies towards universal access. While a wide variety of different technologies is available in the environment, the underlying goal is to hide their presence from users or to smoothly integrate them within the surrounding context as augmented physical objects, rather than technological gadgets. The pervasiveness of technology fosters the elaboration of new interaction concepts beyond current user interfaces (e.g., the desktop metaphor), through embedding interaction in everyday objects and seamlessly integrating the physical and digital worlds.

However, the benefits of AAL environments can only be fully achieved and availed if such technologies can guarantee seamless accessibility for diverse functional limitations due to age. Changes in abilities due to aging affect the way older people use ICTs, thus preventing their acceptance. Requirements are imposed not only by the physical characteristics of this section of the population but also by their unfamiliarity with technology.

Accessibility in the context of individual applications and services has been defined as follows: for each task a user has to accomplish through an interactive system, and taking into account specific functional limitations and abilities, as well as other relevant contextual factors, there is a sequence of input and output actions which leads to successful task accomplishment [4]. However, the accessibility of AAL systems and environments poses different problems and is more complex than currently available approaches to the accessibility of desktop or web applications and services, as such environments do not simply introduce a new technology, but an integrated set of technologies.

The variety of applications and the diversity of target devices make the situation even more complex. Different levels of accessibility may be distinguished. A first level concerns the accessibility of individual devices. Such devices need to be accessible in the first place to their owners according to their needs, but basic accessibility features should be provided also for other users with potentially different needs. A second level concerns the accessibility of the environment that should enable an equivalent access to content and functions for users with diverse characteristics, not necessarily through the same devices, but through a set of interaction options integrated in the environment in a dynamic configuration/ensemble.

Therefore, AAL environments cannot be made accessible through common methods such as the use of conventional assistive technologies or the development of ad-hoc adaptations. In this respect, multimodality and the availability of alternative means of interaction is a key feature towards facilitating the provisioning of smart environments that will be accessible by users with functional limitations and varying abilities and preferences [5].

Interaction adaptation and personalization, which are intrinsic features of intelligent environments, are also anticipated to play a crucial role towards ensuring accessibility in the context of AAL. The notion of automatic adaptation reflects the capability of software to adapt during runtime based on the individual end-user, as well as the particular context of use, by delivering a most appropriate user interface instance [4].

So far, no systematic and comprehensive approach is available to inclusive accessibility in AAL environments for a wide variety of functional limitations brought about by age or disabilities. Towards elaborating such an approach, it becomes particularly important to investigate new accessibility interaction techniques, designed for all, beyond the spectrum of the conventional computational devices, and elaborate an inclusive interaction model, addressing the needs of everyday living in smart environments, encompassing both emerging and standardized techniques. This chapter proposes some basic steps towards a new dimension of accessible user interaction in AAL environments for their older and disabled inhabitants. The main idea behind the proposed approach is to make easily available, usable and "pluggable" in the intelligent environment mature and emerging assistive technology solutions as well as alternative multimodal interaction techniques.

Against this background, this chapter presents a scenario of an envisaged accessible AAL environment, and subsequently discusses the emerging requirements, the current state of the art in the related domains, as well as the emerging research and development challenges which need to be addressed towards making the scenario a reality.

18.2 An accessible AAL environment scenario

Jenny is a 71-year-old retired lady who lives in an apartment in a recently constructed block and needs some assistance in her everyday activities. Jenny has vision and hearing decrease due to age and some motion restrictions due to arthritis. The apartment where Jenny lives, as part of an innovative smart housing project, is equipped with an accessible AAL system.

Jenny has just finished her lunch and decides to sit in her favorite living room corner and have a look at some photos. She sits on an armchair and indicates that she wishes to interact with her smart television, through an appropriate speech command. The environment informs her that the selected device is the TV. Jenny starts navigating the television multimedia menu, in order to open the photo album. Navigation in the various menus and submenus is available through specific "next" and "previous" gestures. Once the photo album application is displayed, she starts looking at the pictures using the "next" and "previous" gestures (e.g., moving a hand to the right or to the left), while through a "zoom-in" gesture (e.g., carrying out an embracing gesture) she can focus on specific parts of a picture. When she first used this gesture interaction feature, she thought it would be a funny thing to do; after a few days however, this interaction modality became her favorite, since it turned out that it is quite easier for her than using touch screens, buttons, and remote controls. What is more is that there is a small set of simple gestures, very easy to remember, which she can execute easily and through which she can interact with all the smart devices of her home. Furthermore, although the gestures' set is small in order to be easily remembered, gestures themselves are personalized according to her personal dexterity and strength.

Her smart wearable assistant, a special wrist clock able to monitor her physiological signs, realizes that Jenny is a little bit stressed, so it sends a message to the smart home to dim the lights and starts playing her favorite relaxing music, while the smart television starts displaying photos of happy moments of her and her family. Launching family photos and playing this type of music has been indicated by Jenny herself as a preferred way for relaxing. A 3D virtual avatar is also displayed between the photos and talks to her. The volume of the voice is adjusted according to her hearing abilities, but also the music playing in the room. Also, the smart wearable assistant informs her that it is time to take her blood pressure pill prescribed by her doctor. The message is spoken but also displayed on the television, including also the correct dosage.

Jenny takes the medicine. When the system tells her that her pressure is back to normal values she decides to go check her mail. As soon as she stands up, the lights are turned on. She sits at her smart secretaire and opens her letters. The first letter she opens is from her bank. Unfortunately she cannot read it, since it is written in very small letters. However, this is not a problem in her smart home. She uses the reading assistant of her secretaire, by placing the letter on the predefined desk area. Right away the assistant announces the letter title and asks her if she would like to have it fully described. Jenny says "Yes, please" and the requested information is presented to her visually in large fonts and high color contrast as well as through speech.

Few months ago, Jenny had actively participated with interest and curiosity to the process of designing and fine tuning her personal AAL environment. Although at the beginning she was a little skeptical, she was happy to be invited to visit the local AAL environment simulator, where she could try different applications and interaction solutions, provide her feedback, select the ones she wanted, easily provide her data to configure the system to be installed in her apartment, and practice using her new technological environment. It was a memorable experience.

The above scenario, depicted in Figure 18.1, illustrates the potential of providing accessibility to older people in a seamless way throughout different essential activities such as health monitoring and taking medications, reading, as well as leisure and relax. The scenario combines speech and gesture interaction as well as a mix of interconnected ambient and wearable devices, and illustrates the role of the user modeling and interaction ontologies in fine-tuning the available modalities to the user's abilities, as well as the capability of the environment to provide to the user accessible multimodal information on various devices, and even making nonaccessible print information accessible.

Although many of the technologies described in the scenario, for example, voice and gesture interaction, are currently sufficiently mature and available on the market, many research challenges are still open towards making the above scenario a reality. This chapter discusses the main arising challenges, and namely:

- Section 18.3 discusses the need to pursue, adopt and evolve deeply human-centered approaches to the development of AAL environments, allowing for direct and active user involvement in all phases of the process from user requirements collection and analysis to deployment.

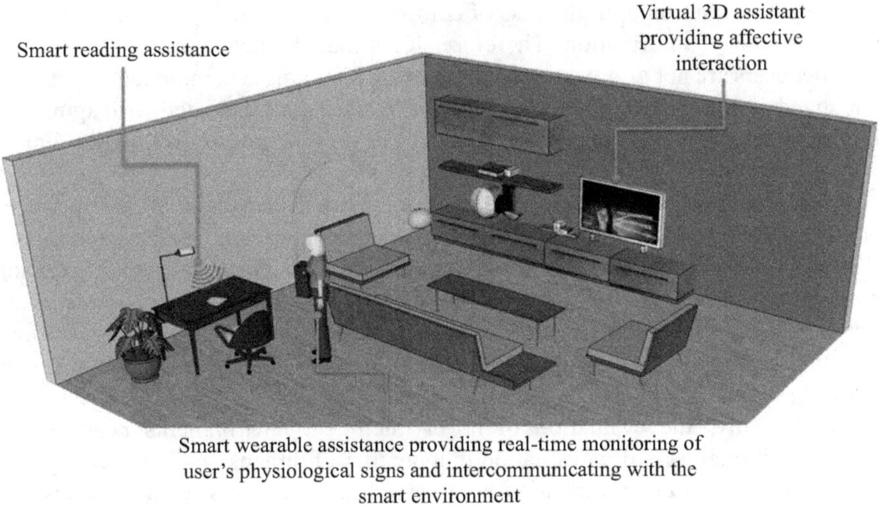

Figure 18.1 A scenario illustrating an accessible AAL environment for older users

- Section 18.4 elaborates on approaches to model knowledge of user requirements and of the appropriateness of different interaction solutions for different combinations of user characteristics/functional limitations and environment characteristics/functions towards ensuring accessibility in AAL environments.
- Section 18.5 discusses the need to seamlessly integrate a variety of accessibility solutions supporting alternative interaction techniques for various combinations of user abilities/functional limitations, addressing the needs of everyday living in smart environments, and encompassing both emerging and standardized techniques, for example, voice, touch, gestures and head recognition, and eye-tracking.
- Section 18.6 addresses the need to develop an architecture supporting user-centric interaction mode, addressing the personalized needs of potential individual users of smart environments, and providing targeted smart interaction solutions in the user's environment.

18.3 Human-centered design of accessible AAL environments

Designing technology applications to support older people in their homes is becoming more and more necessary as the developed world is aging. Although older people are not generally considered to have disabilities, the natural aging process carries some degenerative ability changes, which can include diminished vision, varying degrees of hearing loss, psychomotor impairments, as well as reduced attention, memory, and learning abilities. All of these changes affect the way older people use ICT, which ultimately also contribute to the loss of confidence and in difficulties in orientation and absorption of information. Therefore, ICTs must be accommodated to ensure that older users are not disadvantaged when using ICT. This accommodation can only be realized after a thorough understanding of the changes associated with aging and their impact on the needs of older people concerning the interaction with technical systems.

Principles for providing accessibility to older users in conventional user interfaces include improved contrast, enlargement of information presented on the screen, careful organization of information, choice of appropriate input devices, avoiding relying on memory, and design simplicity [6]. In the context of AAL, user requirements are likely to be more subjective, complex, and interrelated than in previous generations of technology [5], constituting thus necessary the adoption of human-centered design approaches.

Additionally, the evaluation of novel intelligent environments needs to go beyond traditional usability evaluation in a number of dimensions, concerning both the qualities of the environment to be assessed and the assessment methods [5]. Performance-based approaches are not adequate for such environments, as they were developed for single user, desktop applications, and are usually applied in laboratory evaluations. In addition, laboratory-based evaluations are inappropriate for AAL environments, since it is difficult to specify tasks that capture the complexity of everyday activities, while a more subjective view of the user experience is necessary. Qualities which may need to be taken into account in evaluating AmI technologies and environments include highly subjective cognitive and emotional factors. Therefore, evaluation should target the overall user experience and the emotional response to technology [7] rather than traditional usability. However, the concept of user experience needs to be further articulated in order to derive measurable qualities.

In particular, AAL requests a high willingness of users to accept new forms of assistance and to rely on technology in very sensitive activities. Also, user diversity is a serious issue, as increasingly more, and especially older users will have to master technical devices in the near future. Research has shown that the sole consideration of ease of use and the perceived usefulness is not sufficient, and there is a need for integrating additional aspects like trust in technology, privacy protection, system reliability, and fun [8]. Accessibility is another fundamental dimension of user experience in intelligent environments [5], and a necessity for AAL environments. Currently, there are two major options for evaluating accessibility in AAL environments, namely, user-based evaluation and evaluation through simulation tools.

AAL challenges traditional user-based evaluation methods, because the context of use can be difficult to recreate in a laboratory setting, and also because longer-term experiments are required to assess the impact of the environment on everyday life activities. This suggests that the evaluation of the user experience should take place in real-world contexts. However, evaluation in real settings also presents difficulties, as there are limited possibilities of observing and monitoring users and their activities. The emergence of research facilities where AAL environments can be tested under real-life conditions such as, for example, Georgia Tech's Aware

Home[1], MIT's House_n[2], Fraunhofer-Gesellschaft's inHaus[3], and the ICS-FORTH Ambient Intelligence Research Facility[4], represents a solution to the above problem, providing adequate settings for prototyping, validating and refining AAL environments and applications with the involvement of users of various ages with diverse functional limitations, degrees of expertise, and attitudes to technology. Such facilities offer an appropriate experimental environment, combining user experience in context with the availability of the necessary technical infrastructure for monitoring and studying the users' behavior over extended periods of time [9, 10]. However, methods and tools to assess accessibility of AAL environments still present many open research questions.

Finally, simulation software based on user and environment models can be very helpful for the initial accessibility assessment of AAL solutions. In Reference 11, a framework is described for automatic simulated accessibility and ergonomic testing of virtual prototypes of products using virtual user models. The virtual user modeling framework describes virtual humans focusing on the elderly and people with disabilities. Geometric, kinematic, physical, behavioral, and cognitive aspects of the user are addressed in order to represent people with various functional limitations. Furthermore, hierarchical task and interaction models are introduced, in order to describe the user's capabilities at multiple levels of abstraction. The use of alternative ways of task execution, exploiting different modalities and assistive devices, is supported by the proposed task analysis. The framework has been integrated in simulation environments and has been tested in the evaluation of various designs.

As described in the scenario of Section 18.2, a combination of these approaches in the design of AAL environments and applications can have twofold benefits: on the one hand, it allows target users to come to direct contact with AAL technologies and familiarize with them, while on the other hand it allows collecting a wealth of data about user's behavior and feedback in AAL environments.

18.4 User and interaction models in AAL

Towards ensuring the accessibility of AAL environments, it is of vital importance to capture user requirements into models capable of supporting the appropriate matching of user requirements with accessibility solutions, as well appropriate adaptations of the selected interaction modalities.

Research efforts on user modeling have focused on models that support intelligent environments to capture and represent information about users and contexts so as to enable the environment to adapt to both [12]. The most common approaches to user modeling for pervasive computing and smart environments are ontology-based. For example, Heckmann *et al.* in Reference 13 propose a general user model ontology

[1] http://www.awarehome.gatech.edu/, last accessed: March 2016
[2] http://web.mit.edu/cron/group/house_n/, last accessed: March 2016
[3] http://www.inhaus.fraunhofer.de/en.html, last accessed: March 2016
[4] http://www.ics.forth.gr/ami/about, last accessed: March 2016

for the uniform interpretation of distributed user models in intelligent semantic web-enriched environments. Basic user dimensions represented in the ontology include ability and proficiency, personality, emotional state, physiological state, mental state, nutrition, and facial expression. De Carolis *et al.* in Reference 14 describe a multiagent architecture in which users and environments are represented by agents that negotiate task execution and generate results according to user in context features. Two interacting entities are involved: an agent representing the user and an agent representing physical and logical place in which various services are available. User modeling includes four main sections: IDENTITY (with identification data such as the user name, sex, id, password, and email), MIND (background knowledge, interests, and know-how), BODY (disabilities or preferences in using a body part during interaction), and PERSONALITY (personality traits and habits).

Some user models have also been developed which focus on human functional limitations. For example, Casas *et al.* [15] discuss a model of older or disabled users leading to the creation of personas for design purposes, while Kaklanis *et al.* [11] developed an open library of various categories of virtual user models, including VR models, covering a wide range of population groups and especially focusing on groups in risk of exclusion, for example, older people, people with functional limitations (vision, hearing, speech, motor), targeted to support automatic simulated accessibility and ergonomic testing of virtual prototypes of products.

In the context of AAL, Castillejo *et al.* in Reference 16 review the existing models and identify some challenges to be addressed towards dynamic interaction adaptations in AAL. In particular, they discuss the need to model physiological parameters in order to obtain a very accurate model of individual abilities. Another important issue concerns the quality and reliability of data, which can be achieved by using an ontology and a reasoning engine capable of working with ambiguity and uncertain data. Finally, domain dependency of the solution is a relevant problem, which hinders the reusability of solutions in different domains. Ontology-based modeling is also helpful in this respect.

A user model for the adaptation of AAL services provided through Next-Generation Networks is proposed in Reference 17. The adopted ontology extends the approach of Reference 13 to include user preferences for various aspects of everyday activities, as well as user abilities and demographics. User's state is also modeled, including emotional state, location, mental state, mood, motion, and physical environment.

Finally, research work has also been conducted towards systematically modeling interaction in AAL. Petzold *et al.* developed a domain-specific taxonomy for the design of user interaction in AAL systems [18]. The taxonomy includes concepts such as number of available devices, obtrusiveness, interactivity, intention, user experience, adaptability, and modality.

A suitable approach to user modeling towards accessible AAL environments should be based on a generic ontology framework for representing an object model of abstract, language-independent specification of knowledge, using ontology-based scheme. This ontology-based framework can provide a simplified and highly abstract model of ontology which is independent of a specific ontology representation language and operates with ontologies on a conceptual, rather than syntactic level. This approach allows for applications to operate with ontologies described in diverse language forms and gives significant advantages in simplicity of software development. Based on such framework, specific ontology schemes for user profiles and interaction solutions need to be developed. The user-profile ontology should encompass the necessary concepts and relations to define the main characteristics, abilities and preferences of users, facilitating the provision of highly personalized services in the context of accessible AAL environments. The interaction ontology should represent the features, the functionality, and the capabilities supported by various interactive solutions, while the knowledge captured by this ontology should refer to hardware and software capabilities, as well as device-supported interaction modalities. Such knowledge needs to be encapsulated in an inclusive ontology-based interaction model, i.e., a model which will match user profiles, application tasks, and interaction techniques, addressing the needs of everyday living in smart environments, encompassing both emerging and standardized interaction techniques. The models will provide the infrastructure for the implementation and deployment of personalized smart assistive solutions/services that will be able to transform everyday living spaces into smart accessible environments supporting disabled and older persons.

The user-profile and interaction ontologies constitute an essential requirement of a context-aware system allowing context information to be exchanged and used between different entities such as users, devices and services in a same semantic understanding, supporting, among others, semantic interoperability which enables the common schemas to be shared between different entities.

An additional issue concerns the collection of individual information for filling in the user's profile. As depicted in the scenario of Section 18.2, a mix of methods can be adopted, including gathering data from the user herself, from her carers, as well as from environmental monitoring.

18.5 Accessibility solutions

Multimodality and the availability of alternative means of interaction is a key feature towards facilitating the provisioning of a personalizable smart environment that will be accessible by users with functional limitations and varying abilities and preferences. Different modalities and interaction techniques can be used concurrently, to increase the quantity of information made available or present the same information in different contexts, or redundantly, to address different interaction channels, both to reinforce a particular piece of information or to cater for the different abilities of users.

In the context of accessible AAL environments, the assistive technologies that are currently used to make interactive applications and services accessible to disabled users will continue to constitute the basis for users' interaction, but they will need to be extended and enriched in terms of functionality and use in order to provide multimodal and personalized accessible interaction beyond conventional ICT environments. Different modalities can be selected and combined based on the user and interaction models discussed in Section 18.3.

Personalized assistive solutions are required to provide alternative input, output, and information-rendering modalities capable of addressing each individual's needs. Such solutions can be designed and developed building upon state-of-the-art mainstream interaction techniques, including, for example, user interface visual adaptation, voice-based interaction, scanning-based interaction, touch and gesture interaction, haptic feedback, eye-tracking technologies, and other techniques. An innovative aspect relates to the seamless integration of different applications and devices in the environment. Furthermore, it is necessary to provide harmonized interoperability of assistive technologies and conventional computational devices (PCs, smart phones, etc.) with innovative high-end technology artifacts in the context of accessible assistive environments. The following sections discuss the state of the art regarding the most prevalent accessibility technologies and assistive solutions aimed at addressing the needs of older and disabled people in AAL environments.

18.5.1 Visual layout adaptations

Methods and techniques for user interface adaptation meet significant success in modern interfaces. Some popular examples include the desktop adaptations in modern operating systems (e.g., Microsoft Windows, MAC OS, Linux distributions), offering, for example, the ability to hide or delete unused desktop items, to add helpful animations, transparent glass menu bars, live thumbnail previews of open programs and desktop gadgets (like clocks, calendars, weather forecast, etc.). However, adaptations integrated into commercial systems need to be set manually, and mainly focus on aesthetic preferences. In terms of accessibility and usability to people with disabilities or older people, only a limited number of adaptations are available.

Research efforts in the past decades have elaborated comprehensive and systematic approaches to user interface adaptations in the context of universal access and design for all [4]. Such approaches were conceived and applied as a vehicle to efficiently and effectively ensure, through an adaptation-based approach, the accessibility and usability of UIs to users with diverse characteristics, supporting also technological platform independence, metaphor independence and user-profile independence. In such a context, automatic UI adaptation seeks to minimize the need for a posteriori adaptations and deliver products that can be adapted for use by the widest possible end-user population (adaptable user interfaces). This implies the provision of alternative interface manifestations depending on the abilities, requirements, and preferences of the target user groups, as well as the characteristics of the context of use (e.g., technological platform, physical environment). The main objective is to ensure that each end-user is provided with the most appropriate interactive experience at run time.

UI adaptations for older users are addressed in various research efforts. For example, Leonidis *et al.* present a development library of self-adapting interactive components specifically designed for older users, and a related prototyping tool allowing to view the resulting adaptations at design time [19]. This approach has been adopted in the development of games [20] and mobile health applications for older users [21].

From a user requirement point of view, defining adaptations of older users is not straightforward, as research is still needed on how to optimally employ adaptations in order to improve interaction for older users. Adapted user interfaces for older users are discussed in the context of interactive TV in Reference 22 and in the context of ubiquitous services in Reference 23.

In AAL, visual layout adaptations are extended to address adaptation in a multimodal context [24, 25].

In the context of a systematic approach to AAL accessibility, visual layout adaptations are fundamental to address diverse levels of visual but also cognitive functioning, as well as the diverse characteristics and sizes of the available screen devices.

18.5.2 Voice interaction

In the accessibility and assistive technologies fields, the speech communication channel is considered as one of the most important modalities to benefit older and disabled persons. Voice interactions allow users to communicate with computers or computer-related devices without the use of a keyboard, mouse, buttons, or other physical interaction devices, leveraging a skill that is mastered early in life, and have the potential to be more natural than other devices such as the keyboard. Additionally, speech is a compelling input alternative when the user's hands are busy with another task (e.g., driving a car, conducting medical procedures) and the traditional keyboard and mouse may be inaccessible or inappropriate.

Various applications of voice technologies (i.e., speech recognition, speech synthesis and spoken dialogue systems) have been reviewed for disabled users in general [26], and more specifically for blind and low vision people [27]. Among them, screen-readers, text-to-speech systems, voice commands for device control, speech-based dictation, automatic audio description, talking devices, automated reading devices, voice browsers, voice portals, voice-based mobility aids, digital talking books are some of the technologies that provide or augment access to printed or electronic information, daily or social activities, as well as private or public facilities for a variety of target user groups, including the elderly and people with visual disabilities, but also in the context of use where visual interaction is not possible (e.g., while driving). Speech input to support universal access is reviewed in Reference 28.

In AAL, voice-based interaction has been proposed for home control [29], reminder and calendar systems [30], and everyday activities assistance [31].

While speech recognition systems have now reached good performances with close talking microphones (e.g., head-set), the main challenge of speech interaction in AAL is that recognition performance decreases significantly in a home environment where the microphone is moved away from the mouth of the speaker (e.g., it is set in the ceiling), and background noise, such as TV, radio, and other devices, interferes with recognition. Research efforts are therefore focusing on how to improve speech recognition in the home environment [32].

In accessible AAL environments, speech interaction for both input and output can be particularly useful in order to interact with applications which are out of reach (e.g., lights) or sight (e.g., a misplaced device). Such modalities need to be fine-tuned and personalized according to the user's hearing functioning and preferred voice characteristics.

18.5.3 Scanning

Scanning is an interaction method addressing the needs of users with hand motor impairments [33]. The main concept of this technique is to eliminate the need for interacting with a computer application through the traditional input devices, i.e., mouse and keyboard. In scanning, all the interactive objects composing a graphical user interface (GUI) are being sequentially focused and highlighted (e.g., by a colored marker), while users can select to interact with the object currently having the focus by activating a switch. Activation switches can vary from hand, finger, foot, tongue, or head switches to breathe-controlled switches or eye-tracking switches. In order to eliminate the need for using a keyboard to type in text, an on-screen keyboard is usually provided.

During scanning, the focus marker scans the interface and highlights interactive objects sequentially, in a predefined order (e.g., from left to right and from top to bottom). Scanning can be either automatic or manual. In the first case, the marker automatically moves from one interface element to the next, after a predefined time interval of user inactivity (i.e., not pressing the activation switch), while the time interval can usually be customized according to user needs. In manual scanning, the user moves the focus marker to the next interface element, whenever she/he wishes, with the use of a switch.

Applications with embedded scanning are developed so as to support scanning in the first place and are accessible to people with motor impairments. Such applications may vary and include, for example, web browsers [34], web browser add-ons [35], and games [36].

Scanning tools enable users to operate the graphical environment of the operating system, thus eliminating the need of using various specialized applications for carrying out everyday tasks. Keyboard and mouse emulation programs with embedded scanning are popular among scanning tools. Tools which scan interface elements based on their screen location or their place in the hierarchy of objects comprising the active window [33], provide switch access to windows-based systems for motor-impaired users.

More recent advancements refer to scanning-based accessibility of mobile devices. In general, scanning accessibility of iOS-based applications relies upon the VoiceOver technology [37], and therefore all the applications which are VoiceOver compatible are also accessible through scanning.

Finally, another domain where the scanning technique is employed is for environmental control systems [38, 39]. Lately, applications with embedded scanning for artifacts of smart environments have also been developed [40]. In the context of AAL, scanning can be employed as a modality for controlling the overall environment in combination with other interaction modalities (e.g., head pose tracking, eye-tracking, voice-based interaction), according to the needs of each user and the current context of use, in order to achieve a more efficient interaction.

18.5.4 Touch

Touch is a very natural and intuitive way for people to interact and, as a result, touch-based interaction has been widely exploited in user interfaces in recent years, especially on mobile devices, evolving as a fundamental interaction technique. Multi-touch is the natural evolution of the single touch interaction, introducing novel challenges and perspectives.

The use of touch interfaces of mobile devices has been extensively researched. Although generally easy to use for older people, touch-based interfaces present typical issues such as small targets, cumbersome gestures and confusing layouts [41]. It has also been found that, while older adults are significantly slower than younger adults in general, the touchscreen reduces this performance gap in comparison to the desktop and mouse [42]. Based on user studies, Culén and Bratteteig propose to include in the design of touch-based user interfaces for the elderly an automatic customization procedure where the system adapts to individual user and their particular way of touching the screen [43].

Lee and Kuo in Reference 44 present a study related to the usability of small touch-screens among young adults, middle-aged adults and older adults. Grammenos *et al.* introduce a touch-based interface for controlling home lights and carry out a user-based evaluation with older and disabled users [40]. The appropriateness of a touch-screen mobile phone as a user interface element allowing older people to choose their meals to be delivered by means of a home care service was evaluated by the authors in Reference 45.

Touch-based interaction has now advanced from tablets and screens to much larger surfaces and walls, using computer-vision methods [46, 47]. Systems employing touch in AAL range from health monitoring [48] and medication management [49], to games [50], and reminder applications [51].

An additional new dimension of touch-based interaction, which is rapidly developing in the context of intelligent environments, is tangible interaction with physical objects. Klompmaker *et al.* in Reference 52 present an approach for touch detection and object interaction which supports multi-touch interaction with arbitrary physical objects. The approach is vision-based and uses images from a depth-sensing camera, while tracking of user's fingers and of the palm of hands, is combined with object interaction such as grasping, grouping, and stacking.

18.5.5 Haptics

Haptic interfaces include hardware and software systems that are able to provide physical signals which can act as adequate haptic stimuli. There is a rich potential for utilizing the haptic channel to complement or substitute regular audiovisual communication: it can be used to produce cues, such as single taps on the skin to grab attention, directional vibration guiding a person in wayfinding, and continuous signals that enhance the sense of presence in remote interpersonal communication. Low-tech aids such as the long cane, Braille and embossed pictures are examples of haptic solutions often used by the visually impaired.

Today, haptic interaction is common in the vibratory tactile interfaces of smartphones and wearable devices such as smart watches. One of main issues in haptic interaction is the limited quantity of information which can be transmitted through contact with the skin, as well as the difficulty in remembering haptic vibration patterns. Kobayashi and Nakano have investigated vibration patterns easy to understand and learn for older people [53]. Haptic interaction has also been investigated in assistive robots for the elderly [54], as well as for movement feedback in physical exercising systems [55].

In the context of AAL, haptic and particularly vibration-based interaction can constitute a useful and unobtrusive secondary modality, if appropriately used in combination with speech, touch or gestures, taking also into account that aging often brings about a decline in sight. For example, vibratory interfaces can provide natural ways of staying in contact with friends and relatives at a distance. However, its use is limited and further research is needed to integrate haptics in AAL environment in meaningful ways.

18.5.6 Gestures

Gestures have become a highly popular interaction technique in many environments. For example, they are widely used in computer games, supported by mainstream devices such Wiimote, Move and Kinect. Wiimote is based on accelerometer and gyroscope, Kinect is based on computer-vision techniques, while Move combines both techniques.

Gesture-based control can be very efficient in proper settings, since it allows natural interaction and, in some cases, hands-free control of technology. From a user perspective, both Wiimote and Move use a handheld device, while Kinect provides a completely hands-free solution. From a content viewpoint, game developers have used gestures in a range of settings. Sport-related games and programs are most popular. Gestures are used, for example, to play golf, table tennis, fight with swords. Another popular subgenre of gesture-controlled or motion-controlled games include fitness applications, which sometimes use additional devices to capture movements as accurately as possible. Gesture control is also used in the computing context, in combination with multi-touch technologies, and independently as well. For example, Kinect has been used already to control well-known media center applications (such as XBMC and Boxee).

Gestures can bring many benefits for users with functional limitations as well. For example, in the context of control of everyday home appliances (such as televisions and all kind of remote controlled devices), a gesture-controlled handheld device can decrease the need of using physical buttons, and make it possible to control appliances without seeing the interface [56].

Gestures can be roughly categorized as natural movements, which are intuitive in the current context, and those requiring learning. In practice, most gesture-controlled applications require learning. Thus, complex gesture sets should be avoided in most cases, unless there are no other ways for control. Instead, rather simple and few gestures can be highly efficient when they are combined with other modalities, such as speech and haptics, especially for older users. Furthermore, the heterogeneity of the elderly user group implies various restrictions when designing and developing gestural interaction modalities for the elder users. Individual persons have different individual needs based on their different age-related health impairments. Such impairments can affect the person's mobility capabilities and cognitive functions, which results in mobility restrictions in different body parts, as well as difficulties in remembering the specified gestures and the optimal way of performing them. These limitations should be taken under consideration when designing the parameterization properties of such gesture recognition engines.

In AAL, gesture-based interaction has been investigated in the context of home mobility [57], while the design of gestures for older users in AAL environments is discussed in Reference 58.

Recent advances in technology allow exploiting not only hand gestures but also the users' head movements as a means for providing information to computer systems. Head movements and head pose estimation have been used in recent research efforts, as indicators of the user's attention, assisting, for example, automotive applications and conversations with embodied agents. Tracking of head and body movements can be achieved either with a wearable device or through vision technologies [59, 60]. In the context of AAL, head pose estimation has been studied so far mainly to achieve human behavior and understanding.

18.5.7 Eye-tracking

Eye-tracking enables communication and computer control by gaze. An eye-tracking device records the eye movements and a computer program maps the gaze points into screen coordinates. The most common way to control technology by gaze is to use direct pointing. For example, to enter text by gaze, the user first looks at the desired letter on an on-screen keyboard. The focused item can then be selected either by using a separate switch, eye blink, frown, or by gaze alone using "dwell time," a protocol that refers to the use of a prolonged gaze over the focused item [61].

Another way to give commands by gaze is to use gaze gestures. The user initiates a command by making a sequence of "eye strokes" in a certain order. Gestures can offer a fast way to give simple commands, since they do not require dwell time. Gaze gestures are especially useful with small screens and in mobile situations involving mobile trackers [62] and they could well be applied into the control of smart house environments.

Until recently, communication aids based on eye tracking technology have been in use by a limited number of users. The technology has mostly been utilized by people for whom gaze-based communication is a necessity, for example, people with locked-in syndrome, or people suffering from degenerative conditions, such as amyotrophic lateral sclerosis. During the last few years, advances in the technology and the appearance of commercial applications have extended the potential user groups remarkably [63].

The improvement in the quality of systems and the increase in user population give great motivation for research on the applications that benefit from eye-tracking technology. The range of possible applications is wide, from gaze-based communication, computer control and entertainment applications to full environmental control or even the control of mobile phones [63].

There are several different ways to exploit gaze control in smart accessible environments. Eye-tracking enables control by gaze alone, or can be combined with other modalities such as speech, where gaze could be used to point at the object of interest and activate it for commands received via speech. Furthermore, the implementations can be indirect, for example, remote control of house electronics via a gaze-operated control panel, or direct, allowing the user to directly gaze the actual real-world objects [64]. Sometimes, simply detecting the presence of the eyes is enough for useful domestic applications. Smart house appliances that are equipped with such eye-aware sensors can detect if a user is attending to them. Thus, the user can address commands directly to the device, for example, by using speech or gestures. Such smart "eyepliances" include, for example, sensitive lights that know when they are talked to, or an attentive television that will automatically pause if the user no longer watches it [65].

18.5.8 Augmented reality

Augmented reality (AR) is defined as a technology which combines real and virtual imagery, is interactive in real time, and registers the virtual imagery with the real world [66]. Displays for viewing the merged virtual and real environments can be classified into the following categories (Azuma *et al.*, 2001): head worn, handheld, and projective. Head-worn displays (HWDs) are mounted on the user's head, providing imagery in front of their eyes. Two types of HWDs exist: optical see-through and video see-through. Optical see-through displays are those that allow the user to see the real world with their natural eyes and which overlay graphics onto the users' view by using a holographic optical element, half-silvered mirror, or similar technology. On the other hand, video see-through displays are those in which the user has a video view of the real world with graphics overlaid on it. In projection-based displays, the desired virtual information is projected directly on the physical objects to be augmented. Finally, handheld displays are a good alternative for AR applications, particularly because they are minimally intrusive, socially acceptable, readily available, and highly mobile.

Tracking techniques can be sensor-based, exploiting sensors such as magnetic, acoustic, inertial, optical and/or mechanical, vision-based, using image-processing methods to calculate the camera pose relative to real-world objects, or hybrid which combines sensors and vision techniques [67]. Interaction techniques with AR can be classified as follows: tangible, in which the physical manipulation of real-world objects can provide a way to interact with the virtual content; collaborative, in which AR can be used to enhance a shared physical workspace and interface for computer-supported collaborative work; and hybrid interfaces which combine a variety of different but complementary interfaces.

Both device-based and projection-based AR applications have been investigated in AAL. An infrastructure for supporting elderly people needs by simple interactions with the environment through augmented-reality using handheld devices such as mobile phones is presented in Reference 68. Authors in Reference 68 propose a general and adaptive model to transform physical information of objects in the environment into a virtual representation through accelerometers and digital compasses. On the other hand, authors in Reference 69 present a design of a user interface on a projection tabletop system for the elderly. The system is targeted to support daily living activities and uses projection technologies to display visual information on physical surfaces in the home environment.

18.5.9 Multimodal interaction in AAL

In today's pursuit of more transparent, flexible, and efficient human–computer interaction, a growing interest in multimodal interface design has emerged [70]. The goals are twofold: to achieve an interaction closer to natural human–human communication, and to increase the robustness of the interaction by using redundant or complementary information.

Multimodal systems are radically different from standard GUIs, largely because of the nature of human communication, and their basic architecture reflects these differences. Whereas input to GUIs is atomic and certain, machine perception of human input such as speech and gesture is uncertain, so any recognition-based system's interpretations are probabilistic. This means that what were formerly basic events in a GUI, such as object selection, now are events that require recognition and are subject to misinterpretation. Second, whereas standard GUIs assume a sequence of discrete events, such as keyboard and mouse clicks, multimodal systems must process two or more continuous input streams that frequently are delivered simultaneously [71].

Designing and developing multimodal application for the elderly user group can be very challenging due to the vast heterogeneity that characterizes it. Different users are facing different difficulties in their everyday lives depending on their level of cognitive and psychosocial functioning decline due to aging. In AAL, Anastasiou *et al.* have integrated speech and gesture modalities for multiple device control [72].

Towards the development of accessible AAL environment, multimodality plays a fundamental role, by integrating logic into the system making it capable of selecting at run-time the most appropriate modality based on the profile of the user and the environmental situation. The work in Reference 31 provides both visual and aural interaction in the context of a night time assistance system. The modality is selected based on the user profile, and as a result the system displays interface content visually, including textual messages, buttons for interaction and pictures, or when auditory interaction has been selected, speech output and speech recognition are used. Taking a step further from modality selection, it is also important to ensure the appropriate parameterization so as to best suit each selected modality to the needs of each specific user. Finally, more modalities can be active concurrently in the environment. This necessitates suitable modality integration approaches by fusing the semantics of the various involved input signals. Marinc *et al.* address modality configuration through a rule-based approach and modality integration through input/output handlers to support different input and output devices and combinations thereof [73].

18.6 AAL Architectures

In AAL environments, where interactions are realized by the confluence of different interconnected computing systems, the organization of the overall system architecture into a well-defined set of distributed software entities is crucial in order to achieve accessibility and multimodality. Centralized approaches where all software entities run on a monolithic computing platform, despite being easier to implement, are neither scalable nor flexible. On the contrary, a distributed approach allows for (a) flexible, dynamic extension of the overall system with novel functionality, (b) enhanced system scalability, by sharing computation demands among different computers, (c) enhanced robustness, by isolating potential failures of individual software entities, and (d) unambiguous and straightforward modularization of the system's architecture.

The services' architecture in smart environments is considered as a collection of interconnected distributed services, i.e., a collection of software entities that run on different machines, and are able to communicate with each other in order to provide to the infrastructure all the required functionality for sensing, drawing conclusions, and responding to the needs of its users. Essential requirements that a smart environment's services architecture should address are heterogeneity integration, synchronous and asynchronous communication, resilience, security and ease of use [74].

Recent research efforts have suggested several architectures for smart environments. For instance, Janse *et al.* developed a middleware that dynamically integrates heterogeneous systems to achieve interoperability between services and devices in intelligent home networks [75]. The architecture consists of a base middleware component, a programming and deployment framework and a series of legacy services named "Intelligent User Services." On the other hand, Eisenhauer *et al.* in Reference 76 present a middleware which aims to support mostly mobile devices and provide a common layer over different communication technologies (e.g., Bluetooth, Wireless LAN, GSM, GPRS). In addition to the low-level communication facilities, the framework offers higher level methods that on the one hand provide localization and context data to applications and on the other hand perform specialized data filtering depending on personalization information. Finally, the framework also provides a content delivery application that can display XHTML data according to user- and context-specific preferences. A communication middleware specifically designed for the requirements of pervasive computing is proposed in Reference 77. To address the high degree of heterogeneity of platforms and networking technologies, it is based on a layered design and provides a common set of APIs for different programming languages (Java, C++, Python) on a wide range of different devices. Honkola *et al.* developed an interoperability platform (IOP) for smart environments that follows the Smart-M3 concept and supports multidomain, multiplatform, and multivendor software development [78]. Finally, Hanke *et al.*, based on the input from related projects, focus on interoperability and standardization to ensure a broad range of applicability and to develop an open platform to make it technically feasible and economically viable to develop AAL applications [79]. To this purpose, they propose a layered architecture consisting of a Middleware, a Generic Platform Services, AAL Platform Plug-Ins, and AAL Applications and Services.

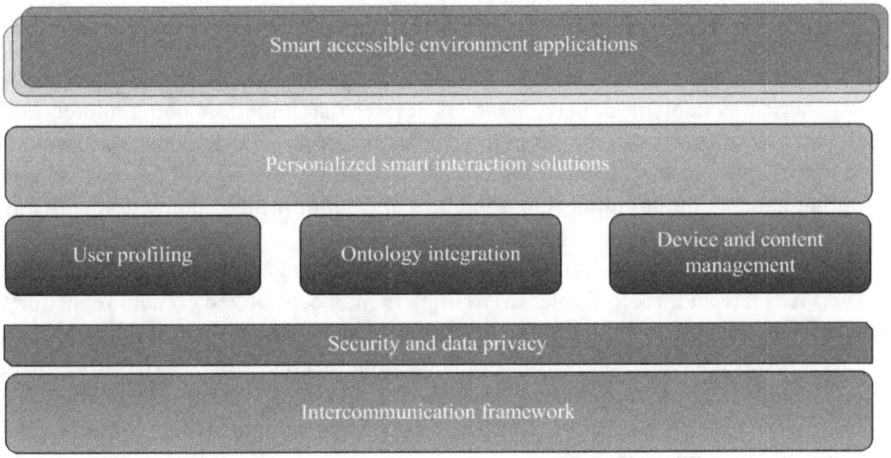

Figure 18.2 An open service architecture for accessible AAL environments

Figure 18.2 illustrates an open service architecture based on seamless connectivity regardless of the communication technology used. An innovative aspect of the proposed architecture is that of supporting the need to interact accessibly with dynamic ensembles of hardware and software that comprise a smart environment, and accommodates smart interaction solutions allowing their intercommunication over different application domains in a transparent way. The intercommunication framework (middleware) dynamically integrates heterogeneous systems over the network in order to achieve interoperability between services and devices in smart environments. Different wireless sensor network technologies are transparently integrated. Standards (such 802.15.4, Bluetooth, Zigbee, and others) can be widely used to build such a hierarchical heterogeneous communication network, extending them when needed. The horizontal layer provided over the intercommunication framework ensures security and users' data privacy and integrity.

Fundamental services for the development of smart accessible applications, which facilitate user profiling, ontology-based user and interaction techniques model, as well as the smart environments' devices and applications' content management, are also incorporated into the framework. The user profiling service facilitates all the necessary functionality needed from the system, in order to keep users' data about their preferences and behavior. This service is fundamental for personalization support, enabling designers and developers to enhance the services and the applications that will be developed with user personalization features. Another fundamental service is the ontology integration service that accommodates an integrated mechanism for web-based intercommunication between the services, tools, and applications. The main aim of this service is to provide in a seamless and transparent way, over the intercommunication framework, the ability to the services and applications to acquire semantic information concerning properties and relations of the users which are using them, the assistive solutions that they include and relevant characteristics of the environment where they are hosted.

The device and content management services provide web services that facilitate a generic mechanism for heterogeneous devices manipulation, both remotely and locally by any interaction solution in the environment. Any device that is accommodated in the smart accessible environment should also include a local device manager agent that handles input/output functionalities and support their interaction with any smart accessible application over the network.

Additionally, web services that facilitate content management are included. These web services have a threefold role: (a) to implement a centralized content repository, providing transparent content access and management by any service, application, and tool, as if it was a local resource (providing caching mechanisms), (b) to encapsulate a content classification mechanism, providing the necessary content-related rationale to data mining procedures, and (c) to encapsulate a sophisticated filtering mechanism for personalized content delivery.

18.7 Conclusions and future work

Although AAL environments have been in the focus of research during the past years and despite the fact that several interaction technologies have reached a maturity stage where they can be exploited for accessibility purposes, the issue of developing truly accessible AAL environments still presents many challenges. In this context, a fundamental move is needed beyond accessibility in human–computer interaction towards accessibility in human–environment interaction. To this end, it is necessary to elaborate systematic approaches providing all the building blocks and tools for the realization of smart environments which exhibit inclusion as an intrinsic feature. This chapter has proposed the following steps towards achieving AAL environments accessible and adaptable by design:

- The elaboration and adoption of deeply human-centered design approaches directly and actively involving target users in all development phases. Such approaches may be facilitated by physical as well as virtual AAL environments simulators.
- The elaboration and adoption of ontology-based user and interaction models, capturing of user requirements and of the appropriateness of different solutions for different combinations of user characteristics/functional limitations and environment characteristics/function.
- The provision and seamless integration of ready-to-use accessibility solutions supporting alternative interaction techniques suitable for various combinations of human abilities and contextual factors in a pervasive, natural, and nonobtrusive fashion.
- The development of a reference architectural model that addresses user needs for inclusive design in smart environments, allowing for the provision of accessible multimodal interaction in a natural way.

In the context of ongoing work, a user model based on the International Classification of Functioning, Disability and Health (ICF[5]) is elaborated, allowing the detailed profiling of inhabitants of AAL environments and the context-sensitive matching of their abilities and skills with a number of interaction modalities. Additionally, a software framework for developing multimodal user interfaces in AAL environments is currently under development including touch- and speech-based interaction, gestures, tangible interaction with smart objects, and AR. The proposed framework supports efficient development, adaptation, and configuration of multimodal user interfaces for a variety of application scenarios involving personal and ambient devices in AAL environments.

Acknowledgments

This work is carried out in the context of the ongoing ICS-FORTH Ambient Intelligence Programme.

References

[1] United Nations, "World Population Aging 2013," United Nations: New York; 2013.

[2] AAL Programme, "Active and Assisted Living Programme: ICT for Aging Well," 2015. [Online]. Available: http://www.aal-europe.eu/about/objectives/ [Last accessed 6 July 2016].

[3] M. Pieper, M. Antona, and U. Cortes, "Editorial: Ambient Assisted Living," *ERCIM News, Special Theme: Ambient Assisted Living*, no. 87, pp. 18–19, 2011.

[4] A. Savidis and C. Stephanidis, "Unified user interface design: Designing universally accessible interactions." *International Journal of Interacting with Computer*, vol. 16, no. 2, pp. 243–270, 2004.

[5] C. Stephanidis, "Human factors in ambient intelligence environments" in G. Salvendy (Ed.), *Handbook of Human Factors and Ergonomics (4th edn.)*. Hoboken, New Jersey, United States: John Wiley and Sons; 2012. pp. 1354–1373.

[6] S. Kurniawan, "Age-related differences in the interface design process" in Stephanidis, C. (Ed.). *The Universal Access Handbook*. Boca Raton, FL: Taylor & Francis; 2009. pp. 8.1–8.8.

[7] M. Hassenzahl and N. Tractinsky, "User experience – a research agenda," *Behaviour & Information Technology*, vol. 25, no. 2, pp. 91–97, 2006.

[8] M. Ziefle and W. Wilkowska, "Why traditional usability criteria fall short in ambient assisted living environments.," *Proceedings of the 8th International Conference on Pervasive Computing Technologies for Healthcare*, ICST

[5]http://www.who.int/classifications/icf/en/, last access: March 2016

(Institute for Computer Sciences, Social-Informatics and Telecommunications Engineering), ICST, Brussels, Belgium; Oldenburg, Germany — May 20–23, 2014.

[9] C. Stephanidis, M. Antona, and D. Grammenos, "Universal access issues in an ambient intelligence research facility" in C. Stephanidis (ed.)." *Universal Access in Human-Computer Interaction: Ambient Interaction* – Volume 6 of the Proceedings of the 12th International Conference on Human-Computer Interaction (HCI International 2007), Beijing, P.R. China; Berlin Heidelberg: Lecture Notes in Computer Science Series of Springer, 2007. pp. 208–217.

[10] C.T.A. Pereira and M. Oliveira e Silva, "Live evaluation within ambient assisted living scenarios.," *Proceedings of the 7th International Conference on Pervasive Technologies Related to Assistive Environments*, ACM, New York, NY, USA: Island of Rhodes, Greece; May 27–30, 2014.

[11] N. Kaklanis, P. Moschonas, K. Moustakas, and D. Tzovaras, "Virtual user models for the elderly and disabled for automatic simulated accessibility and ergonomy evaluation of designs," *Universal Access in the Information Society,* vol. 12, no. 4, pp. 403–425, 2013.

[12] A. Jameson and A. Krüger, "Preface to the special issue on user modelling in ubiquitous computing.," *User Modeling and User-Adapted Interaction,* vol. 15, no. 3–4, pp. 193–119, 2005.

[13] D. Heckmann, T. Schwartz, B. Brandherm, M. Schmitz, and M. von Wilamowitz-Moellendorff, "Gumo: The general user model ontology," in *User Modelling* 2005, 10th International Conference, UM 2005, Edinburgh, Scotland, UK: Springer, July 24–29, 2005. pp. 428–432.

[14] B. De Carolis, S. Pizzutilo, and I. Palmisano, "D-Me: Personal interaction in smart environments," *Proceedings of User Modeling '03*, 9th International Conference, UM 2003 Johnstown, PA, USA: Springer, June 22–26, 2003.

[15] R. Casas, R. Blasco Marín, A. Robinet, *et al.*, "User modelling in ambient intelligence for elderly and disabled people" in *Computers Helping People with Special Needs*, Proceedings of the 11th international conference on Computers Helping People with Special Needs (ICCHP '08), K. Miesenberger, J. Klaus, W. Zagler, and A. Karshmer (Eds.). Springer-Verlag, Berlin, Heidelberg, 2008. pp. 114–122.

[16] E. Castillejo, A. Almeida, D. López-de-Ipiña, and L. Chen, "Modeling users, context and devices for ambient assisted living environments," *Sensors*, vol. 14, no. 3, pp. 5354–5391, 2014.

[17] P.A. Moreno, M.E. Hernando, and E.J. Gómez, "AALUMO: A user model ontology for ambient assisted living services supported in next-generation networks," L.M.R. Romero, (Ed.). *XIII Mediterranean Conference on Medical and Biological Engineering and Computing*, 2013, Volume 41 of the series IFMBE Proceedings, MEDICON 2013, 25–28 September 2013, Seville, Spain; 2014. pp. 1217–1220.

[18] M. Petzold, F. Barbabella, J. Bobeth, D. Kern, C. Mayer, and M. Morandell, "Towards an ambient assisted living user interaction taxonomy.," *CHI '13*

Extended Abstracts on Human Factors in Computing Systems (CHI EA '13), ACM, New York, NY; 2013. pp. 49–54.

[19] A. Leonidis, M. Antona, and C. Stephanidis, "Rapid prototyping of adaptable user interfaces," *International Journal of Human-Computer Interaction*, vol. 28, no. 4, pp. 213–235, 2012.

[20] M. Foukarakis, S. Leonidis, I. Adami, M. Antona, and C. Stephanidis, "An adaptable card game for older users," *Proceedings of the 4th International Conference on Pervasive Technologies Related to Assistive Environments (PETRA 2011)*, Crete, Greece, 25–27 May 2011. pp. 7 pages. 2011.

[21] A. Mourouzis, A. Leonidis, M. Foukarakis, M. Antona, and N. Maglaveras, "A novel design approach for multi-device adaptable user interfaces: Concepts, methods and examples.," C. Stephanidis (Ed.), *Universal Access in Human-Computer Interaction. Design for all and eInclusion* – Volume 5 of the combined Proceedings of the 14th International Conference on Human-Computer Interaction (HCI International 2011), Orlando, FL, USA, 9–14 July, 2011. pp. 400–409.

[22] R. Edlin-White, S. Cobb, M. D'Cruz, A. Floyde, S. Lewthwaite, and J. Riedel, "Accessibility for older users through adaptive interfaces: Opportunities, challenges and achievements," *Human-Computer Interaction. Towards Mobile and Intelligent Interaction Environments*, Volume 6763 of the series Lecture Notes in Computer Science, 2011. pp. 483–489.

[23] R. Miñón and J. Abascal, "Supportive adaptive user interfaces inside and outside the home.".

[24] N. Carbonell, "Ambient multimodality: towards advancing computer accessibility and assisted living," *Universal Access in the Information Society*, vol. 5, no. 1, pp. 96–104, 2006.

[25] M. Blumendorf and S. Albayrak, "Towards a framework for the development of adaptive," In C. Stephanidis (Ed.). Multimodal User Interfaces for Ambient Assisted Living Environments. *Universal Access in HCI. Intelligent and Ubiquitous Interaction Environments*, Volume 5615 of the series Lecture Notes in Computer Science, 5th International Conference, UAHCI 2009, Held as Part of HCI International 2009, San Diego, CA, USA, July 19–24, 2009. Proceedings, Part II, 2009. pp. 150–159.

[26] K. Fellbaum and G. Koroupetroglou, "Principles of electronic speech processing with applications for people with disabilities," *Technology and Disability*, vol. 20, no. 2, pp. 55–85, 2008.

[27] D. Freitas and G. Kouroupetroglou, "Speech technologies for blind and low vision persons," *Technology and Disability*, vol. 20, no. 2, pp. 135–156, 2008.

[28] J. Feng and A. Sears, "Speech input to support universal access" in Stephanidis, C. (Ed.). *The Universal Access Handbook*. Boca Raton, FL: Taylor & Francis; 2009, pp. 30.1–30.16.

[29] M. Alessandrini, G. Biagetti, A. Curzi, and C. Turchetti, "A speech interaction system for an ambient assisted living scenario" in S. Longhi, P. Siciliano, M. Germani, A. Monteriù, (Eds.), *Ambient Assisted Living*. Berlin: Springer; 2014, pp. 233–239.

[30] S. Goetze, N. Moritz, A.J.M. Meis, C. Bartsch, and J. Bitzer, "Acoustic user interfaces for ambient-assisted living technologies," *Informatics for Health and Social Care*, vol. 35, no. 3–4, pp. 125–143, 2010.

[31] J. McNaull, J.C. Augusto, and M.M.P. Mulvenna, "Flexible context aware interface for ambient assisted living," *Human-centric Computing and Information Sciences*, vol. 4, no. 1, 2014. pp. 1–41.

[32] M. Vacher, F. Portet, B. Lecouteux, and C. Golanski, "Speech analysis for ambient assisted living : Technical and user design of a vocal order system" in F. Hu (Ed.)., *Telhealthcare Computing and Engineering: Principles and Design*. New York, USA: CRC Press; 2013, pp. 607–638.

[33] S. Ntoa, A. Savidis, and C. Stephanidis, "Automatic hierarchical scanning for windows applications" in Stephanidis, C. (Ed.). *The Universal Access Handbook*. Boca Raton, FL: Taylor & Francis; 2009, pp. 35-1–35-16.

[34] C. Stephanidis, A. Paramythis, M. Sfyrakis, and A. Savidis, "A case study in unified user interface development: The AVANTI web browser" in C. Stephanidis (Ed.). *User Interfaces for All – Concepts, Methods, and Tools*. Mahwah, NJ: Lawrence Erlbaum Associates; 2001, pp. 525–568.

[35] S. Ntoa, G. Margetis, and C. Stephanidis, "FireScanner: A browser scanning add-on for users with motor impairments," in C. Stephanidis, (Ed.), *Universal Access in Human-Computer Interaction – Applications and Services* – Volume 7 of the Proceedings of the 13th International Conference on Human-Computer Interaction (HCI International 2009), San Diego, CA, USA, 19–24 July, Berlin Heidelberg: Lecture Notes in Computer Science Series of Springer. 2009. pp. 755–763.

[36] D. Grammenos, A. Savidis, and C. Stephanidis, "Designing universally accessible games.," *ACM Computers in Entertainment magazine*, vol. 7, no. 1, pp. 8:1–29, 2009.

[37] Apple, "VoiceOverfor OS X.," 2012. [Online]. Available from: http://www.apple.com/accessibility/voiceover. [last accessed 6 July 2016]

[38] C. Tao, X. Zhang, and X. Wang, " Research of environmental control systems for disabled people," in Y. Peng, X. Weng (Eds.) *7th Asian-Pacific Conference on Medical and Biological Engineering*, 22–25 April 2008 Beijing, China. 2008.

[39] T. Felzer and S. Rinderknecht, "3dScan: An environment control system supporting persons with severe motor impairments," *Proceedings of the 11th International ACM SIGACCESS Conference on Computers and Accessibility*, (Assets '09). ACM, New York, NY, USA; October 26–28, 2009. Pittsburgh, PA, USA, 2009.

[40] D. Grammenos, S. Kartakis, I. Adami, and C. Stephanidis, "(2008). CAMILE: Controlling AmI lights easily," *Proceedings of the 1st ACM International Conference on Pervasive Technologies Related to Assistive Environments*, (PETRA 2008), 15–19 July 2008, Athens, Greece. [CD-ROM]. New York: ACM Press. 2008.

[41] M. Kobayashi, A. Hiyama, T. Miura, C. Asakawa, M. Hirose, and T. Ifukube, "Elderly user evaluation of mobile touchscreen interactions," Proceedings of

the 13th IFIP TC 13 international conference on Human-computer interaction – Volume Part I (INTERACT'11), P. Campos, N. Nunes, N. Graham, J. Jorge, and P. Palanque (Eds.), Vol. Part I. Springer-Verlag, Berlin, Heidelberg, 2011. pp. 83–99.

[42] L. Findlater, J.E. Froehlich, K. Fattal, J.O. Wobbrock, and T. Dastyar, "Age-related differences in performance with touchscreens compared to traditional mouse input," *Proceedings of the SIGCHI Conference on Human Factors in Computing Systems (CHI '13)*. ACM, New York, NY, USA, 2013. pp. 343–346.

[43] A.L. Culén and T. Bratteteig, "Touch-screens and elderly users: A perfect match?" *Changes*, vol. 7, no. 15, 2013.

[44] C.F. Lee and C.C. Kuo, "Difficulties on small-touch-screens for various ages," *4th International Conference on Universal Access in Human-Computer Interaction – Part 1*, UAHCI 2007, Held as Part of HCI International 2007, Beijing, China, July 22–27, 2007. pp. 968–974.

[45] J. Häikiö, A. Wallin, M. Isomursu, H. Ailisto, T. Matinmikko, and T. Huomo, "Touch-based user interface for elderly users.," *Proceedings of the 9th International Conference on Human Computer Interaction with Mobile Devices and Services*, (MobileHCI '07). ACM, New York, NY, USA, 2007. pp. 289–296.

[46] J. Han, "Multi-touch interaction wall," *Proceedings of ACM SIGGRAPH 2006 Emerging technologies Conference, (SIGGRAPH '06)*. ACM, New York, NY, USA, Article 25. 2006.

[47] D. Michel, A. Argyros, D. Grammenos, X. Zabulis, and T. Sarmis, "Building a multi-touch display based on computer vision techniques," *Proceedings of the IAPR Conference on Machine Vision Applications*, (*IAPR MVA 2009*), Keio University, Yokohama, Japan, May 20–22, 2009. pp. 74–77.

[48] S. Beul, L. Klack, K. Kasugai, *et al.*, "Between innovation and daily practice in the development of AAL systems: Learning from the experience with today's systems," in M. Szomszor, and P. Kostkova (Eds.). *Electronic Healthcare*. Berlin: Springer; 2012. pp. 111–118.

[49] M. Ervasti, M. Isomursu, and I. Idigoras Leibar, "Touch- and audio-based medication management service concept for vision impaired older people," *IEEE International Conference on RFID-Technologies and Applications (RFID-TA 2011)*, Sitges, 2011. pp. 244–251.

[50] J. Wittland, P. Brauner, and M. Ziefle, "Serious games for cognitive training in ambient assisted living environments – A technology acceptance perspective," in J. Abascal, S. Barbosa, M. Fetter, *et al. Human-Computer Interaction – INTERACT 2015*, Volume 9296 of the series Lecture Notes in Computer Science, 15th IFIP TC 13 International Conference, Bamberg, Germany, Proceedings, Part I, September 14–18, 2015. pp. 453–471.

[51] G. Yamamoto, J. Hyry, M. Krichenbauer, *et al.*, "A user interface design for the elderly using a projection tabletop system," *3rd IEEE VR International Workshop on Virtual and Augmented Assistive Technology (VAAT 2015)*, Arles, 2015. pp. 29–32.

[52] F. Klompmaker, K. Nebe, and A. Fast., "dSensingNI: A framework for advanced tangible interaction using a depth camera," *Proceedings of the Sixth*

International Conference on Tangible Embedded and Embodied Interaction (TEI '12), S.N. Spencer (Ed.). ACM, New York, NY, USA, 2012. pp. 217–224.

[53] D. Kobayashi and M. Nakano, "Designing memorable tactile patterns for older adults," *Proceedings 19th Triennial Congress of the IEA*, Melbourne, 9–14 August 2015. Available at http://ergonomics.uq.edu.au/iea/proceedings/Index_files/papers/1733.pdf [last accessed 6 June 2016]

[54] H. Wang and X. Liu, "Design of a novel mobile assistive robot with haptic interaction," *IEEE International Conference on Virtual Environments Human-Computer Interfaces and Measurement Systems (VECIMS)*, Tianjin, 2012. pp. 115–120.

[55] J. Maculewicz, C. Erkut, and S. Serafin, "An investigation on the impact of auditory and haptic feedback on rhythmic walking interactions," *International Journal of Human-Computer Studies,* vol. 85, 2015. pp. 40–46.

[56] M. Turunen, J. Soronen, S. Pakarinen, *et al.*, "Accessible multimodal media center application for blind and partially sighted people," *ACM Computers in Entertainement*, vol. 8, no. 3, 2010. pp. 30 pages.

[57] D. Anastasiou, "Gestures in assisted living environments," in E. Efthimiou, G. Kouroupetroglou, S.-E. Fotinea (Eds.). *Gesture and Sign Language in Human-Computer Interaction and Embodied Communication*, Volume 7206 of the series Lecture Notes in Computer Science, 9th International Gesture Workshop, GW 2011, Athens, Greece, May 25–27 2011. pp. 1–12.

[58] A. Grguric, M. Mosmondor, M. Kusek, C. Stocklow, and D. Salvi, "Introducing gesture interaction in the ambient assisted living platform universaa," K. Pripuzic, (Ed.) *Proceedings of the 12th International Conference on Telecommunications*. Zagreb: University of Zagreb, 2013. pp. 215–222.

[59] S. Malassiotis and M.G. Strintzis, "Robust real-time 3D head pose estimation from range data," *Pattern Recognition*, vol. 38, no. 8, pp. 1153–1165, 2005.

[60] P. Padeleris, X. Zabulis and A. A. Argyros, "Head pose estimation on depth data based on Particle Swarm Optimization," *2012 IEEE Computer Society Conference on Computer Vision and Pattern Recognition Workshops*, Providence, RI, 2012, pp. 42–49.

[61] P. Majaranta and K.-J. Räihä, "Text entry by gaze: Utilizing eye-tracking" in S. MacKenzie, and K. Tanaka-Ishii (Eds.). *Text Entry Systems: Mobility, Accessibility, Universality*. San Francisco: Morgan Kaufmann; 2007, pp. 175–187.

[62] A. Bulling, D. Roggen, and G. Tröster, "It's in your eyes – Towards context-awareness and mobile HCI using wearable EOG goggles," *Proceedings of the 10th international conference on Ubiquitous computing (UbiComp '08)*. ACM, New York, NY, USA, 2008. pp. 84–93.

[63] P. Majaranta, R. Bates, and M. Donegan, "Eye-tracking" in Stephanidis, C. (Ed.). *The Universal Access Handbook*. Boca Raton, FL: Taylor & Francis, Lawrence Erlbaum Associates; 2009, pp. 587–606.

[64] F. Corno, A. Gale, P. Majaranta, and K.-J. Räihä, "Eye-based direct interaction for environmental control in heterogeneous smart environments" in

H. Nakashima, H. Aghajan, and J. C. Augusto (Eds.). *Handbook of Ambient Intelligence and Smart Environments*, Springer, 2010, pp. 1117–1138.

[65] J.S. Shell, R. Vertegaal, and A.W. Skaburskis, "EyePliances: Attention-seeking devices that respond to visual attention," in *CHI '03 Extended Abstracts of Human Factors in Computing Systems (CHI EA '03)*. ACM, New York, NY, USA, 2003. pp. 770–771.

[66] R. Azuma, Y. Baillot, R. Behringer, S. Feiner, S. Julier, and B. MacIntyre, "Recent advances in augmented reality," *Computer Graphics and Applications*, vol. 21, no. 6, pp. 34–47, 2001.

[67] F. Zhou, H.L. Duh, and M. Billinghurst, "Trends in augmented reality tracking, interaction and display: A review of ten years of ISMAR," *Proceedings of the 7th IEEE/ACM International Symposium on Mixed and Augmented Reality*, (ISMAR '08). IEEE Computer Society, Washington, DC, USA, 2008. pp. 193–202.

[68] R. Hervás, A. Garcia-Lillo, and J. Bravo, "Mobile augmented reality based on the semantic web applied to ambient assisted living." in J. Bravo, R. Hervás, and V. Villarreal (Eds.). *Ambient Assisted Living Third International Workshop*, IWAAL 2011, Held at IWANN 2011, Torremolinos-Málaga, Spain, June 8–10, 2011. pp. 17–24.

[69] G. Yamamoto, J. Hyry, M. Krichenbauer, *et al.*, "A user interface design for the elderly using a projection tabletop system," *3rd IEEE VR International Workshop on Virtual and Augmented Assistive Technology (VAAT 2015)*, Arles, 2015. pp. 29–32.

[70] S. Oviatt, "Multimdal interfaces" in J.A. Jacko, and A. Sears (Eds.). *The Human-Computer Interaction Handbook*. Mahwah, NJ: Lawrence Erlbaum; 2003, pp. 286–304.

[71] S. Oviatt and P. Cohen, "Perceptual user interfaces: Multimodal interfaces that process what comes naturally," *Communications of the ACM*, vol. 43, no. 3, pp. 45–53, 2000.

[72] D. Anastasiou, C. Jian, and D. Zhekova, "Speech and gesture interaction in an ambient assisted living lab," *Proceedings of the 1st Workshop on Speech and Multimodal Interaction in Assistive Environments (SMIAE '12)*. Association for Computational Linguistics, Stroudsburg, PA, USA, 2012. pp. 18–27.

[73] A. Marinc, C. Stocklöw, A. Braun, C. Limberger, C. Hofmann, and A. Kuijper, "Interactive personalization of ambient assisted living environments" in M.J. Smith, and G. Salvendy (Eds.). *Human Interface and the Management of Information – Interacting with Information*, Held as Part of HCI International 2011, Orlando, FL, USA, July 9–14, 2011. pp. 567–576.

[74] Y. Georgalis, D. Grammenos, and C. Stephanidis, "Middleware for ambient intelligence environments: Reviewing requirements and communication technologies." in C. Stephanidis, (Ed.), *Proceedings of the 13th International Conference on Human-Computer Interaction, (HCI International 2009)*, San Diego, CA, USA, Berlin Heidelberg: Lecture Notes in Computer Science Series of Springer, 9–24 July, 2009. pp. 168–177.

[75] M. Janse, P. Vink, and N. Georgantas, “Amigo architecture: Service oriented architecture for intelligent future,” in M. Mühlhäuser, A. Ferscha, and E. Aitenbichler (Eds.). *In-Home Networks. Constructing Ambient Intelligence – AMI 2007 Workshops*, AmI 2007 Workshops Darmstadt, Germany, November 7–10, 2007 Revised Papers, 2008. pp. 371–378.

[76] M. Eisenhauer, P. Rosengren, and P. Antolin, “A development platform for integrating wireless devices and sensors into ambient intelligence systems,” *6th Annual IEEE Communications Society Conference on Sensor, Mesh and Ad Hoc Communications and Networks Workshops*, Rome, 2009. pp. 1–3.

[77] E. Aitenbichler, J. Kangasharju, and M. Mühlhäuser, “MundoCore: A light-weight infrastructure for pervasive computing,” *Pervasive and Mobile Computing*, vol. 3, no. 4, pp. 332–361, 2007.

[78] J. Honkola, H. Laine, R. Brown, and O. Tyrkkö, “Smart-M3 interoperability platform,” *Proceedings of the IEEE Symposium on Computers and Communications* (ISCC ’10). IEEE Computer Society, Washington, DC, USA, 2010. pp. 1041–1046.

[79] S. Hanke, C. Mayer, O. Hoeftberger, *et al.*, “UniversAAL – An open and consolidated AAL platform” in R. Wichert, and B. Eberhardt (Eds.). *Ambient assisted living*. Berlin, Germany, Springer; January 25–26, 2011, pp. 127–140.

Chapter 19
Privacy and ethical issues

Francisco Florez-Revuelta[2], José Ramón Padilla-López[1], John Dinsmore[3], and Barbara Pierscionek[2]

19.1 Introduction

Ambient Assisted Living, commonly abbreviated as AAL, plays a major role in promoting health and in enhancing wellbeing and safety, or to help to maintain the independent living of older people. Additionally, AAL may also help formal and informal carers to provide support in assisting a person cared. AAL systems usually make use of an intelligent system that provides a means to interact between a user (carer and cared for person) and the system itself that can elaborate sensory information to detect events and incidents in a controlled private or public environment. Ambient assisted spaces make use of a collection of sensors that collect information from the environment or its dwellers. Binary sensors are employed to detect or record on/off status, temperature, motion, pressure or lighting level. These embedded sensors make it possible to obtain the state or context of the environment. Detection and tracking of people can be implemented using infrared ceiling sensors, sensors embedded in the carpet, and radio-frequency identification (RFID) tags or other sensors worn by the users. Other more complex systems employing audio and video can also be used. These sensors monitor the persons and their interactions with other people or objects and allow the recognition of activities of daily living, the learning of routines, or the detection of unexpected events, e.g. falls.

This continuous or sporadic collection of data from the users and their environment, as well as the intelligent processing of that data, is often seen as intrusive and as violating rights to privacy. This potential loss of privacy can be exacerbated by the use of particular sensors, e.g. cameras, as the acquired data may be easily understandable by unauthorised users. Therefore, designers and developers of technologies and services for AAL need to address this important issue in order to facilitate adoption and increase users' acceptance.

[1]Dpto. de Tecnología Informática y Computación, Universidad de Alicante, Apdo. de Correos Nº 99, E-03080 Alicante, Spain
[2]Faculty of Science, Engineering and Computing, Penrhyn Road, Kingston upon Thames, Surrey KT1 2EE, UK
[3]Trinity Centre for Practice and Healthcare Innovation, School of Nursing & Midwifery, Trinity College Dublin, 24 D'Olier Street, Dublin 2, Ireland

The remainder of this chapter is organised as follows: Section 19.2 presents the privacy, ethical and legal issues associated to healthcare technologies. Next, Section 19.3 introduces the principles that need to be followed to design and develop privacy-aware technologies. Then, Section 19.4 reviews some works on how users perceive that their privacy is affected by AAL systems. Some of these systems employ cameras to monitor people. Section 19.5 analyses this particular case, by reviewing different approaches to ensure visual privacy and presenting an approach to offer different visualisations according to the user's context. Finally, Section 19.6 presents come conclusions and discussion.

19.2 Privacy, ethical and legal issues

Aspects of ethics appear in a vast range of disciplines and are relevant to all issues that involve human behaviour and human interactions. Yet the meaning of ethics is not clear nor is it without ambiguity. This may be because although ethics dates back over thousands of years, there is no singular definition of ethics and meanings may have been altered with time and implementation.

The Western concept of ethics dates back to Greek civilisation where it was deemed as describing good deeds, good habits and duty. This was supported by philosophers such as Kant who expressed ethical behaviour as emanating from the notion of duty which is linked to reason.

If ethics and ethical behaviour are grounded in reason, it would suggest that there should be a set of ethical principles that are universally applicable and reasonable to apply. However, ethical principles cannot be strict or prescriptive because ethical principles are guidelines the application of which would vary depending on the situation. In simple terms, ethical principles exist to help make decisions and to justify the actions that arise from these decisions.

Decisions can be easy or complex, they can have consequences on few or on many and they can have short-lived or lasting effects. All of these needs to be considered when applying ethics to technology.

19.2.1 Ethics in healthcare and technology

The ethical principles derive from healthcare and are based on the Hippocratic Oath. Hippocrates who lived around 460–380 BC was believed to be part of a cult of physicians who adhered to Asclepius, the god of medicine and healing. The legends pertaining to Asclepius indicate a profound understanding of ethics by the ancient Greeks. The very reason cited for Asclepius' death, that he was killed by Zeus for resurrecting a dead man was to indicate that this was a transgression of mortal power into the realm of gods. This has resonance today when there is a fear that doctors, scientists and technologists may be 'playing God'. Each technological advance brings with it the prospect of greater powers to those who have devised the technology and those who have the authority to apply it.

The Hippocratic Oath survives to this day, a modern version is cited by medical graduates and it still applies to ethics in medical practice and research. Following the Second World War and because of the atrocities committed, the World Medical Association (WMA) expressed concerns that standards of ethics had been eroded. This led to a unifying international oath adopted in 1948, which incorporated the Hippocratic Oath, and is known as The Declaration of Geneva.

The basic principles are based on doing good, avoiding or minimising harm, respecting autonomy and the right for mentally competent adults to make their own choices about anything that will directly or indirectly affect them and the right to confidentiality and privacy. The latter principle has a direct parallel in law with the Human Rights and Data Protection Acts.

Laws are more prescriptive than ethics because they regulate what can and cannot be done for societies and nations and are determined by governments and legislators. There are obvious links between ethics and laws because like ethics, laws are based on a moral code of what is perceived to be right and wrong. Ethics has no strict regulations but rather a number of basic concepts that can serve as an aid to dealing with situations and making decisions and judgements. The basic ethical concepts are: virtues, values, rights, respect, dignity, standards, duties, responsibility, accountability and principles.

19.2.2 Ethical principles

A major ethical principle, which dates back to the Hippocratic Oath, is about doing good. Although this may appear to be so obvious and simple to implement, it raises two major complexities: how good is good enough and how far should the consequences of good spread in terms of people affected and in time for the effects to be felt. The lack of limits means that the consequences and future implications need to be considered very broadly. Technology, by its very nature, looks to the future and to improving life quality. It can be difficult to determine how far and for whom that improvement will be and as with all novel and untapped applications, this has to be balanced against potential harm. A part of the Hippocratic Oath states that above all there should be avoidance of harm. Technologies, particularly those concerned with healthcare, will never actively seek to harm. What needs to be considered are the risks and adverse consequences that may arise and although, unlike the doing of good, avoidance of harm has limits, these principles work together in a risk/benefit analysis.

Another core principle in ethics is respect for autonomy. This means respecting that mentally competent adults can make decisions for themselves and can understand the consequences of those decisions and that their decisions must be respected. Children and adults who lack mental competence are not deemed to be able to decide for themselves and require others to make decisions for them. This can make ethical issues and subsequent applications complicated. It requires recognising that there are individuals who are more vulnerable than others. These individuals and their rights should be protected and not exploited and at all times their dignity must be preserved. There is some legal guidance on gaining of consent from vulnerable and non-autonomous individuals; e.g. in British law the basis is consideration of best interests of that individual. The notion of best interests is another form of doing good and so is fundamental to ethics. What needs to be taken into account is how best interests of the individual are best served, short and long-term best interests and how these are balanced against the best interests of a greater number of individuals and ultimately, societies.

Confidentiality and protection of privacy has both ethical and legal bases. If at any time these conflict, the law will always have precedence and this is more tightly regulated. This should not create the impression that confidentiality is purely a legal concept. There are aspects of confidentiality and protection of privacy in emerging areas that utilise new technologies, which legislators may not have considered as when the laws were being written or updated, the technologies did not exist. The uses of Big Data and telemedicine are two examples of this. In such instances, the ethical approach should be to consider what the law requires and then apply the principles of doing good, avoiding harm and respecting autonomy to outstanding aspects that the law may not clearly cover.

Communities flourish and thrive because of the goodwill and the respectful behaviour individuals have for one another. This extends well beyond the law. Without realising it, this is a fundamental demonstration of ethics. When testing or trialling new treatments, modalities, technologies or instruments that require human subjects, the ethics manifested in daily living become more focussed on the specific objectives of the test or trial and requires greater scrutiny. It often brings into the open aspects of behaviour not considered previously because they do not form a core part of daily life. The wearing and monitoring of devices that collect medical data and transport this remotely, is one example of this; ambient assisted living is another. The ethical approaches to dealing with this will need a great deal of thought and may require delving into the philosophical theories to understand the fundamental principles that have been postulated as underpinning human thought and action. They should be grounded in moral reasoning and seen as an evolution in ethical thinking that brings the application of ethics from ancient times to the modern world.

19.3 Privacy-by-design and privacy-enhancing technologies

The term Privacy by Design [1] was created in the 1990s by Ann Cavoukian, Information & Privacy Commissioner of Ontario, Canada. Privacy by Design refers to the philosophy and approach of embedding privacy into the design, operation and management of information processing technologies and systems.

Her office developed the seven Foundational Principles of Privacy by Design, which are nowadays a reference in the design of technologies. These principles are:

1. *Proactive not reactive; preventative not remedial*: The Privacy by Design approach is characterised by proactive rather than reactive measures. It anticipates and prevents privacy-invasive events before they happen. Privacy by Design does not wait for privacy risks to materialise, nor does it offer remedies for resolving privacy infractions once they have occurred—it aims to prevent them from occurring. In short, Privacy by Design comes before the fact, not after.

2. *Privacy as the default setting*: We can all be certain of one thing – the default rules! Privacy by Design seeks to deliver the maximum degree of privacy by ensuring that personal data are automatically protected in any given IT system or business practice. If an individual does nothing, their privacy still remains intact. No action is required on the part of the individual to protect their privacy—it is built into the system, by default.
3. *Privacy embedded into design*: Privacy is embedded into the design and architecture of IT systems and business practices. It is not bolted on as an add-on, after the fact. The result is that it becomes an essential component of the core functionality being delivered. Privacy is integral to the system, without diminishing functionality.
4. *Full functionality—positive-sum, not zero-sum*: Privacy by Design seeks to accommodate all legitimate interests and objectives in a positive-sum "win-win" manner, not through a dated, zero-sum approach, where unnecessary trade-offs are made. Privacy by Design avoids the pretence of false dichotomies, such as privacy vs. security, demonstrating that it is possible to have both.
5. *End-to-end security—full lifecycle protection*: Privacy by Design, having been embedded into the system prior to the first element of information being collected, extends throughout the entire lifecycle of the data involved, from start to finish. This ensures that at the end of the process, all data are securely destroyed, in a timely fashion. Thus, Privacy by Design ensures cradle to grave, lifecycle management of information, end-to-end.
6. *Visibility and transparency—keep it open*: Privacy by Design seeks to assure all stakeholders that whatever the business practice or technology involved, it is in fact, operating according to the stated promises and objectives, subject to independent verification. Its component parts and operations remain visible and transparent, to users and providers alike. Remember, trust but verify.
7. *Respect for user privacy—keep it user-centric*: Above all, Privacy by Design requires architects and operators to keep the interests of the individual uppermost by offering such measures as strong privacy defaults, appropriate notice, and empowering user-friendly options. Keep it user-centric.

The application of Information and Communication Technologies (ICT) to protect privacy has become extensively known as Privacy-Enhancing Technologies (PETs). They have been described in different ways but the most adopted definition was made in the PISA European Project

> "Privacy-Enhancing Technologies is a system of ICT measures protecting informational privacy by eliminating or minimising personal data thereby preventing unnecessary or unwanted processing of personal data, without the loss of functionality of the information system" [2]

These technologies are very relevant nowadays. For instance, the European Commission considers that PETs should be developed and more widely used, in particular where personal data is processed through ICT networks. The Commission considers that wider use of PETs would improve the protection of privacy as well as help fulfil data protection rules. The use of PETs would be complementary to the existing legal framework and enforcement mechanisms [3].

PETs incorporate Fair Information Practices (FIPs), following for instance the basic principles developed by the Organisation for Economic Co-operation and Development [4]:

Collection limitation: There should be limits to the collection of personal data and any such data should be obtained by lawful and fair means and, where appropriate, with the knowledge or consent of the data subject.

Data quality: Personal data should be relevant to the purposes for which they are to be used, and, to the extent necessary for those purposes, should be accurate, complete and kept up-to-date.

Purpose specification: The purposes for which personal data are collected should be specified not later than at the time of data collection and the subsequent use limited to the fulfilment of those purposes or such others as are not incompatible with those purposes and as are specified on each occasion of change of purpose.

Use limitation: Personal data should not be disclosed, made available or otherwise used for purposes other than those specified in accordance with the Purpose Specification Principle except (a) with the consent of the data subject or (b) by the authority of law.

Security safeguards: Personal data should be protected by reasonable security safeguards against such risks as loss or unauthorised access, destruction, use, modification or disclosure of data.

Openness: There should be a general policy of openness about developments, practices and policies with respect to personal data. Means should be readily available for establishing the existence and nature of personal data, and the main purposes of their use, as well as the identity and usual residence of the data controller.

Individual Participation: An individual should have the right (a) to obtain from a data controller, or otherwise, confirmation of whether or not the data controller has data relating to him; (b) to have communicated to him, data relating to him within a reasonable time; at a charge, if any, that is not excessive; in a reasonable manner; and in a form that is readily intelligible to him; (c) to be given reasons if a request made under subparagraphs (a) and (b) is denied, and to be able to challenge such denial; and (d) to challenge data relating to him and, if the challenge is successful to have the data erased, rectified, completed or amended.

Accountability: A data controller should be accountable for complying with measures, which give effect to the principles stated above.

19.4 Users' perception of privacy

The user is the main stakeholder affected by the use of monitoring technologies in AAL systems. There have been different recent studies analysing their opinions about what, when, and how data can be collected and processed, and who can be the recipient of that data.

As Zagler *et al.* remark in [5], one of the main concerns of users about the use of technologies and services for AAL is related to the possibility of collecting data. Therefore, even if users may recognise the benefits of using AAL systems, they may reject them because of privacy concerns or complexity of use. This is the reason why many products and research projects avoid the use of invasive sensors, as cameras or microphones, which recordings may be easily understandable by others. Another alternative usually considered is to avoid transferring raw data to a central location, where various decision-makings are made. Even if this requirement most likely increases the equipment costs at the assisted person's premises, all data collection and processing is done on a local level and raw data does not leave the home. Later, processed information from the raw data is allowed to transfer outside the home to inform a second party, e.g. a care centre or an informal carer, but via a secured channel. This is especially true when a critical situation is detected and the resulting alert should be sent to a third party.

Wilkowska and Ziefle [6] studied users' perceptions regarding data security and privacy of medical assistive technologies. They highlight that users do not wish to share their e-health usage and measurements with others as well as wearing a visible device that could stigmatise them. However, privacy requirements were lowered by users with poor health status. Additionally, women were more reluctant than men regarding the anonymous and intimate way of using medical technology.

These works show the complexity of dealing with the different perception of privacy that each individual has. These privacy and ethical issues, of using sensors to monitor the status of the user, are intensified when using cameras in a private space. Ziefle *et al.* [7] analysed the attitudes of users towards video-based monitoring systems for long-term care of elderly or disabled people in smart home environments. Participants in the study would probably not let medical personnel monitor their home. However, they would probably accept cameras if they would really help them. Data protection was paramount for all the participants. They also had serious concerns about system malfunctions, data delivery without consent and illegal access.

Clark and McGee-Lennon [8] studied the current barriers in the United Kingdom to the uptake of assisted living technologies and telecare, and discussed how each of these barriers might be overcome. Eleven focus groups involving older users, informal carers, social care and healthcare professionals, technologists, installation technicians, and policy makers participated in this study. Emerging themes across the focus groups were acceptance and ethical, legal and privacy issues. Potential privacy issues were highlighted regarding the use of improved range and resolution of data (such as video streaming and still images) and the improved technology that enables a person's location and activity to be tracked. As in other studies, there were also concerns about the storage of health data electronically, and about the robustness of home care technology systems and who had the legal responsibility for ensuring that the equipment was working correctly and safely.

Fisk [9] proposed seven principles for the use of surveillance technologies in care homes, which are as follows:

Principle 1: Any reasonable level of surveillance (including cameras) is appropriate for common or public areas in care homes.

Principle 2: Care homes should be able to provide or should be willing to permit or facilitate, the use of surveillance technologies (including cameras) within a resident's room or other private areas.
Principle 3: The location of surveillance technologies should be carefully considered. They should be visible or otherwise clearly known to be present.
Principle 4: Staff should be fully aware of their responsibilities in relation to surveillance technologies.
Principle 5: Access to data, images, audio or video footage should be restricted only to authorised persons or agencies in particular, defined circumstances.
Principle 6: Ownership of data, images, audio or video footage.
Principle 7: Minimising intrusion.

Fisk's paper argues that the use of surveillance technologies can be ethically justified if these principles are followed.

All these studies show important concerns about the use of cameras. However, some of these issues seem to be changing in some countries. For instance, in the UK, recent cases of abuses to older people by their carers, either at care homes or family homes, are leading to an increase in the acceptance of video-based monitoring. HC-One, a care provider in the United Kingdom, ran in 2014 a consultation [10] on the use of visible cameras in care homes. It is significant because of the high number of responses: 7,330 from staff, 3,371 from relatives, and 1,536 from residents. The overall support for cameras was 47% among residents, 87% among relatives, and 63% among staff. It seems that cameras may be a useful safeguarding tool for the different users. Similarly to other studies, there are differences in the acceptance on their use in different areas (communal rooms—51%, only residents' bedrooms—20%, no cameras at all—29%). The main concerns are related to the privacy for the residents, who has access or watches the footage, and where footage is used and how secure it is.

Even the Care Quality Commission, the independent regulator of health and adult social care in England, published in 2014 guidelines on the use of overt and covert surveillance in care homes for both care home providers [11] and the general public [12]. This has opened the debate about the compulsory use of CCTV systems in care homes. Recently, some companies have started to install such monitoring systems in UK care homes following those guidelines, and national and European regulations.

For further research studies on these topics, a literature review on privacy or risk and personal health systems can be found in [13], where Mittelstadt *et al.* reviewed works addressing data privacy (control, security and trust) as well as personal privacy (physical, social and decisional).

19.4.1 Users' views in the BREATHE project

The BREATHE project,[1] funded by the AAL Joint Programme and some National funding bodies, aims at providing an ICT-based solution to support and relieve the burden experienced by informal carers and improve the quality of life of both the carer and the person who is being cared for. The platform incorporates a group

[1] http://www.breathe-project.eu (last accessed in March 2016).

Table 19.1 Outcomes of the initial BREATHE survey on the use of video-based monitoring for remote care

		Assisted person (AP) (%)	**Informal carer (IC) (%)**
Do you find it acceptable or useful for the IC to watch real time video?		90	89
When should the IC have access to real-time video?	At any time	50	21
	When alert	45	68
When access should be available to the IC to real time video?	Spain	75	38
	UK	33	20
	Ireland	33	0
Should AP have control over being observed?	Spain	50	75
	UK	100	100
	Ireland	100	100
Do you have concerns about the privacy of the AP in the context of accessing real-time/live video?	Spain	25	75
	UK	50	80
	Ireland	67	83

of cameras and an array of sensors installed at the assisted person's home and is responsible for collecting information on a daily basis and in an automatic way about the activities and behaviour of the assisted person. At the start of the project, a survey was carried out with a reduced number of users (assisted persons and informal carers) in Ireland, United Kingdom and Spain. This study showed the differences in the privacy requirements of the two different types of users and among countries in the perception of privacy when using video-based monitoring systems, as it is presented in Table 19.1. In Spain the participants were more tolerant to have cameras at home than their British or Irish counterparts. There is an interesting outcome regarding the concerns on accessing and controlling the video streaming: the carers (observers) were more reluctant than the assisted persons (observed) on these issues.

After using the video-based monitoring system during in the pilots carried out in those three countries, the majority of assisted persons (80%) and informal carers (68%) found it acceptable and usable to have cameras installed and to allow a remote broadcast of video to facilitate care-related tasks [14], but this would depend on:

- the room in which the camera is located,
- who can access the realtime video,
- the video should not be stored, and
- the assisted persons should have control over when they can be observed.

19.5 Example of application—protection of visual privacy

Video cameras are being employed more frequently in AAL solutions. Computer vision gives the possibility to monitor 24/7 an environment and report on visual information, which is commonly the most straightforward and natural way of describing an event, a person, an object, actions and interactions. Advances produced in the last decades, regarding increased computational power and reduced costs, have contributed to this. Furthermore, computer vision advances have given video cameras the ability of 'seeing', becoming smart cameras. This has enabled the development of video-based intelligent monitoring systems, which are able to extract valuable information from visual data to analyse actions, activities and behaviours [15]. Video cameras are being increasingly used in order to offer novel telecare services that rely on visual information, e.g. monitoring of activities of daily living (ADL), fall detection, warning system, and many others. As was previously mentioned, they are also been considered to evaluate the quality of care, by care providers [10] and regulators [11, 12].

Traditionally, cameras have been used in public spaces for surveillance services in streets, parking lots, banks, airports, train stations, shopping centres, museums, sports installations and many others. Generally the use of video cameras in public places has been tolerated or accepted by people, whereas their use in private spaces has been rejected. There may be several reasons to explain this difference. On the one hand, the perceived public-safety benefits favour the usage of cameras in public places for crime prevention, fight against terrorism and so on. On the other hand, there is a widespread belief that while staying in public environments, people's sensitive information will not be exposed. However, they are not commonly used in private environments due to people's concerns about privacy as video can convey an enormous amount of information, which can be qualified as sensitive, that can be easily understood by anyone. Intelligent vision-based monitoring systems jeopardise individuals' right to privacy because they can acquire, store and show significant information about people habits, relationships, and the like. This evolution in people's monitoring may be seen as approaching an Orwellian Big Brother environment, as people may have the feeling of being constantly monitored. Therefore, privacy awareness becomes essential for video-based AAL applications. Thus, if individuals' privacy can be guaranteed by using smart cameras, acceptance of those applications would be increased.

19.5.1 Privacy-by-context

Ubiquitous computing, context awareness, intelligence and natural interaction are the main properties of ambient intelligence and smart environments. Context is "any information that can be used to characterise the situation of entities (i.e. whether a person, place or object) that are considered relevant to the interaction between a user and an application, including the user and the application themselves. Context is typically the location, identity and state of people, groups, and computational and physical objects [16]." Context-aware computing refers to an application's ability to adapt to changing circumstances and respond according to our context [17].

Recently, a privacy-by-context paradigm was introduced for remote visual monitoring of assisted persons [18]. This paradigm proposes a level-based scheme to access video data (although it could be extended to any personal data) in order to protect privacy. Each level establishes the way in which the video stream is modified and displayed and, therefore, which protection degree is provided. In this scheme, the appropriate level is dynamically selected according to the context, modifying the raw video before it is displayed, as originally proposed in the "Good Brother" proposal [19].

The context should provide enough information to empower the assisted persons to adapt privacy to their preferences, in such a way that they can decide who, how, and when their carers can watch them. In an assisted living environment the context may be described by:

1. the identity of the subject, in order to retrieve the privacy profile;
2. the appearance, e.g. dressed;
3. the location, e.g. kitchen, bathroom, bedroom;
4. the on-going activity, e.g. having a shower, watching TV, sleeping, receiving a visit; and
5. the identity of the observer: the closeness of the assisted person with a particular carer.

Using these variables, an individual, i.e. the assisted/observed person, can then select in advance a visualisation level to any situation that could be modelled by the context, i.e. giving different permissions to access the video stream to the various different carers in different ways according to what is happening at the assisted person's home.

19.5.2 Visual privacy

In order to ensure privacy in video-based AAL solutions, some systems take the approach of avoiding the acquisition of images, which are easily understandable by an observer. Instead they acquire other "visual" information, e.g. depth maps [20, 21] or infrared images [22], in order to analyse the scene. Some other methods, although they acquire images they process them locally [23, 24], only transmitting processed data and avoiding the transfer of the images.

However, in case that images could be either transmitted or stored, there are two approaches in order to protect and preserve individual's privacy. These depend on when privacy protection is applied: before the image is acquired (image acquisition time), or after (image processing time). On the one hand, it is possible to prevent others of successfully capturing images in which an individual appears. On the other hand, once an image has being acquired and is going to be distributed, shared, or visualised; some parts from it, which are considered private, can be filtered or removed.

Current methods for visual privacy preservation include [25]:

Intervention: Intervention methods deal with the problem of preventing someone to capture private visual data from the environment. They can be as simple as to empower the monitored person to turn off the monitoring. There are also automatic methods that physically intervene camera devices to avert the acquisition of an image by means of a specialised device that interferes with the camera optical lens.

Blind vision: Blind vision addresses privacy-related processing issues by means of secure multi-party computation techniques that are applied to vision algorithms.

Secure multi-party computation is a subfield of cryptography that enables multiple parties to jointly compute a function in such a way that their inputs and the function itself are not revealed.

Secure processing: This approach is similar to blind vision but it is not based on secure multi-party computation. These methods process visual information in a privacy respectful way, e.g. working with images in a domain that does not reveal visual information, rejecting visual information that is not necessary for the algorithm to work. Recently, some systems employ depth/range sensors that allow working only with silhouettes avoiding using any other visual data; and

Redaction: Image and video modification or redaction methods are the most common visual privacy protection methods. They modify the sensitive regions of an image (faces, bodies, objects, number plates, etc.) to conceal private information concerning the subjects appearing on it. Redaction methods include ad-hoc distortion/suppression methods, image encryption, face de-identification, and in-painting based algorithms. Computer vision algorithms are used in order to locate the sensitive regions in the image in which the redaction method must operate. This is one of the main difficulties, as those private regions must be correctly detected at every moment.

Hudson and Smith [26] state that when an image is protected, there should be a balance between privacy protection and intelligibility, i.e. the resulting image should be intelligible. Therefore, if a redaction method is applied, i.e. the image is modified to ensure privacy, the resulting image may have lost some visual features relevant for its analysis. Therefore, a trade-off between privacy preservation and intelligibility is required.

Is it possible to protect privacy while, at the same time, provide video-based AAL services? This question is based on the fact that if privacy is fully protected so that no information could be obtained from the image, the access to the video would be meaningless. The lack of information (minimum intelligibility) would not allow any analysis of the scene, for instance to assess the situation after a domestic accident. Therefore, the answer is very important because if there is no way to protect visual privacy and, simultaneously, to work with the captured images, many of the AAL video-based applications would not be feasible or, if so, users would have to make a choice between privacy and security/safety.

A redaction method combined with a privacy-by-context approach is considered by Padilla *et al.* [18]. The system implements eight different visualisation methods to transform the image (Figure 19.1). First, the silhouette of the person needs to be obtained, as it defines the sensitive area to be protected. Then, visual filters are applied to the sensitive region to conceal the person, replace the individual with a different representation, or even remove the person from the image. The privacy preservation of some of these methods is similar. Therefore, they can be grouped into five privacy levels: (1) minimum, no privacy is provided (raw image), (2) low, distorted colours, textures and shapes (blur, pixelating and emboss), (3) medium, suppressed colours and textures but shapes are preserved (silhouette), (4) high, suppressed colours, textures and shapes but posture is preserved (skeleton and 3D avatar), and (5) maximum,

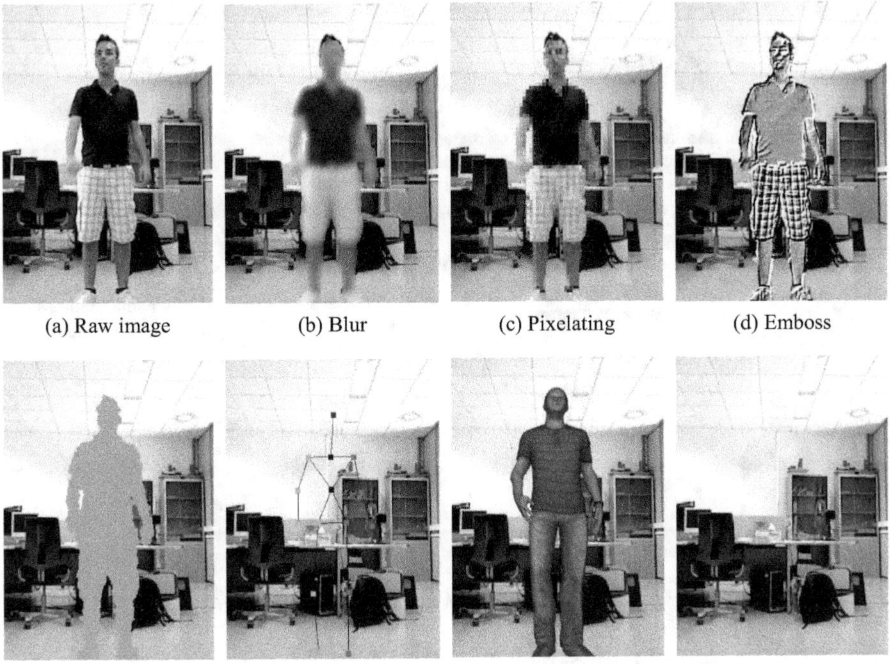

(a) Raw image (b) Blur (c) Pixelating (d) Emboss

(e) Silhouette (f) Skeleton (g) 3D avatar (h) Invisibility

Figure 19.1 Implemented visualisation models ordered from lower to higher protection

suppression of all the visual information (invisibility). Then, each specific context is associated with a particular visualisation level. By adapting the level of privacy depending on the nature of the context/detected event privacy protection can be maximised and yet have enough information to perform the task or assess the situation. Therefore, in order to obtain robust privacy solutions they need to consider the context.

19.5.3 Survey about privacy filters

In order to evaluate the proposed visualisation models, a survey with 28 participants was conducted. A sequence of eight images were shown to each participant where the actual image was previously modified. They were asked to answer several questions about what they were able to see in the modified images regarding appearance (visual identity, nudity, facial expression, clothes, accessories) and activity.

Figure 19.2 shows that some visual attributes are more difficult to be protected in general, such as visual identity, activity and nudity, and that the performance of a visualisation model is not stable across different types of visual attributes. For instance, blur or pixelation achieve significantly better results on visual identity or facial expression than on other attributes. Maybe, this and the reduced processing required are the reasons why they are employed by media to conceal people's identity

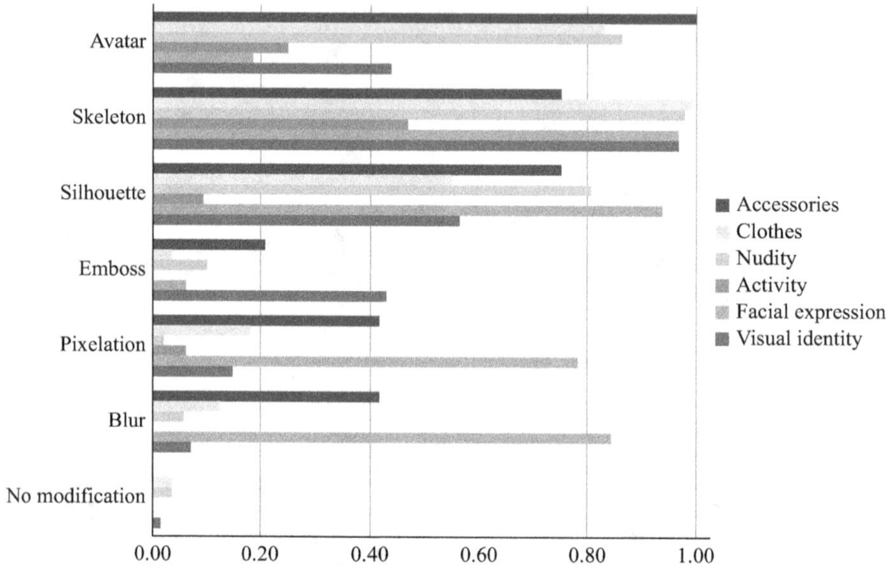

Figure 19.2 Average protection degree of the visualisation models for different visual attributes

in images and videos. On the other side, avatar stands out to hide the accessories worn by the person. It is worth mentioning, that this is the desired result of this visual privacy protection scheme, since the proposed visualisation models allow to set different levels of protection according to the context. Overall (Figure 19.3), it can be observed that silhouette, skeleton and avatar tend to obtain the best protection, presenting nonetheless a sensitivity to activity. However, the avatar obtains poor results on visual identity and facial expression compared to the silhouette and the skeleton, which leads to the thought that users misinterpreted the questions, evaluating the avatar itself instead of the substituted person.

Regarding the visualisation models, users were asked to score how difficult it is for them to see visual features of a person using the same image, in this case, with all the visualisation models. Figure 19.4 details that the invisibility obtains the highest score as expected, but silhouette and skeleton also reach very high scores. Surprisingly, the avatar is below blur and pixelation, even though formally the same information is shown as in the skeleton. Again, the human appearance of the avatar may have misled the user's judgement about what can be seen and what is protected with this visualisation model. This is demonstrated, somehow, by the huge variance in users' replies regarding the score with the avatar filter.

19.5.4 Application in the BREATHE project

This scheme following a privacy-by-context approach and different visualisation has been employed in the BREATHE project. The examples shown in Figure 19.1 were obtained using an RGB-D device, which provides depth information together with

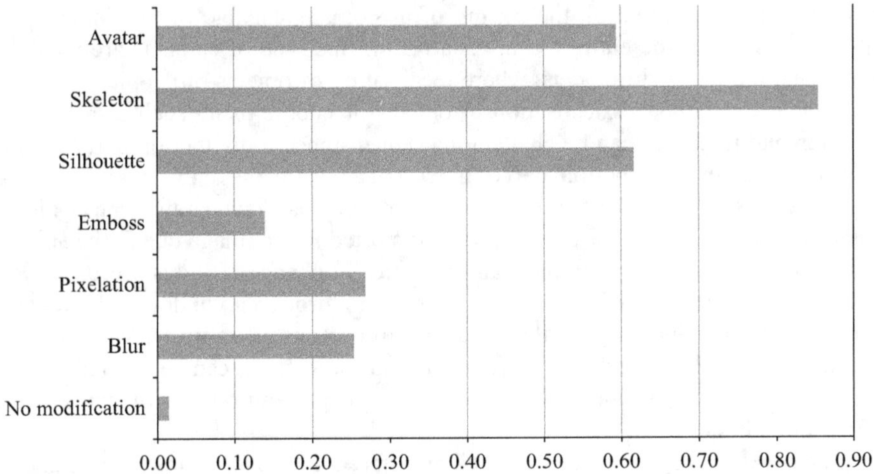

Figure 19.3 Overall average protection degree of the visualisation models considering all the attributes

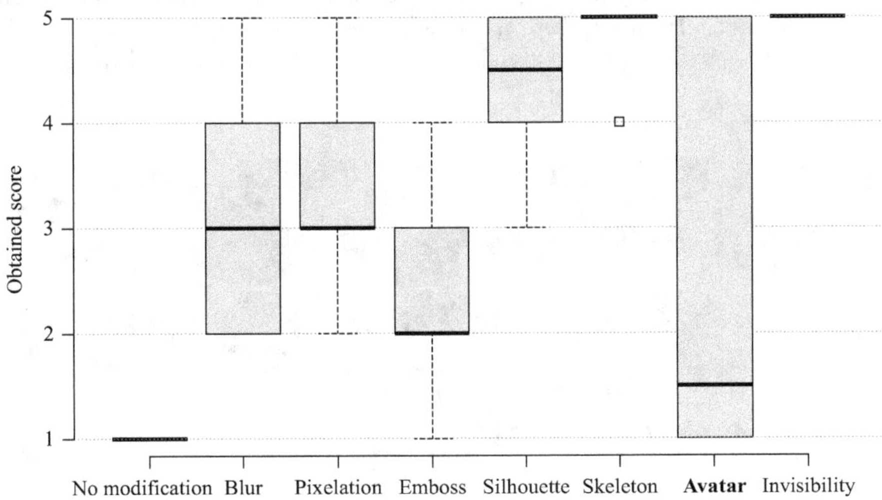

Figure 19.4 Overall average protection degree of the visualisation models considering all the attributes

the regular colour image. Depth data allows an easy location and segmentation of the person/s in the image. These devices provide a horizontal perspective of a monitored environment, their field of views are however limited due the focal length and view angle. In order to analyse the whole room with a single camera, the BREATHE system uses a wide-angle fisheye RGB camera to obtain a top-view of the scene [25]. Such a camera, usually installed on the ceiling of a room, is capable of providing up to 360° view angle of the entire room. As a result, this type of cameras are ideal for monitoring indoor environments, where localisation of objects or people of interest is the main vision task. Localisation of objects or people using such cameras (i.e. detection and tracking) can be challenging. For instance, only the top part (e.g. the head and shoulders) and bottom part (e.g. feet) of a person are visible at most frames; subjects are often very close to the cameras, then changes of appearance can be highly drastic and frequent; and the scene appears distorted in the image due to the fisheye lens. Moreover, image conditions can dramatically change if lights are turned off or on, or blinds are opened and the sunlight enters through a window. Additionally, cluttered environments may create occlusions of persons and objects. Motion based detection and tracking methods, e.g. background subtraction, can be applied to such scenarios where shapes and appearances of the people and objects of interests are difficult to model.

The developed system was validated in different households, installing cameras in the living rooms and kitchens in order to monitor the assisted persons throughout the day. Wide-angle fisheye cameras were mounted on the ceiling of the rooms, which provided a 360° field of view of their entirety. By identifying the location and activity of the assisted person in real-time, the system can automatically adjust, according to the context, the privacy level for other parties to access the corresponding visual data (Figure 19.5). However, there are some of the proposed visualisation methods in Figure 19.1, i.e. skeleton and avatar, that are not currently feasible because they

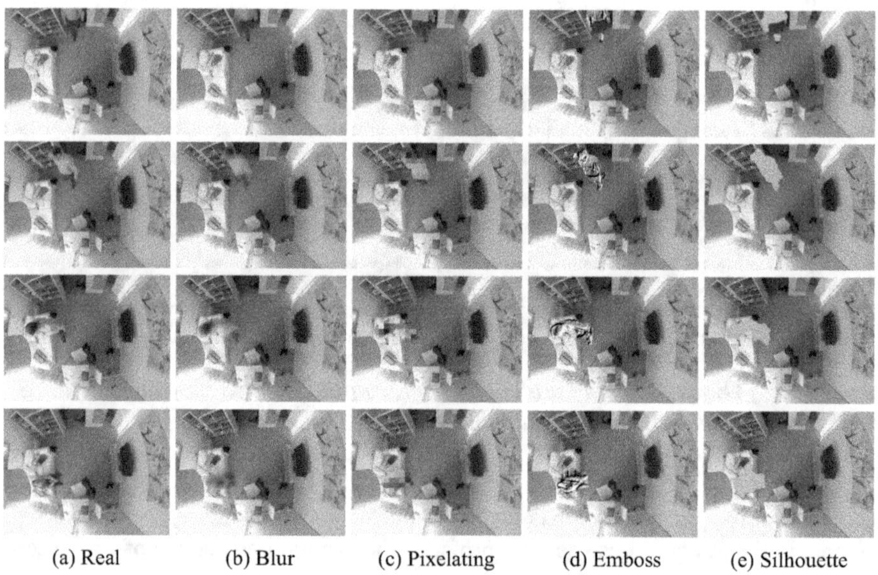

Figure 19.5 A sequence of images where the real image (a) has been protected using some of the visualisation models

require obtaining person's 3D posture, what is difficult using a single RGB camera. Invisibility may be obtained by directly showing the background obtained during the subtraction process.

In addition, visual activity analysis process a visual understanding of the behaviour of the assisted person over time, which is employed to recognise activities of daily living, and which can be potentially coupled with an anomaly detection system to raise alarms on suspicious behaviour and potential dangers.

19.6 Discussion

As we move towards a community based, person-centred model of ageing well at home, invasive sensor technologies such as cameras present significant opportunities to enhance self-management and caregiving processes but at the same time raise concern regards data protection and privacy for end users.

Adhering to fundamental ethical principles (including; integrity, objectivity, beneficence, respect for autonomy, fairness, non-maleficence, truthfulness and justice) when developing and implementing invasive sensor technologies, particularly within health and community care is pivotal to their adoption. Regional (e.g. across EU member states) legal and data protection differences, must be addressed by industry and research programmes to ensure that the human rights are protected. A key question we must consistently address is: "Does the technology introduced do good and improve the quality of life for the end user?" To effectively answer this, developers and researchers must carefully weigh risks and benefits of any proposed technology carefully, considering all issues (e.g. user mental/cognitive capacity) to protect the most vulnerable in society.

With rapidly evolving invasive sensor technologies and related research fields (e.g. 'big data', 'wearables' and the 'Internet of Things (IoT)'), paralleled by ever changing ethical and legal processes to their use, it is important to encourage best practice design processes. As such technology, policy, regulatory and research funding bodies are increasingly endorsing the use of Privacy Enhancing Technologies (PETs) and Fair Information Practices (FIPs). Fundamentally derived from the 'privacy by design' approach their aim is to ensure the strongest protection possible for users of invasive sensor technology.

Previous studies have shown that use of invasive sensor technology in private settings may increase negative user perceptions towards the technology use. However this can be offset if the technology is perceived to help those in most need. Outside of issues concerning the robustness of the technology, data protection is the paramount issue of concern for end users. Therefore embracing the use of PETs and FIPs may better address user fears to personal privacy and negative use of data collected (e.g. to drive up health insurance costs or track and exploit social behaviours) by companies providing the technology, third parties and/or research teams.

While invasive technologies can collect a wide range of user data, that which raises significant concern is visual data. With camera technology now wide spread via the use of smart mobile devices the conversation to the use of visual data is of international concern. A key topic which regards ageing well communities is the use of cameras and visual data to support caregiving at home or in care facilities. Acceptance fluctuates, with relatives of those receiving care most in favour of their use. However with the recepients of care and their caregivers responses mixed, further research is required to understand more about user preferences to aid the acceptability of visual data. This is supported by recent studies such as BREATHE, which suggest that as privacy around camera technology improves so too does user perception of their added value and acceptability of use.

Overall advances in computer vision with reduction to the cost of technology has enabled cameras to become intelligent monitoring healthcare systems. Transitioning concepts for use in healthcare from areas such as personal security, entertainment and public safety provides exciting future opportunities for research and development. However a constant barrier to be addressed is that of personal privacy from the data collected. Thus advancing approaches to data visualisation in areas such as privacy-by-context and visual privacy (before or after images are acquired) is important. The fundamental principle governing this research is find a balance between protecting images and ensuring those images still remain intelligible to the key outcomes they are intended to support (e.g. when used in a care setting that they may mask the identity a visitor but it can be clearly determined a visitor is present). Finally future research to explore and improve the ability of cameras to work in parallel with additional sensing systems (e.g. wearables) may provide exciting promise to advancing their overall use and support for ageing well communities.

References

[1] A. Cavoukian, *Privacy by design: The 7 foundational principles*. Information and Privacy Commissioner of Ontario, Canada, 2009.

[2] PISA Consortium, *Handbook of Privacy and Privacy-Enhancing Technologies—The case of Intelligent Software Agents*, G. Van Blarkom, J. Borking, and J. Olk, Eds. College bescherming persoonsgegevens, 2003.

[3] Commission of the European Communities, *COM(2007) 228, Communication from the Commission to the European Parliament and The Council on Promoting Data Protection by Privacy Enhancing Technologies (PETs)*, 2007.

[4] Organisation for Economic Co-operation and Development, *The OECD Privacy Framework*, 2013.

[5] W. L. Zagler, P. Panek, and M. Rauhala, "Ambient assisted living systems – the conflicts between technology, acceptance, ethics and privacy," in *Assisted Living Systems—Models, Architectures and Engineering Approaches*, ser. Dagstuhl Seminar Proceedings, A. I. Karshmer, J. Nehmer, H. Raffler, and G. Tröster, Eds., no. 07462. Dagstuhl, Germany: Schloss Dagstuhl – Leibniz-Zentrum fuer Informatik, Germany, 2008.

[6] W. Wilkowska and M. Ziefle, "Privacy and data security in e-health: Requirements from the user's perspective," *Health Informatics Journal*, vol. 18, no. 3, pp. 191–201, 2012.

[7] M. Ziefle, C. Rocker, and A. Holzinger, "Medical technology in smart homes: Exploring the user's perspective on privacy, intimacy and trust," in *2011 IEEE 35th Annual Computer Software and Applications Conference Workshops (COMPSACW)*, July 2011, pp. 410–415.

[8] J. Clark and M. McGee-Lennon, "A stakeholder-centred exploration of the current barriers to the uptake of home care technology in the UK," *Journal of Assistive Technologies*, vol. 5, no. 1, pp. 12–25, 2011.

[9] "Surveillance technologies in care homes: seven principles for their use," *Working with Older People*, vol. 19, no. 2, pp. 51–59, 2015.

[10] HC-One Limited, "HC-One's Consultation on an Opt In Visible Camera Scheme," http://www.hc-one.co.uk/media/image/Visible_Camera_Consultation_Report_Oct_2014_LATEST.PDF, 2014, (accessed in March 2016).

[11] Care Quality Commission, "Using surveillance information for providers of health and social care on using surveillance to monitor services," http://www.cqc.org.uk/sites/default/files/20150617_provider_surveillance_information.pdf, 2015, (accessed in March 2016).

[12] Care Quality Commission, "Thinking about using a hidden camera or other equipment to monitor someone's care?" http://www.cqc.org.uk/sites/default/files/20150212_public_surveillance_leaflet_final.pdf, 2015, (accessed in March 2016).

[13] B. Mittelstadt, N. B. Fairweather, N. McBride, and M. Shaw, "Privacy, risk and personal health monitoring," in *ETHICOMP Conference*, 2013.

[14] BREATHE Consortium, "Final Project Report," http://www.breathe-project.eu/gallery/16/D63_-_Final_Project_Report_v10.pdf, 2016, (accessed in March 2016).

[15] A. A. Chaaraoui, P. Climent-Pérez, and F. Flórez-Revuelta, "A review on vision techniques applied to human behaviour analysis for ambient-assisted living," *Expert Systems with Applications*, vol. 39, no. 12, pp. 10 873–10 888, 2012.

[16] A. K. Dey, G. D. Abowd, and D. Salber, "A conceptual framework and a toolkit for supporting the rapid prototyping of context-aware applications," *Human–computer Interaction*, vol. 16, no. 2, pp. 97–166, 2001.

[17] N. Bricon-Souf and C. R. Newman, "Context awareness in health care: A review," *International Journal of Medical Informatics*, vol. 76, no. 1, pp. 2–12, 2007.

[18] J. R. Padilla-López, A. A. Chaaraoui, F. Gu, and F. Flórez-Revuelta, "Visual privacy by context: Proposal and evaluation of a level-based visualisation scheme," *Sensors*, vol. 15, no. 6, p. 12959, 2015.

[19] J. R. Padilla-López, F. Flórez-Revuelta, D. N. Monekosso, and P. Remagnino, "The "Good" Brother: Monitoring People Activity in Private Spaces," in *Distributed Computing and Artificial Intelligence*. Springer, 2012, pp. 49–56.

[20] R. Planinc and M. Kampel, "Introducing the use of depth data for fall detection," *Personal and Ubiquitous Computing*, vol. 17, no. 6, pp. 1063–1072, 2013.

[21] Z. Cheng, L. Qin, Y. Ye, Q. Huang, and Q. Tian, "Human daily action analysis with multi-view and color-depth data," in *European Conference in Computer Vision (ECCV). Workshops and Demonstrations*. Florence, Italy: Springer, Berlin Heidelberg, 2012, pp. 52–61.

[22] S. Tao, M. Kudo, and H. Nonaka, "Privacy-preserved behavior analysis and fall detection by an infrared ceiling sensor network," *Sensors*, vol. 12, no. 12, pp. 16920–16936, 2012.

[23] S. Fleck and W. Straßer, "Smart camera based monitoring system and its application to assisted living," *Proceedings of the IEEE*, vol. 96, no. 10, pp. 1698–1714, Oct 2008.

[24] T. Winkler and B. Rinner, "Sensor-level security and privacy protection by embedding video content analysis," in *18th International Conference on Digital Signal Processing (DSP)*, Santorini, Greece. IEEE, 2013, pp. 1–6.

[25] J. R. Padilla-López, A. A. Chaaraoui, and F. Flórez-Revuelta, "Visual privacy protection methods: A survey," *Expert Systems With Applications*, vol. 42, no. 9, pp. 4177–4195, 2015.

[26] S. E. Hudson and I. Smith, "Techniques for addressing fundamental privacy and disruption tradeoffs in awareness support systems," in *Proceedings of the 1996 ACM Conference on Computer Supported Cooperative Work (CSCW)*. New York, NY, USA: ACM, 1996, pp. 248–257.

Chapter 20
Human-centred design with older adults: examples and recommendations for research, ideation and testing

Ana Correia de Barros and Ana Vasconcelos**

20.1 Introduction

As the population ages, the European Union (EU) seeks technological solutions that will be able to assist older adults in leading healthy, active, joyful and dignified lives. In its call to action the EU has also, namely through the Active Assisted Living Programme (AALP)[1], encouraged the groups engaged in the development of these new solutions to involve end-users in the design process. Literature is profuse in examples of the application of human-centred design (HCD) principles, participatory design or co-creation with different types of end-users. On the other hand, it might be difficult for researchers and developers to find concentrated examples of, and reflexions on, the application of these methods and techniques with older adults.

Older adults are an extremely heterogeneous group and there are specific aspects to take into account when engaging in these design approaches with old participants. In this chapter we guide the reader through the principles and practice of HCD with older adults by revising the existing literature in the field, providing examples of its application and laying out recommendations for the different phases of research, ideation, testing and refinement.

More than just descriptions of HCD methods, older adults as end-users and the subsets thereof, this chapter intends to serve as a companion to practitioners and researchers in the field of Active Assisted Living (AAL) to support all stages of the HCD process.

Publicly funded research programmes are requesting proposals to include users in the development of Information and Communication Technology (ICT) solutions [1]. Notably, in the AAL field, AALP specifically requests proponents to present details on the perspective of end-users, the pilot application and the exit strategy, denoting concerns with the development of useful, usable and acceptable technology, and also a concern with the ethics principles of involving users in the design and validation

*Associação Fraunhofer Portugal Research, Rua Alfredo Allen 455, 4200-135 Porto, Portugal

[1] Formerly known as the Ambient Assisted Living Joint Programme (AAL JP).

of the solutions. This approach is a people-centred one: the system will be shaped according to users' requirements, aspirations, needs, abilities and opinions.

> *'Determining actual user needs instead of merely guessing or generalising can make the difference between a* ***real innovation*** *for users or simply an interesting technical development for the shelf.'*[2]

The AALP wishes to foster the creation of useful technologies tapping the challenge of an ageing society in Europe. To that end, it is focused on supporting research projects that result in market-ready solutions. Based on the conviction implied by the quotation above, the AALP has commissioned the development of documents[3] to assist organisations partaking in the programme in the involvement of end-users throughout the project. The system, the services and the product itself need to efficiently mediate the relationship of the user with technology. The goal of a user-centred interface, as explained by Donald Norman [2], is to provide 'intelligent, understandable tools that bridge the gap between people and systems: convivial tools' (p. 55).

Involving users in the design process is part of a long-standing tradition in design, namely *participatory* or *cooperative* design, with roots in Scandinavia [3]. According to Carroll and Rosson [3], there are two propositions in participatory design: the *moral* and the *pragmatic* one: the former claims that users 'have a right to be directly included' (p. 243), while the latter claims that 'users' input will increase the chances of a successful design outcome' (p. 243). This is particularly easy to imagine in the case of people with disabilities: designers can only go so far in trying to put themselves into users' shoes and many aspects of a life with a disability cannot be imagined or enacted by a person who is not disabled in the same sense. Indeed, Gregor, Newell and Zajicek [4] have noted that, even in the cases where products are designed to accommodate the needs and abilities of people with impairments, these are often focusing on young people with impairments and, therefore, these products, too, fail to accommodate the diverse needs of older users. Failing to consider the widest possible range of end-users, regardless of age, height, abilities, culture or other distinctive aspects, will lead to the creation of products and services that create exclusion. Placing users at the centre is a step in the direction of creating inclusive products and services. This is the philosophy behind *Inclusive Design* [5] and the specifics of the ageing process are what motivated Gregor, Newell and Zajicek [4] to further advocate for *User Sensitive Inclusive Design*.

When designing a product, there are three types of users one might consider: primary, secondary and tertiary. In AAL-related systems, we could consider respective examples to be: a senior, a caregiver and a care institution. In AAL, beyond considering the widest possible range of users, there is a context that must be taken into account and several stakeholders that will, directly or indirectly be affected by the design solution. This is why, in the course of this chapter, we will favour the use of the

[2]Active Assisted LivingProgramme (n.d.), Why do we care so much about users? Available at: http://www.aal-europe.eu/support-to-projects/end-users/, last access: June 2016.
[3]Active Assisted Living Programme (n.d.), *End users*. Available at: http://www.aal-europe.eu/support-to-projects/end-users/ (highlight in the original), last access: March 2016.

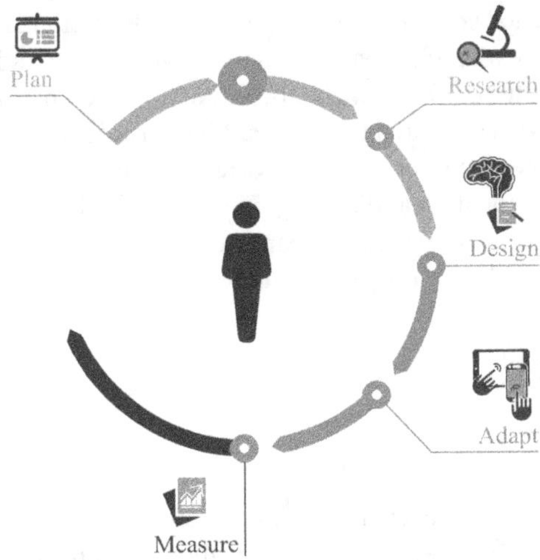

Figure 20.1 Human-centred design process

term 'human-centred design' (HCD), as we consider it to be more wide-ranging than 'user-centred design'.

The process of HCD is a flexible one, based on frequent iterations and cycles. It includes user research, ideation, testing and implementation and, ideally, involves users at all stages. There is always the possibility to return to any previous stage (Figure 20.1).

Apart from the documents provided by the AALP, there are toolkits available to guide researchers, designers and developers through the process of HCD. One comprehensive resource is the *IDEO HCD toolkit*.[4] In a book format, the toolkit is built with practical examples of the application of each method and is available for download free of charge. The same design consultancy company that published the toolkit has recently employed 90-year-old Barbara Beskind[5] in their design team and has created an online repository of methods to be used in HCD, called *Design kit*.[6]

Despite the available toolkits, in relation to specific user groups, as is the case of older people, there are aspects one should take into account when adopting HCD and its methods. This chapter will try to assist the reader in relation to these particular aspects. It will do so by following a structure that maps to the natural cycle of HCD and by compiling the recommendations and experiences from several researchers in the field, as well as the authors' own experience in HCD within a user network – COLABORAR[7] – created for the purpose of designing new technologies with and for older people.

[4]IDEO HCD Toolkit: http://www.ideo.com/images/uploads/hcd_toolkit/IDEO_HCD_ToolKit.pdf, last access: March 2016.

[5]Barbara Beskind's profile on IDEO's website: http://www.ideo.com/people/barbara-beskind, last access: March 2016.

[6]Design kit: http://www.designkit.org/methods, last access: March 2016.

[7]This user network is an initiative of Fraunhofer Portugal AICOS: http://colaborar.fraunhofer.pt/, last access: March 2016.

The intention is not so much to list the available methods and describe them in detail, which the reader may easily find elsewhere, but rather to pinpoint the specificities of using this approach with older adults. We begin by addressing the issue of recruitment and logistics related to getting older participants together for activities. Then we move to three different sections corresponding to sequential phases in the process: user research, ideation and testing. Finally, we highlight some characteristics that should be present in the research team, as well as some reflexions and ethics concerns that have been laid out by other authors.

20.2 Practicalities

20.2.1 Characteristics of old age

When it comes to designing user interfaces for older people, while there are many questions left unanswered, there are relevant guidelines in the literature regarding general recommendations for mobile devices [6, 7], design patterns recommendations [8], gestures on mobile devices [9], device manipulation [10, 11], navigation and information architecture [12, 13], graphic design [10–12, 14], and content [15]. These guidelines are mostly derived from the acknowledgement that there are motor, cognitive and sensory changes that are likely to come about with old age.

Despite being very relevant and useful, taken alone, guidelines for designing user interfaces for older users may also divert designers from the most fundamental issues. Lindsay and co-workers [1] argue that focusing on what makes older adults different from younger adults may bring designers away from 'considering older people as complex individuals with their own set of social and emotional needs and desires'.

Nevertheless, the characteristics that motivated the development of guidelines are also likely to impact the engagement of older people in design activities. Common challenges found have to do with recruitment, maintaining participants' focus during sessions, getting people to envision abstract or intangible concepts [1], accommodating characteristics of people with dementia [16] or finding common ground for people with diverse interests, motivations and backgrounds [17]. Some of these will be looked at in more detail in the following sections.

20.2.2 Recruitment and logistics

Different authors point out that recruitment techniques traditionally used with younger participants (e.g. public advertisements) were unsuccessful when used to recruit older adults. This is likely to be related to trust issues, older people's motivation and increased time spent in isolation [18–22].

Precisely due to trust, recruiting participants through institutions, such as associations, clubs or residences, may be a good starting point. We can explain our purposes to the staff or person responsible at the institution, the gatekeeper [23], who already has a relationship of trust with the potential participants and can then easily establish the bridge. From our experience and from reports in the literature [22], researchers will be requested to present their team and their goals to the group of older people.

Furthermore, we also know that, once an institution has agreed to participate, the first activity with older adults should be carried out within a short time frame [24], otherwise institutions and older adults might consider that their agreement or the work is not being taken seriously. At the beginning of the activities, the purpose of the research, its relevance and expected outcomes should be clearly explained to participants. Furthermore, it is advisable to share an estimate of when the results will be available.

Activities towards the recruitment of new participants must start well ahead of the first participatory activity for the project. On the one hand, until final authorisation for participation is granted, there might be bureaucracy issues to be dealt with, such as the signature of collaboration protocols between the research institution and the institution granting access to older adults. Then, there is the time needed to build trust between the researcher/practitioner and the informant/older person. Ellis and Kurniawan [19] report that they held a 12-week workshop related to ICT skills for older adults before they first requested older adults to partake in the participatory design process for their project.

There are also differences in recruiting for a single project or for a long-term relationship between institutions that will serve multiple projects. In the latter case, it is advisable to do an initial assessment of the group of older people to identify who would be willing to participate and to identify important characteristics (e.g. literacy level, cognitive and communication abilities, motor abilities or illnesses) [18]. Indeed, pre-screening before any research activity is advisable [23]. One aspect mentioned in the literature is the difficulty in recruiting a large number of older participants. When doing recruitment for a user network serving many research projects, the initial assessment is also useful in this sense. For instance, if a project requires older people with rheumatoid arthritis, there will already be a starting point to identify the institutions that have access to people with such characteristics.

When recruitment for multiple projects is the option, a trust person from the side of the research institution is also recommended in order to establish the bridge to old participants. That is to say, participants may be accustomed to a group of researchers due to a given project, but be unacquainted with a different group that comes afterwards for another project. Finally, when creating these larger user networks, it is important to manage the network in order for it to be ready for demands. A practical example: in developing solutions for older users, often we require novice smartphone users to assess how easy it is for them to interact with a system for the first time or learn how to use it; once a group of users has interacted with smartphones through usability tests, they will no longer be eligible for this type of tests, thus requiring the user network to, at all times, have a group available in these conditions which are frequently required for the projects.

Chisnell and co-workers [18] recommend reminding the participants of the session's right before they take place: participants should be reminded of the date and place, the test conditions (e.g. participants will be recorded), they should be reminded of bringing their glasses or relevant daily living aids, and they should also be reminded of eating first. When scheduling the sessions, different arrangements must be made depending on where the sessions will take place. It is possible that participants are expecting food to be available if the session is a group activity held at the research institution's facilities. When preparing something to eat, bear in mind food restrictions, chronic illnesses and prepare healthy food. Another four tips were given by Chisnell and co-workers [18] regarding preparation and logistics: prepare the space to receive participants' spouses or caregivers, be prepared to receive participants ahead of time, as they tend to arrive early, avoid setting up sessions on rush hour, and consider having the sessions in the morning, as this is the time of the day when older people are likely to be more attentive.

Within the facilities, the rooms where the activities will take place should also be highly accessible, i.e. not requiring participants to climb stairs or walk long distances [23]. When available, the session may be conducted in an informal setting to make participants more comfortable and at ease. A room with too much equipment that is not familiar to the participants or with many people can create a stressful environment and hinder their performance. As an example, a room within a living lab would be a good setting for a session. Along with the common pack with information about the research project, participants should be given clear instructions about where the activities will take place [23]. Ideally, the instructions should contain bus routes and information about parking. We have found that some people have difficulties reading a map; therefore, it is advisable to quickly test the instructions to make sure they are clear.

A basic recommendation for planning sessions with older participants is to keep them short. If the protocol cannot be shortened, consider splitting it into two or more sessions [25]. Of course this will have implications on the calendar and project budget, so it must be thought of in terms of compromises that must be made. Another issue that may affect the project budget is insurance. If you are planning to bring older participants over to the lab for validation (e.g. activity monitoring, 'exergames'), you might need to consider insurance for participants and this should have been planned in advanced on the project budget. In order to keep things simple, we are leaving a short checklist for this phase on Table 20.1. Nevertheless, a more comprehensive table may be found on Dickinson and colleagues' publication [26, p. 10].

A consent form should be prepared and should be read out loud to participants, allotting enough time for questions and answers. Often researchers would like visual materials gathered during the activities to illustrate their reports. Consent to use

Table 20.1 Checklist for preparation of participatory sessions

Authorisation needed from ethics committees?
Authorisation required from public data protection agency?
Insurance needed?
Specific type of users (hard to come by?)
Does the involvement in previous sessions bias the participants in any way?
Need to provide transportation?
Informed consent for data collection?
Consent for the use of images for dissemination purposes
Duration (splitting into 2 or more sessions?)
Task complexity (reduced mental burden and reduced functional ability)

participants' image should also be requested. In our particular case, we use two different consent documents.

Regardless of the stage in the HCD one is in, there are aspects that should be taken into account. Attention should be given to the comfort of participants and researchers should look for places with reduced background noise and interference from others. When talking to older people, both 'elderspeak' [27] and difficult technology jargon should not be used. Researchers should also bear in mind that what seems straightforward to them may be alien to another person with less experience in technology:

> *'Older people spent their formative years when a chip was a piece of wood or cooked potato, hardware was nuts and bolts, a window was made of glass, a monitor was a school prefect and software was not even a word.'* [28]

Beyond being open to embrace a different vocabulary, researchers should also respect participants' pace, respect people's time (the duration of the test should not exceed what participant were told it would last), they should explain how participants' inputs are valuable and promote participants' self-confidence [25, 29]. The last item has something to do with trust. In a researcher-participant relationship there is a power relation and, being less knowledgeable about technology or design possibilities, older participants may refrain from participating as openly as they and the researchers would like to [29]. Hands-on experiences and scenarios have shown to be useful in overcoming the initial difficulties in communication: 'Inspiring the participants is as important as eliciting information…' [29].

When it comes to communication, one should adjust the volume and repeat (or paraphrase) as many times as necessary, give participants time to think, respect people's opinions and listen to people's stories [25]. There is a tendency to associate older adults with what is known as 'off-target verbosity' [30], which consists of talking outside of focus or topic. While it is part of the stereotype, it is not common to all older people [30]. However, should this happen, researchers should listen to participants' stories and carefully lead them back on track. Finally word-finding problems are also to be expected, as they increase in old age [30].

Finally, in any research and design activity, participants should be given a chance to witness the outcomes. This can be done by preparing a specific demonstration or workshop or by inviting participants to events of results' demonstration.

There are useful strategies to recruit and maintain a group of older participants for design and research activities, such as celebrating anniversaries or creating leisure activities. These and other strategies were described in more detail elsewhere [24].

20.3 User research

Examples of methods that can be used in order to understand users, the context of product/system use, the various stakeholders and users' needs, hopes or aspirations are questionnaires, focus groups, observations, contextual inquiry, participant observation, user reported data (e.g. diary studies) or interviews.

Having contact with real potential users at the stage of requirements gathering is of great value, as designers would likely fail to consider relevant aspects by themselves. For instance, Eisma and colleagues [29] report that, while asking a stroke survivor to compare two mobile phones, the researchers were expecting the interviewee to comment on button size, but she commented on phone size instead – it had to be small enough to fit in her pocket because she could not use a handbag due to hemiplegia. In another example, Dickinson and colleagues [31] report that only by witnessing how an old interviewee went about picking up her phone from the floor using the walking stick, did the researchers understand that the phone had to resist strong blows. This close contact can reveal people's motivations to accept or reject technology. Yoxall and colleagues [32] have reported that social dynamics may play important roles which designers alone could fail to consider in the design of a lid opener – having to request somebody for help to open a jar lid may create a social situation which is given a higher value by the older person than being able to do it herself. In research about stroke survivors Correia de Barros and co-workers [33, 34] found a wide range of examples of how people and their relatives created spontaneous inventions to cope with everyday challenges in activities of daily living.

It is important to be aware of the gap that may exist between researchers and older people being researched in relation to knowledge of the world and, in particular, knowledge of technology: this influences the vocabulary people use, the references and the ability to imagine certain concepts. It works both ways: both researchers and participants have a unique knowledge of the world that may not be shared by the other. This is why some authors suggest planning activities to create a common ground [29] or just to build a 'cohesive group' [35]. Examples of these activities involve having group members on a session to expose prejudices about each other [35], having hands-on activities with technology – which are much more efficient in describing the technology than just talking about it [29] – or taking the time to engage in discussions with older interviewees beyond what was planned [36].

One common way of gathering insights about participants' real life without the presence of researchers is through self-reported instruments (such as the diary studies mentioned earlier). However, these may have some drawbacks with older participants in the end because participants did not understand the instructions, had limited literacy, forgot, or did not know which aspects were relevant to report [23]. One of the authors of this chapter once created diaries which resembled games for older participants and handed them out. To her surprise when she went to collect the diaries, one of the participants had given hers to her grandson because she thought it would be more fun for him to fill in.

Detailed examples and advice to use focus groups, surveys or interviews are given in Fisk and co-workers [14] and Pak and McLaughlin [12]. Park and Schwarz [2000, cited in (31)] say that older people need a higher 'threshold of certainty' before selecting an option on a questionnaire, hence the higher rate of 'don't know' answers found. Other pieces of practical advice for questionnaires are: to use 14-point font in a printed questionnaire, to repeat scale headings if they break from one page to the next, to pre-test the questionnaire and to pay attention to the order or the questions [12].

We would like, however, to draw the readers' attention to qualitative in-home user research. The reason for this is two-fold: (1) we have seen little reflexion on this type of approach in AAL, and (2) experience and literature suggest that this approach brings very rich insights into user requirements. The particular approach we are referring to is ethnographic research, which comes from anthropology practice and consists mainly of fieldwork to allow for detailed documentation about people in their natural environment. For years it has been used in design. The researchers using it for design purposes highlight its benefits [31, 37–39], even though its use is not free of controversy.

To begin with, getting into an old participant's house inverts the power roles. In this case, researchers are guests and this may contribute to participants' confidence [31] which, in turn, may provide for a richer interaction and more relevant data to analyse. Secondly, it is a very useful approach to tap the types of issues described at the beginning of this section: observing people in their natural environment, having the chance to pose questions in relation to tangible artefacts or requesting people to demonstrate how they go about performing any given activity are examples of things that are not possible when user research is conducted at the research centre's facilities. Furthermore, being surrounded by familiar objects may prompt participants' memories [31].

Notwithstanding, the access to this type of rich data should not be taken for granted, as it also depends on the interlocutors. As with questionnaires, questions must be posed carefully. Asking a person what she usually does will retrieve a completely different reply than asking a person to guide the researcher through a normal day's activities from the time she gets up to the time she goes to sleep. Abeele and Van Rompaey [40] exemplify how hidden true meanings of respondents' motivations can only be unveiled through ethnographic techniques. In order to engage in ethnographic-like fieldwork, researchers may not just count on observations. As clearly described by Hirsch and colleagues [38] in their work for the ELDER project, in-home observations complemented with a literature review, self-documentation by seniors (see also, for example, Abeele and Van Rompaey [40]), and lifestyle interviews with both elders and caregivers may complete a rich picture that can unveil potential users' emotional concerns, aspirations, fears and ideas about what 'quality of life', 'independence', 'engagement' and 'care' mean. An example of cases summarised after ethnographies of older adults may be seen on project ATHENE website.

The involvement of other stakeholders during the user research phase can enrich the gathered information and provide different insights on a particular subject. Often, when designing for users with cognitive impairments (e.g. dementia or people with Alzheimer's disease) caregivers are the main source of information with patients not always being directly engaged. While caregiving is generally a full time responsibility that compels caregivers to immerse themselves into the care receiver's life [41] their point of view over specific aspects may be affected by caregiver bias [16]. As an example many assistive technologies are safety-oriented as a consequence of being designed with the caregivers' agenda in mind, when in turn care receivers could have given more focus to other aspects such as social interaction [42]. All things considered, caregivers' input should be analysed critically and faced as an addition that should not replace the primary users' own opinions.

20.4 Ideation

In ideation phases, trust is also to be taken into account. Trust can be built prior to the design activities [19], by exercises at the beginning of the sessions [35], during the group activities [20] and also guaranteed by recruitment itself. For instance, Rogers and colleagues [43] recruited groups of friends to collaborate with each other in co-creation sessions. When bringing together different stakeholders to design an AAL solution, a single group may contain end-users, nurses, engineers or others. Points of view and jargon may be different. This is why, also in the ideation phase, it is important to engage in activities to work on team relations and trust, as explained in Section 20.3.

Groups can be used for ideation and usually include between 6 and 10 people [Morgan (1997), cited in Goodman *et al.* (44)], however, researchers who have engaged in co-creation groups with old people recommend splitting participants into smaller groups [20, 44, 45], keeping sessions short [1] and making use of breaks. One common strategy is to have the groups work separately and later bringing them together to discuss the ideas they have generated or to engage in other activities altogether [23, 44]. There are reports in the literature of groups including only seniors [43], seniors and caregivers [45] or seniors and designers [40]. Another strategy to make the most out of the session and leave participants at ease is to pair abilities (e.g. person who can draw with another who has hand tremor, person who can read with another who has a low literacy level), as reported by Massimi and colleagues [20].

Finally, there are several reports of problems in keeping the group's attention throughout the session, even in sessions that took 1.5 h including break [46]. Some authors have noted that participants were easily distracted, that the group would be silent after a question or that participants could become fixed on a topic and struggle to shift attention to the current discussion topic [46]. This could have to do with memory and attention impairments, difficulties in communication, because of unfamiliar people and/or environment or just because the topics and/or the session are not motivating for the participants. Not only the way in which the session is planned, but also the moderator, play an important role in creating a favourable, comfortable and enjoyable environment, as explained in Section 20.6.

In order to conduct the ideation sessions, different stimuli can be used, such as photographs [20, 44], cards [47], scenarios [20, 45], videos [1] or theatre [48, pp. 129–143] – the advantages and disadvantages of which are discussed by different authors, even though all are recommended to be used with older participants.

The process of co-design follows the same steps as in any design process. Therefore, after the user research phase, several authors engage in a series of steps that will gradually lead to the final design. Common reports in the literature are to have the following steps: brainstorming, low-fidelity prototyping, medium-fidelity prototyping and high-fidelity prototyping [21]. Low-fidelity prototypes are very much used as a means to validate the positioning of elements on the user interface, the legibility and meaningfulness of the icons, and to stimulate discussion [21, 49]. However, most of the examples in the literature report researchers coming to users with preliminary drawings for discussion rather than drawing from scratch with the end-users. Rice and Carmichael [50] have tried having older participants draw, but even though it was found useful to support group discussion, the authors encountered difficulties: some people would prefer to write down their ideas and others, about a half of the participants, were reluctant to draw. Davidson and co-workers, on the other hand, have used drawing and made no comments on challenges faced by choosing this approach [17].

When it comes to the input, different authors note that it seems easier for older participants to criticise existing things and situations than to propose new solutions [20]. However, critiquing existing products before engaging in creative activity to create a product in the same category has been found to be harmful in the sense that it primed participants and led to lower levels of creativity [17]. In their study of participatory design with older people, Davidson and Jensen [17] found that prior experience with smartphones had no negative implication on the level of creativity that participants could bring to the design of a smartphone application, having indeed found that the most creative idea for an application, as evaluated by an external panel through a creativity scale, came from a group with no prior experience with smartphones and with no critiquing activity prior to the creation phase itself.

An example of a method specifically designed for participatory design with older people is the OASIS (Open architecture for Accessible Services Integration and Standardization) approach. This approach was designed to take into account the challenges mentioned in the literature and presents a specific task list, along with respective duration, that tries to move away from abstract concepts, focus older participants on everyday life and leave space for younger researchers and designers to easily find insights for requirements [1]. Using previously created video prompts, the approach involves two stages: (1) 'Exploratory meetings' divided into information gathering, prompt viewing, scenario generation, claims analysis, feature envisioning and scenario envisioning – all adding up to little over 2 h [1], and (2) 'Low-fidelity prototyping' using creation work based on the PICTIVE process [Muller (1993), cited in (1)]. The success of the latter stage, the authors say, largely depends on the conditions for running the session. Too large a group, wrong motivation or personality clashes may hinder the flow of the session and the quality of the outcomes [1]. PICTIVE [51] uses office materials (post-it, markers) and other low-fidelity representations for conjoint collaboration in the design of user interfaces. It was created to bridge the gap between the lay user and the developer in participatory design. It furthermore uses video to record the session. Beyond the account from Lindsay and colleagues [1], other authors have reported on the use of the method with old participants [50].

Despite being preferably arranged into short sessions, the involvement in participatory design activities can be quite intense and extend for a long time. Therefore, it is likely that the group and relationships within it will grow along with the project. This brings about ethical issues (exit strategy, participants' expectations) and also pragmatic issues in relation to the quality of the work that is coming out [35]. For instance, after continuous work on co-design sessions, Clemensen and co-workers [35] found that conducting the pilot tests with the same group of users might not be advisable because they might at this point lack critical ability.

20.5 Testing

In this section we focus on short evaluation tests, long-term trials and in-home testing. The modalities differ in time, effort and conditions required.

20.5.1 Short evaluation tests

The most common of the modalities is short evaluation tests. These are mostly reported in the literature as tests to assess usability of a given ICT solution. These tests can be conducted either on-site or off-site. While inviting participants to the lab has the advantage of a better controlled environment and flexibility in terms of equipment to use, sometimes the option might not be viable either because there is no budget available to pay for participants' travels, because it takes longer than meeting the participants off-site, or other reasons. The steps needed for the arrangement of the sessions are already described in Section 20.2.

In usability tests the researcher is in a power position. There are anecdotal accounts of older participants being more reserved when facing more than one researcher at a time [52]. One important aspect to bear in mind is to take special care to guarantee that participants understand the test [21] and then relate the test to the participants' world [25]. For instance, while with proficient technology users one can adopt a standard approach to a usability test and ask people to perform certain tasks, with novice technology users it might be beneficial to provide context for the task. For instance, one may want to know whether or not the participant can read the daily step goal on an activity monitoring graphic. Rather than asking what the daily goal is, it is probably better to take the time before to explain that the smartphone has the ability to count the user's steps and that there is a minimum number of daily steps considered to be healthy. In much the same way, role-playing can help bridge the gap between the technology presented during the test and participants' reality [25]. Given the numerous accounts of older people, who are novice technology users, believing that they do not and will not have the ability to interact with new technologies [18], it is especially important that researchers facilitating the tests do not forget to make it clear to the participants that they are not being tested [23, 25].

Novice technology users tend to take longer in completing the tasks than more proficient users. This means that researchers conducting the test have to give participants time to think [25] and it also means that the number of tasks in a test protocol must be controlled, because, in practice, the test may easily extend beyond what was planned. While pre-testing the protocol, take into account not only the strict time for task completion, but bear in mind that participants may take the opportunity to engage in conversations which are not directly related to the test [18, 25, 52]. In order to capture more details about how participants are making their decisions while trying to complete a task, it is common to ask participants to verbalise their thoughts while engaged in the task. This is called 'Think aloud' and it is often encouraged; still, there are reports of authors being unable to use it because it was too difficult for participants [21, 23, 26, 53]. However, there are completely different results depending on the background of participants, as shown by Dickinson and colleagues [26], who report how a retired surgeon was remarkable in his ability to engage in think aloud due to his professional experience.

Finally, one course of action that may be advisable is to pre-test the system with more technology-proficient users than the target group. Sometimes the usability problems are not related to a specific medical condition or impairment – they are general design problems. Moffatt and co-workers [21] tested the design with a different group other than the target audience to eliminate these general design flaws before taking back the system to their target users for testing.

20.5.2 Long-term trials and in-home testing

Long-term and in-home trials are useful not only because they provide the chance for the researchers to evaluate the performance of the system in a natural environment and its suitability to users' everyday life, but also because it allows the opportunity to evaluate the system in the context of social dynamics which are difficult to grasp in short usability tests. As we mentioned in the beginning, Section 20.1, considering different types of users for a single solution is of utmost importance in AAL, as noted by Ballegard and co-workers [54]. In long-term trials, which are common in research projects, representative users may be requested to have the AAL system installed in their homes. Naturally, the system will be a strange element which can, at times, not be very well accepted by the participants [54]. Furthermore, this type of testing allows for the placement of the product within the ecologies [26] which should be embraced in the evaluation, as they represent a closer to natural environment where the system will be used. This means that other stakeholders should be accounted for. Does the person receive visitors? Will the visitors interact with the technology? Are there other stakeholders who can remotely interact with the participant? Neves and co-workers [53] report, for instance, that the system they were testing was supposed to allow the interaction with a caregiver, but during the trial they found that one of the caregivers did not know how to interact with the technology, even though she possessed the device.

When installing or delivering the equipment to participants it is fundamental that they understand the purpose of the evaluation and have the right tools to begin the experiment. These can be in the form of a brief explanation of the product given by the facilitators or written guides and manuals made available to the participants. It is also important to have users experiment with the devices demystify any fears that may arise. Will the equipment break while I use it? Am I responsible for it? By the end of an initial introductory session, facilitators should make sure that there are no questions left unanswered and that participants are able and comfortable to proceed with the trial.

Naturally, when an AAL system is undergoing a long-term trial, support to the users must be available at all times. The support may be technical or social. Rather than leaving people on their own for months, it is courteous to have regular contact and ask about how the test is going. Participants may also take the opportunity to ask about technical issues which, otherwise, they would not ask for fear of bothering researchers or just because they thought it was not worth the trouble.

Different motivations underlie the decision of older people to take part in long-term trials of technology, e.g. a desire to contribute for new AAL technology, curiosity, loneliness, amongst others [Grönvall and Kyng (2012), cited in Novitzky *et al.* (55)]. However, as explained by Dickinson and co-workers [31], 'the stability of the home environment is of particular importance' and badly arranged trials or the introduction of 'inappropriate technology' [31] may not only be unfruitful for research but, most importantly, cause serious harm to the participants. The harm can come in the form of physical, psychological or social harm. Therefore, careful strategies must be planned and put in place to accommodate for these and other relevant ethics aspects that may arise, especially when the target group involves fragile users [55].

Long-term trials naturally raise several ethics issues, for which there are several guidelines available. A comprehensive review was recently published by Novitzky and colleagues [55]. One important aspect to take into account is that laws may differ from one country to the next, thus being advisable to refrain from adopting a standardised approach.

20.6 The design and research team

In a study published in 2013, the authors report young researchers' challenges, experiences and strategies in engaging with HCI research with older people [52]. The results emerged from interviews with a group of young researchers who had built all their professional experience in research and development for and with older users. A summary of the themes emerged may be found in Table 20.2.

Because of the heterogeneity found amongst the user group of older adults and due to cultural differences, many of the aspects have to rely on common sense [18] and ability to improvise during the sessions. Reports on the importance of having a well-trained and informed research and facilitator team are common in relation to HCD with older users [17, 19–21, 35, 46]. Therefore, we advise the research teams to invest in this specific training with enough time in advance, passing on the knowledge from the senior researchers to the younger ones. A poorly designed

Table 20.2 Difficulties, experiences and main strategies of young researchers working in HCI related to old users

Difficulties	Experience	Main strategies
User management	Older adults' behaviour	Time-saving
Introducing the test sessions	Bias	Results-improving
Conducting the test	Time	
Test planning	First few minutes	

protocol or script, as well as a poorly trained researcher, may render the data useless and, worse than that, it may cause discomfort to participants. In the results from the interviews to young researchers [52] we have found that, without being instructed to do so, researchers would naturally engage in behaviour that would take their time and not produce relevant data in favour of participants' comfort. For instance, when conducting tests off-site (e.g. day-care centre) it is often the case that non-eligible old people want to participate. If this is the case, tests (if not potentially harmful) should be conducted with these participants as well and later discarded.

We did not delve into the topics of projects for especially fragile target audiences, such as people with specific illnesses or severe impairments, nevertheless, we have pointed towards authors who have discussed these matters [16, 21]. In this section we would like to remind the reader that logistics may need to be rethought and specific variations of the methods might be needed in these cases. For instance, one of the characteristics of Parkinson's disease is 'On–Off' fluctuation. People are 'On' when medication is taking effect and 'Off' under no medication or when medication starts to wear off. In the Off period, symptoms show more intensely and may be severely disabling. Thus, when working with people with Parkinson's disease, sessions must be planned to be extremely short in order to accommodate for the duration of every participant's 'On' period, which may change from one person to the next. The best time for the session should be agreed upon with all participants so that all can participate while in 'On' state. Hand tremor and micrography are common characteristics in Parkinson's disease. Therefore, sessions should avoid the need for participants to write. These and other characteristics of the disease should be taken into account in order to design the activities' plan so as to create the best possible conditions to gather data and make participants comfortable in their participation.

Contrary to what would be expected, it is not only seniors who may be afraid of discussing or of trying out new technology – in our previous study, we have found that young researchers often dreaded their first encounter with older participants because they said they did not know how to engage in conversation with the participants [52]. Communication is, indeed, one of the fundamental aspects to work on. Authors in the literature report that enough time should be planned for in order to build the sensitivity of researchers and fine tune the discourse up to the level where it is efficient [21]. There should be enough time allotted to building trust and to explain the project to participants [19].

Moderators also have to be skilful in leading the discussions, in eliciting participants' opinions or in engaging participants in the co-creation activities. Clemensen and colleagues [35] suggest that researchers should not be afraid of putting themselves in uncomfortable positions so as to act as the role model for participants – one example is in case of role playing, where researchers can also be involved. Researchers should also be tactful to maintain the attention of the group and ensure that all participants are given equal opportunity to intervene – this may require bringing off-target verbosity back on track or encouraging participants to express their true opinions, rather than what they might think the researcher wants to hear. This summarises in controlling the breadth, depth of the discussion and, as highlighted by Clemensen and colleagues [35], in being open-minded about the contributions, and avoiding being paternalistic [19].

Dickinson and co-workers [26] alert to the possibility that old participants' social skills may lead participants to involve the researcher in the experiment. This is similarly pointed out by Chisnell and co-workers [56], who alert to facilitators having to resist the temptation of helping participants with task completion during usability testing.

While we may define in our projects that we will consider older adults to be people above 55, above 60 or 65, our definitions may be far away from people's view of themselves. Placing the discussion within scenarios may help avoid misunderstandings or terms that people may find to be offensive or that they do not relate to.

We would like to end this section by giving special attention to an aspect brought about by Massimi and colleagues [20], which is whether the discussion in participatory sessions should be guided as 'design for me', 'design for us' or 'design for them' (p. 161). The first happens when participants think about their own needs, the second when they consider the needs of the group in the session and the third when they are considering users in general, for whom the project is to be designed. Massimi and colleagues [20] say that participants in the same group may be considering different stands on these options and it might be the facilitator's role to opt for one option to guide the discussion, depending on the requirements for the project.

The assumptions behind the discourse of designers and researchers are very important, as recently noted by Vines and co-workers [57]. While analysing the discourse in HCI publications related to ageing, the authors noticed negative discourses and generalisations. We would like to point out to the recommendations of the authors, as they could frame the work carried out using HCD principles methods that we have described in this chapter. Vines and co-workers' [57] recommendations include avoiding a unique biomedical vision of the person, considering diversity and changes within a single person through time, and finally to carefully think about what are failures and what successes and finding metrics to evaluate these.

20.7 Summary of human-centred design methods and tools

In this chapter we have sought to summarise the current knowledge about HCD with older adults, with a special focus on AAL solutions. As noted in the Introduction, Section 20.1, there are several repositories where the reader can look for further information on the existing methods and tools for each stage in the HCD process.

Nevertheless, to ease the search and to summarise some of the recommendations laid out throughout this chapter, we are leaving the readers with a non-exhaustive list of common methods and tools, along with pointers towards their use with older people in Table 20.3.

Table 20.3 Examples of human-centred design methods and tools

Name	Notes	Method	Tool	Requires user involvement	User research	Ideation	Testing
Focus group	Duration. Trust. Breadth and depth of discussion. Controlling drifts from topic. Comfort. Language.	x		x	x	x	x
Survey	Language. Accessible graphic design and content.	x		x	x	x	x
Participant observation	Trust. Social rules (e.g. accepting cup of tea)	x		x	x		x
Interview	Duration. Bringing focus to the topic of conversation.	x		x	x		x
Diary study	Explain scope and goals. Avoid writing. Keep it simple.	x		x	x		x
Guided tour	Attention to mobility impairments. Trust.	x		x	x		x
Shadowing	Trust. Mobility.	x		x	x		x
Empathy tools	Inclusive design (e.g. glasses, gloves, movement restrainers)	x			x		x
Persona	Inclusive design	x	x		x	x	
Scenario	Situations and stakeholders	x	x		x	x	
Storyboard	Situations and stakeholders. Avoid abstract concepts.	x	x		x	x	
Mood board	Hopes and joys beyond just needs.		x		x	x	
Use cases	Situations and stakeholders. Avoid abstract concepts.		x			x	
User story	Situations and stakeholders. Avoid abstract concepts.		x			x	
Brainstorming	Duration. Trust. Breadth and depth of discussion. Controlling drifts from topic. Comfort. Language. Control pace.	x				x	
Analogy	Avoid abstract concepts	x				x	
Sandpit	Pairing people according to abilities and personalities. Small groups.	x		x	x	x	x

(*Continues*)

Table 20.3 (Continued)

Name	Notes	Method	Tool	Requires user involvement	User research	Ideation	Testing
Card sorting	Avoid abstract concepts. Provide context. Accessible graphic design and content.	x				x	x
Affinity diagram	Avoid abstract concepts. Provide context.	x	x			x	
Paper prototype	Pay attention to scale. Accessible graphic design and content. Easy to manipulate and less frightening for novice technology users.		x		x	x	x
Mid- to high-fidelity prototypes	Pay attention to scale. Accessible graphic design and content.		x		x	x	x
Service blueprint	Consider all the relevant stakeholders and situations	x	x		x	x	x
User journey	Consider all the relevant stakeholders and situations	x	x		x	x	x
Mindmap	If done collaboratively, assess participants' willingness and ability to produce a mind map	x	x		x	x	
Role play	If working with people with specific illnesses, consider whether characteristic of the disease will impact the ability to do role play. Allot time to break the ice. Have facilitators/moderators give the example.	x			x	x	x
Sketching	Some people might not be able, might be embarrassed or might not like to draw. Assess this in advance.	x			x	x	
Face-to-face moderated test	Pay attention to moderators' position in relation to the participant. Resist offering help outside what is predicted in the protocol.	x			x		x
Think aloud	Some people might not be able to efficiently engage in dual-tasking	x			x		x

Acknowledgements

The authors would like to thank the team at Fraunhofer Portugal AICOS and the many students who did their dissertations work with us. A special thanks to Sílvia Rêgo who does the daily management of the user network COLABORAR and to all the people and institutions that collaborate with Fraunhofer Portugal AICOS in designing useful, fun and usable technology.

References

[1] Lindsay S, Jackson D, Schofield G, Olivier P. Engaging older people using participatory design. In: *Proceedings of the SIGCHI Conference on Human Factors in Computing Systemshi'12*. Austin, Texas: ACM; 2012; p. 1199–208.

[2] Norman DA. User centered system design. In: Norman DA, Draper SW, editors. *User centered system design: new perspectives on Human-Computer Interaction*. Hillsdale, New Jersey: Lawrence Erlbaum Associates Publishers; 1986.

[3] Carroll JM, Rosson MB. Participatory design in community informatics. *Des Stud*. 2007;28(2007):243–61.

[4] Gregor P, Newell AF, Zajicek M. *Designing for dynamic diversity: interfaces for older people*. ACM Press, New York, 2002. p. 151.

[5] Clarkson PJ, Coleman R, Keates S, Lebbon C. *Inclusive design: design for the whole population*. London: Springer; 2003.

[6] Correia de Barros A, Leitão R, Ribeiro J. Design and evaluation of a mobile user interface for older adults: navigation, interaction and visual design recommendations. *Procedia Comput Sci*. 2014;27:369–78.

[7] Holzinger A, Searle G, Nischelwitzer A. On some aspects of improving mobile applications for the elderly. In: Stephanidis C, editor. *Universal access in human computer interaction coping with diversity*. Springer Berlin Heidelberg; 2007. p. 923–32.

[8] Leitão R, Silva PA. Target and spacing sizes for smartphone user interfaces for older adults: design patterns based on an evaluation with users. In: *Conference on Pattern Languages of Programs*. Arizona, United States; 2012.

[9] Leitão R, Silva PA. A study of novice older adults and gestural interaction on smartphones. In: *MOBACC 2013: Mobile Accessibility Workshop at CHI 2013*. Paris, France: ACM; 2013.

[10] Hawthorn D. Possible implications of aging for interface designers. *Interact Comput*. 2000;12(5):507–28.

[11] Kobayashi M, Hiyama A, Miura T, Asakawa C, Hirose M, Ifukube T. Elderly user evaluation of mobile touchscreen interactions. In: *INTERACT 2011, Part I (LNCS 6946)*; 2011. p. 83–99.

[12] Pak R, McLaughlin A. *Designing displays for older adults*. Boca Raton: CRC Press; 2011.

[13] Ziefle M. Information presentation in small screen devices: the trade-off between visual density and menu foresight. *Appl Ergon*. 2010 Oct;41(6): 719–30.

[14] Fisk AD, Rogers WA, Charness N, Czaja SJ, Sharit J. *Designing for older adults: principles and creative human factors approaches*, Second Edition. Boca Raton: CRC Press, 2009.

[15] Czaja SJ, Lee CC. Older adults and information technology: opportunities and challenges. In: *The Human-Computer Interaction Handbook: Fundamentals, evolving technologies, and emerging applications*. Third Edition. Boca Raton: Taylor & Francis; 2012. p. 825–40.

[16] Lindsay S, Brittain K, Jackson D, Ladha C, Ladha K, Olivier P. Empathy, participatory design and people with dementia. In: *Proceedings of the SIGCHI Conference on Human Factors in Computing Systems*. New York, NY, USA: ACM; 2012. p. 521–30.

[17] Davidson JL, Jensen C. Participatory design with older adults: an analysis of creativity in the design of mobile healthcare applications. In: *Proceedings of the 9th ACM Conference on Creativity & Cognition*. New York, NY, USA: ACM; 2013. p. 114–23.

[18] Chisnell D, Lee A, Redish J. Recruiting and working with older participants in usability studies [Internet]. AARP; 2004 [cited 2015 Aug 14]. Available from: http://www.redish.net/images/stories/PDF/recruiting_and_working_with_older_participants.pdf

[19] Ellis RD, Kurniawan SH. Increasing the usability of online information for older users: a case study in participatory design. *Int J Hum-Comput Interact*. 2000;12(2):263–76.

[20] Massimi M, Baecker RM, Wu M. Using participatory activities with seniors to critique, build, and evaluate mobile phones. In: *Proceedings of the 9th International ACM SIGACCESS Conference on Computers and Accessibility*. New York, NY, USA: ACM; 2007. p. 155–62.

[21] Moffatt K, McGrenere J, Purves B, Klawe M. The participatory design of a sound and image enhanced daily planner for people with aphasia. In: *Proceedings of the SIGCHI Conference on Human Factors in Computing Systems*. New York, NY, USA: ACM; 2004. p. 407–14.

[22] Müller C, Wan L, Stein M, Neufeldt C. Experience of giving and receiving – living lab-based technology design with elderly people. In: *SIGCHI Conference on Human Factors in Computing Systems (CHI'12)*. Austin, Texas: ACM; 2012.

[23] Damodaran L, Olphert CW, Sandhu J. *A toolkit for engaging older people in research, design and development of ICT based products and services*. Loughborough: Loughborough University; 2013.

[24] Correia de Barros A, Rêgo S, Antunes J. Aspects of human-centred design in HCI with older adults: experiences from the field. In: Sauer S, Bogdan C, Forbrig P, Bernhaupt R, Winckler M, editors. *Human-Centered software engineering*. Springer Berlin Heidelberg; 2014. p. 235–42.

[25] Silva PA, Nunes F. 3 x 7 Usability testing guidelines for older adults. In: *Proceedings of the 3rd Human-Computer Interaction, Usability Testing, Older Adults*. Universidad Politécnica de San Luis Potosí; 2010. p. 1–8.

[26] Dickinson A, Arnott J, Prior S. Methods for human-computer interaction research with older people. *Behav Inf Technol*. 2007;26(4):343–52.

[27] Williams K, Kemper S, Hummert ML. Improving nursing home communication: an intervention to reduce elderspeak. *The Gerontologist*. 2003; 43(2):242–7.

[28] Eisma R, Dickinson A, Goodman J, Syme A, Tiwari L, Newell AF. Early user involvement in the development of information technology-related products for older people. *Univers Access Inf Soc*. 2004;3(2):131–40.

[29] Eisma R, Dickinson A, Goodman J, Mival O, Syme A, Tiwari L. *Mutual inspiration in the development of new technology for older people*. London: The Helen Hamlyn Research Centre; 2003.

[30] Kemper S. Language production in later life. In: Gerstenber A, Voeste A, editors. *Language development: the lifespan perspective*. Amsterdam, Philadelphia: John Benjamins B.V.; 2015. p. 58–76.

[31] Dickinson A, Goodman J, Syme A, Tiwari L, Mival O, Newell AF. Domesticating technology. In-home requirements gathering with frail older people. In: *Proceedings of the 10th International Conference on Human-Computer Interaction*. Crete, Greece; 2003. p. 827–31.

[32] Yoxall A, Langley J, Musselwhite C, Rodrigues-Falcon EM, Rowson J. Husband, daughter, son and postman, hot-water, knife and towel: assistive strategies for jar opening. In: *Designing inclusive interactions: inclusive interactions between people and products in their contexts of use*. London: Springer; 2010. p. 187–96.

[33] Correia de Barros A, Duarte C, Cruz JB. "I made it myself" – DIY in the design of assistive devices for stroke patients. In: *Proceedings of the 7th International Conference on Design & Emotion*. Chicago, USA; 2010.

[34] Correia de Barros A, Duarte C. Assistive devices: stroke patients' design. In: *Proceedings of the Seventh ACM Conference on Creativity and Cognition*. New York, NY, USA: ACM; 2009. p. 79–86.

[35] Clemensen J, Larsen SB, Kyng M, Kirkevold M. Participatory design in health sciences: using cooperative experimental methods in developing health services and computer technology. *Qual Health Res*. 2007;17(1):122–30.

[36] Goodman J, Syme A, Eisma R. Age-old question(naire)s. In: *Include 2003*. London: The Helen Hamlyn Research Centre; 2003.

[37] The FOOD Project Consortium. Deliverable 2.1 – User requirements [Internet]. 2013. Available from: http://www.food-aal.eu/index.php/downloads/UNA/Deliverable-2.1/

[38] Hirsch T, Forlizzi J, Hyder E, Goetz J, Stroback J, Kurtz C. The ELDer project: social and emotional factors in the design of eldercare technologies. In: *Proceedings of the Universal Usability Conference*. London: ACM; 2000. p. 72–9.

[39] Morris M, Lundell J, Dishman E, Needham B. New perspectives on ubiquitous computing from ethnographic study of elders with cognitive decline. In: Dey AK, Schmidt A, McCarthy JF, editors. *UbiComp 2003: Ubiquitous Computing*. Springer Berlin Heidelberg; 2003. p. 227–42.

[40] Abeele V v. d., Van Rompaey V. Introducing human-centered research to game design: designing game concepts for and with senior citizens. In: *CHI '06 Extended Abstracts on Human Factors in Computing Systems*. New York, NY, USA: ACM; 2006. p. 1469–74.

[41] Chen Y, Ngo V, Park SY. Caring for caregivers: designing for integrality. In: *Proceedings of the 2013 Conference on Computer Supported Cooperative Work*. New York, NY, USA: ACM; 2013. p. 91–102.

[42] Morris M, Lundell J, Dishman E. Catalyzing social interaction with ubiquitous computing: a needs assessment of elders coping with cognitive decline. In: *CHI '04 Extended Abstracts on Human Factors in Computing Systems*. New York, NY, USA: ACM; 2004. p. 1151–4.

[43] Rogers Y, Paay J, Brereton M, Vaisutis KL, Marsden G, Vetere F. Never too old: engaging retired people inventing the future with MaKey MaKey. In: *Proceedings of the 32nd Annual ACM Conference on Human Factors in Computing Systems*. New York, NY, USA: ACM; 2014. p. 3913–22.

[44] Goodman J, Dickinson A, Syme A. Gathering requirements for mobile devices using focus groups with older people. In: MA SK, CEng JCM, MIEE, BA PL, CEng PRM, editors. *Designing a more inclusive world*. Springer London; 2004. p. 81–90.

[45] Wherton J, Sugarhood P, Procter R, Hinder S, Greenhalgh T. Co-production in practice: how people with assisted living needs can help design and evolve technologies and services. *Implement Sci*. 2015;10(1):75.

[46] Barrett J, Kirk S. Running focus groups with elderly and disabled elderly participants. *Appl Ergon*. 2000;31(6):621–9.

[47] Seale J, McCreadie C, Turner-Smith A, Tinker A. Older people as partners in assistive technology research: the use of focus groups in the design process. *Technol Disabil*. 2002;14(1):21–9.

[48] Newell AF. *Design and the digital divide: insights from 40 years in computer support for older and disabled people*. Morgan & Claypool Publishers; 2011.

[49] Correia de Barros A, Cevada J, Bayés À, Alcaine S, Mestre B. User-centred design of a mobile self-management solution for Parkinson's disease. In: *Proceedings of the 12th International Conference on Mobile and Ubiquitous Multimedia*. New York, NY, USA: ACM; 2013. p. 23:1–23:10.

[50] Rice M, Carmichael A. Factors facilitating or impeding older adults' creative contributions in the collaborative design of a novel DTV-based application. *Univers Access Inf Soc*. 2011;12(1):5–19.

[51] Muller MJ. PICTIVE – an exploration in participatory design. In: *Proceedings of the SIGCHI Conference on Human Factors in Computing Systems*. ACM, New York; 1991. p. 225–31.

[52] Correia de Barros A, Leitão R. Young practitioners' challenges, experience and strategies in usability testing with older adults. In: Encarnação P, Azevedo L, Gelderblom GJ, Mathiassen N-E, editors. *Assistive technology: from research to practice – AAATE 2013*. Amsterdam: IOS Press; 2013. p. 787–92.

[53] Neves BB, Franz RL, Munteanu C, Baecker R, Ngo M. "My Hand Doesn'T Listen to Me!": adoption and evaluation of a communication technology for the "Oldest Old." In: *Proceedings of the 33rd Annual ACM Conference on Human Factors in Computing Systems*. New York, NY, USA: ACM; 2015. p. 1593–602.

[54] Ballegaard SA, Bunde-Pedersen J, Bardram JE. *Where to, Roberta? Reflecting on the role of technology in assisted living*. Oslo: ACM; 2006. p. 373–6.

[55] Novitzky P, Smeaton AF, Chen C, *et al*. A review of contemporary work on the ethics of ambient assisted living technologies for people with dementia. *Sci Eng Ethics*. 2015;21(3):707–65.

[56] Chisnell D, Lee A, Redish J. Recruiting and working with older participants in usability studies [Internet]. AARP; 2004 [cited 2015 Aug 14]. Available from: http://www.redish.net/images/stories/PDF/recruiting_and_working_with_older_participants.pdf

[57] Vines J, Pritchard G, Wright P, Olivier P, Brittain K. An age-old problem: examining the discourses of ageing in HCI and strategies for future research. *ACM Trans Comput-Hum Interact*. 2015;22(1):2:1–2:27.

Chapter 21

Design and implementation of a smart home technological platform for the delivery of AAL services: from requirements to field experience

Susanna Spinsante[1], *Ennio Gambi*[1], *Adelmo De Santis*[1], *Laura Montanini*[1], *Giovanni Pelliccioni*[1], *Laura Raffaeli*[1], *and Giorgio Rascioni*[2]

21.1 Introduction

The design of a technological platform able to support the provisioning of Ambient Assisted Living (AAL) services has to face the complexities of a multi-disciplinary field, in which Information and Communication Technologies (ICT) are used to address and counteract the effects of growing elderly population. Requirements expressed by target users of AAL platforms may be strongly heterogeneous and difficult to settle, as a consequence technology should be personalized, adaptive, anticipatory, and able to provide high quality of service. At the same time, the design of an AAL platform typically foresees the integration of existing technologies, consumer devices, and applications, to offer solutions for the benefit of different stakeholders (end-users, caregivers, policy-makers, and business entities). This poses relevant difficulties, related to the achievable interoperability, usability, security and accuracy features, needed to facilitate the transition from research prototypes to real, mass-market products.

This chapter addresses the aforementioned aspects, by providing first an overview of research and development projects aimed at the implementation of technological platforms, to be used at home by elderly people aiming to prolong and improve their independence and quality of life. Most of the projects presented were developed in the framework of the European Ambient Assisted Living Joint Platform (AAL JP) calls; they were selected and reviewed here, based on their common target of a technological platform for home-based AAL service provisioning. Following the projects overview, from which some shared technological approaches and solutions

[1]Dipartimento di Ingegneria dell'Informazione, Universita' Politecnica delle Marche, Via Brecce Bianche 12, I-60131, Ancona (Italy)
[2]ArieLAB Srl, Via Casine di Paterno snc, 60131 Frazione Paterno, Anacona (Italy)

may be identified, a specific home platform is described, which was developed within a research project carried out by the authors. The peculiarity of the platform herein presented relies on the fact that it was the result of a cooperative design involving several Small Medium Enterprises (SMEs), each contributing its specific technology, product, or service, that have been integrated within the platform. Consequently, how to get interoperability and integration of pre-existing components to be harmonized in a common system for the delivery of new conceived services was the primary driver in the design of the platform. From this specific experience, it was possible to derive some general remarks about the motivations that may underpin specific design solutions, due to the need of facing issues appearing when approaching the real-world implementation of AAL systems, based on the integration of different and often pre-existing components.

The experimental results obtained from the development of the home platform are then presented, addressing the issues related to: technology, referred to the implementation of the Proof of Concept (PoC), data management and analysis, concerning the AAL services enabled by the platform, and to the user-interaction. Corresponding tests and outcomes are discussed. A discussion on the future directions in AAL platforms design is presented, and finally, the main conclusion of the chapter are provided.

21.2 Background

A rich amount of research and development projects related to technological platforms for smart environments supporting AAL services may be found, that is reviewed in this section to gather a comprehensive knowledge of the most commonly adopted approaches.

Within the European context, a reference source of information is provided by the AAL JP [1] catalogues of projects, issued once per year, collecting basic details about the co-funded projects approved in response to several calls for proposals for research, development and innovation projects in the field of ICT for Active and Healthy Ageing.

The different calls for proposals issued by the AAL JP deal with specific targets. Among all the calls issued up to now, the following ones are closer to technologies for smart environments:

- Call 1, which funded projects starting in 2009, was focused on ICT based solutions for prevention and management of chronic conditions of elderly people.
- Call 5, with projects starting in 2013, was dedicated to ICT-based solutions for self-management of daily life activities of older adults at home.

In the AAL JP view and approach towards research and technology development, users are always involved in the design, evaluation and test phases, and participate in defining the expected solutions.

The ALADDIN project [2] aimed to adopt state of the art in ICT to develop an integrated solution and innovative tools for the self-management of dementia patients.

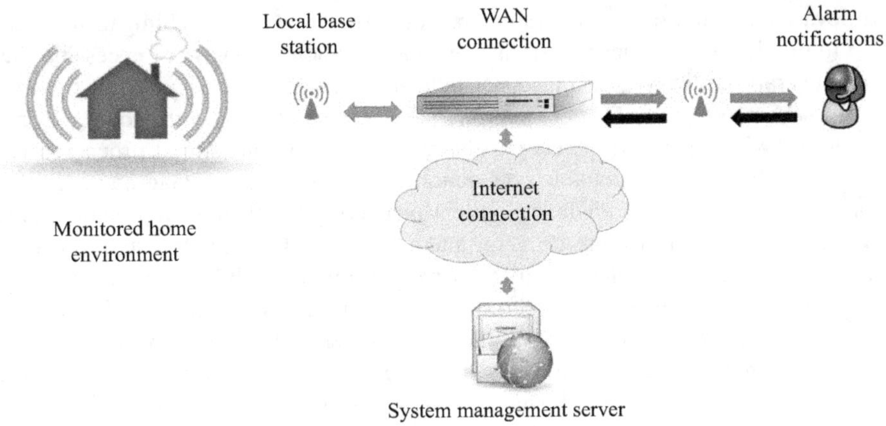

Figure 21.1 CARE project platform architecture

The solution substantiates in an integrated platform for distant monitoring of patient status, thus facilitating personalized intervention and adaptive care. The platform was designed to support users in maintaining health and functional capability, through risk assessment and early detection of deterioration symptoms, and distress signs of the caregiver. Through the provision of user-friendly ICT tools for frequent and unobtrusive monitoring, the self-management of chronic conditions was enabled.

The CARE project [3] focused on an intelligent monitoring and alarming system for independent living of elderly persons, according to the architecture shown in Figure 21.1. Specifically, this project targeted the automated recognition and alarming of critical situations (like falls) using a visual sensor and real-time processing, while preserving the privacy, and taking into account system dependability issues, ensuring reliability, availability, security, and safety from a holistic point of view. A biologically inspired, neuromorphic vision sensor was integrated into the alarm, security, and monitoring system for seamless analysis and tracking of elderly persons at home [4], providing a real-time information exploited for incident detection (e.g., fall detection, motionless person), and to instantly deliver alarms to the concerned parties. The system mainly addressed single individuals living in their own private homes.

Project eCAALYX (Enhanced Complete Ambient Assisted Living Experiment) [5] was designed to address several objectives: health monitoring of elderly persons with multiple chronic conditions, at home and on the move; improving the quality of life of elderly persons by increasing their freedom and safety, through prompt detection and control of decompensation episodes, in order to prolong their independent life at home, and reduce hospitalization or admission in nursing homes for longer periods; preventing deterioration of the patient condition by providing continuous support, guidance, and relevant health education.

The Health at Home project (H@H) [6] targeted societal problems related to healthcare services for elderly citizens affected by Chronic Heart Failure (CHF), providing them with wearable sensor devices for monitoring of pathophysiological cardiovascular and respiratory parameters and, at the same time, enabling the medical staff to monitor their situations at distance and take action in case of necessity. The H@H overall hardware platform consists in the Home Monitoring Sensor System, the Home Gateway and a Remote Server. Patient's physiological data acquired by the sensors are wireless transmitted to the Gateway, where they are stored, processed and sent, through a safe and reliable connection, to the Server. The Gateway acts also as a Reminder, in order to help the patient respecting the therapy and planning the measures, as a System Failure Detector and as an Extra-Protocol Measure Manager. The clinician can access data from the server (through a web-based user interface) and view the information coming from the measurements, but he can also modify the measurement protocol or the pharmacological therapy according to the symptoms coming out from the data monitoring. The Bluetooth technology was chosen for the communication between the sensors and the gateway, due to the wide diffusion of available devices on the market. On the other side, ADSL was chosen as the primary transmission channel for data communication between the home gateway and the remote server, while a mobile broadband connection (GSM/GPRS/UMTS) was used as secondary data channel.

A technical validation of the system was implemented involving 30 patients with CHF disease in NYHA class III and IV, with an average age of 62 years and recently hospitalized for HF. The minimum period of monitoring was one month. A specific testing protocol and a questionnaire were developed to gather patients, caregivers, and physicians feedbacks and validate the system. The results showed a very limited number of activity misses (<3%), mostly in the first days of monitoring, confirming also the property of the system to improve the therapy compliance. Moreover, the number of false positive alarms was less than 5%. No connectivity and transmission problems, including data lost, occurred. All end-users reported a positive feedback and good satisfaction level in the final questionnaire. Physicians reported that the use of this platform did not load up in a significant way their regular activity, but represented a valid means to control at distance the evolution of the followed patients thanks to the high quality of acquired signals and alarms detection capability.

The Happy Ageing project [7] promoted the design and development of a system to counteract the incidence of chronic conditions, composed by three main modules: a lifestyle monitor for recording main activities in the home and reminding the user about activities to perform; a navigation assistant, to support user's mobility in the nearby environment; and a personal assistant, to support actions reminding and search for personal objects [8]. All the developmental phases were performed in accordance with the user-centred design paradigm: first, a survey was conducted in the countries involved in the field trial, in order to collect information on the needs of the elderly, attitudes towards technology, and system requirements. This was followed by focus groups with stakeholder representatives in each country, to analyse the system opportunities for the future. Then, rapid testing techniques were carried out on an early version of the system. Finally, a field trial was conducted to test the user satisfaction and acceptance of the system.

The HERA project [9] addressed the provision of a platform with cost-effective specialized assisted living services for the elderly people suffering from Mild

Cognitive Impairment (MCI), or mild/moderate Alzheimer's Disease (AD), or other diseases (diabetes, cardiovascular) with identified risk factors, to improve the quality of their home life, extend its duration and at the same time reinforce social networking. The HERA platform focuses on the following main categories of services: cognitive reinforcement services; physical reinforcement services; patient specific home care services; general home care services for elderly. HERA applies technological solutions for aiding users managing their daily lives; it employs agent technology and a human–machine interface approach based on the user's TV set and a set-top-box, along with Bluetooth enabled sensors in the user's home. Thus, the equipment used are familiar, and the ambient intelligence functions support the user in performing daily tasks [10].

The HOPE project [11] proposed an integrated, smart platform that enables the elderly people with Alzheimer's disease to use innovative technology for a more independent life, easy access to information, monitor their health, and serve as a source of inspiration for users as well as for people working with assistive devices. HOPE consists of a self-operated, self-adjusted, innovative intelligent IP-Based Universal Control Box (UCB) that uses intelligence to manage the various connected subsystems and devices within a residence of the elderly people. The HOPE system can be split into two main blocks: the Server Block and the Home Block, which represent the main agent and every subsystem at each elderly user's home respectively. The Home Block covers functionalities associated to each person's environment monitoring, indicating alarms when necessary. The Server Block is responsible of the decision-making functionalities of the system, including storing all information from any Home System, alarm delivery in case of a fall detection to caregivers and relatives, evaluation of information and provision of rules for estimating the most appropriated scenario, synchronization with all home databases, interfaces to relatives and doctor.

The recently started BREATHE project [12] addresses a technological platform which provides daily guidance and support for the informal caregiver in the long-term care of elderly people. Cameras and other types of binary sensors are included in the BREATHE system for the intelligent monitoring of AAL environments; the system enables interaction among different stakeholders, as schematically shown in Figure 21.2. The project follows a privacy-by-design approach in order to ensure the privacy of user data and information. The level of privacy requested, based on the context, is assessed by evaluating where data are acquired and transmitted, according to a number of parameters that account for the identity and state of the assisted person, the caregiver, and the environment.

The NITICS project [13] aims at designing and building an expandable holistic platform that offers advanced ICT services, like monitoring and navigational support, for the mobility of elderly and disabled persons in their home, during their daily activities. Furthermore, it also brings suitable services for elderly and people with diseases or disabilities to preserve as long as possible their cognitive capabilities. A list of the available services is graphically shown in Figure 21.3.

The platform will rely on a set of basic and task oriented services: localization of personal objects; localization and analysis of movement patterns of elderly and disabled people inside their homes; a multimedia bi-directional platform

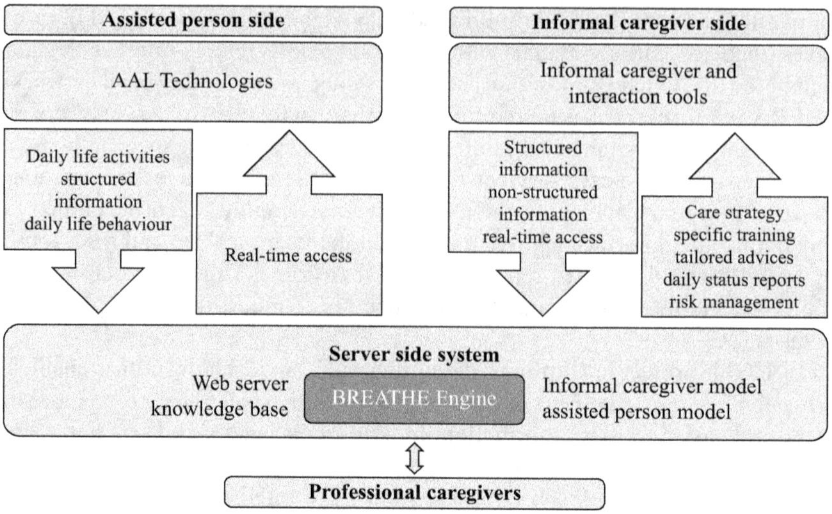

Figure 21.2 Overall BREATHE system functionalities and stakeholders

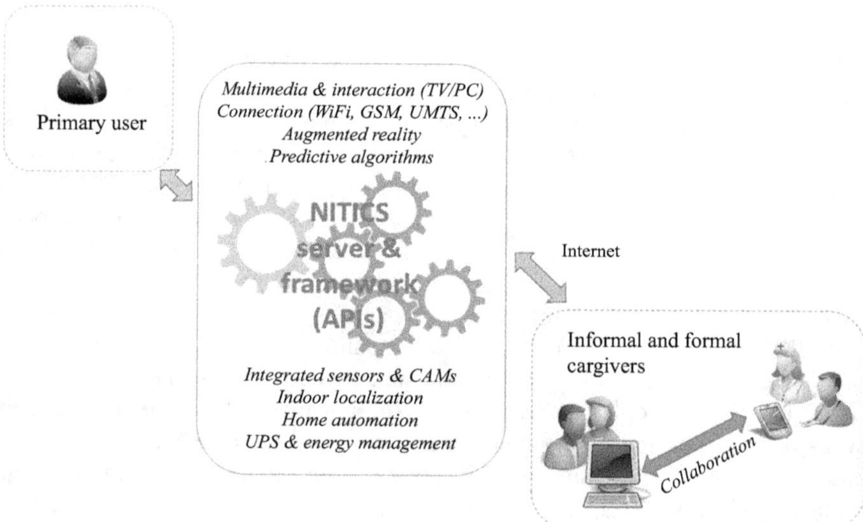

Figure 21.3 Schematic representation of the NITICS architecture, services, and stakeholders

(TV/PC/Smartphone) to ease, stimulate and support daily activities; an augmented-reality system to assist users in finding objects. NITICS will enable disabled persons to create, participate and continue their social activities not only via an Internet connection but also by using localization technology inside their homes, supporting an active social life. The localization technology is exploited also to detect unpredicted or abnormal behaviours, lack of movement or erratic behaviour, and to trigger actions by care providers in case of need. The NITICS platform will also allow integration with standardized Hospital Information Systems and Emergency Services. The capability of remote configuration, sensing and user interaction offered by the NITICS platform will improve existing services and enable novel ones, directly improving the quality of life of elderly and people with disabilities. The system will be developed starting from existing technologies (whenever possible), prioritizing open systems interconnectivity and open source philosophy in general.

To further emphasize the importance of evaluating new proposed technologies through experiments, it is worth mentioning that several projects developed in the framework of the AAL JP calls, such as PERSONA [14], CogWatch [15], and others, exploited Living Labs as a testbed to allow real users to interact in a realistic manner with the AAL environment. Living Labs have been often used also to promote co-design of AAL technologies and validation with relevant stakeholders (users, caregivers, relatives, etc.). A Living Lab may be seen either as a physical facility to support the technology development by means of infrastructures simulating a realistic environment, and as a tool to support the design methodology, by enabling user involvement, experimentation, and identification of requirements [16].

Looking outside the European context, several AAL projects and research activities are being carried out in those countries that are currently experiencing, or are expected to experience issues with population ageing in the medium/short term. Among them, countries like Japan or Korea are pushing the development of AAL technologies as a mean to sustain the expected impact of an increased rate of aged citizens on their welfare systems and resources. Kim *et al.* [17] in 2010 presented the U-Health Smart Home project, focused on building a smart home along with an autonomic system to monitor the home as well as the inhabitants to provide intelligent support and assistance in any situation, at any time. Non-intrusive and non-invasive monitoring enable the delivery of assistance to the elderly directly in their own home. Most of the research efforts are devoted to developing robotics technology to assist ageing people at home, and support them in Activities of Daily Living (ADLs).

In the United States, research in AAL is mainly devoted to support and assist people affected by chronic diseases as Alzheimer's and Parkinson's Disease [18, 19].

21.3 The TRASPARENTE platform

The objective of the project developed by the authors is the design of an integration platform dedicated to active ageing. The project aimed at extending the independent living of elderly living at home alone, therefore it considers aspects such as home security, environmental control, behaviour and health monitoring [20]. Specifically, the following domains have been considered: home automation system, load monitoring, behaviour detection, telemedicine, and system-user interface.

Figure 21.4 shows a representation of the system architecture. The home automation system includes a set of devices connected by means of a two-layers architecture. The CAN bus forms the lower layer, and enables to connect and manage the set

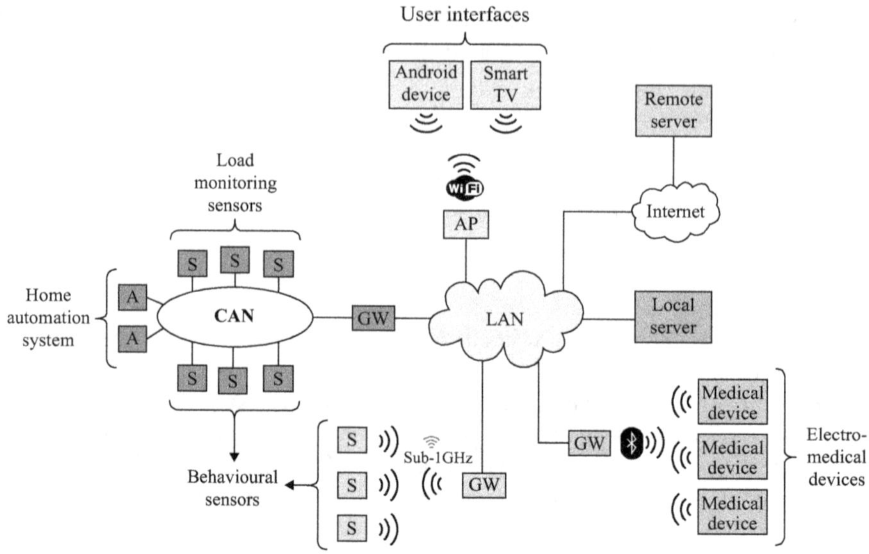

Figure 21.4 Representation of the TRASPARENTE platform architecture

of sensors and actuators. The higher layer is represented by the Ethernet infrastructure, where proper devices act as gateways to interface networks implemented by means of different communication technologies. Actuators are generally wired devices, while for example some magnetic sensors are connected in wireless mode, in order to facilitate their installation. For this reason, some gateways are equipped with a radiofrequency interface (Sub-GHz frequency ranges). Each gateway owns its IP address and each node connected to it has its identifier (ID): this way, the IP address-ID pair identifies a unique device. Actuators enable to control lights and window fixtures, while sensors include Passive Infra-Red presence sensors (PIRs), magnetic sensors applied on doors and windows, gas, smoke and water leaks sensors. An actuator could receive a command from its reference gateway installed on the Ethernet network. Otherwise, each sensor is connected to a home automation node that transmits the acquired data to the reference gateway.

By exploiting the same infrastructure, some meter nodes have been installed, dedicated to power load detection and monitoring. Meter nodes can be applied to a single load or connected to a master switch, in order to monitor the power consumption of the whole supply line. They provide essentially two functions. The first one is to measure the consumption at predefined intervals and sends this information to the server. The second one refers to the load control: if the power absorption exceeds a threshold value for a certain time interval, the node sends a notice to the server which could decide to disable a load, by sending the proper command.

As for the health monitoring, the user can exploit a set of electro-medical devices, i.e. a scale, a blood pressure meter and an oximeter. Each time the user performs a new measurement, the device automatically transmit the data via Bluetooth to the local gateway, which forwards it over the IP network to a remote platform so that the doctor/caregiver can monitor the patient's health condition anytime. The platform chosen and the associated biomedical devices have been certified as medical equipment, in order to satisfy the expected requirements for a telemedicine system.

Another aspect of interest when dealing with elderly people living alone is the behaviour analysis, with the aim to study the user's habits, and to detect and evaluate any change. This activity could be very complex, so it is necessary to first decide which features are to be considered, and then design the system properly. In the authors' proposal, the behaviour analysis includes:

- an integrated smart shoe, whose insole is equipped with force sensing resistors, that allows posture and balance evaluation;
- pressure sensors placed on the bed and on the sofa, to localize the person and evaluate their level of activity;
- a set of wearable sensors (e.g. accelerometers, gyroscopes, temperature sensors) to measure the user's parameters, that are collected by a central node named Body Sensor Network Concentrator (BSNC).

Data collected by these heterogeneous sensors are stored in a database and are available for processing. Each algorithm extracts only the data relevant to infer the specific desired information for behaviour analysis.

The description of the system provided so far shows that it is quite complex in terms of devices, technologies, and functionalities. Therefore, the architecture also comprises a central control unit that coordinates the entire system, representing an interconnection layer among the sub-systems. The importance of introducing a central server will be described later, in Section 21.4.

The other elements that complete the architecture are: a local database to store raw and processed data, and different types of interfaces for the user-system interaction. User interfaces enable both input and output functions, such as the control functionalities of the home automation system, the visualization of the current status of the devices, the power consumption and so on. Multimodal interaction by means of touchscreen portable devices and Smart TVs has the objective to better fulfil the needs and capabilities of the users. Finally, by accessing the remote platform, it is possible to show the time trend of the user's medical parameters, and of some features related to the behaviour analysis.

21.4 Motivations for design solutions

21.4.1 Issues in real-world implementation

A significant amount of research organizations and industry entities are being active in the field of AAL since several years, as the amount of co-funded projects available in the literature confirms (see Section 21.2). Despite the high number of recent research and development activities, and the extensive research efforts, however, the real-world usage of the results obtained appears to be quite poor. Only a few devices and standards have fulfilled the pilot-study level and moved to a more mature market-oriented engineering step. Looking for motivations to this trend, it is possible to state that a quite remarkable gap still remains between the capabilities of currently available solutions and technologies, and the requirements emerging from a real-world AAL scenario. Relevant resources (technical, economical, and organizational) are needed to move from research prototypes developed within pilot studies to marketable products; further, the amount of experiments performed during pilot studies is typically limited and not enough significant to provide a sound motivation to the investments needed during the go-to-market phase.

The variety of user experience, socio-technical requirements, and needs expressed by different user populations creates additional obstacles to the effective transposition of prototypes into real products or services available for purchase from a mass market. Recently introduced methodologies, such as user-centred and participatory design, are supporting a better identification of user requirements thus making AAL systems become more aligned to real-world needs and scenarios; reliability and dependability of AAL systems are improving, thanks to the adoption of formal modelling and validation tools. In the Happy Ageing project, for example, the development stage followed a user-centred design: as a first step, they have collected information on the user's needs and the system requirements; moreover, after the development, a survey has been conducted to evaluate the effective users' satisfaction about the field trial. A validation stage was performed also in the Health at Home project, in this case involving both patients and medical personnel with the aim to receive feedbacks from two distinct points of view. When moving from research prototype design of an AAL system to its real-world implementation, some apparently obvious issues shall be taken into due account [16]. Device failures may happen, and these events may not only affect prototypes (as it could be reasonably expected), but also legacy devices, infrastructures, or cables. As a consequence, it is of basic importance to foresee the use of monitoring services when implementing an AAL system, possibly operating at different levels, from the transmission media to the network devices, to the software applications, in order to enable a quick and reliable intervention in the case of problems. Related to the availability of monitoring services is the recommendation of dealing with the integration of heterogeneous devices, technologies, and applications, since the early stage of the system design. Sometimes, even if devices or applications declare compliance to a given standard or protocol, their effective interoperability with other devices or systems may be not easy to obtain, due to slight variations in the real implementation of the standard or protocol. Checking for true interoperability and integration of every new added element to the system should be a basic requirement. The issue of integration is highly considered in the NITICS project, in fact they plan to adopt existing technologies and follow an open source philosophy and open systems interconnectivity, with the aim to improve the current services and to offer new ones. In addition to a proper design, in order to ensure a correct integration of the different components in a AAL system, and the correct execution of all the envisioned functionalities, tests should be performed not only by a suitable pool of end users, but primarily from developers, that can implement a first verification of the system and identify the critical technical issues to be addressed. The technical evaluators should be able to generate specific (i.e. not generic) feedback to correct and improve the system, by means of proper bug tracking tools.

21.4.2 Integration of different components in TRASPARENTE

As described in the previous paragraph, one of the most important aspects to consider when developing complex systems is the interoperability among the various components. As for the NITICS project, in fact, the platform described herein is not a project from scratch, but it tries to put together diverse, existing and already in use solutions. This explains the variety of technologies exploited in the system. In order to let the different components communicate, we foresee the use of gateways interfacing heterogeneous communication technologies. However, this is not enough and from the early stages of implementation we have realized the need for a device to put together all the information coming from different parts and to manage the entire system. We decided, therefore, to introduce a central server in the system architecture. The purpose of the server is, precisely, to collect the data acquired by behavioural sensors, power consumption monitoring system and electro-medical devices, in order to implement algorithms for sensor fusion. It also handles the home automation system and controls the state of the whole infrastructure. Moreover, the server takes in charge the management of the requests made by the application interfaces. The client/server communication employs the Representational State Transfer (REST) technology, and a specific "protocol" has been defined in order to recognize the different types of queries, that mainly correspond to information requests or commands (for example, the application can ask for the state of the sensors or need to send commands to the home automation system). Thanks to this architecture, each interface acting as a client follows the same syntax to query the server, which replies regardless the requesting device.

As soon as the physical installation of the sensors was completed, the system started generating data, originated by the actions performed by people accessing the Laboratory. Once the set up was completed, it became evident that the amount of information collected every day is quite large. This led to the decision to store the raw data in a local database (DB) within the Lab, and send to the remote platform aggregated information only once a day, with the aim of regulating the data flow. As will be described in Section 21.5, the server applies data fusion algorithms to the data collected during the day by the different components of the system, and generates three reports to be sent remotely. These daily reports contain information referred to the system status and the user's behaviour, such as energy consumption, user's activities and presence/absence in the house.

As already stated, during the implementation of the project specifications, the need of local processing functions aimed at integrating all the platform components has emerged. These functions can be considered as logical entities implementable in a distributed way. Specifically, the platform offers diverse operations, some of them are allocated in a server machine, which deals with the DB management; other ones, instead, are handled by a proper gateway. The electro-medical devices and the BSNC directly interface this gateway, which is responsible both for generating the reports to be sent to the remote platform, and forwarding data to the local server for the storage

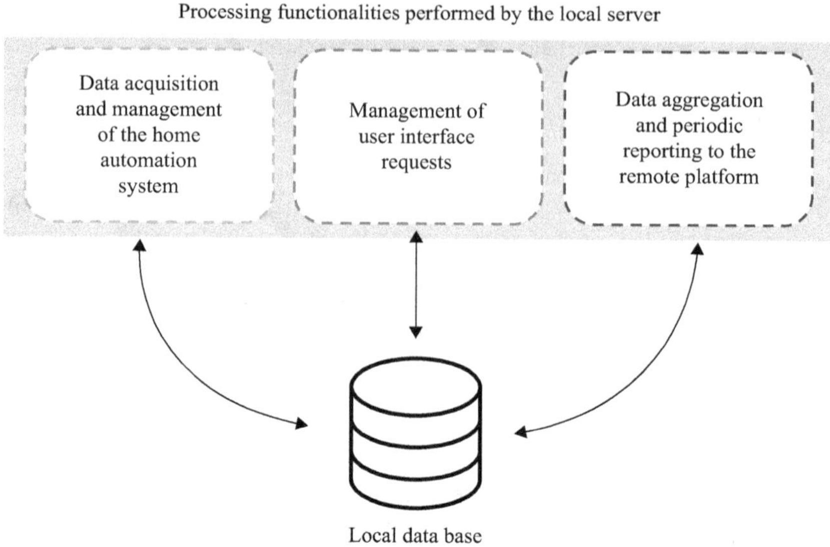

Figure 21.5 Representation of the server modules, extracting data from the same DB

on the DB. This apparent duplication is due to two motivations. Firstly, it is necessary to send data directly from the medical sensors to the remote platform without altering them, in order to not compromise the "medical certification" of the service. Thus, for the local storage a different transmission is needed. Secondly, the BSNC may detect a fall of the user, which should be handled as a priority: the direct transmission of this alarm event allows an immediate notification over the remote platform. Other user's parameters collected by the BSNC are saved in the local DB. The server and the gateway use both a common technology: the REST architectural pattern. Adopting a common technology to implement the processing functionalities allows to differently combine and interchange the tasks performed by the local server and by the gateway, also in preparation for future developments. Moreover, as depicted in Figure 21.5, some of the functions performed by the local server run independently from the others and, thus, could be executed by different devices. This seems apparently in contrast with the need of putting together all the information acquired by sensors in a unique collection point. Actually, even though the execution of each software modules is autonomous, they draw on the same information pool, i.e. the local database.

21.5 Experimental results

21.5.1 Proof of concept

With the aim of describing the effectiveness of the solutions adopted by the authors to solve the tricky issues and problems emerged so far, the outcomes obtained in the experimental phase will be presented in the following. The assistive home technology proposed by the authors and described in the previous sections is already available as a PoC in a laboratory environment. Some aspects of the platform, such as infrastructures and data acquisition, are more consolidated; other ones, such as the ADL recognition algorithms, could be improved, exploiting, for example, more sophisticated methods based on sensor fusion and machine learning techniques. The Laboratory has been set up with sensors connected to objects that are typically present in a home environment, e.g.:

- magnetic sensors connected to the refrigerator door, to the entrance door and to the window;
- power meters connected to the TV, the fan, the microwave, and upstream of the power line;
- PIR sensors placed over the entrance door and in the centre of the room;
- a force sensor on the armchair;
- flux meters on the cooker and on the power supply;
- gas and smoke detection sensors.

This set of devices forms a monitoring network in which both wired and wireless (CAN bus and Sub-GHz) communication technologies are employed. The platform also includes some electro-medical devices that communicate via Bluetooth with the gateway. Then, it forwards the information received to the remote server, using a certified communication. Moreover, such information is also sent to the local server in order to collect all data in a unique database, in the perspective to integrate more sensor fusion algorithms. The user can display the various measurements, such as pressure and weight, by querying the gateway directly by means of the responsible interfaces.

Within the Laboratory environment, the students perform some common daily home activities, such as cooking at lunchtime, sitting, watching TV, turning on the fan and so on. Such users do not fit the target user, since they are typically young people, but their contribution has been very useful to improve algorithms and interfaces, and to obtain a feedback on the system reliability and effectiveness. From a functional point of view, the solution provided meets all the preset objectives. The data collected from the various sensors are sent successfully through the different communication interfaces to the central server which stores and processes them.

21.5.2 Data management

The storage of all the raw data generated by the system allows to perform a large variety of analysis. On the other hand, it requires a lot of memory space. The following table (Table 21.1) provides an evaluation of the daily and average amounts of DB rows generated by the devices of the AAL platform (environmental sensors and power meters).

This analysis refers to two specific tables of the database, concerning data collected by the set of sensors included in the platform, and by the power meters installed in the Lab. The dates have been selected randomly over a period of four months, and refer to days of the week from Monday to Friday. In fact, when the Lab is closed,

Table 21.1 Average number of DB rows generated by the sensors and by the power meters per day

	Sensor data	Power meter data	Total
Average DB rows	4,353	202	4,555

i.e. during the weekend or holidays, around 300 rows are collected daily, due to alive messages generated by the sensors. The sensors table includes events detected by:

- three presence sensors (PIR);
- three magnetic sensors (applied on door, window, refrigerator door);
- two force sensors (sofa, bed);
- gas and smoke detection sensors;
- water and gas flux meters.

Gas, smoke and flux sensors transmit a few data, while the number of activations of force and magnetic sensors is around 50 per day. Consequently, almost all the collected data are generated by the presence sensors, which transmit an activation message every time a presence is detected, and a deactivation message after that. In terms of memory space, if we consider only the relevant information (data and timestamp), the alive messages alone, produced during the weekends, take approximately 20 kB of memory space per day, compared to an average amount of 300 kB on the other days, due to the activations of the AAL sensors. If all the records of the table are considered, the amount of required memory increases to 30 kB and 400 kB, respectively. The power meters table contains the values detected by five power meter nodes: four of them are connected to wall sockets, and one is applied on the general power supply line. Each node transmits the measured power value every minute, which is recorded in the database only if it is different from the previous one. Due to this operating principle, the amount of row generated by the power meters can vary widely, depending on the activities performed in the Lab, and on the specific load connected to each socket. This behaviour is clearly demonstrated by the variability of the number of rows as reported in Table 21.1. Every night at a set time, the server application executes some aggregation tasks, with the aim to process the data collected the day before, and to save the results in other tables of the local database. By supposing to delete raw data and keep only aggregated data, it is possible to obtain a large gain in terms of disk space every day. Once the data have been aggregated, the server generates three reports formatted as JSON files, containing information referred to:

- activity performed during the day;
- presence in/out of home;
- energy consumption.

A more detailed analysis on the type of aggregated data and on the database structure may be found in [21].

21.5.3 Data analysis

Thanks to the data collected by the sensors, it is possible to monitor and analyse the user's habits. Each of the three files generated every day contains a different behavioural information identified by the platform: energy consumption, presence and activities. While the first concerns only the energy consumption and thus may be derived directly from the observation of the data acquired from the power meters, without using more challenging algorithms, the last two types exploit more complex analysis methods. The presence detection at home is a very important aspect because it allows to identify outliers or alarm and to implement any security measures. To this aim, only few sensors have been used, such as PIRs, magnetic and force sensors. At the present development stage, the algorithm is able to detect the user's status, i.e. present or absent, and its degree of reliability, exploiting the user's location at home. Such localization is related to the data generated by the sensors: whenever a sensor event occurs, e.g. opening/closing the door, enabling/disabling the PIR, etc., the system is able to roughly determine the user's position. For example, if the user is outdoor and the door is closed, just few minutes without the occurrence of any other event are sufficient to assert that the user is absent; conversely, when the user is in the house and the door is closed, much more time is necessary before the system can notify the absence. From tests carried out so far, the algorithm is able to correctly recognize the user's status. Through environmental sensors, it is also possible to recognize the type of activities performed by the user. In this regard, first elementary algorithms have been implemented to understand if he is cooking, sitting or sleeping. Such information can be derived using respectively a flux meter on the cooker, a pressure sensor under the cushion of the sofa and another one under the mattress of the bed. While the first two actions have been tested by real people, the sleeping activity has just been simulated, although it is expected to use a sensor identical to the one exploited in the sitting activity. The activation and deactivation of the various sensors allow to recognize the activities carried out by the subject, however further improvements foresee the implementation of more refined methods and machine learning algorithms in order to predict the elder's attitude, relying on observations of behavioural data over a long period. The results of the behavioural analysis are stored in the database and periodically sent to the remote telecare platform.

21.5.4 User interfaces

As stated in Section 21.3, the system provides both a web interface to access a remote platform and local application interfaces, available for multiple devices.

As for the web interface, the doctor or caregiver can access it to monitor both behavioural data and measurements acquired from electro-medical devices, allowing him to assess deterioration or improvement in the subject's health status in the long term. Such information can be also visualized in graph format, as shown in Figures 21.6, 21.7, and 21.8, respectively.

Through the local interfaces, the user can control the home automation system. It facilitates some tasks, such as opening or closing windows and shutters, and turning on or off lights, providing a safe and easy system to interact with the dwelling.

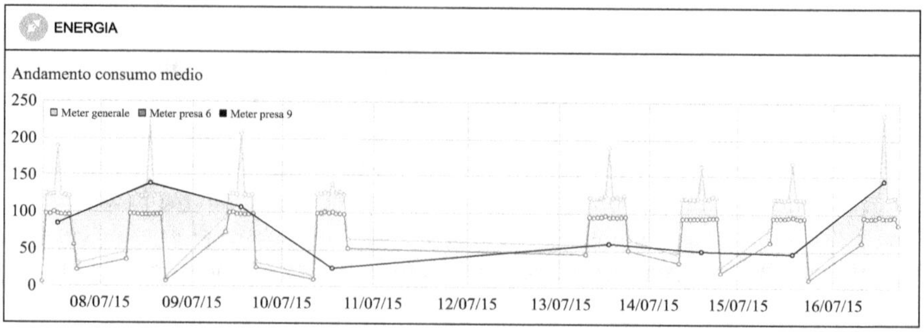

Figure 21.6 Screenshot of the remote telecare platform showing the average power consumption in watts: "generale" is the whole line, "presa 6" is the fan, and "presa 9" is the microwave consumption, respectively

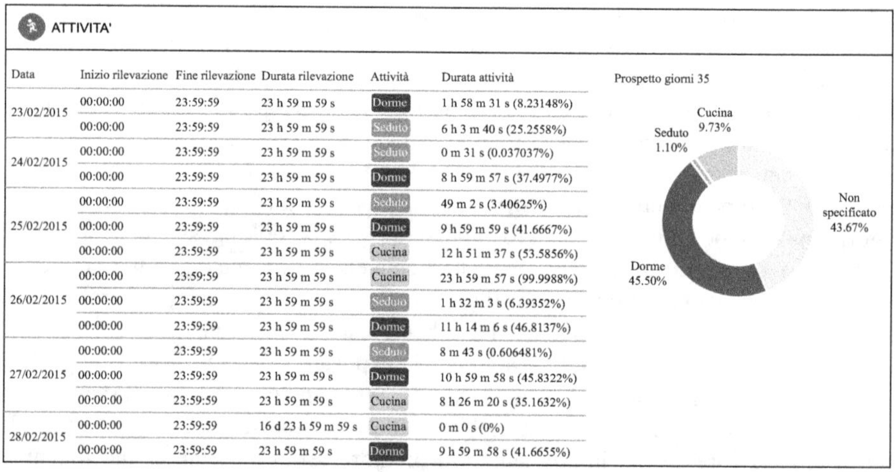

Data	Inizio rilevazione	Fine rilevazione	Durata rilevazione	Attività	Durata attività
23/02/2015	00:00:00	23:59:59	23 h 59 m 59 s	Dorme	1 h 58 m 31 s (8.23148%)
	00:00:00	23:59:59	23 h 59 m 59 s	Seduto	6 h 3 m 40 s (25.2558%)
24/02/2015	00:00:00	23:59:59	23 h 59 m 59 s	Seduto	0 m 31 s (0.037037%)
	00:00:00	23:59:59	23 h 59 m 59 s	Dorme	8 h 59 m 57 s (37.4977%)
25/02/2015	00:00:00	23:59:59	23 h 59 m 59 s	Seduto	49 m 2 s (3.40625%)
	00:00:00	23:59:59	23 h 59 m 59 s	Dorme	9 h 59 m 59 s (41.6667%)
	00:00:00	23:59:59	23 h 59 m 59 s	Cucina	12 h 51 m 37 s (53.5856%)
26/02/2015	00:00:00	23:59:59	23 h 59 m 59 s	Cucina	23 h 59 m 57 s (99.9988%)
	00:00:00	23:59:59	23 h 59 m 59 s	Seduto	1 h 32 m 3 s (6.39352%)
	00:00:00	23:59:59	23 h 59 m 59 s	Dorme	11 h 14 m 6 s (46.8137%)
27/02/2015	00:00:00	23:59:59	23 h 59 m 59 s	Seduto	8 m 43 s (0.606481%)
	00:00:00	23:59:59	23 h 59 m 59 s	Dorme	10 h 59 m 58 s (45.8322%)
	00:00:00	23:59:59	23 h 59 m 59 s	Cucina	8 h 26 m 20 s (35.1632%)
28/02/2015	00:00:00	23:59:59	16 d 23 h 59 m 59 s	Cucina	0 m 0 s (0%)
	00:00:00	23:59:59	23 h 59 m 59 s	Dorme	9 h 59 m 58 s (41.6655%)

Figure 21.7 Screenshot of the remote telecare platform showing ADL graphs: each label represents the average time spent for the respective activity ("Dorme" for sleeping, "Cucina" for cooking, "Seduto" for sitting, "Non specificato" for undefined)

In addition, in order to make these operations even faster, some scenarios have been implemented, allowing the activation/deactivation of groups of different actuators. The user interfaces are already available in a prototype version, they have been developed for both Smart TV and Android touchscreen devices. Aside from the features already mentioned, they allow to monitor the status of sensors, power consumption and the user's ADL history. Additionally, the system is able to support user's drug therapy, sending notification messages to the various interfaces, to reminding drug intake at the proper time. With the exception of data gathered from medical devices,

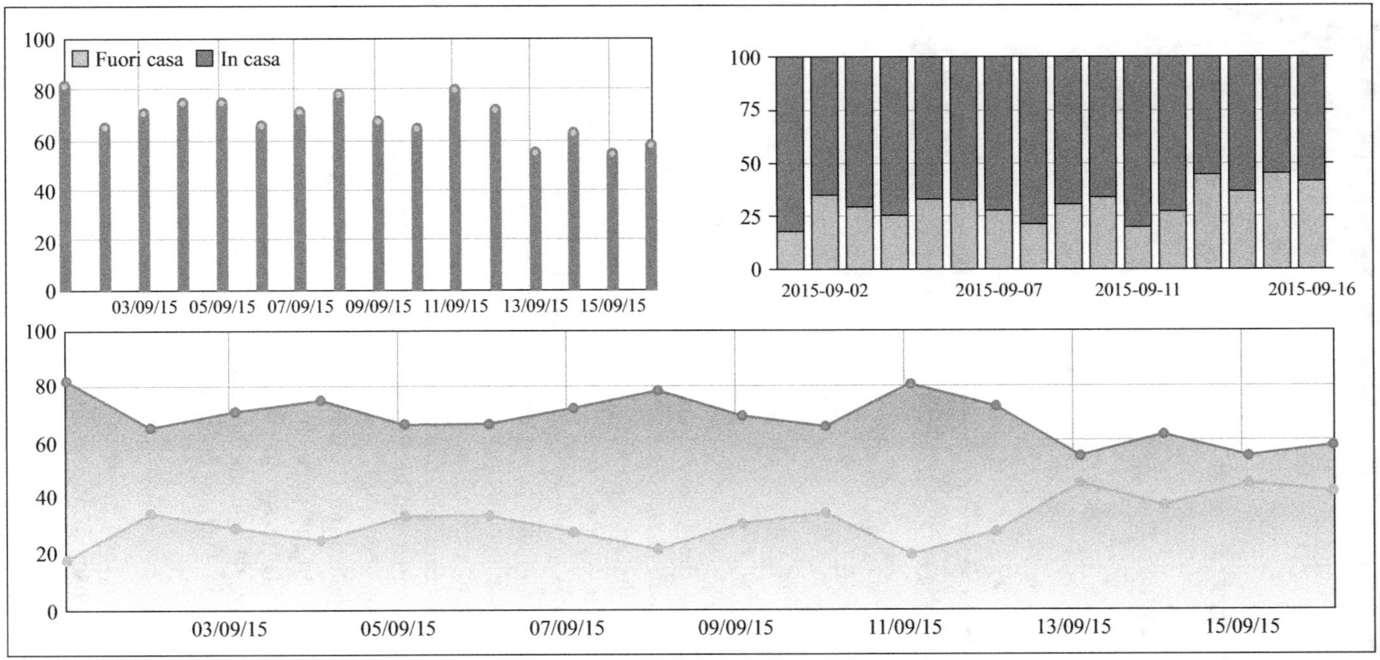

Figure 21.8 *Screenshots of the remote telecare platform showing the presence graphs: the light and dark grey portions represent respectively the absence and presence percentage time*

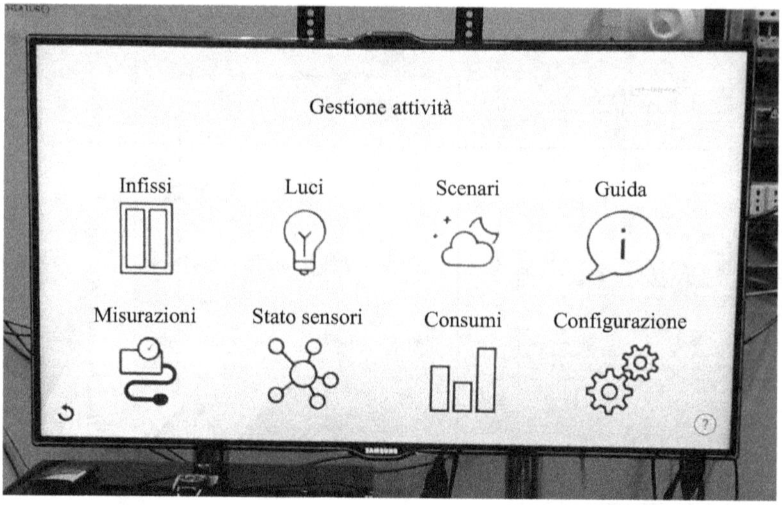

Figure 21.9 Main menu of the smart TV application prototype (Icons from "icons8.com")

the local server provides all the services and information requested by the user interfaces, according to the classic client-server model, using the standard JSON format. During the design phase, the developers have considered some rules ensuring a greater acceptability and usability. Jakob Nielsen [22] has provided a list of 10 "heuristic" principles for the interaction design. Such a list is quite general and establishes the guidelines to follow in order to provide a tool as easy to use as possible, and having a good impact on the user from the beginning. In addition to these general rules, other more specific principles for the different interfaces have been followed, also taking into account the needs and characteristics of the target users. As for the smart TV, for example, the key factors to respect when designing a user interface have been summarized by Samsung in [23]. They mainly concern: simplicity, clarity, consistency, feedback, aesthetic considerations and user control. The developed prototype application is characterized by a correct and adequate organization of content within the page, in order to facilitate the understanding and reading. Thanks to the large screen size, the developers implemented an application having a shallow navigation hierarchy and a few subsections in the main menu (as shown in Figure 21.9), in order to simplify content navigation through the remote control.

Differently from Smart TV, the touchscreen interface is a technology unfamiliar for the elderly, for this reason, to be careful in the design phase has been a necessary factor. Graphics chosen for the prototype application are very simple and easily understandable, there are no animations and the buttons are large sufficiently. Moreover, different layouts have been chosen to best fit the size of the screen as shown in Figures 21.10 and 21.11.

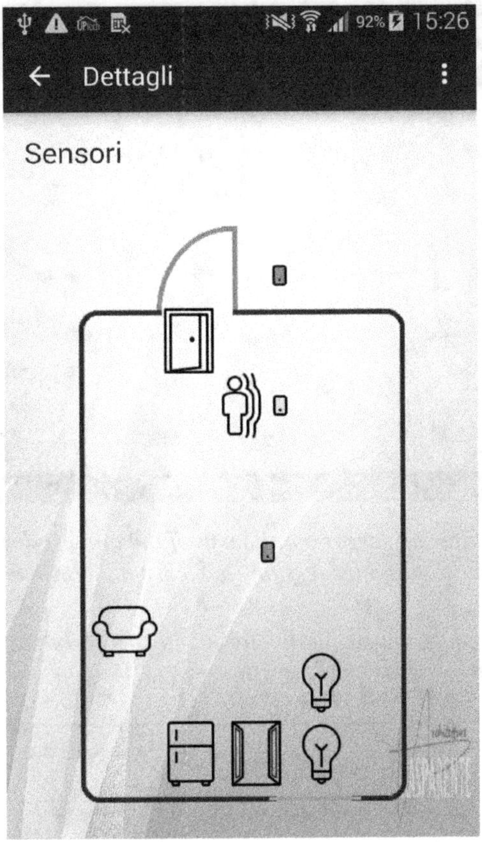

Figure 21.10 Interface for sensor monitoring of the touchscreen prototype application: smartphone version (Icons from "icons8.com")

21.5.5 Conclusive tests

In order to ensure a complete test of all the system, we first have created specific checklists for each single functionality. Secondly, it has been necessary to verify the correct integration of all the devices composing the architecture. Some problems emerged during the first tests have been solved. After 6 months of running, the final version of the platform can be considered quite stable, since no significant malfunctions occurred. As already stated, people not involved in the project have tested the system in the Laboratory. Although they do not fit the target users perfectly, they provide an initial positive response. Moreover, in order to verify the functioning, usability and effectiveness of the interface prototypes, a survey has been conducted: 15 users among students and technical staff rated the applications and provided some useful suggestions to improve them. Each volunteer was interviewed, asking an overall judgment on: visual appearance, intuitiveness, ease of navigation, efficiency and feedback. Moreover, two further questions were added, asking a personal opinion on

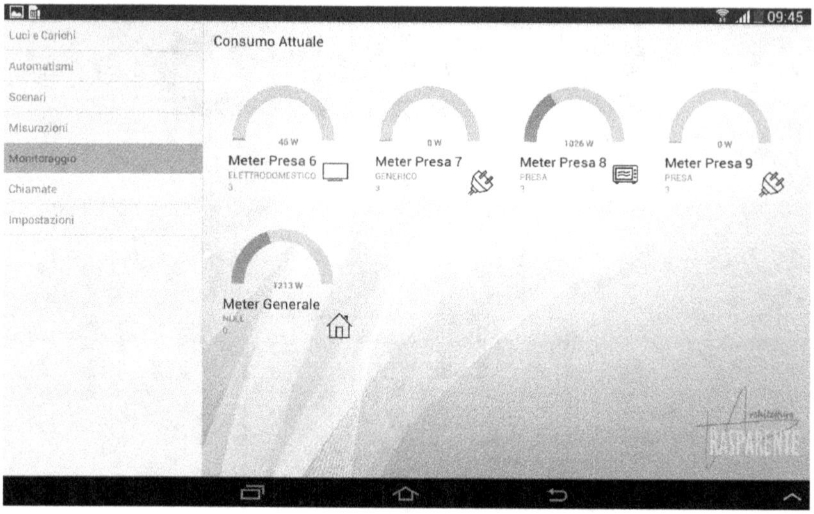

Figure 21.11 Interface for sensor monitoring of the touchscreen prototype application: tablet version (Icons from "icons8.com")

Table 21.2 Key points of the evaluation form and corresponding scores, voting in a range from 1 (very bad) to 10 (very well)

Evaluated aspect	**Smart TVs**	**Touchscreen devices**
Look and feel: clarity, visual impact, coherence among the different sections	9.2	8.7
Intuitive use: easy to learn how to interact with the application	9	8.4
Immediate detection of features, ease of navigation	8.9	8.8
Efficiency: level of user control, ability to reach the objectives	9.2	8.7
Feedback, notifications, error handling	8.2	7.6
Estimated usability by disabled or elderly users	7.7	7.2
Estimated acceptability and impact on users	8	7.5

the level of usability of the application when used by elderly people. The results can be considered very good, as the average score is between 7.6 and 9.2 for all the five main features (see Table 21.2).

Regarding the estimated acceptability and usability by older users, the opinion of respondents gives a positive outcome, with scores ranging from 7.2 to 8. The people interviewed have notified a few marginal problems: this is not unusual when testing prototype interfaces. However, the overall operation is correct and this survey helped the developers to improve the applications.

21.6 Discussion and open issues

The term *smart home* is generally referred to a house in which a number of interconnected electronic devices, home appliances, and applications are controlled or programmed to perform certain actions, in order to ensure safety, comfort, and security of the inhabitants. Such a concept evolved from the original design of home automation systems, already in use in their basic form for decades, in many modern homes (e.g. programming a heater for a specific time and temperature, or opening an automated garage door by a remote control). Modern home automation systems are, however, much more advanced. Their key element relies in the *integration* of various smart devices and smart appliances, and their control, through a central interface, typically accessible remotely over the internet. This paradigm is related to an idea first envisioned in 1999, and now becoming a reality: the *Internet of Things* (IoT)—a huge network of smart objects which work together in collecting and analysing data, and autonomously perform certain actions, through machine-to-machine (M2M) communication. The smart home scenario is now the one in which the connected devices in the home are all interconnected, have the potential to share information with one another, and to cooperate in fulfilling predefined tasks: they have become part of IoT.

A clear and easy example of how IoT may already impact our daily home life is given by the security alarm, which uses sensors to detect movement in the house and the opening or breaking of doors and windows. Unless a specific code is entered into the central system panel, the system might sound an alarm to distract burglars and/or alert the owner, the police or a security company. In a smart home, the security alarm is integrated into a more complex system, in which other actions can be performed, such as blocking exits or enabling remote access to surveillance cameras via the internet. The system could be programmed to turn on lights, the TV or radio in order to prevent burglaries, by simulating the appearance of an occupied home. Carbon monoxide or water leaks alarms, as well as fire detection, can also be additionally integrated in this system. This simple example shows how the integration and interconnection of devices and applications that were considered not correlated in the past, can pave the way to the design of new automated services, in which the user intervention may even be not requested, or at least not necessary to generate a reaction to external events.

IoT will necessarily influence and impact the design of future AAL solutions and systems, as it is leading to smart objects being capable of identifying, locating, sensing and connecting, and thus opening possibilities for new forms of communication between people and things, and things themselves. Using IoT for AAL should consequently simplify the interaction between diverse healthcare devices, and healthcare centre/person. IoT provides all the main characteristics necessary for an AAL environment including heterogeneous devices, and shares with AAL some basic pillars in the development of related technology, such as: demand for reliable energy-efficient designs and energy harvesting, including low-power circuits and communication protocols; layers of intelligence, to transform the massive amount of data generated by pervasive systems into wisdom; essential privacy and security requirements.

Despite the fact that technology shows a lot of promising potential in improving the quality of life for the elderly and disabled, there are possible negative consequences. Some are wary about loneliness or isolation resulting from the use of certain devices that replace human caretakers, which may be the user's only regular social contact. Others are concerned with privacy issues surrounding biometrics and smart home systems, which collect personal information about users and monitor their activities. Discrimination is yet another concern, especially for those with well-managed, chronic conditions. Some technologies, such as fall alert systems, wearable biometric monitors, or mobility devices are difficult to hide, and can make a person's disability very obvious. This can be stigmatizing, making the user uncomfortable, even unwilling to use the device. The massive use of technology in AAL requires a judicious approach with an emphasis on empathy. The market of customers with the greatest potential to benefit from AAL are retirees but many of these people will have had limited exposure to technology. As a consequence, the adoption and acceptance of technology-based AAL will require the development of novel approaches to User Interface Design (UID), and an in-depth appreciation of this market's requirements for the User Experience Design (UXD).

As a final remark, it is possible to state that AAL research has not sufficiently focused on solving the most critical problems of interoperability, usability, security, reliability, and the quality of the user experience. Instead, the research most often addressed isolated aspects of AAL, resulting in a lack of innovation and focus for achieving usable and affordable AAL systems [24]. The overall vision of producing heterogeneous, open, and reusable infrastructure platforms and components, consisting of diverse software and hardware elements spanning from reasoning algorithms to specific healthcare and assisted living devices, is still to be attained.

21.7 Conclusion

This chapter focused on the issues of designing feasible smart home technological platforms for the delivery of Ambient Assisted Living services, providing a review of the most relevant research projects in the field, and discussing in details the real-world experiences performed by the authors in a similar scenario. Matching the complexity of user requirements to the available consumer devices and technologies means a hard work to be performed on integration of different components, harmonization of services and applications.

As understood by the authors, through personal experience, the approach to the real-world implementation often brings out different aspects and problems with respect to the results of a laboratory or experimental development. Referring to the main issues to overcome, some general considerations have been made, such as: the advantages of a user-centred design, the need of monitoring the correct operation of the system, and the importance of checking the real interoperability among different components. All these factors should have influence on the design solutions adopted, with the aim to bring benefits to all the involved stakeholders, from final users to designers and developers, to business entities.

References

[1] The Ambient Assisted Living Joint Programme; 2015. Available from: http://aal-europe.eu/about.

[2] ALADDIN Project: A technology pLatform for the Assisted living of Dementia elDerly INdividuals and their carers; 2009. Available from: http://aladdin-project.eu.

[3] CARE Project: Safe Private Homes for Elderly Persons; 2009. Available from: http://care-aal.eu.

[4] Belbachir AN, Litzenberger M, Schraml S, Hofstatter M, Bauer D, Schon P, *et al.* CARE: A dynamic stereo vision sensor system for fall detection. In: 2012 IEEE International Symposium on Circuits and Systems (ISCAS); 2012. p. 731–734.

[5] Kamel Boulos MN, Lou RC, Anastasiou A, Nugent CD, Alexandersson J, Zimmermann G, *et al.* Connectivity for healthcare and well-being management: examples from six European projects. International Journal of Environmental Research and Public Health. 2009;6(7):1947. Available from: http://www.mdpi.com/1660-4601/6/7/1947.

[6] Fanucci L, Saponara S, Bacchillone T, Donati M, Barba P, Sanchez-Tato I, *et al.* Sensing devices and sensor signal processing for remote monitoring of vital signs in CHF patients. IEEE Transactions on Instrumentation and Measurement, 2013 March;62(3):553–569.

[7] HAPPY AGEING Project: A Home based APProach to the Years of AGEING; 2009. Available from: http://www.aal-europe.eu/projects/happy-ageing/.

[8] Marcellini F, Bevilacqua R, Stara V. HAPPY AGEING (AAL-2008-1-113) Project Overview: Solutions for Independent Living and Technology Acceptance. In: Proceedings of the 4th International Conference on PErvasive Technologies Related to Assistive Environments. PETRA '11. New York, NY, USA: ACM; 2011. p. 61:1–61:5. Available from: http://doi.acm.org/10.1145/2141622.2141695.

[9] HERA Project: Home sERvices for specialised elderly Assisted living; 2009. Available from: http://www.aal-europe.eu/projects/hera/.

[10] Spanoudakis N, Grabner B, Kotsiopoulou C, Lymperopoulou O, Moser-Siegmeth V, Pantelopoulos S, *et al.* A Novel Architecture and Process for Ambient Assisted Living – The HERA approach. In: Proceedings of the 10th IEEE International Conference on Information Technology and Applications in Biomedicine. ITAB10. New York, NY, USA: IEEE; 2010. .

[11] HOPE Project: Smart HOme for the elderly PEople; 2009. Available from: http://www.hope-project.eu/about.asp.

[12] BREATHE Project: Because you have the right to feel good and breathe. Project Reference: AAL-JP-2012-5-045; 2013. Available from: http://www.breathe-project.eu/.

[13] NITICS Project: Networked InfrasTructure for Innovative home Care Solutions, 2012-5-255; 2013. Available from: http://nitics.eclexys.com/.

[14] Tazari MR, Furfari F, Ramos JP, Ferro E. The PERSONA Service Platform for AAL Spaces. In: Nakashima H, Aghajan H, Augusto J, editors. Handbook of Ambient Intelligence and Smart Environments. Springer US; 2010. p. 1171–1199. Available from: http://dx.doi.org/10.1007/978-0-387-93808-0_43.

[15] Pastorino M, Fioravanti A, Arredondo MT, Cogollor JM, Rojo J, Ferre M, *et al.* Preliminary evaluation of a personal healthcare system prototype for cognitive eRehabilitation in a living assistance domain. Sensors. 2014;14(6):10213. Available from: http://www.mdpi.com/1424-8220/14/6/10213.

[16] Colomer JBM, Salvi D, Cabrera-Umpierrez MF, Arredondo MT, Abril P, Jimenez-Mixco V, *et al.* Experience in evaluating AAL solutions in living labs. Sensors. 2014;14(4):7277. Available from: http://www.mdpi.com/1424-8220/14/4/7277.

[17] Kim J, soo Choi H, Wang H, Agoulmine N, Deerv MJ, Hong JWK. POSTECH's U-Health Smart Home for Elderly Monitoring and Support. In: 2010 IEEE International Symposium on a World of Wireless Mobile and Multimedia Networks (WoWMoM), 2010 p. 1–6.

[18] Hartin P, Nugent C, McClean S, Cleland I, Tschanz J, Clark C, *et al.* Encouraging Behavioral Change via Everyday Technologies to Reduce Risk of Developing Alzheimer's Disease. In: Pecchia L, Chen L, Nugent C, Bravo J, editors. Ambient Assisted Living and Daily Activities. Lecture Notes in Computer Science, vol. 8868. Springer International Publishing; 2014. p. 51–58. Available from: http://dx.doi.org/10.1007/978-3-319-13105-4_9.

[19] Kaye J. Unobtrusive measurement of daily computer use to detect mild cognitive impairment. Alzheimer's & Dementia: The Journal of the Alzheimer's Association. 2013;10(1):10–17.

[20] De Santis A, Gambi E, Montanini L, Pelliccioni G, Raffaeli L, Rascioni G, *et al.* Smart Homes for Independent and Active Ageing: Outcomes from the TRASPARENTE Project. In: 2015 Italian Forum on Ambient Assisted Living; 2015.

[21] Spinsante S, Gambi E, Montanini L, Raffaeli L. Data Management in Ambient Assisted Living Platforms Approaching IoT: a Case Study, 2015 IEEE Globecom Workshops (GC Wkshps); 2015. p. 1–7. doi: 10.1109/GLOCOMW.2015.7414079

[22] Nielsen J. 10 Usability Heuristics for User Interface Design. Nielsen Norman Group, Internet: http://wwwnngroupcom/articles/ten-usability-heuristics/. 1995.

[23] Principles for Designing Applications for Samsung Smart TV; 2014. Available from: https://www.samsungdforum.com/UxGuide/2014/01_principles_for_designing_applications_for_samsung_smart_tv.html#ux-01.

[24] Memon M, Wagner SR, Pedersen CF, Beevi FHA, Hansen FO. ambient assisted living healthcare frameworks, platforms, standards, and quality attributes. Sensors. 2014;14(3):4312. Available from: http://www.mdpi.com/1424-8220/14/3/4312.

Chapter 22

Technologies and applications for active and assisted living – What's next?

Francisco Florez-Revuelta[1] *and Alexandros Andre Chaaraoui*[2]

Twenty-five years ago, Mark Weiser published his paper *The computer for the 21st century* [1], seminal work on ubiquitous computing. Weiser and his colleagues at the Xerox Palo Alto Research Centre conceived a new way of thinking about computers, taking into account the human world and allowing the computers to vanish into the background. They thought about rooms with hundreds of computers interconnected in a ubiquitous network, of different sizes suiting particular tasks, which would come to be invisible to common awareness, and that people will simply use them unconsciously to accomplish everyday tasks. That paper finishes with the sentence "Machines that fit the human environment instead of forcing humans to enter theirs will make using a computer as refreshing as taking a walk in the woods," as an analogy of the problem of information overload. But it can also be understood as a goal to locate the user/person as the centre/goal of all this ubiquitous technology.

Following and extending Weiser's ideas, some years later, the term *Ambient Intelligence (AmI)* was coined by Philips. Early references to it appear in several publications, providing different definitions highlighting diverse aspects of the technologies and services involved:

> "In the near future our homes will have a distributed network of intelligent devices that provides us with information, communication, and entertainment. Furthermore, these systems will adapt themselves to the user and even anticipate on user needs. These consumer systems will differ substantially from contemporary equipment through their appearance in peoples environments, and through the way users interact with them. Ambient Intelligence is the term that Philips uses to denote this new paradigm for in-home computing and entertainment. (2000)" [2]

[1]Faculty of Science, Engineering and Computing. Penrhyn Road, Kingston upon Thames, Surrey KT1 2EE, UK
[2]Google, Inc., Zurich, Switzerland

"AmI provides a vision of the Information Society where the emphasis is on greater user-friendliness, more efficient services support, user-empowerment, and support for human interactions. People are surrounded by intelligent intuitive interfaces that are embedded in all kinds of objects and an environment that is capable of recognising and responding to the presence of different individuals in a seamless, unobtrusive and often invisible way. (2001)" [3]

"Ambient intelligence will become a network of hidden intelligent interfaces that recognise our presence and mould our environment to our immediate needs. (2002)" [4]

"...a new paradigm in information technology is emerging, in which people are served by a digital environment that is aware of their presence and context, and is responsive to their needs, habits, gestures and emotions: ambient intelligence. (2003)" [5]

"An AmI system requires the use of distributed sensors and actuators to create a pervasive technological layer, able to interact transparently with a user, either passively by observing and trying to interpret what the user actions and intentions are, but also actively, by learning the preferences of the user and adapting the system parameters (applied to sensors and actuators, for instance) to improve the quality of life and work of the occupant. (2005)" [6]

From these definitions, AmI is based on four key aspects:

Ubiquitous computing: Embedded processing into everyday objects like furniture, clothes or toys; i.e. there is an interconnection of embedded systems that are hidden in the environment. This includes computers, communication systems, cameras, microphones, and pressure light, motion, and temperature sensors. There is also a possibility of communication among objects and between objects and the user.

Awareness: The system must know at all times the state of the environment and its dwellers, being able to locate and identify objects and people.

Intelligence: The system must be able to adapt to new data and establish connections with the existing knowledge.

Natural interaction: The basic requirements must be able to be transmitted and interpreted without the user performing strange or unnatural movements or orders. This suggests the development based on the most common means of interaction for human interfaces, such as speech and gestures. Besides, interfaces must be mixed with the environment so that the user does not feel them present.

In 2001, Ducatel *et al.* [3] foresaw how ambient intelligence was going to be used in 2010, presenting several case scenarios. This is a relevant document, now that we have long passed that date and we can compare those expectations with the actual achievements. Regarding technologies, those scenarios required:

- very unobtrusive hardware: self-generating power and micro-power usage of devices/objects; new displays, smart surfaces and paints; active devices, nanodevices, embedded computers;
- seamless mobile/fixed web-based communications infrastructure: complete integration of mobile and fixed and radio and wired networks, optical networks, and seamless and dynamic reconfiguration;
- dynamic and massively distributed device networks: plug-and-play solutions, multi-domain networking, embedded intelligence, distributed data management and storage systems, middleware;
- natural feeling human interfaces: intelligent agents, adaptive interfaces, multimodal (multi-user, multilingual, multi-channel and multipurpose) user interfaces for speech, gesture, and pattern recognition; and
- dependability and security: self-testing and self-organising software, secure ID authentication, micropayment systems and biometrics.

They envisioned that, in 2010, all the required computers and interfaces would be embedded in a wrist band, identification though immigration and customs would be automatic using that wearable device, mobile audio systems, cars had to be driven but they would use a navigation system, rooms would automatically adapt to the preferences of the dweller, video conference systems would be projected on the walls, files would be stored remotely, people would be connected through social networks, each one of us would have a Digital-Me who learns from the user and interacts with others, people would share transport, the fridges would be smart, people would do their shopping electronically being delivered to home or to local kiosks, cars would auto-pay when entering specific city areas, their speed would be automatically reduced by the city authorities due to pollution, cars would sense other vehicles and persons around them, people would work from home to reduce carbon print, and people would share face-to-face/remote meetings using virtual reality.

A couple of years later, the Institute for Prospective Technological Studies (IPTS) and the European Science and Technology Observatory (ESTO) network, published a thorough report [7] proposing a roadmap for the design and development of ambient intelligence technologies and services. Friedewald *et al.* predicted that, in 2015, some technologies would be widely used, for instance: beyond 3G mobile networks, GPS-like systems, integration of all types of wireless networks, body area networks, in-body sensors, emotion measurement, bio-sensors, molecular electronics, high-density storage, nano-systems that can move inside the body for clinical applications, self-organising and self-repairing software, large-scale distributed systems, embedded systems, new and open standards, use of ontologies, semantic web, context-aware systems, natural real-time language translation, multi-users–environment—language—mood speech recognition, computers surpassing human logic and learning abilities, autonomous robots, most software written by a machine, "affective" robots, artificial noses, natural speech-based user interaction, multi-modal interfaces, augmented reality with see-through displays, brain–computer interaction, personalisation in commerce in physical stores, rollable displays, electronic ink, holographic displays, large (>50″) displays, PCs using a small battery with more than one year of duration, ambient power, privacy and anonymity protection, secure transactions and payments, and reliable and fault-tolerant systems.

The rationale to mention all these different technologies and services, which were forecast more than 10 years ago, is to show how difficult it is to predict which new developments will be in the market and adopted by the general public in the future, even the near one. The authors of those two documents stressed that either these were the views of a relatively small group, or they were pure wishes, mind-opening vision or science fiction. As it is explained in [3], there a number of socio-political factors, and business and industrial models that produce these differences in developments and early adoption by the users. However, in those scenarios anyone can see current companies as Amazon, Google, Skype, Facebook, Dropbox and FitBit, to mention just some of them. But, there are many others that were not foreseen and some other models that are far from being accomplished. An interesting paper, by Gunnarsdóttir and Arribas-Ayllon [8], critically reviews the evolution of AmI and how many of those initial expectations have not been achieved.

Ambient Assisted Living (AAL), or Active and Assisted Living as it is the term currently employed in Europe, arises from the application of the Ambient Intelligent paradigm to support the well-being, health and care of older people. In [7] this was expressed as prevention, cure, care, and health administration. Similarly to the achievements and unfulfillments for AmI, the expected progress in AAL during these years has had diverse degrees of accomplishment.

Therefore, the remainder of this chapter will critically analyse where the achievements in technologies and applications for AAL that have been presented in this book stand with respect to the foreseen progress. Next, some forecast challenges for the next decade are introduced based on the aforementioned analysis and the summarised book chapters' conclusions. Finally, this chapter and with it this book are concluded.

22.1 Foreseeing the future of AAL

There are also some recent roadmaps that address specifically the technologies and services for active and assisted living. The AALIANCE2 Coordination Action is aimed at building consensus upon research priorities in AAL. As part of their work, the AALIANCE2 Consortium published in 2014 an AAL Roadmap and Strategic Research Agenda for the upcoming decades [9]. They presented ten different scenarios for the use of AAL technologies:

1. Prevention of early degeneration of cognitive abilities
2. Healthy living
3. Management of chronic diseases
4. Age-friendly and safe environments
5. Fall prevention
6. Management of daily activities and keeping control over own life
7. Keeping social contact and having fun

Table 22.1 Key enabling technologies for AAL (adapted from [9])

Sensing	Smart sensors, Micro-Electro-Mechanical System (MEMS), lab-on-a-chip, biosensors, vision sensors, acoustic sensors, olfactory sensors, environmental sensors, pervasive sensing and smart environments, in-body and on-body sensors, quantum sensors, and energy harvesting
Reasoning	Context awareness and sensor data fusion, artificial (quasi-human) intelligence, advanced controls for AAL robotics and devices, dependability and maintainability, and "Smart" everywhere! – miniaturisation of processors
Acting	Service robotics, smart mobility, smart actuators, green technologies, neurorobotics, wearable robotics, and cloud robotics
Interacting	Smart multimedia appliances and apps, smart wearable accessories, sensorial interfaces, spatial interfaces, natural language interfaces, multi-modal interfaces, neural interfaces and brain-computer interfaces, and service integrations
Communicating	Standardisation and certification, middleware, semantic technologies, self-configuration, self-optimisation, self-healing, self-protection, data protection regulations, body area network (BAN), personal area network (PAN), local area network (LAN), and wide area network (WAN)

8. Outdoors mobility (i.e. pedestrians, public transport and private cars)
9. Avoiding caregivers isolation
10. Senior citizens at work,

analysing the needs and opportunities for the different stakeholders (i.e. older persons, informal carers, formal carers, service providers, local communities, and healthcare systems) and the key enabling technologies involved (see Table 22.1).

The German Commission for Electrical, Electronic & Information Technologies of the German Institute for Standardisation (DIN) and the Association for Electrical, Electronic & Information Technologies (VDE) published a roadmap for standardisation in AAL systems, with special attention to interoperability [10]. This document mentions all the available standards for sensors, actuators and house buses, for user interfaces, and for middleware/services/runtime platforms (including clinical standards). A repository of standards in technology related to AAL was also compiled by the AALIANCE2 project.[1]

An interesting document [11] was published in 2015 by Nesta, the innovation foundation, about their vision of the UK National Health System in the medium-term future. Although this report focuses its attention to the challenges that the health systems will address due to the ageing population, it is relevant as it proposes to empower people to track and analyse their own health data using ubiquitous technology for a personalised healthcare, and to share this data with others for promoting healthy living, prevention and management of chronic conditions.

As it can be observed from these examples of recent analysis of the state of the art and the future of AAL, and as it has been seen throughout this book, most of the

[1] http://nero.offis.de/projects/aaliance2/start (Accessed in March 2016)

technologies and services are similar to those that were proposed one decade ago. Are we still at the same point? No. Have we reached those predictions? Neither.

22.2 What have we reached so far and what is next?

This book provides a review of the current developments in technologies and applications for AAL, which can be compared with those expectancies that some early researchers in the area had some decades ago. This section summarises the current and expected advances in AAL.

There are many important new players in the market, however, global adoption by the users is going much slower than it was expected some years ago. In many Western countries, only the use of gas or smoke detectors, or HVAC controllers, are mainstream. In most cases, far from being "smart."

While the use of ambient sensors seems to be limited to public environments, with the exceptions just mentioned, wearable sensors either embedded in mobile phones or worn by the user (e.g. smart watches) are currently getting the interest of people. There are a variety of mobile applications that allow the monitoring of vital signs or physical activity in order to provide solutions to support a healthy living. In the near future, research and innovation should focus on miniaturisation, lower energy consumption and comfortability.

Among ambient sensors, video-based devices stand out, as they can acquire a lot of information from the environment for a variety of services (safety, security, recognition of activities of daily living) as it is mentioned in several chapters. However, current systems for video analytics are not robust enough for a reliable use under unconstrained conditions, and there are privacy concerns that need to be addressed in order to increase user acceptance. However, recent devices (e.g. RGB-D devices as the Microsoft Kinect) may contribute to address these issues.

All these sensors offer different functionalities, are able to acquire different information from the environment and the user, and have different degrees of reliability. Therefore, instead of working in isolation, there is a need for information fusion to reduce uncertainty and develop more robust AAL solutions. There are current projects working on this fusion, as well as on the integration of other data sources, e.g. social networks, integration at different scales (home, city, transport), or user profiles.

An area where research is getting huge advances is that of assistive robots, which are able to support older people in their everyday activities, to monitor them, or to allow remote carers to interact with the assisted people [12, 13]. Relevant developments are also expected in exo-skeletons or wearable prosthetics for rehabilitation and support to people with motor impairment [14, 15].

Anyway, these sensors would need to gather and integrate as much information as possible about the status of the environment and the user/s in order to offer intelligent, proactive, context-aware and adaptive AAL solutions, which could be valid for the changeable health, physical, sensory, and emotional conditions of older people. These large amounts of data would need to be analysed to extract valuable information for the users. We are entering the era of the Big Data [16] and the quantified self [17].

Currently, fall detection is one of the main AAL applications related to safety. Systems such as those that have been presented in this book employ inexpensive accelerometers and gyroscopes, which are embedded in mobile phones or in specific devices (e.g. pendant alarms). These are wearable systems that can be employed ubiquitously. However, they will not detect any fall if the user forgets to wear it, what can regularly happen with older people or people with dementia. Therefore, some other systems using radars or cameras in the environment are currently being developed. Nonetheless, they are not robust enough in real conditions, for instance, in a home where the environment can be cluttered with furniture.

Frailty is very related to fall detection and, in particular, fall prevention. A measure of frailty can be obtained by analysing human gait. Systems for gait analysis usually employ the same sensors than those for fall detection. Only in specialised facilities other devices such as sensorised walkways are employed. However, they are too expensive to be used in a home. Further research in the near future should be focused on making the systems more robust, and designing systems that allow long-term gait analysis in order to find variability over time.

Another application where long-term analysis and continuous learning would be required is in the recognition of activities of daily living (ADLs). These systems would be able to learn users' routines and behaviour, and detect deviations from those routines. There are still many challenges that AAL systems have to address for an effective recognition of ADLs: multiple occupancy of the environment, context-awareness, recognition of complex activities, fusion of data captured with environmental and wearable sensors, etc. Although common to all the AAL services provided at home, one of the main issues with recognition of ADLs is the requirement of sensorising the usual environment of a person, the home. Installation of multiple sensors and actuators in every room is, in many cases, not accepted by the users as it requires retrofitting the house, wiring all those devices. Although there have been advances in wireless devices, most of them require being connected to mains power. The autonomy of battery-powered devices is still very short. Therefore, research needs to be carried out on lower-consumption or self-powered devices.

Regarding systems to support mobility of older people, either outdoors or indoors, great advances have been obtained in the last years. Navigation in outdoor environments is integrated in most smart phones, while systems for indoors navigation are improving. Efforts may be directed towards the support of mobility of people with disabilities, e.g. visually-impaired people. Another area where mobility will be improved in the near future is the use of autonomous cars [18, 19], as they will not require attention from the driver, facilitating their use by older and disabled people.

While some of these AAL solutions will not require any active input by the users, other systems will involve some person-environment interaction. Following the AmI paradigm, this interaction should be natural using gestures, speech and touch as input; and using images, voice synthesis, scents [20, 21] and haptics [22] as output. Current WIMP (windows, icons, menus, point-and-click devices) [23], and even tablet devices, are not natural to use, and this is exacerbated with older people, who are less accustomed to Information and Communication Technologies. Some other more advanced interfaces, as virtual and augmented reality [24], brain-computer interfaces [25], and affective interfaces [26, 27] will be very relevant in AAL.

However, in order to actually fulfil a complete integration, not only of sensors but also of actuators, as well as of processing and communication systems, there is a need for standards and effective interoperability among them. Since some years ago, there have been different initiatives to reach that interoperability, however this has not been obtained yet, limiting the global deployment of the Internet of Things [28], smart homes and technologies for AAL.

22.3 Conclusion

Last but not least, in this section, we want to clarify the objective of bringing together the broad fields of technologies and applications involved in AAL in this book. *Active and Assisted Living: Technologies and Applications* is aimed at being a handbook for researchers, as well as technical and healthcare professionals from any of the related fields who may need to acquire or refresh their knowledge of the recent state of the art and obtain a well-balanced in-depth and broad know-how of the field's technologies and applications.

This goal is pursued by covering all major technologies used in the field, including but not limited to smart homes, environmental and wearable sensors, visual monitoring, information fusion, standards and interoperability, reasoning systems and person-environment interaction. On the other hand, AAL applications are important also to healthcare professionals and caregivers. In this part, the book has reviewed tele-care and tele-health, as well as gait analysis, fall detection, support of ADLs, outdoor and indoor mobility, well-being and social interaction and decision support systems. Finally, associated issues such as accessibility, privacy and human-centred design; and study cases for smart cities and smart homes are presented.

In this sense, this book gives a unified view of the significant amount of scientific fields that are emerging from and coming together in AAL. As such, it can provide not only a way of precise and direct initiation in the field, but it may also serve as reference for similar state-of-the-art reviews and retrospections in the future.

References

[1] M. Weiser, "The computer for the 21st century," *Scientific american*, vol. 265, no. 3, pp. 94–104, 1991.

[2] E. Aarts and R. Harwig, "Ambient Intelligence. A new user experience," CeBit 2000, http://www.research.philips.com/technologies/projects/ami/vision.html, (Accessed in March 2016).

[3] K. Ducatel, M. Bogdanowicz, F. Scapolo, J. Leijten, and J.-C. Burgelman, *Scenarios for ambient intelligence in 2010*, 2001.

[4] J. Horvath, "Making friends with Big Brother?" Telepolis, http://www.heise.de/tp/artikel/12/12112/1.html, 2002, (Accessed in March 2016).

[5] B. J. A. Kröse, J. M. Porta, A. J. N. Breemen, K. Crucq, M. Nuttin, and E. Demeester, *Ambient Intelligence: First European Symposium, EUSAI 2003, Veldhoven, The Netherlands, November 3–4, 2003. Proceedings*. Berlin, Heidelberg: Springer Berlin Heidelberg, 2003, Lino, the User-Interface Robot, pp. 264–274.

[6] P. Remagnino, H. Hagras, N. Monekosso, and S. Velastin, *Ambient Intelligence: A Novel Paradigm*. New York, NY: Springer New York, 2005, Ambient Intelligence, pp. 1–14.

[7] M. Friedewald, O. Da Costa *et al.*, "Science and technology roadmapping: Ambient intelligence in everyday life (AmI@ Life)," *IPTS Report*, vol. 73, 2003.

[8] K. Gunnarsdóttir and M. Arribas-Ayllon, "Ambient intelligence: a narrative in search of users (discussion paper)," 2011.

[9] AALIANCE2 Consortium, "AALIANCE2 AAL Roadmap 2014," http://www.aaliance2.eu/system/files/drafts/AALIANCE2_Roadmap2014.pdf, (Accessed in March 2016).

[10] DKE German Commission for Electrical, Electronic & Information Technologies of DIN and VDE, "The German AAL Standardization Roadmap," http://www.vde.com/en/dke/std/Documents/German%20AAL%20Standardization%20Roadmap%20[1].pdf, 2012, (Accessed in March 2016).

[11] J. Bland, H. Khan, J. Loder, T. Symons, and S. Westlake, "The NHS in 2030. A vision of a people-powered, knowledge-powered health system," Nesta, http://www.nesta.org.uk/sites/default/files/the-nhs-in-2030.pdf, 2015, (Accessed in March 2016).

[12] J. Broekens, M. Heerink, and H. Rosendal, "Assistive social robots in elderly care: a review," *Gerontechnology*, vol. 8, no. 2, pp. 94–103, 2009.

[13] T. S. Dahl and M. N. K. Boulos, "Robots in Health and Social Care: A Complementary Technology to Home Care and Telehealthcare?" *Robotics*, vol. 3, no. 1, pp. 1–21, 2013.

[14] H. S. Lo and S. Q. Xie, "Exoskeleton robots for upper-limb rehabilitation: State of the art and future prospects," *Medical Engineering & Physics*, vol. 34, no. 3, pp. 261–268, 2012. [Online]. Available: http://www.sciencedirect.com/science/article/pii/S1350453311002694

[15] A. J. Veale and S. Q. Xie, "Towards compliant and wearable robotic orthoses: A review of current and emerging actuator technologies," *Medical Engineering & Physics*, vol. 38, no. 4, pp. 317–325, 2016. [Online]. Available: http://www.sciencedirect.com/science/article/pii/S135045331600031X

[16] V. Mayer-Schönberger and K. Cukier, *Big data: A revolution that will transform how we live, work, and think*. Houghton Mifflin Harcourt, 2013.

[17] M. Swan, "The quantified self: Fundamental disruption in big data science and biological discovery," *Big Data*, vol. 1, no. 2, pp. 85–99, 2013.

[18] H. Bradshaw-Martin and C. Easton, "Autonomous or 'driverless' cars and disability: a legal and ethical analysis." *European Journal of Current Legal Issues*, vol. 20, no. 3, 2014.

[19] J. Yang and J. Coughlin, "In-vehicle technology for self-driving cars: Advantages and challenges for aging drivers," *International Journal of Automotive Technology*, vol. 15, no. 2, pp. 333–340, 2014.

[20] J. J. Kaye, "Making Scents: aromatic output for HCI," *interactions*, vol. 11, no. 1, pp. 48–61, 2004.

[21] M. Obrist, A. N. Tuch, and K. Hornbaek, "Opportunities for odor: Experiences with smell and implications for technology," in *Proceedings of the SIGCHI Conference on Human Factors in Computing Systems*, CHI '14. New York, NY, USA: ACM, 2014, pp. 2843–2852.

[22] M. I. Tiwana, S. J. Redmond, and N. H. Lovell, "A review of tactile sensing technologies with applications in biomedical engineering," *Sensors and Actuators A: Physical*, vol. 179, pp. 17–31, 2012.

[23] A. van Dam, "Beyond wimp," *Computer Graphics and Applications, IEEE*, vol. 20, no. 1, pp. 50–51, 2000.

[24] M. Ma, L. C. Jain, and P. Anderson, *Virtual, augmented reality and serious games for healthcare 1*. Springer, 2014.

[25] L. F. Nicolas-Alonso and J. Gomez-Gil, "Brain computer interfaces, a review," *Sensors*, vol. 12, no. 2, p. 1211, 2012.

[26] R. W. Picard, *Affective computing*. MIT press, Cambridge, 1997, vol. 252.

[27] C. Mühl, B. Allison, A. Nijholt, and G. Chanel, "A survey of affective brain computer interfaces: principles, state-of-the-art, and challenges," *Brain–Computer Interfaces*, vol. 1, no. 2, pp. 66–84, 2014.

[28] J. Gubbi, R. Buyya, S. Marusic, and M. Palaniswami, "Internet of things (IoT): A vision, architectural elements, and future directions," *Future Generation Computer Systems*, vol. 29, no. 7, pp. 1645–1660, 2013.

Index

www.ingramcontent.com/pod-product-compliance
Lightning Source LLC
LaVergne TN
LVHW011859200926
848484LV00002B/485